Electronic Structure of Polymers and Molecular Crystals

NATO ADVANCED STUDY INSTITUTES SERIES

A series of edited volumes comprising multifaceted studies of contemporary scientific issues by some of the best scientific minds in the world, assembled in cooperation with NATO Scientific Affairs Division.

Series B: Physics

Volume 1 – Superconducting Machines and Devices
edited by S. Foner and B. B. Schwartz

Volume 2 – Elementary Excitations in Solids, Molecules, and Atoms (Parts A and B)
edited by J. Devreese, A. B. Kunz, and T. C. Collins

Volume 3 – Photon Correlation and Light Beating Spectroscopy
edited by H. Z. Cummins and E. R. Pike

Volume 4 – Particle Interactions at Very High Energies (Parts A and B)
edited by David Speiser, Francis Halzen, and Jacques Weyers

Volume 5 – Renormalization and Invariance in Quantum Field Theory
edited by Eduardo R. Caianiello

Volume 6 – Interaction between Ions and Molecules
edited by Pierre Ausloos

Volume 7 – Low-Dimensional Cooperative Phenomena
edited by H. J. Keller

Volume 8 – Optical Properties of Ions in Solids
edited by Baldassare Di Bartolo

Volume 9 – Electronic Structure of Polymers and Molecular Crystals
edited by Jean-Marie André and János Ladik

Volume 10 – Progress in Electro-Optics
edited by Ezio Camatini

The series is published by an international board of publishers in conjunction with NATO Scientific Affairs Division

A	**Life Sciences**	Plenum Publishing Corporation
B	**Physics**	New York and London
C	**Mathematical and Physical Sciences**	D. Reidel Publishing Company Dordrecht and Boston
D	**Behavioral and Social Sciences**	Sijthoff International Publishing Company Leiden
E	**Applied Sciences**	Noordhoff International Publishing Leiden

Electronic Structure of Polymers and Molecular Crystals

Edited by

Jean-Marie André
Laboratoire de Chimie Théorique Appliquée
Facultés Universitaires de Namur
Namur, Belgium

and

János Ladik
Lehrstuhl für Theoretische Chemie
Technische Universität München
München, BRD

with the cooperation of

Joseph Delhalle
Laboratoire de Chimie Théorique Appliquée
Facultés Universitaires de Namur
Namur, Belgium

PLENUM PRESS • NEW YORK AND LONDON
Published in cooperation with NATO Scientific Affairs Division

Library of Congress Cataloging in Publication Data

NATO Advanced Study Institute on Electronic Structure of Polymers and Molecular Crystals, Facultés Universitaires de Namur, 1974.
Electronic structure of polymers and molecular crystals.

(NATO advanced study institutes series: Series B, physics; v. 9)
Includes bibliographical references and index.
1. Polymers and polymerization–Congresses. 2. Molecular crystals–Congresses.
I. André, Jean Marie. II. Ladik, János. III. Title. IV. Series.
QD380.N37 1974 547'.84 75-12643
ISBN 978-1-4757-0321-4 ISBN 978-1-4757-0319-1 (eBook)
DOI 10.1007/978-1-4757-0319-1

Lectures presented at the 1974 NATO Advanced Study Institute on Electronic Structure of Polymers and Molecular Crystals, held at the Facultés Universitaires de Namur, Belgium, September 1–14, 1974

Softcover reprint of the hardcover 1st edition 1975

A Division of Plenum Publishing Corporation
227 West 17th Street, New York, N.Y. 10011

United Kingdom edition published by Plenum Press, London
A Division of Plenum Publishing Company, Ltd.
Davis House (4th Floor), 8 Scrubs Lane, Harlesden, London, NW10 6SE, England

Preface

The NATO Advanced Study Institute on "Electronic Structure of Polymers and Molecular Crystals" was held at the Facultés Universitaires de Namur (F.U.N.) from September 1st till September 14th, 1974. We wish to express our appreciation to the NATO Scientific Affairs Division whose generous support made this Institute possible and to the Facultés Universitaires de Namur and the Société Chimique de Belgique which provided fellowships and travel grants to a number of students.

This volume contains the main lectures about the basic principles of the field and about different recent developments of the theory of the electronic structure of polymers and molecular crystals.

The school started with the presentation of the basic SCF-LCAO theory of the electronic structure of periodic polymers and molecular crystals (contributions by Ladik, André & Delhalle) showing how a combination of quantum chemical and solid state physical methods can provide band structures for these systems. The numerical aspects of these calculations were also discussed. Lectures by Mahan have shown how optical properties of molecular crystals can be interpreted on the basis of the exciton theory. Little has reviewed the present status of the hypothesis about excitonic superconductivity and the different approaches to synthesize a superconductive polymer. McCubbin in his first series of lectures has given a very precise group theoretical treatment of the symmetry properties of polymers. Atkins' lectures have covered both the theoretical aspects of X-ray structure determination and its application to different polymers. In a subsequent lecture series, Coppens has shown how the electronic density function of a molecular crystal can be determined from X-ray data and has given a critical comparison of the experimental and theoretical electron densities. Clark in his lectures has reviewed the fundamentals of the ESCA technique, the theoretical models developed for the interpretation of the measurements and the application of the method to polymers

The question of electronic correlation in the ground state of polymers and molecular crystals was treated by Calais using both the alternant molecular orbital method and Green's function technique. The correlation in excited states of crystals was discussed by a series of lectures by Collins showing that by starting from ab initio SCF-LCAO band structures one can get really good agreement for the band widths with experiment if one takes into account separately the long-range and the short-range correlations and the relaxation due to the excitation. These techniques seem to be very promising also for the treatment of excited states of polymers.

Harris has shown in his lectures how one can calculate the total energy of a crystal taking into account all neighbors' interactions using the method of Fourier transforms. Santry approached the problem of total energy and band structure of a polymer or molecular crystal from the aspect that starting from the wave functions of the monomer and using perturbation theory including third order terms, one obtains a good approximation. Rein has treated in his lectures the problem of interactions between polymer constituents with the aid of perturbation theory showing the decomposition of the interaction energy into physically different terms. Simonetta has discussed how force field calculations using empirical potential functions (molecular mechanics) can be applied to organic molecules and to molecular crystals built up from them.

An alternative approach to the treatment of electronic structure of polymers and molecular crystals was proposed by Johnson, Herman and Kjellander who suggested the application of the SCF-Xα scattered wave method.

In a second series of lectures, McCubbin discussed the different theoretical attempts to treat possible kinds of disordered systems (for instance, geometrical disorder, compositional disorder, etc.). Ladik reviewed the semi-empirical SCF-LCAO crystal orbital methods and their applications to the calculation of the band structures of different periodic DNA and protein models. In a final lecture, Laki and Ladik gave some considerations on the structure of different proteins which act as energy converters.

The lectures of the Institute were kept on a rather high level and they really covered (perhaps with the exception of transport phenomena in polymers) the important aspects of the field.

Our sincere thanks are due to the Academic Authorities of the Facultés Universitaires de Namur - espcially Prof. J. Denis, rector, who provided accomodation and very valuable organizational support. The secretarial task of the Institute was extremely heavy. We are much obliged to Dr. J. Delhalle and Mrs. S. Delhalle for their outstanding contributions to the practical organization and to the editing of the lectures. Messrs. G. Kelner and D. Vercauteren did a wonderful job in arranging all practical details and splendid trips to the Ardennes. Special thanks are due to our hostesses: M.C. André, C. Derouane, N. Renkin, and A. Lacroix who took such good care of everyone during the Institute. N. Renkin and P. Lonnoy handled the conference correspondence and contributed actively to the preparation of this volume.

Last but not least, our sincere gratitude to the lecturers for their active collaboration and for having prepared their manuscripts in good time.

Jean-Marie André

Janos Ladik

December 1974

Contents

AB INITIO AND SEMI-EMPIRICAL BAND STRUCTURE CALCULATIONS ON POLYMERS

J.M. André

Laboratoire de Chimie Théorique Appliquée

Facultés Universitaires de Namur

61, rue de Bruxelles

5000 NAMUR (Belgium)

I. INTRODUCTION

An important group of chemicals known as polymers can be characterized by a possible periodicity in their conformational structure. Up to now, polymer chemistry was mainly concerned with the study and understanding of its own chemical synthesis and mechanical properties. Besides their important role as plastics and in biological processes, polymers now appear to be more and more interesting from an electronic and magnetic point of view. They will presumably exibit special characteristics and it is hoped that more emphasis will be given to the synthesis of those polymers which present specific features.

The aim of this lecture is to illustrate the general theory, actually used, for studying the electronic properties of regular macromolecules. Since the bonds in a polymer are not different from usual bonds in organic molecules, the methods already proved successful in the study of the electronic structure of molecules can be used for describing the electronic properties of polymers. It means that we shall use the tight binding approximation in the sense in which it is used in the LCAO approximation in molecular quantum chemistry. The LCAO approximation is merely a mathematical expansion of monoelectronic functions, although,in solid state physics, the tight binding approximation is more a physical method of "atoms in molecules, solids or crystals". In other words, what is called "tight binding approximation" is generally considered as a

LCAO development with only those atomic orbitals used in the ground states of isolated atoms. During the last five years, we realized that the LCAO approximation gives better and better results if the LCAO basis is very extended giving rise to considerable improvements in total energies but loosing a part of the "intuitive" chemical sense of analysis in terms of atomic orbitals.

This allows us to point out another interesting feature in polymer theoretical chemistry. Indeed, it appears that polymer quantum chemistry turns out to be the ideal link between solid state physics and molecular quantum chemistry. There exists a common problem in discussions concerning orbital energies, orbital symmetry and atomic gross charges between chemists who have a clear understanding in this work and solid state physicists who use somwhat more obscure terms as first Brillouin zone, dependence of "wave-function" (in fact the *monoelectronic wave-function* is an orbital for the chemist) with respect to wave-vector k, Fermi surfaces, Fermi contours and density of states.

Actually, both scientists use the same quantum mechanical background, the same approximations and even the same degree of sophistication so that each can understand the other. An interesting feature of polymer theoretical chemistry is that its scientific language is partly that of quantum chemistry and partly that of solid state physics. As a consequence, this field has, in my opinion, a considerable impact in teaching solid state concepts as first Brillouin zone, energy bands or density of states to chemists with a certain knowledge of basic quantum chemistry.

This lecture is divided into four parts ; part one gives the summary of the main theorems to be used for systems with periodical properties allowing us to indicate the general LCAO periodical form. Part two summarizes how we obtained band structures from the knowledge of basic atomic orbitals. Part three is a sample calculation at the Hückel level. It explains band theory as applied to simple systems. Part four is a review of the main calculations in theoretical polymer chemistry.

II. BLOCH FUNCTIONS OF INFINITE PERIODICAL SYSTEMS

In usual expansion techniques, the functions to be developed are linear combinations of well-defined basis functions. The expansion coefficients are unknown and must be determined. In quantum mechanics, polyelectronic wave-functions and monoelectronic wave-functions (or orbitals) have no physical meaning by themselves. Only the square of the wave-functions or of the orbitals are meaningful and represent a density probability. In a periodic system, since the electron density and, as a consequence, the density probability are periodic, there must exist some relationship between the expansion coefficients of the polymer orbitals in the basis of the fixed atomic orbitals. Bloch's theorem defines the relation between the value of the polymer orbital at a point of direct space and its value at a point translated by a translation A_n, i.e. n times a primitive translation vector a. If introducing the translation operator $T(A_n)$ such that :

$$T(A_n)\, f(r) = f(r+A_n)$$

we easily note that the monoelectronic potential function $V(r)$ and the electronic density $\rho(r)$ are eigenfunctions of the translation operator with eigenvalues equal to 1 due to the periodic properties of the infinite lattice:

$$T(A_n)\, V(r) = V(r+A_n) = 1.\, V(r)$$

$$T(A_n)\, \rho(r) = \rho(r+A_n) = 1.\, \rho(r)$$

Formally, the eigenfunctions of the translation operator can be represented as

$$T(A_n)\, \phi(r) = \phi(r+A_n) = \lambda_n\, \phi(r)$$

By noting the periodic conditions of the electron density of the polymer orbitals

$$T(A_n)|\phi(r)|^2 = |\phi(r+A_n)|^2 = \lambda_n^2\, |\phi(r)|^2 = |\phi(r)|^2$$

we deduce that the eigenvalue is a complex number of modulus equal to unity :

$$\lambda_n = \exp(i\theta A_n)$$

with $\theta(A_n)$ being a scalar of A_n :

$$\theta(A_n) = kA_n$$

We remark that since A_n lies in the direct space (L-dimension), k must be defined in the reciprocal space (L^{-1} -dimension). The most convenient representation of the polymer orbitals in these conditions turns out to be :

$$\phi(k,r+A_n) = \exp(ikA_n)\, \phi(k,r)$$

The precedent equation allows us to see that only some values of k give non-redundant information. In particular, if comparing two polymer orbitals related by a translation Kn in the reciprocal space, we see that :

$$T(An)\ \phi(k,r) = \exp(ikAn)\ \phi(k,r)$$

$$T(An)\ \phi(k+Kn,r) = \exp(i(k+Kn)An)\ \phi(k+Kn,r)$$

$$= \exp(ikAn)\ \phi(k+Kn,r)$$

Therefore, Bloch functions which differ by a reciprocal lattice vector admit the same eigenvalue for any translation operator. This means that we can define a reduced wave vector k lying from 0 to, at most, a translation in reciprocal space. The lenght bound by the maximum of all possible reduced vectors is called the *first Brillouin zone* of the polymer.

The Bloch form ϕ of a single atomic orbital χ is thus obtained as :

$$\phi(k,r) = \Sigma(n)\ c_n\ \chi(r+An)$$

$$= N^{-1/2}\ \Sigma(n)\ \exp(ikAn)\ \chi(r-An)$$

where $N^{-1/2}$ is the normalisation factor for a N-cells polymer.

The LCAO Bloch form which describes the delocalized polymer orbital $\phi_n(k,r)$ as a periodic combination of functions centered at the atomic nuclei of polymers is then

$$\phi_n(k,r) = \Sigma(j)\ \Sigma(p)\ c_{npj}\chi_p(r-Aj)$$

$$= N^{-1/2}\ \Sigma(j)\ \exp(ikAj)\Sigma(p)\ c_{np}(k)\ \chi_p(r-Aj)$$

where n is the band index and k the position vector in the first Brillouin zone.

III. L.C.A.O. METHOD FOR POLYMERS

The aim of the present part is to deduce the LCAO equations for the case of infinite periodic one-dimensional polymers. This is an interesting case of a one-dimensional periodicity although the wave-functions are truly three-dimensional.

The one-electron theory (which means the band theory in solid state physics) defines a set of polymeric orbitals to represent the wave-functions of a single electron in a periodic potential produced by nuclei and other electrons. The optimal set of polymer orbitals for a given basis set is constructed in the usual way by solving a set of Hartree-Fock equations. The SCF monoelectronic operator has an explicit form :

$$h(i) = -1/2 \ \nabla^2(i) - \Sigma(h)\ \Sigma(p)\ Z_p/|r_i-R_p-R_h| + \Sigma(n')\ \Sigma(k')\ \{2\ J_{k'n'}(i) - K_{k'n'}(i)\}$$

with the coulomb and exchange terms being detailed as

$$J_{k'n'}(i)\ \phi_n(k,i) = \int \phi^*_{n'}(k',j)\ 1/r_{ij}\ \phi_{n'}(k',j)\ dv_j\ \phi_n(k,i)$$

$$K_{k'n'}(i)\ \phi_n(k,i) = \int \phi_{n'}(k',j)\ 1/r_{ij}\ \phi_n\ (k,j)\ dv_j\ \phi_{n'}(k',i)$$

The preceding terms express respectively the kinetic energy operator, the attraction of a single electron with all nuclei (p) centered in all cells (j), the averaged electrostatic potential of all electrons and the averaged exchange interaction.

From Bloch's theorem, we know that a polymeric orbital has the LCAO form:

$$\phi_n(k,r) = N^{-1/2}\ \Sigma(j)\ \Sigma(p)\ \exp(ikAj)\ c_{np}(k)\ \chi_p(r-Aj)$$

By forming the expectation value of the monoelectronic Hartree-Fock operator and by using the variational procedure for the LCAO coefficients $c_{np}(k)$, we obtain for the whole polymer the system of equations:

$$\Sigma(p)\ c_{np}(k)\ \{\ \Sigma(j)\ \exp(ikAj)\ h_{pq}(j)\} - \varepsilon_n(k)\{\ \Sigma(j)\ \exp(ikAj)\ S_{pq}(j)\} = 0$$

The compatibility condition of the preceding system equation:

$$|\Sigma(j)\ \exp(ikAj)\ h_{pq}(j) - \varepsilon(k)\ \Sigma(j)\ \exp(ikAj)\ S_{pq}(j)| = 0$$

will produce in the reduced scheme the band structure $\varepsilon(k)$ as a multivalued function of k. We note that $h_{pq}(j)$ is a matrix element of the monoelectronic operator h between the atomic orbital χ_p centered in the origin unit cell and the atomic orbital χ_q centered in cell j. $S_{pq}(j)$ has the same meaning for the unit operator. Both matrix elements decrease exponentially with the distance between the orbitals giving rise to a natural convergency of the summation over cells appearing in the secular systems and determinants. It is to be noted that the dimensions of the matrix equations to be solved are equal to the number of atomic orbitals per unit cell, the effect of the infinite lattice being included in the naturally converging but formally infinite sums.

The preceding equations show that in a general way a polymer can be considered as a large molecule. In this case, the usual methods of quantum chemistry can be used to investigate the electronic structure of polymers. As a consequence, we are allowed to introduce the well-known approximations of the molecular quantum chemistry into the formalism of polymeric orbitals. For instance, we can either restrict the number of electrons to be considered or approximate several electronic integrals or group of integrals. In this way, we will speak of "ab initio" or "non-empirical" techniques if considering all electrons and calculating all the necessary integrals. On the other hand, we will use semi-empirical methods if reducing the number of electrons is to be considered and as a consequence if the necessary integrals are to be evaluated from experimental data. Table I represents the different degrees of sophistication which can be obtained.

In the simple Hückel method, only π-electrons of purely conjugated organic molecules are taken into account and all interactions except for the nearest neighbours are neglected. This method obtained a considerable success by its simplicity and by the good validity of the involved approximations. Around 1950, Mulliken, Wolfsberg and Helmolz suggested a very simple type of Hückel parametrization which is easily extended to σ-bonded systems. Hoffmann took up this method and applied it to a large variety of saturated and conjugated organic molecules. This is by now well known as the Extended Hückel method. More sophisticated methods no longer have their justification on a entire evaluation of the matrix elements but more precisely on a close analysis of the energy terms. The matrix elements are splitted into their kinetic, attraction and repulsion parts and some integrals are either neglected or approximated by careful procedures. This is the case of the Pariser-Parr-Pople type methods for π-electrons and CNDO (Complete Neglect of Differential Overlap) type methods when considering all-valence electrons. Fundamentally, those procedures attempt to reproduce by means of one-electron models the one-electron results of the many-electrons Hartree-Fock theory. Sometimes, however, the fit to experimental data results in an implicit inclusion of a part of the correlation effects although always in a very indirect way.

TABLE I: Methods of Quantum Chemistry

Restrictions on the number of electrons	Global Evaluation of Matrix elements (Non-iterative procedures)	Approximations in the calculation of matrix elements (Iterative Procedures)
π electrons	Hückel Method	Pariser-Parr-Pople Method
All-valence electrons	Extended Hückel Method	CNDO -type Methods
All electrons	-	**Ab Initio Methods**

The Hartree-Fock theory obtained its considerable success with the ab initio calculations on molecules of "chemical" sizes. In these calculations, all the electrons were taken into account and all the necessary integrals were explicitly computed for a given basis of atomic orbitals. Minimal basis (i.e. atomic orbitals which are occupied in the ground states of isolated atoms, for instance, 1s, 2s, 2px, 2py, 2pz-type orbitals for the carbon atom) are only used when simulating the behaviour of electrons in the molecule as compared to isolated atoms. Extended basis are to be used if trying to obtain best quantitative results near the Hartree-Fock limit.

Non-empirical methods are dependant on the fourth power of the number of orbitals considered while empirical methods are more dependant on square power. In fact, diagonalizations which appear for both non-empirical and semi-empirical approximations are more n^3 procedures and in some non-empirical calculations, heavy n^5 or n^6 transformation procedures are sometimes found.

At the present time, all the above cited methods are used to calculate energy band structure of some polymers. A review of the calculations is given in the last part of this paper, but it is now important to have an estimation of the computational effort needed. From the expression of the converging expansion of matrix elements in secular equations, it is trivial to observe that we need to calculate N times n^2 two-center matrix elements. If rewriting the monoelectronic system equations as

$$F(k)\ C_n(k) = S(k)\ C_n(k)\ E_n(k)$$

we see that the Fock and overlap matrices F(k) and S(k) between Bloch orbitals depend on the k-position in the first Brillouin zone but can be represented as converging sums of square matrices independent of k and noted below as F(j) and S(j):

$$F(k) = \Sigma(j)\ \exp(ikAj)\ F(j)$$

$$S(k) = \Sigma(j)\ \exp(ikAj)\ S(j)$$

The precedent matrices collect the Fock and overlap integrals between the set of atomic orbitals centered in the origin unit cell and the set of atomic orbitals centered in cell j. We thus deduce the minimum number of integrals to compute and to store in a computer. If N is the number of cells for which all the overlap and Fock elements are greater than a given threshold (for instance, 10^{-5}) and n the size of the basis set centered in the unit cell, we have to calculate Nn^2 Fock integrals and Nn^2 overlap integrals. The exact number is $Nn(n+1)/2$ since parts of the integrals are redundant. The repulsion operator by its two-electron character gives rise to much more complex matrices. Those matrices (of number N^3) contain $n^2(n+1)/4$ integrals as first approximation. As a consequence, the time-consuming part of an "ab initio" program is of a general $N^3n^4/8$ time dependence. Taking into account the realistic possibilities and the availability of important machine time, it turns out that ab initio computations for polymers are actually practicable in the framework of nearest or next nearest cell approximation only. Extended Hückel or CNDO band calculations have a much easier Nn^2 dependence for the integral calculations. The time-consuming process is then the diagonalisation and, as a consequence, those type of calculations are feasible on normal computers in realistic computation times. Since extended Hückel calculations avoid SCF iterative process, one might think that the computing time ought to be less than the time needed for the CNDO method. However, the former needs heavy matrix transformations in order to treat the nonorthogonal atomic and Bloch basis while the atomic and consequently the Bloch basis is assumed orthogonal in the CNDO-type methods. As a result, computing times are very similar for both methods.

IV. SAMPLE CALCULATIONS IN ONE-DIMENSION

One dimension infinite π-electron systems turn out to be a good illustration of the calculation of energy as a function of the reciprocal vector k. We shall choose the physical example of an infinite conjugated chain with two atoms per unit cell and a lattice vector of length a. For labelling purposes, we call the π-atomic $2p_z$ orbital centered on the lower left carbon χ_1 and the other one χ_2.

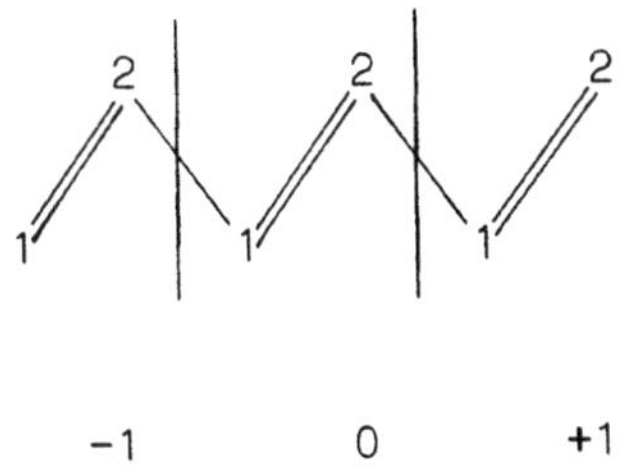

From Bloch's theorem, we deduce that a good trial polymeric orbital for the π-system has the form

$$\phi_n(k,r) = c_{n1}(k)\ b_1(k) + c_{n2}(k)\ b_2(k)$$

or

$$\phi_n(k,r) = N^{-1/2}\ \Sigma(j)\ \exp(ikAj)\ \{c_{n1}(k)\ \chi_1(r-ja) + c_{n2}(k)\ \chi_2\ (r-ja)\}$$

where defining the π-Bloch orbitals:

$$b_1(k) = N^{-1/2}\ \Sigma(j)\ \exp(ikAj)\ \chi_1(r-ja)$$

$$b_2(k) = N^{-1/2}\ \Sigma(j)\ \exp(ikAj)\ \chi_2(r-ja)$$

When using the Hückel approximation (neglect all but the nearest interactions) and defining the symbolic matrix elements:

$$h_{11}(j) = h_{22}(j) = \int \chi_1(r-ja)\ h\ \chi_1(r-ha)\ dv = \int \chi_2(r-ja)\ h\ \chi_2(r-ha)\ dv = \alpha\delta_{hj}$$

$h_{12}(j) = \int \chi_1(r-ja)\ h\ \chi_2(r-ha)\ dv = \beta$ if both the atoms on which χ_1 and χ_2 are centered are nearest neighbours.

$h_{12}(j) = \int \chi_1(r-ja)\ h\ \chi_2(r-ha)\ dv = 0$ otherwise

For the evaluation of the overlap matrix elements, the basis is assumed to be orthogonal with respect to each atomic center :

$$\int \chi_p(r-ja)\ \chi_q(r-ha)\ dv = \delta_{pq}\ \delta_{jh}$$

The matrix elements of the Hamiltonian in the basis of the π-Bloch orbitals are now:

$$H_{11}(k) = H_{22}(k) = \Sigma(j)\ \exp(ikAj)\ h_{11}(j) = \alpha$$

$$H_{12}(k) = H_{21}(k) = \Sigma(j)\ \exp(ikAj)\ h_{12}(j) = \beta(1 + \exp(ika))$$

The system of equations for the π-system of the whole polymer is thus :

$$c_{n1}(k)\ \{\alpha - \varepsilon_n(k)\} + c_{n2}(k)\ \beta\{1 + \exp(ika)\} = 0$$

$$c_{n1}(k)\beta\{1 + \exp(-ika)\} + c_{n2}(k)\{\alpha - \varepsilon_n(k)\} = 0$$

of which the compatibility conditions :

$$\begin{vmatrix} \alpha - \varepsilon(k) & \beta\{1 + \exp(ika)\} \\ \beta\{1 + \exp(-ika)\} & \alpha - \varepsilon(k) \end{vmatrix} = 0$$

yields the two π-energy bands of an infinite polyene:

$$\varepsilon_1(k) = \alpha + \beta\sqrt{2 + 2\cos(ka)} = \alpha + 2\beta \cos(ka/2)$$

$$\varepsilon_2(k) = \alpha - \beta\sqrt{2 + 2\cos(ka)} = \alpha - 2\beta \cos(ka/2)$$

Clearly, $-\pi/a \leqslant k < \pi/a$ is all that is necessary for non-redundant information. The representation of the energy as a function of wave-vector k is shown in Figure 1.

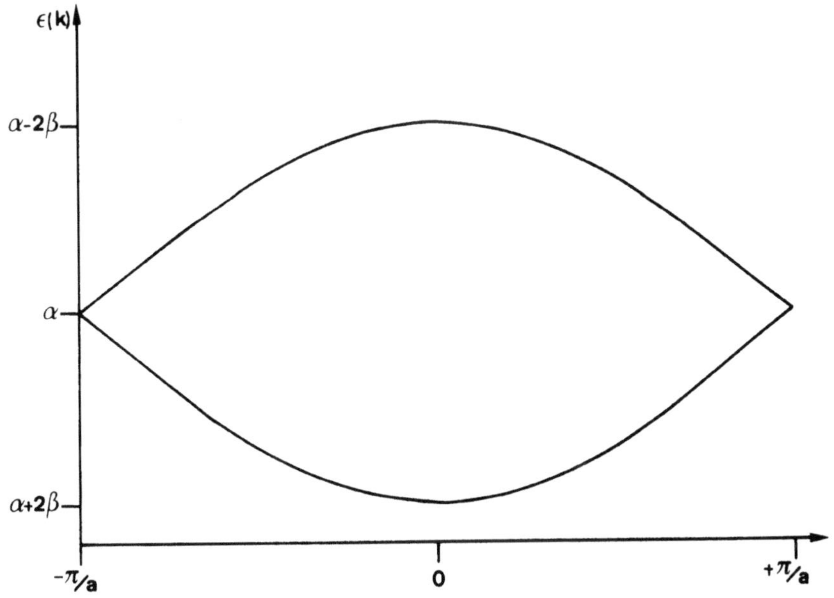

Fig. 1. Band structure of an infinite polyene.

Now, if we differentiate respectively single and double bonds by their matrix elements β and β', the compatibility condition easily becomes :

$$\begin{vmatrix} \alpha - \varepsilon(k) & \beta' + \beta\exp(ika) \\ \beta' + \beta\exp(-ika) & \alpha - \varepsilon(k) \end{vmatrix} = 0$$

with two energy bands

$$\varepsilon_1(k) = \alpha + \sqrt{\beta^2 + \beta'^2 + 2\,\beta\beta'\cos(ka)}$$

$$\varepsilon_2(k) = \alpha - \sqrt{\beta^2 + \beta'^2 + 2\,\beta\beta'\cos(ka)}$$

Figure 2 represents the differences between both situations: a polyene of single and double bond alternation ($\beta \neq \beta'$) and a regular polyene without bond alternation ($\beta = \beta'$). In the former case, there exists current carrying states at infinitesimal distance in energy on top of the occupied levels. There is however no energy gap. This is the typical behaviour of a *metal*. Should an energy gap exist, a finite excitation energy is required to carry the electrons over that gap. If the energy gap is greater than the thermal energy, the polymer would be an *insolator*. If the energy gap is small compared to thermal energy, then a small but non-negligeable density of electrons excited by the thermal energy into the upper band will exist ; the material will have observable electrical conductivity becoming easier at higher temperature. The polymer would then be a *semi-conductor*.

The links between the energy levels of the finite molecular systems and the infinite polymeric chain can easily be visualized by drawing the orbital energies (filled and empty) associated with polyenes of increasing lengths (Ethylene, Butadiene, Hexatriene,...). In a series of homogeneous molecules, the number of energy levels increases with the increase of molecular dimensions and, correspondingly, the distance between their energy levels diminishes. The progressive transformation of distinguishable energy levels into energy bands is schematized in Figure 3 for the case of regular and alternant polyene.

We note the analogy between the occupied levels and the valence bands, the unoccupied levels and the conduction band. Furthermore, the energy of the highest occupied molecular orbital (HOMO) is the first ionization potential of the molecule which is the top of the valence band for a polymer. On the other hand, the energy of the first unoccupied molecular orbital (LUMO) is the electron affinity of the molecule which is nothing but the bottom of the conduction

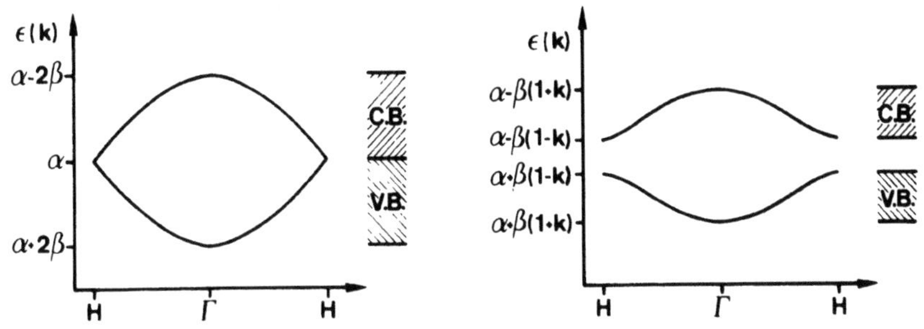

Fig. 2 : Band structures of a regular (LEFT) and an alternant polyene (RIGHT).

band of the polymer. Since, the simple molecular orbital method neglects modifications of repulsion and exchange effects when going from ground states to excited ones, the energy differences between the last occupied and the lowest unoccupied level is a measure of the first UV transition; in polymers, the same difference (i.e. the energy gap) can be estimated as the difference between the ionisation potential and the electron affinity :

$$\Delta E = I - A$$

Besides the representation of molecular levels and its contraction into energy band schemes, the right-hand side of Figure 3 shows a third representation which is the so-called *density of states* functions N(E) of the band. The density function N(E) indicates the variation of the number of allowed energy levels with the energy value within a band i.e. the number of allowed electron states per unit energy range. Since the N states of the polymer are unequivocally described by a length equal to $2\pi/a$ of reciprocal space, the relation between the number of states dN and a length dk of the reciprocal space is

$$dN = (Na/2\pi)\ dk$$

From the above sample calculation on a regular polyene, we

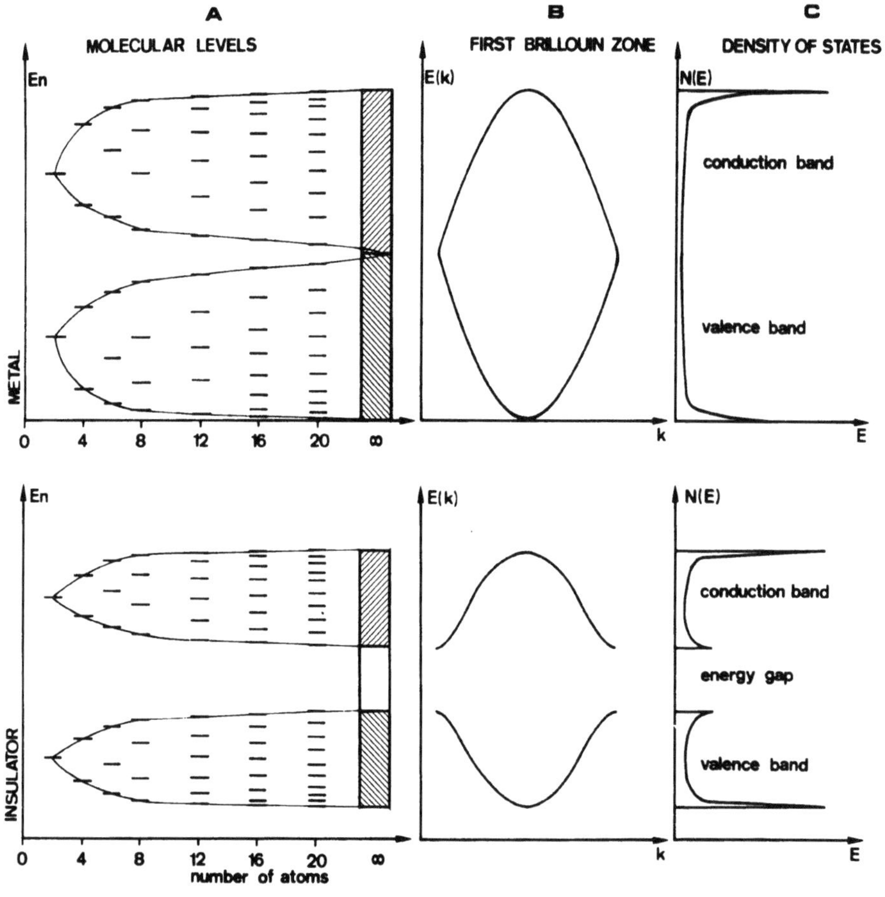

Fig. 3 : Molecular levels (A), band structure (B), density of states (C) for regular (TOP) and alternant (BOTTOM) polyenes.

obtain the following theoretical expression of the valence π-band:

$$\varepsilon(k) = \alpha + 2\,\beta\cos(ka/2)$$

and

$$d\varepsilon = -\,a\beta\,\sin(ka/2)\,dk = -\,a\beta\,\sin(\mathrm{arc}\,\cos(\varepsilon-\alpha/2\beta))$$

$$dN = -\,(N/2\pi\,\sin(\mathrm{arc}\,\cos(\varepsilon-\alpha/2\beta)))\,d\varepsilon$$

The density of states is directly deduced from the preceding expression :

$$N(E) = dN/dE = -\,(N/2\pi)\,(1/\sin(\mathrm{arc}\,\cos(\varepsilon-\alpha/2\beta))$$

N(E) is greater than zero since β is a negative quantity. As expected, this depends on the size of the polymer. This one-dimensional density of states is represented on the right-hand side in Figure 3. All representations a) in molecular levels b) in first Brillouin zone and c) in density of states are quite equivalent as far as the information contained is regarded. But the last two graphs only require a single calculation for an infinite system which makes them more useful for practical purposes. However, the first scheme clearly demonstrates the equivalence between representations of molecular chemistry and solid state physics.

V. LCAO BAND STRUCTURE CALCULATIONS ON POLYMERS

Polymeric orbital theory has followed the general scheme which in turn was followed by molecular orbital theory. In this sense, early applications were mainly concerned with the general understanding of the π-electrons of aromatic or conjugated one- or two-dimensional polymers. The pioneer work in this field is due, independently to Coulson, Wallace, Barriol and Metzger around the fifties for describing the electronic structure of the planar graphite in a Hückel type procedure. Similar work was done for the infinite polyene chain by Lennard-Jones and Kuhn. After the work of these pioneers, both graphite and polyene were studied in approximate theories taking account of the details of electron interactions. The aim of this work was primarely to establish or to demonstrate the existing link between molecular orbital theory and solid state band theory. The generalization of the LCAO methods to all electrons or, at least, to all valence electrons in a polymer is quite recent and the first all valence calculation on a polymer as far as we know is the band structure of polyethylene as calculated by Manne & McCubbin. More recently, advancement in mathematical techniques

TABLE II: Band structure calculations on polyethylene

METHOD	REFERENCES	AUTHORS
Extended Hückel	1	McCubbin & Manne (1968)
	2	Imamura (1970)
	3	André & Delhalle (1972)
	4	André, Kapsomenos & Leroy (1971)
	5	Falk & Fleming (1973)
CNDO/2	6	Morokuma (1970)
	7	Fujita & Imamura (1970)
	3	André, Kapsomenos & Leroy (1971)
	8	Morosi & Simonetta (1971)
	9	Morokuma (1971)
	4	André & Delhalle (1972)
	10	André, Delhalle, Kapsomenos & Leroy (1972)
	11	McAloon & Perkins (1972)
	5	Falk & Fleming (1973)
INDO	12	Beveridge, Jano, Ladik (1972)
MINDO/2	12	Beveridge, Jano, Ladik (1972)
Ab initio	13	André (1970)
	14	André & Leroy (1970)
	15	André & Leroy (1971)
	16	Clementi (1971)
	17	André, Delhalle, Delhalle, Caudano, Pireaux & Verbist (1973)
	5	Falk & Fleming
	18	Delhalle, André, Delhalle, Caudano, Pireaux,Vergist (1974)

1) W.L. McCubbin & R. Manne, Chem.Phys.Lett. 2, 230 (1968)
2) A.Imamura, J.Chem.Phys. 52, 3168 (1970)
3) J.M. André, G.S. Kapsomenos, G. Leroy, Chem.Phys.Lett.8,195(1971)
4) J.M. André, J.Delhalle, Chem.Phys.Lett. 17, 145 (1972)
5) J.E. Falk, R.J.Fleming, J.Phys.C:Solid State Phys.6, 2954 (1973)
6) K.Morokuma, Chem.Phys.Letter. 6, 186 (1970)
7) H.Fujita, A.Imamura, J.Chem.Phys. 53, 4555 (1970)
8) G.Morosi, M.Simonetta, Chem.Phys.Lett. 8, 358 (1971)
9) K.Morokuma, J.Chem.Phys. 54, 1962 (1971)
10) J.M. André, J.Delhalle, G.S. Kapsomenos, G.Leroy, Chem.Phys. Lett. 14, 485 (1972)
11) B.J. McAloon, P.G. Perkins, Faraday Discuss.II,1972, 1121
12) D.L. Beveridge, I.Jano, J.Ladik, J.Chem. Phys. 56, 4744 (1972)
13) J.M. André, Compt.Phys.Commun. 1, 391 (1970)
14) J.M. André, G.Leroy, Ann.Soc.Sci.Brux. 84, 133 (1970)
15) J.M. André, G.Leroy, Chem.Phys.Lett. 5, 71 (1971)
16) E.Clementi, J.Chem.Phys. 54, 2491 (1971)
17) J.M. André, J.Delhalle, S.Delhalle, R.Caudano, J.J.Pireaux, J. J.Verbist, Chem.Phys.Lett. 23, 206 (1973)
18) J.Delhalle, J.M.André, S.Delhalle, J.J.Pireaux, R.Caudano, J.J.Verbist, J.Chem.Phys. 60, 595 (1974).

and programming have reached a point at which moderate size polymers can be studied by ab initio methods avoiding the use of semi-empirical parameters. This allows the band theory of polymers to advance in a clearly defined way. The purpose of this section is to review the current status of all-valence and all-electrons calculations on polymers with particular emphasis to theoretical predictions and complementary information on experimental techniques.

A. POLYETHYLENE and its substitution derivatives

Polyethylene is the best example of recent proliferation of band structure calculations. In the sense, it plays the same role as benzene did in molecular calculations. Polyethylene was extensively described in the literature in extended Hückel, CNDO/2, INDO, MINDO/2 and ab initio methods. The calculations are summarized in Table II. The conformation of polyethylene chains in crystals has benn investigated in the CNDO/2 approximation for the isolated chain energy and an empirical method for the packing energy by Morosi and Simonetta. The most stable conformation appeared to be the usual zig-zag chain (T). All calculations show a large energy gap between the filled and the vacant band in agreement with the experimental fact that polyethylene is a good insulator. Recent papers by Wood, Barber & Hillier, Beveridge & Wun, André, Delhalle, Delhalle, Caudano, Pireaux & Verbist and Falk & Fleming show that band structure calculations of polyethylene are of great interest in identifying photoelectron spectra. The analysis of the valence levels, for instance, produced a very satisfactory interpretation of the ESCA spectrum by comparing extended Hückel or ab initio valence bands and experimental density of states. The position of both theoretical and experimental peaks agree surprisingly well and both fine structures are directly comparable. The excellent agreement is proved by a correlation diagram of theoretical versus experimental positions of the valence bands where the sets of extended Hückel and ab initio values are related to experimental data by straight lines with correlation coefficients of 0.990 and 0.997 respectively.

Substitution derivatives of polyethylene have also been studied as shown in Table III. A conformation study of poly (tetrafluorethylene) was carried out in the semi-empirical CNDO/2 approximation. It shows that the stable helical rotation angle of the chain must lie near 165° in excellent agreement with the experimental results (166 to 168°).

The results of Morokuma thus predict a slightly twisted zigzag conformation. Other recent studies were proved succesfull in indentifying ESCA chemical shifts for fluorine derivatives of polyethylene. A good example is given by cis and trans 1-2 difluoro polyethylene where a different orientation of fluorine atom per unit cell was shown by André & Delhalle to produce a non-negligea-

TABLE III: Band structure calculations on substitution dervations of polyethylene

POLYMER	METHOD	REFERENCE	AUTHOR
-CFH-CH_2-			
-CFH-CFH-			
-CFH-CFH-	CNDO/2 & EH	1	André, Delhalle (1972)
-CH_2-CF_2-		2	Delhalle, Delhalle, André (1974)
-CF_2-CFH-		3	Delhalle (1974)
-CF_2-CF_2-	EH	4	McCubbin (1971)
		1	André, Delhalle (1972)
		2	Delhalle, Delhalle, André(1974)
			Delhalle (1974)
	CNDO/2	5	Morokuma (1971)
		1	André, Delhalle (1972)
		2	Delhalle, Delhalle, André(1974)
		3	Delhalle (1974)
		3	Delhalle (1974)
-CH_2-CCl-	CNDO/2	6	McAloon, Perkins (1972)

1) J.M.André, J.Delhalle, Chem.Phys.Lett. 17, 145 (1972)
2) J.Delhalle, S.Delhalle, J.M.André, Bull.Soc.Chim.Belges.83, 107 (1974)
3) J.Delhalle, accepted for publication in Chemical Physics
4) W.L.McCubbin, Chem.Phys.Lett. 8, 507 (1971)
5) K.Morokuma, J.Chem.Phys. 54, 962 (1971)
6) B.J.McAloon, P.G.Perkins, Faraday Discuss.II, 1972, 1121

TABLE IV: Band structure calculations on polyene

METHODS	REFERENCE	AUTHOR
Extended Hückel	1	André, Kapsomenos,Leroy(1971)
CNDO/2	1	André,Kapsomenos,Leroy (1971)
	2	O'Shea, Santry (1971)
	3	McAloon, Perkins (1972)
	4	O'Shea,Santry (1974)
Ab initio	5	André, Leroy (1970)
	6	André, Leroy (1971)
	1	André,Kapsomenos,Leroy (1971)

1) J.M.André,G.S.Kapsomenos,G.Leroy,Chem.Phys.Lett.8, 195 (1971)
2) S.O'Shea,D.P.Santry,J. Chem.Phys.54, 2667 (1971)
3) B.J.McAloon, P.G.Perkins,Faraday Discuss.II (1972) 1121
4) S.O'Shea, D.P.Santry,Chem.Phys.Lett. 25, 164 (1974)
5) J.M.André, G.Leroy,Ann.Soc.Scien.Brux.84, 133 (1970)
6) J.M.André, G.Leroy,Internat.J.Quant.Chem.5, 557 (1971)

TABLE V: Band structure calculations on particular types of polymers

POLYMERS	METHOD	REFERENCE	AUTHOR
LiF chain	CNDO/2	1	O'Shea,Santry (1971)
Polyglycine	CNDO/2	2	Fujita,Morokuma (1970)
	INDO,MINDO/2	3	Beveridge,Jano,Ladik(1972)
Phosphite Polymer	CNDO/2	4	McAloon,Perkins (1972)
Boron-Nitrogen Polymers	CNDO/2	5	Armstrong,McAloon,Perkins (1973)
Ladder Polymers	E.H.	6	Perkins (1973)

1) S.O'Shea, D.P.Santry, J.Chem.Phys. 54, 2667 (1971)
2) H.Fujita,A.Imamura, J.Chem.Phys. 53, 4555 (1970)
3) D.L.Beveridge, I.Jano, J.Ladik, J.Chem.Phys. 56, 4744 (1972)
4) B.J.McAloon, P.G.Perkins, Faraday Trans.II., 68, 1833 (1972)
5) D.R.Armstrong, B.J.McAloon, P.G.Perkins, Faraday Trans.II, 69, 968 (1973)
6) P.G.Perkins, Rev.Roum.Chimie, 18, 931 (1973)

ble shift of about 0.15 eV on the C_{1s} level of the carbon atom. A study of McAloon & Perkins on the band structure of polyvinylchloride leads to satisfactory interpretation of the stabilization of polyvinyl chloride against chemical degradation.

B. POLYENE

The polyene chain has been studied in the Hückel approximation from a long time. All valence or all electrons calculations are recent and the main emphasis is given on the possibility of bond alternation in the systems. It now appears that in their stable conformation infinite polyene would be in an alternant position. Related to this problem is the question of the existence or non-existence of an energy gap in infinite regular polyene. Although it is generally admitted that the alternant polyene will show a non-zero energy gap, it is now considered that regular (or non-alternant) polyene would not have any energy gap and must therefore be a metallic polymer. This question was treated on a group theory basis by McCubbin. More than the polyethylene chain, the polyene chain study appears to be of an academic interest. Table IV summarized of the main calculations on these chains.

C. OTHER POLYMERS

Several papers have been published on other types of polymers. These are mainly oriented to the investigation of biopolymers or to the study of candidates to superconductivity. Table V presents the main calculations in these fields. May we point out that some works have been carried out on band structures of molecules in molecular crystals. The main work in this field is due to Chojnacki and Santry.

VI. CONCLUSIONS

In our opinion, the recent proliferation and the actual study of band structure calculations raises the question of the value of polymer computation. A realistic answer was given by Clementi who hopes that more and more emphasis will be placed on the synthesis of polymers having unique electronic and magnetic characteristics. For McCubbin, three main areas are isolated in which the values of polymer calculations appear to be reasonably unequivocal : i.e. electrical conductivity, nature of reactive states in solid state chemistry and conformational analysis. As noted by Ladik, polymers are now the candidates to new physical phenomena such as supraconductivity. Our studies show that band structure calculations offer useful perspectives in interpreting results of photoelectron spectroscopy.

We hope that in the near future band structure calculations and correlations with experimental facts will become routine work and achieve predictive ability of band theory for polymers.

DIFFERENT LCAO BAND STRUCTURE CALCULATION METHODS FOR PERIODIC POLYMERS AND MOLECULAR CRYSTALS

J. Ladik

Lehrstuhl für Theoretische Chemie

Technische Universität München

München, BRD

and

Department of Applied Mathematics

University of Waterloo

Waterloo, Ontario, Canada

ABSTRACT

Applying an unitary transformation to block-diagonalize cyclic hypermatrices, *ab initio* SCF LCAO crystal orbital (CO) methods have been formulated for one- and three-dimensional periodic systems (polymers or molecular crystals) with arbitrary number of orbitals in the elementary cell. The corresponding expressions are given for both cases when either a closed-shell molecule, or an open-shell system forms the unit cell and the CO method is formulated also for different orbitals for different spins. Possible applications of the newly formulated "open-shell" *ab initio* SCF LCAO CO method to the TCNQ-TTF system are discussed.

1. INTRODUCTION

Polymers play a very essential role as plactics, biopolymers like proteins and nucleic acids (DNA and RNA) have fundamental importance in life processes and most recently highly conducting polymers (like the TCNQ-TTF system with pseudo one-dimensional stacked chains embedded in a three-dimensional molecular crystal) seem to be the candidates for the discovery of new physical phenomena (superconductive-type behaviour at higher temperatures etc.) To understand their different physical and chemical properties (which underlie in the case of biopolymers their biological functions) one has to get a fair knowledge of their electronic structure.

If a polymer is completely aperiodic, it is very difficult to treat it with present-day methods. Clusters of randomly distributed molecules have been approximately investigated with the aid of the Green's function technique (1), but the problem of an aperiodic large polymer or molecular crystal is generally unsolved (for a detailed discussion of this problem see the paper of W. L. McCubbin (2) in this volume). On the other hand if we are able to define in a polymer or in a molecular crystal an elementary cell, by the repetition of which we can build up the whole system, we can treat them with the aid of a proper combination of quantum chemical and solid state physical methods. This repetition can be a simple translation, but it can be obtained also by a combination of the translation and another symmetry operation (e.g. in the DNA double helix we get from one base pair to the next by the simultaneous translation of 3.36 Å along the long axis of the molecule and by the rotation of 36° around this axis).

The aim of this paper is to present the formalisms of the <u>ab initio</u> SCF LCAO crystal orbital (CO) methods for closed shell and open shell systems and in the different orbitals for different spins (DODS) cases. The application of the closed shell formalism will be discussed in a subsequent paper of J.-M. André (3).

2. AB INITIO SCF LCAO CRYSTAL ORBITAL METHOD FOR CLOSED-SHELL SYSTEMS

2.1. Bloch-diagonalization of the Hamiltonian matrix

Let us suppose that we have a three-dimensional periodic polymer or molecular crystal containing m orbitals in the elementary cell of one or more atoms. For the sake of simplicity let the number of elementary cells in the direction of each crystal axis be equal to an odd number $N_1 = N_2 = N_3 = 2N + 1$. We suppose further that there is an interaction between orbitals belonging to different elementary cells. In that case we can describe, in the one-electron approximation, the delocalized crystal orbitals of the polymer with the aid of the LCAO approximation, in the form

$$\phi_h^{\vec{p}}(\vec{r}) = \sum_{\vec{q}} \sum_{g=1}^{m} C(\vec{p})_{h;\vec{q},g} \, \chi_g^{\vec{q}} \qquad (1)$$

where $\vec{p}=(p_1,p_2,p_3)$, $\vec{q}=(q_1,q_2,q_3)$; the integers p_j and q_j (j=1,2,3) run from $-N_1\ldots,0_1\ldots,N_1$ and $\sum_{\vec{q}}$ is a shorthand notation for $\sum_{q_1=-N}^{N}\sum_{q_2=-N}^{N}\sum_{q_3=-N}^{N}$. Further $\chi_g^{\vec{q}} \equiv \chi_g(\vec{r}-\vec{R}_{\vec{q}}-\vec{r}_{\vec{g}s})$ is the gth AO (belonging to the atom with position vector $\vec{g}_A$) of the cell characterized by the vector $\vec{R}_{\vec{q}} = q_1\vec{a}_1 + q_2\vec{a}_2 + q_3\vec{a}_3$ ($q_j = -N_1\ldots,0_1\ldots,N$), where $\vec{a}_1,\vec{a}_2$ and $\vec{a}_3$ are the three basis vectors of the crystal.

If we form the expectation value

$$\frac{\langle \phi_h^{\vec{p}} \mid \hat{H}_{eff.} \mid \phi_h^{\vec{p}} \rangle}{\langle \phi_h^{\vec{p}} \mid \phi_h^{\vec{p}} \rangle} = \varepsilon(\vec{p})_h \tag{2}$$

of one-electron effective Hamiltonian $\hat{H}_{eff.}$ and perform a Ritz variational procedure for the coefficients in (1), we obtain in the standard way for the whole polymer the matrix equation

$$\underline{\underline{H}}\,\underline{C}\,(\vec{p})_h = \varepsilon(\vec{p})_h\,\underline{\underline{S}}\,\underline{C}\,(\vec{p})_h. \tag{3}$$

The hypermatrix $\underline{\underline{H}}$ of dimension $m(2N+1)^3$ has the submatrices $\underline{\underline{H}}_{\vec{p};\vec{q}}$ of dimension m consisting of interactions between orbitals belonging to the elementary cells characterized by the lattice vector $\vec{R}_{\vec{p}}$ and $\vec{R}_{\vec{q}}$, respectively. The f,gth element $H_{\vec{p},f;\vec{q},g}$ of the matrix $\underline{\underline{H}}_{\vec{p},\vec{q}}$ is then given by

$$H_{\vec{p},f;\vec{q},g} = \langle \chi_f^{\vec{p}} \mid \hat{H}_{eff.} \mid \chi_g^{\vec{q}} \rangle \tag{4}$$

The overlap matrix $\underline{\underline{S}}$ can be partitioned into blocks, similar to the matrix $\underline{\underline{H}}$, with elements

$$S_{\vec{p},f;\vec{q},g} = \langle \chi_f^{\vec{p}} \mid \chi_g^{\vec{q}} \rangle \tag{5}$$

In consequence of the three-dimensional translational symmetry of the polymer and Born-Kármán periodic boundary conditions the matrices $\underline{\underline{H}}$ and $\underline{\underline{S}}$ are cyclic hypermatrices. For the sake of simplicity we show this for the one-dimensional case, the generalization for two- and three-dimensional cases is straightforward. In the one-dimensional case the hypermatrices $\underline{\underline{H}}$ and $\underline{\underline{S}}$ have the form

$$\begin{pmatrix} \boxed{\underline{\underline{0}}} & \boxed{\underline{\underline{1}}} & \boxed{\underline{\underline{2}}} & & & \boxed{\underline{\underline{2N}}} \\ \boxed{\underline{\underline{-1}}} & \boxed{\underline{\underline{0}}} & \boxed{\underline{\underline{1}}} & \boxed{\underline{\underline{2}}} & & \boxed{\underline{\underline{2N-1}}} \\ & & & & & \\ \boxed{\underline{\underline{-2N}}} & & & & \boxed{\underline{\underline{-1}}} & \boxed{\underline{\underline{0}}} \end{pmatrix} \qquad (6)$$

if we take into account the translational symmetry. Here the submatrix $\boxed{\underline{\underline{0}}}$ denotes interactions within the elementary cell, the submatrices $\boxed{\underline{\underline{1}}}$ and $\boxed{\underline{\underline{-1}}} = \boxed{\underline{\underline{1}}^{tr}}$ correspond to first neighbours' interactions and so on (all the submatrices have the dimensions of mXm only). Introducing the Born-von Kármán periodic boundary conditions, we obtain the relations $\boxed{\underline{\underline{2N}}} = \boxed{\underline{\underline{-1}}}$, $\boxed{\underline{\underline{2N-1}}} = \boxed{\underline{\underline{-2}}}$ and so on. Therefore, the hypermatrices (6) will become cyclic hypermatrices. As it is well known, for a cyclic matrix, it is always possible to find a unitary transformation which diagonalizes it. In the same way there exists the unitary hypermatrix with blocks $\underline{\underline{U}}_{p,q}$ ($\underline{\underline{U}}_{p,q}$ is the block of $\underline{\underline{U}}$ belonging to the pth row of blocks and qth column of blocks).

$$\underline{\underline{U}}_{p,q} = \frac{1}{(2N+1)^{1/2}} \exp\left(\frac{i\, 2\pi\, pq}{2N+1}\right) \underline{\underline{1}} \qquad (7)$$

which blockdiagonalizes the cyclic hypermatrices (the unit matrix $\underline{\underline{1}}$ has again the dimensions of mXm only). So we can write

$$\underline{\underline{H}}' = \underline{\underline{U}}^{+} \underline{\underline{H}}\, \underline{\underline{U}} = \begin{pmatrix} \square & & & \underline{\underline{0}} \\ & \square & & \\ & & \square & \\ \underline{\underline{0}} & & & \ddots \; \square \end{pmatrix} \quad \underline{\underline{S}}' = \underline{\underline{U}}^{+} \underline{\underline{S}}\, \underline{\underline{U}} , \qquad (8)$$

where only the diagonal mXm blocks differ from zero. It should be pointed out that the unitary matrix $\underline{\underline{U}}$ defined by (7) blockdiagonalizes any cyclic hypermatrix of dimension m(2N+1) with blocks of dimension m independently of the values of the elements building up the submatrices. Therefore the same matrix $\underline{\underline{U}}$ will blockdiagonalize both $\underline{\underline{H}}$ and $\underline{\underline{S}}$.

Returning to the three-dimensional problem it can be shown in a similar way (4) that the matrices $\underline{\underline{H}}$ and $\underline{\underline{S}}$ are cyclic hypermatrices also in this case and the unitary matrix $\underline{\underline{U}}$ of dimension $m(2N+1)^3$ with mXm blocks

$$\underline{\underline{U}}_{\vec{p},\,\vec{q}} = \frac{1}{(2N+1)^{3/2}} \exp\left(\frac{i\,2\pi\,\vec{p}\,\vec{q}}{2N+1}\right) \underline{\underline{1}} \tag{9}$$

blockdiagonalizes them. With the definition of (9) it is easy to prove that $\underline{\underline{U}}^{+}\underline{\underline{U}} = \underline{\underline{I}}$ where the unit matrix $\underline{\underline{I}}$ has the dimension of $m(2N+1)^3$ and the $\vec{p}$th diagonal blocks of the blockdiagonal matrices

$$\underline{\underline{H}}' = \underline{\underline{U}}^{+}\,\underline{\underline{H}}\,\underline{\underline{U}}\,,\quad \underline{\underline{S}}' = \underline{\underline{U}}^{+}\,\underline{\underline{S}}\,\underline{\underline{U}} \tag{10}$$

have the form

$$\underline{\underline{H}}'(\vec{p}) = \sum_{\vec{q}} \exp\left(\frac{i\,2\pi\,\vec{p}\,\vec{q}}{2N+1}\right) \underline{\underline{H}}(\vec{q}) \tag{11a}$$

$$\underline{\underline{S}}(\vec{p}) = \sum_{\vec{q}} \exp\left(\frac{i\,2\pi\,\vec{p}\,\vec{q}}{2N+1}\right) \underline{\underline{S}}(\vec{q}) \tag{11b}$$

In these expressions the matrices $\underline{\underline{H}}(\vec{q})$ and $\underline{\underline{S}}(\vec{q})$, respectively, are submatrices of the original matrices $\underline{\underline{H}}$ and $\underline{\underline{S}}$ before the unitary transformation. So in the one-dimensional case $q=0$ gives the submatrix $\boxed{\underline{\underline{0}}}$ of (6), $q=1$ and $q=-1$ refer to the blocks $\boxed{\underline{\underline{1}}}$ and $\boxed{\underline{\underline{-1}}}$ (first neighbours interactions) of (6) and so on.

Multiplying equ.-s(3) from the left by $\underline{\underline{U}}^{+}$ and inserting in them after $\underline{\underline{H}}$ and $\underline{\underline{S}}$, respectively, the unit matrix $\underline{\underline{I}}$ in the form $\underline{\underline{I}}=\underline{\underline{U}}\,\underline{\underline{U}}^{+}$, we obtain

$$\underline{\underline{U}}^{+}\,\underline{\underline{H}}\,\underline{\underline{U}}\,\underline{\underline{U}}^{+}\,\underline{C}(\vec{p})_h = \varepsilon(\vec{p})_h\,\underline{\underline{U}}^{+}\,\underline{\underline{S}}\,\underline{\underline{U}}\,\underline{\underline{U}}^{+}\,\underline{C}(\vec{p})_h \tag{12}$$

Introducing the notation $\underline{\underline{U}}^{+}\,\underline{C}(\vec{p})_h = \underline{D}(\vec{p})_h$ and taking into account (10) we can rewrite (12) as

$$\underline{\underline{H}}'\,\underline{D}(\vec{p})_h = \varepsilon(\vec{p})_h\,\underline{\underline{S}}'\,\underline{D}(\vec{p})_h \tag{13}$$

Using the fact that $\underline{\underline{H}}'$ and $\underline{\underline{S}}'$ are blockdiagonal equations (13) can be decomposed to the much simpler equations

$$\underline{\underline{H}}'(\vec{p})\,\underline{d}(\vec{p})_h = \varepsilon(\vec{p})_h\,\underline{\underline{S}}'(\vec{p})\,\underline{d}(\vec{p})_h \qquad (14)$$

where $\underline{\underline{H}}'(\vec{p})$ and $\underline{\underline{S}}'(\vec{p})$ have been defined by (11a) and (11b), respectively.

If $N \to \infty$ we can take the new variables

$$k_1 = \frac{2\pi\, p_1}{a_1(2N+1)}\,,\quad k_2 = \frac{2\pi\, p_2}{a_2(2N+1)}\,,\quad k_3 = \frac{2\pi\, p_3}{a_3(2N+1)} \qquad (15)$$

as continuous ones. Since the p_j-s(j=1,2,3) had the values -N,..., 0,...,N k_1, k_2 and k_3 will have values between $-\frac{\pi}{a_1}$ and $\frac{\pi}{a_1}$; $-\frac{\pi}{a_2}$ and $\frac{\pi}{a_2}$; $-\frac{\pi}{a_3}$ and $\frac{\pi}{a_3}$, respectively. Defining the vector

$$\vec{k} = r_1\vec{b}_1 + k_2\vec{b}_2 + k_3\vec{b}_3 \qquad (16)$$

where $\vec{b}_1$ etc. are the basis vectors of the reciprocal space (by definition $\vec{a}_i\vec{b}_j = 2\pi\,\delta_{i,j}$)[1.)] we can rewrite (14) as

$$\underline{\underline{H}}(\vec{k})\,\underline{d}(\vec{k})_h = \varepsilon(\vec{k})_h\,\underline{\underline{S}}(\vec{k})\,\underline{d}(\vec{k})_h \qquad (h=1,2,\ldots,m) \quad (17)$$

where now

$$\underline{\underline{H}}(\vec{k}) = \sum_{q=-\infty}^{\infty} e^{i\vec{k}\vec{R}_{\vec{q}}}\,\underline{\underline{H}}(\vec{q}) \qquad (18)$$

and

$$\underline{\underline{S}}(\vec{k}) = \sum_{q=-\infty}^{\infty} e^{i\vec{k}\vec{R}_{\vec{q}}}\,\underline{\underline{S}}(\vec{q}) \qquad (19)$$

1.) The first Brillouin zones of a crystal with a given symmetry are usually given in terms of the Cartesian coordinates of $\vec{k}$. On the other hand, we can characterize the elementary cells most easily with the aid of the lattice vectors $\vec{R}_{\vec{q}}$ expressed in terms of the unit vectors $\vec{a}_1$,$\vec{a}_2$ and $\vec{a}_3$ which are not necessarily orthogonal. Therefore, we need to transform the rectangular coordinates k_x, k_y and k_z of $\vec{k}$ into the k_j -s(j=1,2,3). It is easy to show that this can be achieved with the aid of the equations

$$k_j = \frac{1}{2\pi}\,(a_{jx}\,k_x + a_{jy}\,k_y + a_{jz}\,k_z) \qquad (j=1,2,3)$$

where a_{jx}, a_{jy} and a_{jz} are the rectangular coordinates of the basis vector $\vec{a}_j$.

2.2. Elimination of the overlap matrix

To solve equ. -s (17) first we have to eliminate the matrix $\underline{\underline{S}}\,(\vec{k})$. We can do this most simply with the aid of Löwdin's symmetric orthogonalization procedure (5). Multiplying (17) from the left by $\underline{\underline{S}}\,(\vec{k})^{-1/2}$ and inserting $\underline{\underline{S}}\,(\vec{k})^{-1/2}\,\underline{\underline{S}}\,(\vec{k})^{1/2} = \underline{\underline{1}}$ we can write

$$\underline{\underline{S}}\,(\vec{k})^{-1/2}\,\underline{\underline{H}}\,(\vec{k})\,\underline{\underline{S}}\,(\vec{k})^{-1/2}\,\underline{\underline{S}}\,(\vec{k})^{1/2}\,\underline{d}\,(\vec{k})_h =$$

$$\varepsilon(\vec{k})_h\,\underline{\underline{S}}\,(\vec{k})^{-1/2}\,\underline{d}\,(\vec{k})_h \tag{20}$$

or introducing the notations

$$\hat{\underline{\underline{H}}}\,(\vec{k}) = \underline{\underline{S}}\,(\vec{k})^{-1/2}\,\underline{\underline{H}}\,(\vec{k})\,\underline{\underline{S}}\,(\vec{k})^{-1/2} \tag{21}$$

and

$$\underline{b}\,(\vec{k})_h = \underline{\underline{S}}\,(\vec{k})^{1/2}\,\underline{d}\,(\vec{k})_h \tag{22}$$

$$\hat{\underline{\underline{H}}}\,(\vec{k})\,\underline{b}\,(\vec{k})_h = \varepsilon(\vec{k})_h\,\underline{b}\,(\vec{k})_h \tag{23}$$

The square root and inverse square root of $\underline{\underline{S}}\,(\vec{k})$ are easily calculated by diagonalizing it,

$$\underline{\underline{S}}_o\,(\vec{k}) = \underline{\underline{V}}\,(\vec{k})^+\,\underline{\underline{S}}\,(\vec{k})\,\underline{\underline{V}}\,(\vec{k}) \tag{24}$$

where the unitary matrix V $(\vec{k})$ contains as colum $\underline{v}\,(\vec{k})_i$ of the eigenvalue problem

$$\underline{\underline{S}}\,(\vec{k})\,\underline{v}\,(\vec{k})_i = s\,(\vec{k})_i\,\underline{v}\,(\vec{k})_i \tag{25}$$

and the diagonal matrix $\underline{\underline{S}}_o(\vec{k})$ contains as elements the eigenvalues $s(\vec{k})_i$. From (24) one can express $\underline{\underline{S}}\,(\vec{k})$ as

$$\underline{\underline{S}}\,(\vec{k}) = \underline{\underline{V}}\,(\vec{k})\,\underline{\underline{S}}_o\,(\vec{k})\,\underline{\underline{V}}\,(\vec{k})^+$$

and $\underline{\underline{S}}\,(\vec{k})^{1/2}$ and $\underline{\underline{S}}\,(\vec{k})^{-1/2}$, respectively,

as $\underline{\underline{S}}(\vec{k})^{1/2} = \underline{\underline{V}}\,(\vec{k})\,\underline{\underline{S}}_o\,(\vec{k})^{1/2}\,\underline{\underline{V}}\,(\vec{k})^+$, $\underline{\underline{S}}\,(\vec{k})^{-1/2} = \underline{\underline{V}}\,(\vec{k})\,\underline{\underline{S}}_o\,(\vec{k})^{-1/2}$

$$\times\ \underline{\underline{V}}\,(\vec{k})^+ \tag{26}$$

(For the proof of the validity of equations (26) see for instance (6). Since it can be shown that all the eigenvalues of $\underline{\underline{S}}\,(\vec{k})$ are positive, it is easy to calculate $\underline{\underline{S}}_o(\vec{k})^{1/2}$ and $\underline{\underline{S}}_o(\vec{k})^{-1/2}$, respectively.

The most simple way to diagonalize the Hermitian complex matrix $\underline{\underline{S}}(\vec{k})$ defined by (19) is, if we rewrite its eigenvalue equation (25), or

$$(\mathrm{Re}[\underline{\underline{S}}(\vec{k})] + i\,\mathrm{Im}[\underline{\underline{S}}(\vec{k})])\,(\mathrm{Re}[\underline{v}(\vec{k})_i] + i\,\mathrm{Im}[\underline{v}(\vec{k})_i]) = s(\vec{k})_i\,(\mathrm{Re}[\underline{v}(\vec{k})_i] + i\,\mathrm{Im}[\underline{v}(\vec{k})_i]) \tag{27}$$

by separating its real and imaginary parts :

$$\mathrm{Re}[\underline{\underline{S}}(\vec{k})]\,\mathrm{Re}[\underline{v}(\vec{k})_i] - \mathrm{Im}[\underline{\underline{S}}(\vec{k})]\,\mathrm{Im}[\underline{v}(\vec{k})_i] = s(\vec{k})_i \times \mathrm{Re}[\underline{v}(\vec{k})_i] \tag{28a}$$

$$\mathrm{Im}[\underline{\underline{S}}(\vec{k})]\,\mathrm{Re}[\underline{v}(\vec{k})_i] + \mathrm{Re}[\underline{\underline{S}}(\vec{k})]\,\mathrm{Im}[\underline{v}(\vec{k})_i] = s(\vec{k})_i\,\mathrm{Im}[\underline{v}(\vec{k})_i] \tag{28b}$$

Instead of writing the two matrix equations (28a) and (28b) we can write the single matrix eigenvalue equation

$$\begin{pmatrix} \mathrm{Re}[\underline{\underline{S}}(\vec{k})] & -\mathrm{Im}[\underline{\underline{S}}(\vec{k})] \\ \mathrm{Im}[\underline{\underline{S}}(\vec{k})] & \mathrm{Re}[\underline{\underline{S}}(\vec{k})] \end{pmatrix} \begin{pmatrix} \mathrm{Re}[\underline{v}(\vec{k})_i] \\ \mathrm{Im}[\underline{v}(\vec{k})_i] \end{pmatrix} = s(\vec{k})_i \begin{pmatrix} \mathrm{Re}[\underline{v}(\vec{k})_i] \\ \mathrm{Im}[\underline{v}(\vec{k})_i] \end{pmatrix}$$

of a real matrix which has the order of 2m (if the order of the Hermitian complex matrix $\underline{\underline{S}}(\vec{k})$ was m). Each of the always real eigenvalues $s(\vec{k})_i$ of $\underline{\underline{S}}(\vec{k})$ will occur in (29) twice and taking into account (19) we can write

$$\mathrm{Re}[\underline{\underline{S}}(\vec{k})] = \sum_{\vec{q}=-\infty}^{\infty} \cos(\vec{k}\vec{R}_{\vec{q}})\,\underline{\underline{S}}(\vec{q}), \quad \mathrm{Im}[\underline{\underline{S}}(\vec{k})] = \sum_{\vec{q}=-\infty}^{\infty} \sin(\vec{k}\vec{R}_{\vec{q}})\,\underline{\underline{S}}(\vec{q}). \tag{30}$$

We can rewrite in a similar way also the eigenvalue equation (23), or

$$\underline{\underline{\hat{H}}}(\vec{k})\,\underline{\underline{B}}(\vec{k}) = \underline{\underline{B}}(\vec{k})\,\underline{\underline{\varepsilon}}(\vec{k}) \tag{31}$$

(here the matrix $\underline{\underline{B}}(\vec{k})$ contains again as columns the vectors $\underline{b}(\vec{k})_h$ and the diagonal matrix $\underline{\underline{\varepsilon}}(\vec{k})$ has as elements the eigenvalues

$\varepsilon(\vec{k})_h$) as

$$\begin{pmatrix} \mathrm{Re}[\hat{\underline{\underline{H}}}(\vec{k})] & -\mathrm{Im}[\hat{\underline{\underline{H}}}(\vec{k})] \\ \mathrm{Im}[\hat{\underline{\underline{H}}}(\vec{k})] & \mathrm{Re}[\hat{\underline{\underline{H}}}(\vec{k})] \end{pmatrix} \begin{pmatrix} \underline{\underline{B}}_1(\vec{k}) \\ \underline{\underline{B}}_2(\vec{k}) \end{pmatrix} = \begin{pmatrix} \underline{\underline{B}}_1(\vec{k}) \\ \underline{\underline{B}}_2(\vec{k}) \end{pmatrix} \underline{\underline{\varepsilon}}'(\vec{k}) \quad (32)$$

where $\underline{\underline{B}}_1(\vec{k}) = \mathrm{Re}[\underline{\underline{B}}(\vec{k})]$, $\underline{\underline{B}}_2(\vec{k}) = \mathrm{Im}[\underline{\underline{B}}(\vec{k})]$ and the diagonal matrix $\underline{\underline{\varepsilon}}'(\vec{k})$ contains twice each eigenvalue $\varepsilon(\vec{k})_h$. Taking into account ($\overline{2}1$) and the second equation of (26) we obtain

$$\mathrm{Re}[\hat{\underline{\underline{H}}}(\vec{k})] = \mathrm{Re}[\underline{\underline{G}}(\vec{k})]\,(\mathrm{Re}[\underline{\underline{H}}(\vec{k})]\,\mathrm{Re}[\underline{\underline{G}}(\vec{k}) - \mathrm{Im}[\underline{\underline{H}}(\vec{k})]\,\mathrm{Im}[\underline{\underline{G}}(\vec{k})])$$
$$- \mathrm{Im}[\underline{\underline{G}}(\vec{k})]\,(\mathrm{Im}[\underline{\underline{H}}(\vec{k})]\,\mathrm{Re}[\underline{\underline{G}}(\vec{k})] + \mathrm{Re}[\underline{\underline{H}}(\vec{k})]\,\mathrm{Im}[\underline{\underline{G}}(\vec{k})])\quad(33)$$

$$\mathrm{Im}[\hat{\underline{\underline{H}}}(\vec{k})] = \mathrm{Re}[\underline{\underline{G}}(\vec{k})]\,(\mathrm{Re}[\underline{\underline{H}}(\vec{k})]\,\mathrm{Im}[\underline{\underline{G}}(\vec{k}) + \mathrm{Im}[\underline{\underline{H}}(\vec{k})\,\mathrm{Re}[\underline{\underline{G}}(\vec{k})])$$
$$+ \mathrm{Im}[\underline{\underline{G}}(\vec{k})]\,(\mathrm{Re}[\underline{\underline{H}}(\vec{k})]\,\mathrm{Re}[\underline{\underline{G}}(\vec{k})] - \mathrm{Im}[\underline{\underline{H}}(\vec{k})]\,\mathrm{Im}[\underline{\underline{G}}(\vec{k})])\quad(34)$$

with

$$\mathrm{Re}[\underline{\underline{G}}(\vec{k})] = \mathrm{Re}[\underline{\underline{S}}(\vec{k})^{-1/2}] = \underline{\underline{V}}_1(\vec{k})\,\underline{\underline{S}}_o(\vec{k})^{-1/2}\,\underline{\underline{V}}_1(\vec{k})^+ +$$
$$\underline{\underline{V}}_2(\vec{k})\,\underline{\underline{S}}_o(\vec{k})^{-1/2}\,\underline{\underline{V}}_2(\vec{k})^+ , \quad (35a)$$

$$\mathrm{Im}[\underline{\underline{G}}(\vec{k}) = \mathrm{Im}[\underline{\underline{S}}(\vec{k})^{-1/2}] = -\underline{\underline{V}}_1(\vec{k})\,\underline{\underline{S}}_o(\vec{k})^{-1/2}\,\underline{\underline{V}}_2(\vec{k})^+ +$$
$$\underline{\underline{V}}_2(\vec{k})\,\underline{\underline{S}}_o(\vec{k})^{-1/2}\,\underline{\underline{V}}_1(\vec{k})^+ , \quad (35b)$$

$$\underline{\underline{V}}_1(\vec{k}) = \mathrm{Re}[\underline{\underline{V}}(\vec{k})] \quad \underline{\underline{V}}_2(\vec{k}) = \mathrm{Im}[\underline{\underline{V}}(\vec{k})]$$

With the aid of equations (33) and (34) one can always calculate the real and imaginary parts of $\hat{\underline{\underline{H}}}(\vec{k})$ and thus solve equation (32). Having thus determined the matrices $\underline{\underline{B}}_1(\vec{k})$ and $\underline{\underline{B}}_2(\vec{k})$ on the basis of the relation

$$\underline{\underline{B}}(\vec{k}) = \underline{\underline{B}}_1(\vec{k}) + i\,\underline{\underline{B}}_2(\vec{k}) = \underline{\underline{S}}(\vec{k})^{1/2}\,\underline{\underline{D}}(\vec{k}) \quad (36)$$

one can always calculate back the real and imaginary parts of the original vectors $\underline{d}(\vec{k})_h$:

$$\underline{\underline{D}}(\vec{k}) = \underline{\underline{S}}(\vec{k})^{-1/2}\,\underline{\underline{B}}(\vec{k}) ,$$

$$\underline{\underline{D}}_1(\vec{k}) + i\,\underline{\underline{D}}_2(\vec{k}) = (\mathrm{Re}\,[\underline{\underline{S}}(\vec{k})^{-1/2}] + i\,\mathrm{Im}\,[\underline{\underline{S}}(\vec{k})^{-1/2}])\ \mathrm{x}$$

$$(\underline{\underline{B}}_1(\vec{k}) + i\,\underline{\underline{B}}_2(\vec{k}))$$

$$\underline{\underline{D}}_1(\vec{k}) = \mathrm{Re}\,[\underline{\underline{S}}(\vec{k})^{-1/2}]\,\underline{\underline{B}}_1(\vec{k}) - \mathrm{Im}\,[\underline{\underline{S}}(\vec{k})^{-1/2}]\,\underline{\underline{B}}_2(\vec{k}) \quad (37a)$$

$$\underline{\underline{D}}_2(\vec{k}) = \mathrm{Re}\,[\underline{\underline{S}}(\vec{k})^{-1/2}]\,\underline{\underline{B}}_2(\vec{k}) + \mathrm{Im}\,[\underline{\underline{S}}(\vec{k})^{-1/2}]\,\underline{\underline{B}}_1(\vec{k}) \quad (37b)$$

In these equations we can substitute for the real and imaginary parts of $\underline{\underline{S}}(\vec{k})^{-1/2}$ the expressions (35).

2.3. Hartree-Fock-Roothaan Formalism

Until now we have not specified the matrices $\underline{\underline{H}}(\vec{q})$ in (18). To do this let us write down the Hamiltonian of the whole polymer (in atomic units) as

$$\hat{H} = \sum_{\mu=1}^{n_e(2N+1)^3} \hat{H}^N(\mu) + \sum_{\mu<\nu}^{n_e(2N+1)^3} \frac{1}{r_{\mu\nu}} \quad (38)$$

$$\hat{H}^N(\mu) = -\frac{1}{2}\Delta - \sum_{\vec{q}} \sum_{\alpha=1}^{M} \frac{Z_\alpha}{|\vec{r}_\mu - \vec{R}_\alpha^{\vec{q}}|} \quad (39)$$

where the Greek letters μ,ν refer to the electrons, n_e denotes the number of electrons within the elementary cell, Z_α is the nuclear charge of atom α, $\vec{R}_\alpha^{\vec{q}}$ stands for the position vector of the αth nucleus in the cell characterized by $\vec{R}_{\vec{q}}$ and M is the number if atoms in the elementary cell. If we take the one-electron orbitals in the (1) LCAO form, build up with the aid of them a Slater determinant many-electron wave function and perform the same variational procedure on the expectation value $<\hat{H}>$ of (38) (calculated with the mentioned Slater determinant), as Roothaan (7) did for molecules, we obtain the matrix equation (3), where the elements of the hypermatrix $\underline{\underline{H}}$ are defined by (4) if we substitute in it for $\hat{H}_{eff.}$ the Fock operator $\hat{F}$

$$\hat{H}_{eff.} = \hat{F} = \hat{H}^N + \sum_{\vec{p}} \sum_{h=1}^{n^*} [2\,\hat{J}(\vec{p},h) - \hat{K}(\vec{p},h)] \ . \quad (40)$$

Here the operator $\hat{H}^N$ is given by (39) and the Coulomb and exchange operators $\hat{J}(\vec{p},h)$ and $\hat{K}(\vec{p},h)$ are defined, respectively, as

$$\hat{J}(\vec{p},h,\mu)\,\psi(\mu) = <\phi_h^{\vec{p}}(\nu)\,|\frac{1}{r_{\mu\nu}}|\,\phi_h^{\vec{p}}(\nu)>\,\phi(\mu) \quad (41a)$$

$$\hat{K}(\vec{p},h,\mu)\ \phi(\mu) = \langle \phi_h^{\vec{p}}(\nu) | \frac{1}{r_{\mu\nu}} | \phi(\nu) \rangle\ \phi_h^{\vec{p}}(\mu) \qquad (41b)$$

and n^* denotes the number of MO's which would be doubly filled by the n_e electrons coming from one elementary cell ($n^* = \frac{n_e}{2}$).

As we have seen above, if we take into account the translational symmetry of the system and introduce the Born-Kármán periodic boundary conditions our matrix equation (3) reduces to (17). In the Hartree-Fock-Roothaan case the elements of the Fock matrices $\underline{\underline{F}}(\vec{q})$ occurring in the expression

$$\underline{\underline{F}}(\vec{k}) = \sum_{\vec{q}} e^{i\vec{k}\vec{R}_{\vec{q}}}\ \underline{\underline{F}}(\vec{q}) \qquad (42)$$

can be obtained, if we substitute into the expression

$$(\underline{\underline{F}}(\vec{q}))_{r,s} = \langle \chi_r^{\vec{o}} | \hat{F} | \chi_s^{\vec{q}} \rangle = \langle \chi_r^{\vec{o}} | \hat{H}^N + \sum_{\vec{p}} \sum_{h=1}^{n^*} [2\hat{J}(\vec{p},h) - \hat{K}(\vec{p},h)] | \chi_s^{\vec{q}}] \rangle \qquad (43)$$

(41a) and (41b) together with the LCAO form (1) of the crystal orbitals occuring in the operators $\hat{J}$ and $\hat{K}$. We arrive in this way at the formula

$$[\underline{\underline{F}}(\vec{q})]_{r,s} = \langle \chi_r^{\vec{o}} | \hat{H}^N | \chi_s^{\vec{q}} \rangle + \sum_{\vec{p}} \sum_{h=1}^{n^*} \sum_{\vec{q}_1} \sum_{\vec{q}_2} \sum_{u=1}^{m} \sum_{v=1}^{m} C(\vec{p})^*_{h;\vec{q}_1,u}$$

$$\times\ C(\vec{p})_{h;\vec{q}_2,v}\ (2\langle \chi_r^{\vec{o}}\ \chi_u^{\vec{q}_1} | \chi_s^{\vec{q}}\ \chi_v^{\vec{q}_2} \rangle - \langle \chi_r^{\vec{o}}\ \chi_u^{\vec{q}_1} | \chi_v^{\vec{q}_2}\ \chi_s^{\vec{q}} \rangle) \qquad (44)$$

We can introduce the charge-bond order matrix of the three-dimensional polymer as

$$\underline{\underline{P}} = 2 \sum_{\vec{p}} \sum_{h=1}^{n^*} \underline{C}(\vec{p})_h\ \underline{C}(\vec{p})_h^+ \qquad (45)$$

Taking into account the transformation $\underline{C}(\vec{p})_h = \underline{\underline{U}}\ \underline{D}(\vec{p})_h$ (see (12)) and the form (9)) of the unitary matrix $\underline{\underline{U}}$, one obtains for the subblocks $\underline{\underline{p}}(\vec{q}_1,\vec{q}_2)$ of $\underline{\underline{P}}$ the expression

$$\underline{\underline{p}}(\vec{q}_1,\vec{q}_2) = \frac{2}{(2N+1)^3} \sum_{\vec{p}} \sum_{h=1}^{n^*} \underline{d}(\vec{p})_h\ \underline{d}(\vec{p})_h^+$$

$$\times \exp[i\ 2\pi\ \vec{p}\ (\vec{q}_1 - \vec{q}_2)/(2N+1)] \qquad (46)$$

Introducing instead of $\vec{p}$ the continuous variable $\vec{k}$ defined by (15) and (16) and integrating over $\vec{k}$ instead of summing over $\vec{p}$, we can further write

$$\underline{\underline{p}}(\vec{q}_1,\vec{q}_2) = \frac{2}{\omega} \int_{\omega} \sum_{h=1}^{n^*} \underline{d}(\vec{k})_h \, \underline{d}(\vec{k})_h^+ \, e^{i\vec{k}(\vec{R}_{\vec{q}_1} - \vec{R}_{\vec{q}_2})} \, d\vec{k} \quad (47)$$

where ω is the volume of the first Brillouin zone in $\vec{k}$ space.

Substituting (45) into (44) and using (47) one can show (4) that (44) takes the form

$$[\underline{\underline{F}}(\vec{q})]_{r,s} = \langle \chi_r^{\vec{o}} | \hat{H}^N | \chi_s^{\vec{q}} \rangle + \sum_{\vec{q}} \sum_{\vec{q}_2} \sum_{u,v} p(\vec{q}_1 - \vec{q}_2)_{u,v} \times$$

$$(\langle \chi_r^{\vec{o}} \chi_u^{\vec{q}_1} | \chi_s^{\vec{q}} \chi_v^{\vec{q}_2} \rangle - \frac{1}{2} \langle \chi_v^{\vec{o}} \chi_u^{\vec{q}1} | \chi_v^{\vec{q}2} \chi_s^{\vec{q}} \rangle) \quad (48)$$

here $p(\vec{q}_1 - \vec{q}_2)_{u,v}$ is the μ, νth element of the submatrix.

$$\underline{\underline{p}}(\vec{q}_1 - \vec{q}_2) = \underline{\underline{P}}(\vec{q}_1, \vec{q}_2). \quad (49)$$

(The latter equation follows immediately from the form (47) of $\underline{\underline{P}}(\vec{q}_1,\vec{q}_2)$). It should be mentioned that the described <u>ab initio</u>-SCF LCAO crystal orbital method was developed independently and simultaneously by Del Re, Ladik and Biczó (4) and by Andre, Gouverneur and Leroy (8).

In the case of a <u>one-dimensional chain</u> with translational symmetry all the vectors $\vec{k}$, $\vec{q}$, $\vec{q}_1$ and $\vec{q}_2$ become scalar and so (17) reduces to

$$\underline{\underline{F}}(k)\, \underline{d}(k)_h = \varepsilon(k)_h\, \underline{\underline{S}}(k)\, \underline{d}(k)_h \quad (50)$$

with

$$\underline{\underline{F}}(k) = \sum_{q=-\infty}^{\infty} e^{ikqa}\, \underline{\underline{F}}(q) \quad (51a)$$

$$\underline{\underline{S}}(k) = \sum_{q=-\infty}^{\infty} e^{ikqa}\, \underline{\underline{S}}(q) \quad (51b)$$

where a is the elementary translation

$$[\underline{\underline{S}}(q)]_{r,s} = \langle \chi_r^o | \chi_s^q \rangle \quad (52)$$

and the elements of the matrices $\underline{\underline{F}}(q)$ are now defined by

$$[\underline{\underline{F}}(q)]_{r,s} = \langle \chi_r^o | -\frac{1}{2}\Delta - \sum_{q_1=-\infty}^{\infty} \sum_{\alpha=1}^{M} \frac{Z_\alpha}{|\vec{r} - \vec{R}_\alpha^{q_1}|} | \chi_s^q \rangle +$$

$$\sum_{q_1,q_2=-\infty}^{\infty} \sum_{u,v=1}^{m} p(q_1-q_2)_{u,v} (\langle \chi_r^o \chi_u^{q_1} | \chi_s^q \chi_v^{q_2} - \frac{1}{2}\langle \chi_r^o \chi_u^{q_1} | \chi_v^{q_2} \chi_s^q \rangle \quad (53)$$

with

$$p(q_1-q_2)_{u,v} = \frac{a}{2\pi} \int_{-\pi/a}^{\pi/a} \sum_{h=1}^{n^*} d(k)^*_{h,u} \, d(k)_{h,v} \, e^{ika(q_1-q_2)} \, dk. \quad (54)$$

The elimination of matrix $\underline{\underline{S}}(k)$ from equ. (50) can be performed as it was described above for the three-dimensional case. It has to be taken into account, however, both in the three- and in the one-dimensional case that the vectors $\underline{d}(\vec{k})_h$ and $\underline{d}(k)_h$ (which occur in the expressions (46) and (54) of the charge-bond order matrix elements) have to be calculated back with the aid of the expressions

$$\underline{d}(\vec{k})_h = \underline{\underline{S}}(\vec{k})^{-1/2} \, \underline{b}(\vec{k})_h$$
$$\underline{d}(k)_h = \underline{\underline{S}}(k)^{-1/2} \, \underline{b}(k)_h \quad (55)$$

from the vectors $\underline{b}(\vec{k})_h$ and $\underline{b}(k)_h$, respectively (which one obtains from the solution of the eigenvalue problem (31) and from its counterpart for the one-dimensional case), in each iteration step. For this reason and since the dependence of the vector components $d(k)_{h,u}$ on $\vec{k}$ or k, respectively, is usually not known analytically and therefore one has to perform the integrations (46) and (54) numerically, the programming of the described method is not simple.

For the one-dimensional case with finite neighbours' interactions ($q = -N, \ldots 0, \ldots, N$, where N is a small integer number) Andre (9) has written a program. This program, called POLYMOL, uses a Gauussian basis set and takes into account the fact that the matrices $\underline{\underline{F}}(\vec{k})$ and $\underline{\underline{F}}(k)$ are Hermitian complex one also in the *ab initio* SCF case, which follows from the relations

$$\underline{\underline{F}}(\vec{q}) = \underline{\underline{F}}(-\vec{q})^{tr} \quad ; \quad \underline{\underline{F}}(q) = \underline{\underline{F}}(-q)^{tr} \quad (56)$$

The latter relations can easily be proven if we take into account the translational symmetry of all the occurring integrals and the relations $\underline{\underline{P}}(\vec{q}_1-\vec{q}_2) = \underline{\underline{P}}(\vec{q}_2-\vec{q}_1)^{tr}$; $\underline{\underline{P}}(q_1-q_2) = \underline{\underline{P}}(q_2-q_1)^{tr}$ (which follow again from the translational symmetry). With the aid of POLYMOL *ab initio* calculations for some simple one-dimensional polymers (polyene, polyethylene) have already been performed (10).

In the case of nearest neighbours'interactions approximation in the expressions for $\underline{\underline{F}}(\vec{k})$ and $\underline{\underline{S}}(\vec{k})$ we have to sum only for nearest neighbours. Thus we can write in the three-dimensional case

$$\underline{\underline{F}}(\vec{k}) = \sum_{\vec{q}}^{(n.n.)} e^{i\vec{k}\vec{R}_{\vec{q}}} \underline{\underline{F}}(\vec{q}) \tag{57a}$$

$$\underline{\underline{S}}(\vec{k}) = \sum_{\vec{q}}^{(n.n.)} e^{i\vec{k}\vec{R}_{\vec{q}}} \underline{\underline{S}}(\vec{q}) \tag{57b}$$

whereas the expressions for the one-dimensional case will be

$$\underline{\underline{F}}(k) = \underline{\underline{F}}(0) + \underline{\underline{F}}(1)\, e^{ika} + \underline{\underline{F}}(1)^{tr}\, e^{-ika} \tag{58a}$$

$$\underline{\underline{S}}(k) = \underline{\underline{S}}(0) + \underline{\underline{S}}(1)\, e^{ika} + \underline{\underline{S}}(1)^{tr}\, e^{-ika} \tag{58b}$$

To be consistent we have to apply the nearest neighbours' interactions approximation also in the summations occurring in the matrix elements (48) and (53). Thus we have to write for instance in the one-dimensional case

$$[\underline{\underline{F}}(q)]_{r,s} = \langle \chi_r^o \,|-\tfrac{1}{2}\Delta - \sum_{q_1=-1}^{1}{}' \sum_{\alpha=1}^{M} \frac{Z_\alpha}{|\vec{r} - \vec{R}_\alpha^{q_1}|}\,|\, \chi_s^q \rangle + \sum_{q_1,q_2=-1}^{1}{}' \times$$

$$\sum_{u,v=1}^{m} p(q_1-q_2)_{u,v} \left(\langle \chi_r^o\, \chi_u^{q_1} \,|\, \chi_s^q\, \chi_v^{q_2} \rangle - \tfrac{1}{2} \langle \chi_r^o\, \chi_u^{q_1} \,|\chi_v^{q_2}\, \chi_s^q \rangle\right)$$

$$(q=-1,0,1) \tag{59}$$

where the primes after the summation signs according to q_1 and q_1, q_2, respectively, indicate that from the summations all such values of q_1, and q_1,q_2, respectively, have two be excluded which lead to integrals which have AO-s centered in non-nearest neighbour cells. If for instance q=1 in the four-center integral $\langle \chi_r^o\, \chi_u^{q_1} | \chi_s^q\, \chi_v^{q_2} \rangle$ neither q_1,nor q_2 can take the value of -1. We shall call this restriction nearest neighbours'interactions approximation in the strict sense [2.)]

If we apply the nearest neighbours'interactions approximation in a strict sense for a one-dimensional problem, besides the integrals $\langle \chi_r^o\ \chi_u^o\, | \chi_s^o\ \chi_v^o \rangle$ occuring in the single molecule, we have only three other types of integrals describing interactions between

2.) It is easy to show that in the nearest neighbours'interactions approximation the relations (56) are fulfilled only then, when we apply this approximation in the above defined strict sense.

neighbouring molecules in the infinite chain. These are the following :

1) $< \chi_r^0 \; \chi_u^1 \mid \chi_s^0 \; \chi_v^1 >$ (first electron centered in cell q=0, the second one in cell q=1),

2) $< \chi_r^0 \; \chi_u^0 \mid \chi_s^1 \; \chi_v^0 >$ (the first electron has an intercell distribution, while electron two is centered in the q=0 cell) and finally

3) $< \chi_r^0 \; \chi_u^0 \mid \chi_s^1 \; \chi_v^0 >$ (both electrons have intercell distributions).

Taking into account the translational symmetry and the fact that $\underline{F}^{(q)}$ contains all pairs of indices r,s and that in (59) we have to sum over all values of u and v , it is easy to see that no other four-center integrals will occur[3.)]. This decreases the necessary computer time in a considerable amount.

The question how many neighbours have to be taken into account in a calculation is not a trivial one. Semiempirical all-valence-electron band structure calculations (for these methods see (12)) of H-F...H-F... and

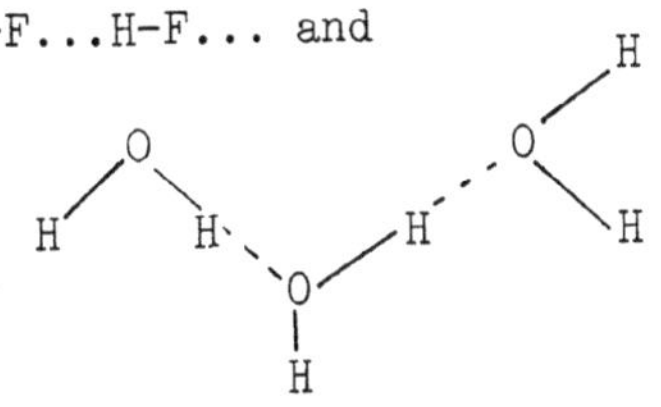

chains (where the dotted lines indicate hydrogen bonds) have shown that if one takes into account interactions until fifth neighbours, the band structures will be consistent (they do not change more in the third decimal of the energy values (in eV-s) if one goes from the fifth to the sixth neighbours) (13). on the other hand one would expect that more neighbours have to be taken into account in a polyene (-CH=CH-CH=CH-) chain to obtain consistent band structures. It should be mentioned that in two- and three-dimensional periodic systems this problem becomes yet more complicated and important. For instance semiempirical all-valence electron band structure calculation of a two-dimensional periodic polyglycine model has shown that the effect of second neighbours (cells with q_1=-1 or 1 and simultaneously q_2=-1 or 1) on the band structures is nearly the same as that of the first neighbours (q_1=-1 or 1 and q_2=0, or q_1=0 and q_2=-1 or 1) (14).

3.) This statement is only then true, when the periodicity is caused by a simple translational symmetry. If the periodicity occurs in consequence of a combined symmetry operation, we may have more integrals. For instance, in the case of a screw axis (DNA helix) it is easy to show (11) that besides $<\chi_r^0 \; \chi_u^0 \mid \chi_s^0 \; \chi_v^0>$ instead of three four additional types of integrals occur in the case of the one-dimensional problem and of the nearest neighbours'interactions approximation (in the strict sense).

The problem of further neighbours' interactions becomes exceedingly important if one wants to calculate the total energy per elementary cell for which in the three-dimensional case the following expression can be derived in a straighforward way

$$\frac{E}{(2N+1)^3} = \frac{1}{\omega} \int_{\omega} \sum_{h=1}^{n^*} [\, I(\vec{k})_{h,h} + \varepsilon(\vec{k})_h]\, d\vec{k} + \sum_{\alpha<\beta=1}^{M} \frac{Z_\alpha Z_\beta}{|\vec{R}_\alpha^{\vec{o}} - \vec{R}_\beta^{\vec{o}}|}$$

$$+ \frac{1}{2} \sum_{q_1 \neq 0} \sum_{\alpha,\beta=1}^{M} \frac{Z_\alpha Z_\beta}{|\vec{R}_\alpha^{\vec{o}} - \vec{R}_\beta^{\vec{q}_1}|} \qquad (60)$$

where

$$I(\vec{k})_{h,h} = \underline{d}\, (\vec{k})_h^+ \ \underline{\underline{I}}\, (\vec{k})\ \underline{d}\, (\vec{k})_h \qquad (61)$$

with

$$\underline{\underline{I}}\, (\vec{k}) = \sum_{\vec{q}} e^{i\vec{k}\vec{R}_{\vec{q}}}\ \underline{\underline{I}}\, (\vec{q}) \qquad (62a)$$

$$[\ \underline{\underline{I}}\ (\vec{q})]_{r,s} = \langle \chi_r^{\vec{o}}\ |\hat{H}^N|\ \chi_s^{\vec{q}} \rangle = \langle \chi_r^{\vec{o}}| - \frac{1}{2}\Delta - \sum_{\vec{q}_1} \sum_{\alpha=1}^{M} \frac{Z_\alpha}{|\vec{r} - \vec{R}_\alpha^{\vec{q}_1}|} |\chi_s^{\vec{q}} \rangle \qquad (62b)$$

In the one-dimensional case (60) modifies to

$$\frac{E}{2N+1} = \frac{a}{2\pi} \int_{-\pi/a}^{\pi/a} \sum_{h=1}^{n^*} [\, I(k)_{h,h} + \varepsilon(k)_h]\, dk + \text{N.R.} \qquad (63)$$

where N.R. stands for the nuclear repulsion terms which are the same as in (60) only now we have to write instead of the vectors $\vec{o}$ and $\vec{q}_1$, the scalars o and q, respectively. The quantity $I(k)_{h,h}$ can be calculated again with the aid of expressions (61) - (62b), if we substitute into them everywhere instead of the vectors the scalars corresponding to the one-dimensional case.

If $N\to\infty$ both the terms N.R. and the second term of $\hat{H}^N$ diverge. It can be shown, however, that the sum of them gives a convergent result (15). On the other hand, if we use a finite neighbours approximation, the electronic energy can decrease with the distance in a different way than the term N.R. providing a wrong result for the total energy. Therefore, the band structure calculation method in its described form should not be used for the calculation of total energy. On the other hand, Harris and Monkhorst (16) making use of Fourier transformations into the reciprocal lattice have worked out a method which makes it possible to sum up to all neighbours in the expressions of the Fock matrix elements and in that of the total energy per unit cell. The method has been applied, however, until now only to a simple cubic atomic hydrogen lattice (16). By the nature of the Fourier transformation the extension of the method to one-dimensional periodic systems is more difficult and has been done until now only for the case of one orbital per elementary cell (17). The formulation of this method for the general case of a one-dimensional periodic polymer with arbitrary numbers of orbitals within the elementary cell is in progress at different places. (For the details of the method see also (18)).

3. THE DIFFERENT ORBITALS FOR DIFFERENT SPINS SCF LCAO CRYSTAL ORBITAL SCHEME

The method of different orbitals for different spins (DODS) is considered as one possible way to take into account a part of the correlation. Using, however, one single Slater determinant build up from different spatial orbitals for the electrons with spin α and β , respectively, (unrestricted Hartree-Fock, UHF method), the difficulty arises that the many electron wave functions will not be an eigenfunction of the operator $\hat{S}^2$ of the total spin. To overcome this difficulty one has to project out with the aid of a projection operator from the Slater determinant the component with the desired multiplicity $2S+1$, annihilating all other "contaminating" components. This can be done either after an already performed calculation (spin-projection after variation, UHF with annihilation), or, as Löwdin (19) has pointed out, one would expect a more negative total energy, if the variation is performed with an already spin-projected Slater determinant [4.] (spin-projection before variation, spin-projected extended Hartree-Fock (EHF) method). If one performs the variation of the expectation value of the Hamiltonian formed with a spin-projected Slater determinant, one gets rather complicated equations (extended Hartree-Fock (EHF) equations) for the one-electron orbitals (20)[5.]. The investigation of these equa-

4.) The reason of this is that a spin-projected Slater determinant is a given linear combination of different Slater determinants. The variation of the expectation value of the Hamiltonian formed with a spin-projected Slater determinant provides thus equations (the so-called EHF equations) whose solutions represent the solution of this particular multiconfigurational SCF problem.

5.) It should be mentioned that the same equations have been derived with the aid of advanded group theory by Goddard (21) for the principal case $M = S$, where M is the quantum number belonging to the eigenvalues of the operator $\hat{S}_Z$.

tions have shown that if the number of electrons $n\to\infty$ and $n \gg s$, the EHF equations go over into the unprojected UHF equation (22). Therefore, the expectation values of all one- and two-electron operators and all one-electron energies and wavefunctions will be the same in the projected and unprojected cases (22). This means that if we wish to apply the DODS method to a molecular crystal or to a polymer we can use its simple unprojected form.

The unprojected UHF equations which we obtain from the variation of the expectation value of the Hamiltonian formed with a DODS Slater determinant, as it is well-known, have the form

$$\hat{F}^{\alpha}\phi_i^{\alpha} = (\hat{H}^N + \sum_{h=1}^{\tilde{n}^{*\alpha}} (\hat{J}_h^{\alpha} + \hat{J}_h^{\beta} - \hat{K}_h^{\alpha})\, \phi_i^{\alpha} = \varepsilon_i^{\alpha}\, \phi_i^{\alpha} \qquad (64a)$$

$$\hat{F}^{\beta}\phi_i^{\beta} = (\hat{H}^N + \sum_{h=1}^{\tilde{n}^{*\beta}} (\hat{J}_h^{\alpha} + \hat{J}_h^{\beta} - \hat{K}_h^{\beta})\, \phi_i^{\beta} = \varepsilon_i^{\beta}\, \phi_i^{\beta} \qquad (64b)$$

Here $\tilde{n}^{*\gamma}$ is the number of filled levels with electrons with spin γ ($\gamma=\alpha$ or β) and the Coulomb operators $\hat{J}_j^{\gamma}$ and exchange operators $\hat{K}_j^{\gamma}$ are defined as

$$\hat{J}_h^{\gamma}(\vec{r}_1)\,\phi(\vec{r}_1) = \int \frac{|\phi_h^{\gamma}(\vec{r}_2)|^2}{|\vec{r}_1-\vec{r}_2|}\, d\vec{r}_2\, \phi(\vec{r}_1) \quad (\gamma=\alpha,\beta) \qquad (65a)$$

$$\hat{K}_h^{\gamma}(\vec{r}_1)\,\phi(\vec{r}_1) = \int \frac{\phi_h^{\gamma}(\vec{r}_2)^{*}\,\phi(\vec{r}_2)}{|\vec{r}_1-\vec{r}_2|}\, d\vec{r}_2\, \phi_h^{\gamma}(\vec{r}_1) \quad (\gamma=\alpha,\beta) \qquad (65b)$$

If we introduce a basis $\{\chi\}$ and express all ϕ_i^{γ} -s as linear combinations of the basis functions,

$$\phi_h^{\gamma} = \sum_{r=1}^{m} c_{h,r}^{\gamma}\, \chi_r \qquad (\gamma=\alpha,\beta) \qquad (66)$$

we obtain in the usual way the matrix equations

$$\underline{\underline{F}}^{\alpha}\, \underline{c}_h^{\alpha} = \varepsilon_h^{\alpha}\, \underline{\underline{S}}\, \underline{c}_h^{\alpha} \quad , \quad \underline{\underline{F}}^{\beta}\, \underline{c}_h^{\beta} = \varepsilon_h^{\beta}\, \underline{\underline{S}}\, \underline{c}_h^{\beta} \qquad (67)$$

Here the elements of the matrices $\underline{\underline{F}}^{\gamma}$ are defined as

$$[\underline{\underline{F}}^{\gamma}]_{r,s} = \langle\chi_r|\hat{H}^N|\chi_s\rangle + \sum_{u,v=1}^{m} [\, (p_{u,v}^{\alpha} + p_{u,v}^{\beta}) \langle\chi_r\, \chi_u|\chi_s\chi_v\rangle - p_{u,v}^{\gamma}\langle\chi_r\, \chi_u|\chi_v\, \chi_s\rangle \,] \qquad (\gamma=\alpha,\beta) \qquad (68)$$

where

$$p^{\gamma}_{u,v} = \sum_{h=1}^{\tilde{n}^{*\gamma}} C^{\gamma *}_{h,u} C^{\gamma}_{h,v} \qquad (\gamma=\alpha,\beta) \tag{69}$$

Let us now consider a three-dimensional periodic system with $(2N+1)^3$ unit cells and with m orbitals within the cell. If we apply again the Born-Kármán periodic boundary conditions the matrices $\underline{\underline{F}}^{\alpha}$, $\underline{\underline{F}}^{\beta}$ and $\underline{\underline{S}}$ become cyclic hypermatrices of order $m(2N+1)^3$. Therefore we can apply again to equ. -s (67) the unitary transformation described in point 2.1. We obtain in this way

$$\underline{\underline{U}}^{+} \underline{\underline{F}}^{\alpha} \underline{\underline{U}}\, \underline{\underline{U}}^{+} \underline{C}^{\alpha}_{h} = \varepsilon^{\alpha}_{h} \underline{\underline{U}}^{+} \underline{\underline{S}}\, \underline{\underline{U}}\, \underline{\underline{U}}^{+} \underline{C}^{\alpha}_{h}, \tag{70a}$$

$$\underline{\underline{U}}^{+} \underline{\underline{F}}^{\beta} \underline{\underline{U}}\, \underline{\underline{U}}^{+} \underline{C}^{\beta}_{h} = \varepsilon^{\beta}_{h} \underline{\underline{U}}^{+} \underline{\underline{S}}\, \underline{\underline{U}}\, \underline{\underline{U}}^{+} \underline{C}^{\beta}_{h} \tag{70b}$$

or

$$\underline{\underline{F}}^{\alpha\prime} \underline{D}^{\alpha}_{h} = \varepsilon^{\alpha}_{h} \underline{\underline{S}}' \underline{D}^{\alpha}_{h} \tag{71a}$$

$$\underline{\underline{F}}^{\beta\prime} \underline{D}^{\beta}_{h} = \varepsilon^{\beta}_{h} \underline{\underline{S}}' \underline{D}^{\beta}_{h} \tag{71b}$$

with $\underline{D}^{\gamma}_{h} = \underline{\underline{U}}^{+} \underline{C}^{\gamma}_{h}$ $(\gamma=\alpha,\beta)$, where the unitary matrix was defined by (7). From equations (71) we obtain in the same way as in point 2.1. the equations

$$\underline{\underline{F}}^{\alpha}(\vec{k})\, \underline{d}^{\alpha}(\vec{k}) = \varepsilon^{\alpha}(\vec{k})_h\, \underline{\underline{S}}(\vec{k})\, \underline{d}^{\alpha}(\vec{k})_h \tag{72a}$$

$$\underline{\underline{F}}^{\beta}(\vec{k})\, \underline{d}^{\beta}(\vec{k}) = \varepsilon^{\beta}(\vec{k})_h\, \underline{\underline{S}}(\vec{k})\, \underline{d}^{\beta}(\vec{k})_h \tag{72b}$$

where the matrices $\underline{\underline{F}}^{\gamma}$ (which have now only the order of m) have the form (23),

$$\underline{\underline{F}}^{\gamma}(\vec{k}) = \sum_{\vec{q}=-\infty}^{\infty} e^{i\vec{k}\vec{R}_{\vec{q}}}\, \underline{\underline{F}}^{\gamma}(\vec{q}) \tag{73}$$

$$[\underline{\underline{F}}^{\gamma}(\vec{q})]_{r,s} = \langle \chi^{\vec{o}}_{r} | \hat{H}^{N} | \chi^{\vec{q}}_{s} \rangle + \sum_{\vec{q}_1,\vec{q}_2} \sum_{u,v} \{ [p(\vec{q}_1-\vec{q}_2)^{\alpha}_{u,v} + p(\vec{q}_1-\vec{q}_2)^{\beta}_{u,v}] \langle \chi^{\vec{o}}_{r} \chi^{q_1}_{u} | \chi^{\vec{q}}_{s} \chi^{q_2}_{v} \rangle - p(\vec{q}_1-\vec{q}_2)^{\gamma}_{u,v} \langle \chi^{\vec{o}}_{r} \chi^{q_1}_{u} | \chi^{q_2}_{v} \chi^{\vec{q}}_{s} \rangle \}$$

$$(\gamma=\alpha,\beta) \tag{74}$$

Here $\hat{H}^{N}$ was defined before (see (39)), the charge-bond order matrix elements $p(\vec{q}_1-\vec{q}_2)^{\gamma}$ are defined as

$$p(\vec{q}_1-\vec{q}_2)^{\gamma}_{u,v} = \frac{1}{\omega}\int_{\omega} \sum_{h=1}^{n^{*\gamma}} d^{\gamma}(\vec{k})^{*}_{h,u}\, d^{\gamma}(\vec{k})_{h,v}\, e^{i\vec{k}(\vec{R}_{q_1}-\vec{R}_{q_2})}\, d\vec{k} \tag{75}$$

and $n^{*\gamma}$ is the number of electrons per unit cell with spin γ (it is easy to see that $\tilde{n}^{*\gamma} = (2N+1)^3\, n^{*\gamma}$). The matrix $\underline{\underline{S}}(\vec{k})$, occurring also in equ.-s (72), was defined also before (see equ. (19)).

The total energy of a molecule can be written in the UHF case as

$$E = \frac{1}{2}\left[\sum_{h=1}^{\tilde{n}^{*\alpha}} (I^{\alpha}_{h,h} + \varepsilon^{\alpha}_{h}) + \sum_{h=1}^{\tilde{n}^{*\beta}} (I^{\beta}_{h,h} + \varepsilon^{\beta}_{h}) + \text{N.R.}\right. \tag{76}$$

For a periodic system we obtain from (76) in a straightforward way for the total energy per unit cell the expression (23)

$$\frac{E}{(2N+1)^3} = \frac{1}{2}\frac{1}{\omega}\int_{\omega} \Big\{ \sum_{h=1}^{n^{*\alpha}} [\, I^{\alpha}(\vec{k})_{h,h} + \varepsilon^{\alpha}(\vec{k})_{h}] + \sum_{h=1}^{n^{*\beta}} [\, I^{\beta}(\vec{k})_{h,h} + \varepsilon^{\beta}(\vec{k})_{h} \,]\Big\}\, d\vec{k} + \text{N.R.} \tag{77}$$

where N.R. was defined before (see (60) and (63)), ω is again the volume of the first Brillouin zone,

$$I^{\gamma}(\vec{k})_{h,h} = \underline{d}^{\gamma}(\vec{k})^{+}_{h}\; \underline{\underline{I}}(\vec{k})\; \underline{d}^{\gamma}(\vec{k})_{h} \qquad (\gamma=\alpha,\beta) \tag{78}$$

and $\underline{\underline{I}}(\vec{k})$ is given by equ.-s (62a) and (62b). Combining again the long range parts of $\underline{\underline{H}}^{N}$ and N.R. it can be shown (24) that they will cancel giving thus a convergent result for the total energy.

The one-dimensional form of the described *ab initio* DODS SCF LCAO crystal orbital scheme can be obtained exactly in the same way from the above described three-dimensional expressions, as we did in the case of the conventional Hartree-Fock scheme (see equ.-s (50) - (54)). We obtain also the corresponding expressions for the nearest neighbours'case in a similar way, as we did in point 2.3.

The described *ab initio* DODS scheme has not been applied yet to any system. Some general considerations based on the total electronic density [6.] (25) and experiences of different calculations seem to indicate that if $n^{\alpha} = n^{\beta}$ (closed-shell ground state) by

6.) Preliminary considerations (25) show that for this a criterion, similar to the Mott's criterion for metal-non-metal transition (26) can be formulated.

usual bond distances in most cases of an ab initio (27), or an all-valence electron calculation (28) one cannot expect a splitting of the levels belonging to electrons with the spin α and β, respectively. The same result was obtained by Calais and Sperber in their Li crystal calculation using the AMO method (29). On the other hand if $n^{\alpha} \neq n^{\beta}$ the experience (30) shows that one does get a splitting and with it a lowering of the total energy. Only after the application of the DODS CO method to some polymers can be decided, whether really no lowering of the total energy can be achieved if $n^{\alpha} = n^{\beta}$. On the other hand, the method will give certainly a splitting of the bands of electrons with spin α and β, respectively, if the number of electrons with spin α per unit cell is different from those with spin β[7].

A further shortcoming of the DODQ crystal orbital method is that it takes into account partially only the correlation within the unit cell (intracell correlation), but neglects completely the electron-correlation between different cells (intercell correlation). If the interaction between the different unit cells is not very strong (like in periodic DNA models, or in other stacked chains), this intercell correlation might be unimportant. On the other hand, however, if the interactions between cells are large (like in a polyene chain) the long range intercell correlations can be very important. A possible way to take them into account in the DODS CO scheme is, to multiply the LACO Bloch orbitals ϕ_h^{α} and ϕ_h^{β} with different standing waves η_h^{α} and η_h^{β} which have wave lengths of several times the elementary translation (suitably chosen for a given polymer) and different phases for electrons with spin α and β (31). In this way we can achieve that the total population of electrons with spin α will be large in a given cell, while that of the electrons with spin β will be small, while in the next two neighbour cells (in the case of a linear chain) the reversed situation will occur and so on. Of course, only numerical calculations performed on some polymers can decide whether the described simple idea is really useful to take into account intercell correlations.

Applying (77) for the one-dimensional case we can write for the total energy per unit cell of a linear chain in the DODS case in the first neighbours'approximation

$$\frac{E}{2N+1} = \frac{a}{2\pi}\frac{1}{2}\int_{-\pi/a}^{\pi/a}\{\sum_{h=1}^{n\alpha^{*}}[I^{\alpha}(k)_{h,h} + \varepsilon^{\alpha}(k)_{h}] + \sum_{h=1}^{n^{*}\beta}$$

7) This statement is true only then, when the interaction between the open shell-molecules (radicals) is not so strong that they merge into a common closed-shell supermolecule. This somewhat unclear, but important problem deserves future investigations.

$$[I^{\beta}(k)_{h,h} + \varepsilon^{\beta}(k)_h]\} \, dk + N.R. \qquad (79)$$

where

$$I^{\gamma}(k)_{h,h} = \underline{d}^{\gamma}(k_h^+) \, \underline{\underline{I}}(k) \, \underline{d}^{\gamma}(k)_h \qquad (\gamma=\alpha,\beta) \qquad (80)$$

$$\underline{\underline{I}}(k) = \underline{\underline{I}}(0) + \underline{\underline{I}}(1) e^{ika} + \underline{\underline{I}}(-1) e^{-ika} \qquad (81)$$

and the elements of $\underline{\underline{I}}(q)$ are given by (62b) if we substitute in it everywhere instead of the vector $\vec{q}_1$ the scalar q_1. Comparing the total energies per unit cell obtained by (79) and (63) one can study the improvement in energy of a linear chain given by the DODS method for the $n^{\alpha} \neq n^{\beta}$ case and (if there is a splitting) also for the $n^{\alpha} = n^{\beta}$ case.

4. "OPEN-SHELL" SCF LCAO CRYSTAL ORBITAL METHOD

4.1. Roothaan's open-shell SCF formalism in LCAO form

For molecules with a single open-shell or for triplet states Roothaan (32) has formulated an open-shell SCF theory. A more general open-shell formalism, applicable also for singlet excited states, was formulated by Hunt *et al.* (33) and by Dahl (34). Though we cannot have an infinite chain (or three-dimensional crystal) in which each unit is in a singlet excited state, we shall formulate the CO method with the aid of this more general procedure. One can show that in the case of a single open-shell this formalism is equivalent to Roothaan's equations for the closed-and open-shell orbitals [8.] (see equ.-s (32) with the definitions (31) of (32)) (34).

In the case of the general formalism, applying the projection operator technique, we can write (34) :

$$\hat{\Theta}_c \hat{F}_c \hat{\Theta}_c \phi_{k'} = \varepsilon_{k'} \phi_{k'} \qquad (k'=1,2,\ldots, n_c^*, n^*+1,\ldots) \qquad (82a)$$

$$\hat{\Theta}_o \hat{F}_o \hat{\Theta}_o \phi_m = \varepsilon_m \phi_m \qquad (m=n_c^*+1,\ldots,n^*, n^*+1,\ldots) \qquad (82b)$$

$$\hat{P}(\hat{F}_c-\hat{F}_o)\hat{P}\phi_l = \varepsilon_l \phi_l \qquad (l=1,2,\ldots,n^*; n^*=n_c^*+n_o^*) \qquad (82c)$$

8) From the point of view of the organization of an *ab initio* program it is more advantageous to use these equations of Roothaan's formalism than the other form of it in which the closed- and open-shell orbitals occur as the solutions of the same eigenvalue problem (see equations (36) - (38) of (32)).

where the projection operators $\hat{\Theta}_c, \hat{\Theta}_o$ and $\hat{P}$, respectively, are defined as

$$\hat{\Theta}_c = 1 - \sum_{m=1}^{n_o^*} |\phi_m\rangle\langle\phi_m| \tag{83a}$$

$$\hat{\Theta}_o = 1 - \sum_{k'=1}^{n_c^*} |\phi_{k'}\rangle\langle\phi_{k'}| \tag{83b}$$

$$\hat{P} = \sum_{k'=1}^{n_c^*} |\phi_{k'}\rangle\langle\phi_{k'}| + \sum_{m=1}^{n_o^*} |\phi_m\rangle\langle\phi_m| \tag{83c}$$

Here n_o^* is the number of singly filled open-shell orbitals (in the case oi the physically most interesting case of polyradicals $n_o^*=1$) and n_c^* that of the doubly occupied closed-shell ones. Clearly, the operators (83a) and (83b) provide the coupling between the closed- and open-shell orbitals in equ.-s (82) and equ. (82c) ensures the fullfilment of the necessary conditions for the Hartree-Fock orbitals (for details see (34)). The Fock-operators for the closed- and open-shell, respectively, are given by

$$\hat{F}_c = \hat{H}^N + 2\hat{J}_c - \hat{K}_c + 2\hat{J}_o - \hat{K}_o \tag{84a}$$

$$\hat{F}_o = \hat{H}^N + 2\hat{J}_c - \hat{K}_c + 2a\hat{J}_o - b\hat{K}_o \tag{84b}$$

where the Coulomb operators $\hat{J}_c$ and $\hat{J}_o$ and exchange operators $\hat{K}_c$ and $\hat{K}_o$, respectively, are defined as

$$\hat{J}_c = \sum_{k'=1}^{n_c^*} J_{k'}, \quad \hat{J}_o = f \sum_{m=1}^{n_o^*} J_m \tag{85a}$$

$$\hat{K}_c = \sum_{k'=1}^{n_c^*} K_{k'}, \quad \hat{K}_o = f \sum_{m=1}^{n_o^*} K_m \tag{85b}$$

and the numerical constants a, b and f depend on the specific case (32).

Introducing a basis $\{\chi\}$ we can write

$$\phi_{k'} = \sum_{r=1}^{n} C_{k',r}\,\chi_r, \quad \phi_m = \sum_{r=1}^{n} C_{m,r}\,\chi_r, \quad \phi_1 = \sum_{r=1}^{n} C_{1,r}\,\chi_r \tag{86}$$

(where n denotes the number of basis functions). Substituting (86) into equ.-s (82) and (83) one can derive in the standard way the matrix equations

$$(\underline{\underline{1}} - \frac{1}{2}\underline{\underline{S}}\,\underline{\underline{P}}^{o})\,\underline{\underline{F}}_{c}\,(\underline{\underline{1}} - \frac{1}{2}\underline{\underline{P}}^{o}\,\underline{\underline{S}})\,\underline{C}_{k'} = \varepsilon_{k'}\,\underline{\underline{S}}\,\underline{C}_{k'} \quad (k'=1,\ldots,n_c^{*},\ldots n^{*}+1,\ldots,n) \tag{87a}$$

$$(\underline{\underline{1}} - \frac{1}{2}\underline{\underline{S}}\,\underline{\underline{P}}^{c})\,\underline{\underline{F}}_{o}\,(\underline{\underline{1}} - \frac{1}{2}\underline{\underline{P}}^{c}\,\underline{\underline{S}})\,\underline{C}_{m} = \varepsilon_{m}\,\underline{\underline{S}}\,\underline{C}_{m} \quad (m=n_c^{*}+1,\ldots,n^{*},\ldots,\ldots n) \tag{87b}$$

$$\frac{1}{2}\underline{\underline{S}}\,(\underline{\underline{P}}^{c} + \underline{\underline{P}}^{o})(\underline{\underline{F}}_{c}-\underline{\underline{F}}_{o})\,\frac{1}{2}\,(\underline{\underline{P}}^{c}+ \underline{\underline{P}}^{o})\,\underline{\underline{S}}\,\underline{C}_{l} = {}_{l}\,\underline{\underline{S}}\,\underline{C}_{l} \quad (l=1,\ldots,n^{*};\; n^{*}=n_c^{*}+n_o^{*}). \tag{87c}$$

Here

$$\underline{\underline{P}}^{c} = 2\sum_{k'=1}^{n_c^{*}} \underline{C}_{k'}\,\underline{C}_{k'}^{+} \tag{88a}$$

$$\underline{\underline{P}}^{o} = 2\sum_{m=1}^{n_o^{*}} \underline{C}_{m}\,\underline{C}_{m}^{+} \tag{88b}$$

and the elements of the matrices $\underline{\underline{F}}_{c}$ and $\underline{\underline{F}}_{o}$, respectively, are given as

$$[\underline{\underline{F}}_{c}]_{r,s} = \langle\chi_r|\,\hat{H}^{N}|\chi_s\rangle + \sum_{u,v=1}^{n} P^{c}_{u,v}\,(\langle\chi_r\chi_u|\chi_s\chi_v\rangle - \frac{1}{2}\langle\chi_r\chi_u|\chi_v\chi_s\rangle) + \sum_{u,v=1}^{n} P^{o}_{u,v}\,f\,(\langle\chi_r\chi_u|\chi_s\chi_v\rangle - \frac{1}{2}\langle\chi_r\chi_u|\chi_v\chi_s\rangle) \tag{89a}$$

$$[\underline{\underline{F}}_{o}]_{r,s} = \langle\chi_r|\hat{H}^{N}|\chi_s\rangle + \sum_{u,v=1}^{n} P^{c}_{u,v}\,(\langle\chi_r\chi_u|\chi_s\chi_v\rangle - \frac{1}{2}\langle\chi_r\chi_u|\chi_v\,\chi_s\rangle + \sum_{u,v=1}^{n} P^{o}_{u,v}\,f\,(a\,\langle\chi_r\chi_u|\chi_s\chi_v\rangle - \frac{b}{2}\langle\chi_r\chi_u|\chi_v\,\chi_s\rangle) \tag{89b}$$

4.2. Block-diagonalization

If we have a linear chain with (2N+1) unit cells and $\tilde{n}$ orbitals within the cell all the matrices occurring in equ.-s (87) will have the dimension $n=\tilde{n}(2N+1)$. Taking into account the translational symmetry (or in the case of a screw axis the periodic symmetry) and introducing again the Born-von Kármán periodic boundary conditions (see after equ. (6) on p.5). we can observe that all the matrices in (87) are cyclic hypermatrices. Therefore with the aid of the unitary matrix $\underline{\underline{U}}$ defined by (7) they can be block-diagonalized. Thus we can write

$$\underline{\underline{U}}^{+} (\underline{\underline{1}} - \frac{1}{2} \underline{\underline{S}}\, \underline{\underline{U}}\, \underline{\underline{U}}^{+} \underline{\underline{P}}^{o} \underline{\underline{U}}\, \underline{\underline{U}}^{+}) \underline{\underline{F}}_{c} \underline{\underline{U}}\, \underline{\underline{U}}^{+} (\underline{\underline{1}} - \frac{1}{2} \underline{\underline{P}}^{o} \underline{\underline{U}}\, \underline{\underline{U}}^{+} \underline{\underline{S}}\, \underline{\underline{U}}\, \underline{\underline{U}}^{+}) \underline{C}_{k'} = \varepsilon_{k'} \underline{\underline{U}}^{+} \underline{\underline{S}}\, \underline{\underline{U}}\, \underline{\underline{U}}^{+} \underline{C}_{k'} \quad (90a)$$

$$\underline{\underline{U}}^{+} (\underline{\underline{1}} - \frac{1}{2} \underline{\underline{S}}\, \underline{\underline{U}}\, \underline{\underline{U}}^{+} \underline{\underline{P}}^{c} \underline{\underline{U}}\, \underline{\underline{U}}^{+}) \underline{\underline{F}}_{o} \underline{\underline{U}}\, \underline{\underline{U}}^{+} (\underline{\underline{1}} - \frac{1}{2} \underline{\underline{P}}^{c} \underline{\underline{U}}\, \underline{\underline{U}}^{+} \underline{\underline{S}}\, \underline{\underline{U}}\, \underline{\underline{U}}^{+}) \underline{C}_{m} = \varepsilon_{m} \underline{\underline{U}}^{+} \underline{\underline{S}}\, \underline{\underline{U}}\, \underline{\underline{U}}^{+} \underline{C}_{m} \quad (90b)$$

$$\frac{1}{4} \underline{\underline{U}}^{+} \underline{\underline{S}}\, \underline{\underline{U}}\, \underline{\underline{U}}^{+} (\underline{\underline{P}}^{c} + \underline{\underline{P}}^{o}) \underline{\underline{U}}\, \underline{\underline{U}}^{+} (\underline{\underline{F}}_{c} - \underline{\underline{F}}_{o}) \underline{\underline{U}}\, \underline{\underline{U}}^{+} (\underline{\underline{P}}^{c} + \underline{\underline{P}}^{o}) \underline{\underline{U}}\, \underline{\underline{U}}^{+} \underline{\underline{S}}\, \underline{\underline{U}}\, \underline{\underline{U}}^{+} \underline{C}_{1} = \varepsilon_{1} \underline{\underline{U}}^{+} \underline{\underline{S}}\, \underline{\underline{U}}\, \underline{\underline{U}}^{+} \underline{C}_{1} \quad (90c)$$

or by introducing the notations

$$\underline{\underline{U}}^{+} \underline{\underline{B}}\, \underline{\underline{U}} = \underline{\underline{B}}' \quad (\underline{\underline{B}} = \underline{\underline{S}}, \underline{\underline{P}}^{c}, \underline{\underline{P}}^{o}, \underline{\underline{F}}_{c}, \underline{\underline{F}}_{o}, \underline{\underline{P}}^{c} + \underline{\underline{P}}^{o} = \underline{\underline{P}}\,,\ \underline{\underline{F}}_{c} - \underline{\underline{F}}_{o}) \quad (91)$$

$$\underline{\underline{U}}^{+} \underline{\underline{C}}_{k'} = \underline{D}_{k'},\ \underline{\underline{U}}^{+} \underline{C}_{m} = \underline{D}_{m},\ \underline{\underline{U}}^{+} \underline{C}_{1} = \underline{D}_{1}$$

$$(\underline{\underline{1}} - \frac{1}{2} \underline{\underline{S}}' \underline{\underline{P}}^{o'}) \underline{\underline{F}}_{c}' (\underline{\underline{1}} - \frac{1}{2} \underline{\underline{P}}^{o'} \underline{\underline{S}}') \underline{D}_{k'} = \varepsilon_{k'} \underline{\underline{S}}' \underline{D}_{k'} \quad (92a)$$

$$(\underline{\underline{1}} - \frac{1}{2} \underline{\underline{S}}\, \underline{\underline{P}}^{c'}) \underline{\underline{F}}_{o}' (\underline{\underline{1}} - \frac{1}{2} \underline{\underline{P}}^{c'} \underline{\underline{S}}') \underline{D}_{m} = \varepsilon_{m} \underline{\underline{S}}' \underline{D}_{m} \quad (92b)$$

$$\frac{1}{4} \underline{\underline{S}}' \underline{\underline{P}}' (\underline{\underline{F}}_{c}' - \underline{\underline{F}}_{o}') \underline{\underline{P}}' \underline{\underline{S}}' \underline{D}_{1} = \varepsilon_{1} \underline{\underline{S}}' \underline{D}_{1} \quad (\underline{\underline{P}}' = \underline{\underline{P}}^{c'} + \underline{\underline{P}}^{o'}) \quad (92c)$$

Taking into account that all the matrices occurring in equ.-s (92) are block-diagonal, assuming that $N \to \infty$ and introducing again the continuous variable k defined by (15), it is easy to show that we can arrive at the matrix equations [9.)].

$$[\underline{\underline{1}} - \frac{1}{2} \underline{\underline{S}}(k)\, \underline{\underline{P}}^{o}(k)]\, \underline{\underline{F}}_{c}(k)\, [\underline{\underline{1}} - \frac{1}{2} \underline{\underline{P}}^{o}(k)\, \underline{\underline{S}}(k)]\, \underline{d}(k)_{k'} = \varepsilon(k)_{k'} \underline{\underline{S}}(k)\, \underline{d}(k)_{k'} \quad (93a)$$

$$[\underline{\underline{1}} - \frac{1}{2} \underline{\underline{S}}(k)\, \underline{\underline{P}}^{c}(k)]\, \underline{\underline{F}}_{o}(k)\, [\underline{\underline{1}} - \frac{1}{2} \underline{\underline{P}}^{c}(k)\, \underline{\underline{S}}(k)]\, \underline{d}(k)_{m} = \varepsilon(k)_{m} \underline{\underline{S}}(k)\, \underline{d}(k)_{m} \quad (93b)$$

9.) In the derivation the fact was used that the product of block-diagonal matrices is also block-diagonal and therefore equ.-s (92) will split off also to matrix equ.-s of order n (the number of orbitals in the unit cell) in the same way, as in point 2.).

$$\frac{1}{4}\, \underline{\underline{S}}(k)\, \underline{\underline{P}}(k) \quad \underline{\underline{F}}_c(k) - \underline{\underline{F}}_o(k) \quad \underline{\underline{P}}(k)\, \underline{\underline{S}}(k)\, \underline{d}(k)_1 = \varepsilon(k)_1\, \underline{\underline{S}}(k)\, \underline{d}(k)_1$$

$$[\, \underline{\underline{P}}(k) = \underline{\underline{P}}^c(k) + \underline{\underline{P}}^o(k)\,] \tag{93c}$$

Here the Hermitian complex matrices

$$\underline{\underline{B}}(k) = \sum_{q=-\infty}^{\infty} e^{ikqa}\, \underline{\underline{B}}(q) \quad (\underline{\underline{B}}(q) = \underline{\underline{S}}(q),\, \underline{\underline{P}}^c(q),\, \underline{\underline{P}}^o(q),\, \underline{\underline{F}}_c(q),$$

$$\underline{\underline{F}}_o(q),\, \underline{\underline{P}}(q) \tag{94}$$

have the order of $\tilde{n}$ only and the matrices $\underline{\underline{B}}(q)$ (which are the corresponding sub-blocks of the original matrices $\underline{\underline{B}}$) are defined as:

$$[\underline{\underline{S}}(q)]_{r,s} = \langle \chi_r^o |\, \chi_s^q \rangle \tag{95a}$$

$$[\underline{\underline{P}}^c(q)]_{r,s} = \frac{a}{2\pi} \int_{-\pi/a}^{\pi/a} 2 \sum_{k'=1}^{\tilde{n}_c^*} d(k)^*_{k',r}\, d(k)_{k',s}\, e^{ikqa}\, dr \tag{95b}$$

$$[\underline{\underline{P}}^o(q)]_{r,s} = \frac{1}{b} \int_{k\ min.}^{k\ max.} 2 \sum_{m=1}^{n_o^*} d(k)_{m,r}\, d(k)_{m,s}\, e^{ikqa}\, dr \tag{95c}$$

(b= k max.- k min.), where k max. and k min. refer to the limiting k-values of the halfly filled band (it is assumed that the chain is antiferromagnetic).

$$[\underline{\underline{F}}(q)_{c,\sigma}]_{r,s} = \langle \chi_r^o | -\frac{1}{2}\Delta - \sum_{q_1=-\infty}^{\infty} \sum_{\alpha=1}^{M} \frac{Z_\alpha}{|\vec{r} - \vec{R}_d^{q_1}|} | \chi_s^q \rangle +$$

$$\sum_{q_1,q_2=-\infty}^{\infty} \sum_{u,v=1}^{\tilde{n}} [\, p^c(q_1-q_2)_{u,v}\, (\langle \chi_r^o\, \chi_u^{q_1} | \chi_s^q\, \chi_v^{q_2} \rangle - \frac{1}{2} \langle \chi_r^o\, \chi_u^{q_1} |$$

$$\chi_v^{q_1}\, \chi^q \rangle) + p^o(q_1-q_2)_{u,v} \times f(\begin{pmatrix}1\\a\end{pmatrix} \langle \chi^o\, \chi_u^{q_1} |\, \chi_s^q\, \chi_v^{q_2} \rangle - \frac{1}{2} \begin{pmatrix}1\\b\end{pmatrix}$$

$$\langle \chi_r^o\, \chi_u^{q_1} \,|\, \chi_v^{q_2} \chi_s^q \rangle)\,] \tag{95d}$$

In the last equation the quantities $\begin{pmatrix}1\\a\end{pmatrix}$ and $\begin{pmatrix}1\\b\end{pmatrix}$, respectively, occurring in the last two terms mean that in the closed-shell (C) case the factor 1 and in the open-shell (o) case the factors a and b, respectively, have to be taken.

In the case of the first neighbours'interactions approximation we have to take only $q=-1,0,1$ terms in (94) and in the summations over q_1 and q_2 also only the same values $(-1,0,1)$ have to be considered taking care that if for instance $q=1$, the $q_1=-1$ and/or $q_2=-1$ terms should be excluded (first neighbours'interactions in the strict sense).

The formalism, described here for a linear chain, can easily be developed also for a three-dimensional periodic system. It can be shown that the resulting expressions are again equ.-s (93) - (95), if we substitute in them everywhere instead of the scalars k,q,q_1 and q_2 the three-dimensional vectors $\vec{k},\vec{q},\vec{q}_1$ and $\vec{q}_2$ and perform in (95) the integrations over the first Brillouin-zone in the three-dimensional $\vec{k}$ space

This ab initio open-shell SCF LCAO CO formalism has not been applied yet to any system, but for a linear chain in the first neighbours'interactions approximation a program of it is under development (35). The main purpose for developing this complicated program is to be able to perform ab initio open-shell SCF LCAO CO calculations for poly ($TCNQ^-$) and poly (TTF^+) chains, since this $TCNQ^-$ - TTF^+ system has shown very high metallic conductivity with a maximum $\sigma\approx 10^4\ \Omega^{-1}\ cm^{-1}$ at $\sim 60°K$ (36). For the detailed theoretical investigation of the assumed paraconductivity (superconductive fluctuation) at this temperature (37) (wich was based on a non-BCS mechanism proposed by Fröhlich back in 1954 (38)), a fair knowledge of the band structures (widths of valence and conduction bands, the gap between them and the form of the k-dependence of the energy of these bands) of these chains containing as unit cells the open-shell $TCNQ^-$ and TTF^+ molecules, respectively, is necessary. After the completion of the program these calculations will be performed (39).

Finally it should be mentioned that if we have an infinite chain (or crystal) of open-shell units, one could expect a halfly filled band, but otherwise such positions of the bands which are around the one-electron levels of the corresponding closed-shell molecule,(because the whole infinite system is a closed-shell one. If the interactions between the different units are strong (like in the N N N N chain which shows also high metallic

S S S S

conductivity (40)), very probably really this will happen. On the other hand, if the interactions are relatively weak (like in the stacked poly ($TCNQ^-$) and poly (TTF^+) chains), the positions of the bands will be probably more or less around the one-electron levels of the corresponding open-shell systems ($TCNQ^-$ and TTF^+, respectively, in our example) and neither around the levels of the corresponding closed-shell neutral molecules (TCNQ and TTF, respectively), nor around those of the ions calculated with the closed-shell formalism (half-electron method). Since the one-electron levels of a molecule-ion calculated with the aid of the closed-shell and open-shell formalism usually differ from each other rather strongly, one would expect rather large differences also in the corresponding band structures (until the interactions between the open-shell subunits are not strong). To give a definite answer to this interesting (and rather important) question, one has to perform, however, first band structure calculations on the same system using both the usual closed-shell CO formalism (with a halfly filled band) and the newly developed open-shell CO formalism, described here.

ACKNOWLEDGMENTS

The author should like to express his sincere gratitude to the NATO Scientific Affairs Division and to the University of Namur, whose sponsorship made it possible to organize the Advanced Study Institute on Electronic Structure of Polymers and Molecular Crystals.

He is further very much indebted to Professor G.L. Hofacker whose interest and support has given him the opportunity to continue his research and to Dr. W. von Niessen for calling his attention to the "direct method" to formulate the open-shell problem.

Last, but not least, he wishes to acknowledge the stimulating discussions with Professors J. Cizek and J. Paldus and their kind hospitality extended to him during his stay at the Department of Applied Mathematics, University of Waterloo. His very pleasant stay in Waterloo, made possible by the support by Professor W. Forbes, Dean of the Faculty of Mathematics, and Professor G. Wertheim, Chairman of the Department of Applied Mathematics, contributed much to the development of methods and ideas described in this paper.

REFERENCES

(1) J. Keller, J. de Physique 33, C3-241 (1972); J. Keller, J. Fritz, and A. Garritz, Proc. of Conference on Disordered Metallic Systems, Strasbourg, 1973.

(2) W.L. McCubbin in "Electronic Structure of Polymers and Molecular Crystals", ed. J.-M. André and J. Ladik, Plenum Press, London,(1975).

(3) J.-M. André in "Electronic Structure of Polymers and Molecular Crystals", ed. J.-M. André and J. Ladik, Plenum Press, London, (1975).

(4) G. Del Re, J. Ladik, and G. Biczó, Phys. Rev. 155, 997 (1967); G. Biczó, unpublished result.

(5) P.-O. Löwdin, J. Chem. Phys. 18, 365 (1950).

(6) J. Ladik, Quantenchemie, Enke Verlag, Stuttgart, 1973, p.124.

(7) C.C.J. Roothaan, Revs. Mod. Phys. 23, 69 (1951).

(8) J.-M. André, L. Gouverneur, and G. Leroy, Int. J. Quant. Chem. 1, 427 and 451 (1967).

(9) J.-M. André, J. Chem. Phys. 50, 1536 (1969).

(10) J.-M. André and G. Leroy, Chem. Phys. Letters 5, 71 (1971).

(11) C. Merkel and J. Ladik, unpublished result.

(12) J. Ladik in "Electronic Structure of Polymers and Molecular Crystals", ed. J.-M. André and J. Ladik, Plenum Press, London, (1975).

(13) A. Karpfen, J. Ladik, P. Rusegger, P. Schuster, and S. Suhai, Theor. Chim. Acta (accepted).

(14) S. Suhai and J. Ladik, Acta Phys. Ac. Sci. Hung. (accepted).

(15) K.F. Berggren and F. Martino, Phys.Rev. 184, 484 (1969).

(16) F.E. Harris and H.J. Monkhorst, in Computational Methods in Band Theory, ed. P. M. Marcus, J. F. Janak and H. R. Williams, Plenum Press, New York, 1971, p. 517.

(17) F.E. Harris, J. Chem. Phys. 56, 4472 (1972).

(18) F.E. Harris in "Electronic Structure of Polymers and Molecular Crystals" ed. J.-M. André and J. Ladik, Plenum Press, London, (1975).

(19) P.-O. Löwdin, Phys. Rev. 97, 1474, 1490, 1509 (1955).

(20) F. Martino and J. Ladik, J. Chem. Phys. 52, 2262 (1970); I. Mayer, J. Ladik, and G. Biczó, Int. J. Quant. Chem. 7, 583 (1973).

(21) W.A. Goddard III, Phys. Rev. 157, 73, 81 (1967); J. Chem. Phys. 48, 450, 5537 (1967).

(22) F. Martino and J. Ladik, Phys. Rev. A3, 862 (1971).

(23) G. Biczó, G. Del Re and J. Ladik (unpublished result); J. Ladik and G. Biczó, Acta Chim. Ac. Sci. Hung. 67, 297 (1971).

(24) See (15) and F.E. Harris and H.J. Monkhorst, Chem. Phys. Letters 23, 1026 (1969); Phys. Rev. Letters 23, 1026 (1969).

(25) F. Martino and J. Ladik, unpublished result.

(26) N.F. Mott and E.A. Davis, Electronic Processes in Non-Crystalline Materials, Clarendon Press, Oxford, 1971, Chap. V (p. 121).

(27) J. Ladik and P. Otto, Chem. Phys. Letters (submitted).

(28) Á. Bartha and J. Ladik, unpublished result.

(29) J.-L. Calais and G. Sperber, Int. J. Quant. Chem. 7, 501 (1973).

(30) W. Meyer, personal communication.

(31) J. Ladik, unpublished result.

(32) C.C.J. Roothaan, Rev. Mod. Phys. 32, 179 (1960).

(33) W.J. Hunt, T.H. Dunning, Jr., and W.A. Goddard III, Chem. Phys. Letters 3, 606 (1969); W.A. Goddard III, T.H. Dunning, Jr. and W.J. Hunt, Chem. Phys. Letters 4, 231 (1969) ; W.J. Hunt, W.A. Goddard III and T.H. Dunning, Jr., Chem. Phys. Lett. 6, 147 (1970).

(34) J.P. Dahl, to be published in the Proc.-s of SRC Atlas Symp. No. 4, Quantum Chemistry, Oxford 1974, ed. V.A. Saunders and J. Brown.

(35) C. Merkel, W. von Niessen, A. Karpfen, G. Stolhoff, and J. Ladik (unpublished).

(36) L.B. Coleman, J.J. Cohen, D.J. Sandman, F.G. Yamagishi, A.F. Garito, and A.J. Heeger, Solid State Comm. 12, 1125 (1973).

(37) J. Bardeen, Solid State Comm. 13, 357 (1973); D. Allender, J. W. Bray, and J. Bardeen, Phys. Rev. 139, 119 (1974).

(38) H. Fröhlich, Proc. Roy. Soc. A223, 296 (1954).

(39) C. Merkel, W. von Niessen, A. Karpfen, G. Stolhoff, J. Ladik, and P. Fulde (to be published).

(40) V.V. Walatka, Jr., M.M. Labes, and J.H. Perlstein, Phys. Rev. Letters 31, 1139 (1973).

NUMERICAL PROBLEMS IN CALCULATING ELECTRONIC STRUCTURE OF ONE-DIMENSIONAL CHAINS

J. Delhalle[+]

Laboratoire de Chimie Théorique Appliquée

Facultés Universitaires de Namur

61, rue de Bruxelles, B-5000 Namur (Belgium)

I. INTRODUCTION

The purpose of this lecture is to provide simple applications on concepts presented in the introductory lectures[1,2] on polymer calculations. It is also an attempt to point out some numerical problems occuring in one-dimensional periodic systems. The three following topics will be considered :

i. The Calculation of the Density Matrix Elements
ii. The Calculation of the First Derivative of the Energy Bands and the Band Indexing Problem.
iii. The Calculation of the Density of Electronic States

Before going through these subjects, it could be useful to define some quantities in the framework of the LCAO-(SCF)-CO formalism together with the notation used in this text, for more details see references 1 and 2.

The polymeric orbitals, $\phi_n(k,\underline{r})$, are expressed as a periodic combination of atomic functions, $\chi_p(\underline{r}-j\underline{a})$:

$$\phi_n(k,\underline{r}) = (2N+1)^{-\frac{1}{2}} \sum_{j=N}^{+N} \sum_{p=1}^{\omega} \exp(ikja)\, C_{np}(k)\, \chi_p(\underline{r}-j\underline{a}) \quad (1)$$

$$. \ k \left[-\frac{\pi}{a}, +\frac{\pi}{a} \right]$$

The coefficients, $C_{np}(k)$, are obtained by solving the generalized eigenvalue problem :

$$\underset{\sim}{F}(k)\ \underline{C}_n(k) = \underset{\sim}{S}(k)\ \underline{C}_n(k)\ E_n(k) \tag{2}$$

where the column vector $\underline{C}_n(k)$ contains the LCAO coefficients, $\{C_{np}(k)\}$, of $\phi_n(k,\underline{r})$.

The Fock and overlap matrix elements, $F_{pq}(k)$ and $S_{pq}(k)$ are expressed as follows :

$$F_{pq}(k) = \sum_j \exp(ikja)\ F_{pq}(j) \tag{3a}$$

and

$$S_{pq}(k) = \sum_j \exp(ikja)\ S_{pq}(j) \tag{3b}$$

$F_{pq}(j)$ and $S_{pq}(j)$ are respectively the Fock and overlap elements between atomic functions.

The density matrix elements, in the direct space, are defined as[2] :

$$D_{pq}(j) = \frac{a}{2\pi} \int_{-\pi/a}^{+\pi/a} \sum_{n=1}^{n_f} C^*_{np}(k) C_{nq}(k) \exp(ikja)\ dk \tag{4}$$

. n_f = number of doubly occupied levels at a specific k-point.

The band structure can also be obtained using the expectation value equation :

$$E_n(k) = \underline{C}^+_n(k)\ \underset{\sim}{F}(k)\ \underline{C}_n(k) \tag{5}$$

Finally, the density of electronic states, which is the number of states in the energy range E to E + dE, for unit "volume" of the polymer, is written as :

$$\mathcal{D}(E) = \frac{1}{\pi} \sum_{n=1}^{n_f} \left(\frac{dk}{dE_n(k)}\right)_{E_n(k) = E} \tag{6}$$

II. THE DENSITY MATRIX

IIa. Definition and Usefulness

Density matrix in k-space is defined as[3] :

$$\underset{\sim}{D}(k) = \underset{\sim}{C}(k)\ \underset{\sim}{C}^+(k) \tag{7}$$

where $\underset{\sim}{C}(k)$ is a $\omega \times n_f$ matrix containing the n_f doubly occupied vectors $\underline{C}_n(k)$. The normalization condition takes the form :

$$\mathrm{tr}\underset{\sim}{D}(k)\,\underset{\sim}{S}(k) = n_f \tag{8a}$$

and in an orthogonal atomic basis set we have :

$$\mathrm{tr}\underset{\sim}{D}(k) = n_f \tag{8b}$$

In the direct space $\underset{\sim}{D}(k)$ transforms in $\underset{\sim}{D}$ according to the next relation :

$$\underset{\sim}{D} = \frac{a}{2\pi}\int_{-\pi/a}^{+\pi/a} \sum_j \underset{\sim}{D}(k)\,\exp(ikja)\,dk = \sum_j \underset{\sim}{D}(j) \tag{9}$$

It is important to compute good values of $\underset{\sim}{D}$ since in SCF methods it varies at each step of the variational loop until a certain level of self consistency has been reached. As the next equation shows it the Fock matrix is directly dependent on $\underset{\sim}{D}$,

$$F_{pq}(j) = F^{nuc}_{pq}(j) + \sum_n \sum_l \sum_r \sum_s D_{rs}(j-l)\,[2(p^{o}q^{j}|r^{h}s^{l}) - (p^{o}r^{h}|q^{j}s^{l})] \tag{10}$$

and obviously numerical errors in $\underset{\sim}{D}(j)$ will affect the band structure[4]. Probably not enough care has been paid to the numerical evaluation of $\underset{\sim}{D}$ when the number of interacting cells increases.

IIb. Application to the Infinite Polyene HCO[1]

In the framework of Hückel method, two interesting quantities derived from the density matrix are given by Q_p, the π-charge density and Q_{pq}, the π-bond order (a quantitative index of π-bonding between two atoms) :

$$Q_p = \sum_{n=1}^{n_f} 2\,C^2_{np} \tag{11a}$$

and

$$Q_{pq} = \sum_{n=1}^{n_f} 2\,C_{np}\,C_{nq} \tag{11b}$$

These quantities may be extended to infinite periodic cases by the following definitions[1,2]

$$Q_p = \frac{a}{2\pi}\int_{-\pi/a}^{+\pi/a} 2\sum_{n=1}^{n_f} C^*_{np}(k)\,C_{np}(k)\,dk \tag{12a}$$

and

$$Q_{pq}(j) = \frac{a}{2\pi} \int_{-\pi/a}^{+\pi/a} 2 \sum_{n=1}^{n_f} C^{*}_{np}(k)\, C_{nq}(k) \exp(ikja)\, dk \quad (12b)$$

They may also be written in terms of the density matrix elements

$$Q_{pq}(j) = \frac{a}{2\pi} \int_{-\pi/a}^{+\pi/a} 2\, D_{pq}(k) \exp(ikja)\, dk \quad (13)$$

Using the results on the infinite polyene already obtained in a previous lecture[1] and summarized in Table I, we are now able to compute Q_1 and $Q_{1,2}(j)$. Expressing the variable k in term of $x = ka$ the coefficients of the occupied π-band are :

$$C_{\pi 1}(x) = \frac{1}{\sqrt{2}} \; ; \; C_{\pi 2}(x) = \frac{1}{\sqrt{2}} \frac{\omega(x)}{|\omega(x)|}$$

$$. \; x \in [-\pi, +\pi] \quad (14)$$

Table I : Summary of the Results on the Infinite Non-Alternant Polyene HCO[1]. $\beta = \beta'$

$\overset{-1}{- C_1 - C_2} \left[\overset{0}{C_1 - C_2} \right] \overset{1}{C_1 - C_2} -$	$- \; k \in [-\frac{\pi}{a}, +\frac{\pi}{a}]$ $- \; C_{\pi 1}(k) = \frac{1}{\sqrt{2}}; \; C_{\pi 2}(k) = \frac{1}{\sqrt{2}} \frac{\omega(k)}{\lvert\omega(k)\rvert}$
$\underset{\sim}{F}(k) = \begin{bmatrix} \alpha & \beta(1+e^{-ika}) \\ \beta(1+e^{ika}) & \alpha \end{bmatrix}$	$- \; E_\pi(k) = \alpha + 2\beta \cos \frac{ka}{2}$ $- \; \frac{d}{dk} E_\pi(k) = - a\beta \sin \frac{ka}{2}$ $- \; \frac{\omega(k)}{\lvert\omega(k)\rvert} = e^{\frac{ika}{2}}$

The π-charge density Q_1 is then :

$$Q_1 = \frac{2a}{2\pi} \int_{-\pi/a}^{+\pi/a} \frac{1}{\sqrt{2}} \cdot \frac{1}{\sqrt{2}}\, dk = \frac{2}{2\pi} \int_{-\pi}^{+\pi} \frac{1}{2}\, dx = 1 \quad (15)$$

and $Q_{1,2}(j)$ is obtained after the following operations :

$$Q_{1,2}(j) = \frac{2}{2\pi} \int_{-\pi}^{+\pi} \frac{1}{\sqrt{2}} \cdot \frac{1}{\sqrt{2}} \frac{\omega(x)}{|\omega(x)|} e^{ijx} dx = \frac{1}{\pi} \int_{-\pi}^{+\pi} e^{\frac{ix}{2}} e^{ixj} dx \quad (16)$$

leading to :

$$Q_{1,2}(j) = \frac{\sin\pi(j + \frac{1}{2})}{\pi(j + \frac{1}{2})} = \frac{(-1)^j}{(j + \frac{1}{2})\pi} \quad (17)$$

Some values of $Q_{1,2}(j)$ are shown in table II.

Table II : π-bond orders between atoms 1 and 2 respectively centered in the origin unit cell and in the cell j.

j	0	-1	1	-2	2
$Q_{1,2}(j)$	.6366	.6366	-.2122	-.2122	.1273

That the values of the density matrix elements can affect the band structure is suggested in the forthcoming section.

IIc. Importance of Good Numerical Values of $\underset{\sim}{D}(j)$

To illustrate the influence of D(j) on the energy bands we have considered the regular alternant infinite polyene in the PPP-SCF-CO approximation[5]. The density matrix has been approximated by a discrete summation (still commonly in use) over n_k k-points :

$$\underset{\sim}{D}(j) = \frac{1}{n_k} \sum_{k}^{n_k} \underset{\sim}{D}(k) \exp(ikja) \quad (18)$$

Two cases have been studied; $N_1 = 2$ and $N_2 = 6$ corresponding to the number of interacting cells in each direction with the origin cell. We have computed the band gap ΔE for different n_k's (starting from the lowest possible value : $n_k = 2N + 1$ to values stabilizing ΔE). Table III contains the results.

It is easy to see that the number of interactions doesn't modify very much the value of the band gap but the effect of $\underset{\sim}{D}(j)$ is important since here it is the only variable factor. As we shall see now good numerical values of D(j) are sometimes difficult to obtain, particularly when j increases ($j \geq 10$).

Table III : Variation of the band gap ΔE of [-CH=CH-] with n_k Units in eV.		
n_k	$\Delta E(N_1 = 2)$	$\Delta E(N_2 = 6)$
5	.4646	——
13	3.9770	.0701
161	4.1049	4.0313
201	4.1049	4.0329

As an example we take the numerical evaluation of $D_{1,2}(o)$ and $D_{1,2}(5)$ of the infinite polyene HCO. They are simply half the value of $Q_{1,2}(0)$ and $Q_{1,2}(5)$ given by equation (17). We have used four methods to compute these values : a discrete summation (Σ), a Simpson quadrature[6,7](S), a Gauss-Legendre quadrature[8,7](G) and a Filon method [7,9 - 11,4](F). Let us consider the results of each method with respect to an increase of n_k (Tables IV and V).

From table IV it comes out that Gauss-Legendre technique is the most interesting compared with Simpson or summation quadratures. But, when j increases (Table V) these three methods require much more k-points, even for Gauss-Legendre method, before reaching the good value. Filon technique we present in the next section is specially designed for handling oscillatory integrands. Effectively it is based on a quadratic interpolation (like Simpson method) and it requires approximatively as many k-points for $D_{1,2}$ (5) as Simpson method needs for $D_{1,2}(0)$. Filon method or similar ones should probably be introduced in our polymer programs in order to avoid numerical uncertainties in band structures and chiefly to reduce the number of diagonalizations. We should bear in mind that $\tilde{D}(k)$ is built after solving the generalized eigenvalue problem, equation (2), at n_k k-points as many times as required to end up with the expected self consistency. Two recent papers reinforce this conclusion; the first by O'SHEA and SANTRY[3] who have pointed out the need of many interacting cells ($j\uparrow$) for cases like polyenes and the second one by KARPFEN and coll.[14] Who had to increase the number of k-points to stabilize the band gap in a water polymer.

Table IV. : Numerical values of $D_{1,2}(0)$ of the polyene HCO from different methods of integration and with respect to n_k. Theoretical value : .31830989.

n_k	Σ	S	G
2	-	-	.30809526
3	.33333333	.31903560	-
4	-	-	.31830738
6	-	-	.31830989
7	.32099708	.31835273	-
8	-	-	.31830989
11	.31939428	.31831826	-
20	-	-	.31830989
23	.31855747	.31831041	-
40	-	-	.31830989
43	.31838069	.31830994	-
64	-	-	.31830989
67	.31833905	.31830989	-

Table V. : Numerical values of $D_{1,2}(5)$ of the polyene HCO from different methods of integration and with respect to n_k. Theoretical value : -0.02893726.

n_k	Σ	S	G	F
6	-	-	-.13372362	-
7	-0.11456040	+.72940450	-	-
10	-	-	-.07083158	-
11	-0.04545455	-.11098211	-	-
14	-	-	-.02896869	-
17	-0.02527878	-.02919049	-	-0.02893786
20	-	-	-.02893726	-
33	-0.02779643	-.02895142	-	-
35	-	-	-	-0.02893725
65	-0.02853833	-0.02893813	-	-0.02893726
129	-0.02878088	-0.02893732	-	-0.02893726
257	-0.02886989	-0.02893726	-	-0.02893726
513	-0.02890629	-0.02893726	-	-0.02893726
1025	-0.02892246	-0.02893726	-	-0.02893726

IId. Filon Method

We have just seen that in polymer calculations exist oscillatory functions. By a rapidly oscillatory integrand we mean one with numerous (generally more than 10) local maxima and minima over the range of integration[15]. Among examples we find the Fourier transform :

$$I(j) = \int_a^b f(x) \exp(ijx)\, dx \qquad (19)$$

What we want are the numerical values of such integrals when oscillations increase. Their general form may be written,

$$I(j) = \int_a^b f(x)\, K(x,j)\, dx \qquad (20)$$

$$.-\infty \leqslant a < b \leqslant \infty$$

with K(x,j) an oscillatory kernel and f(x) the supposed well-behaved part in the sense of a non-oscillatory character. Density matrix elements, $D_{pq}(j)$, are to be included in that class since they may be written in a similar way as in (19) :

$$D_{pq}(j) = \frac{a}{2\pi} \int_{-\frac{\pi}{a}}^{+\frac{\pi}{a}} D_{pq}(k) K(k,j) dk \qquad (21)$$

Numerical integration of oscillatory integrands is beset with difficulties peculiar to it. For example consider the computation of :

$$\int_0^1 x^2 \cos(4\pi x)\, dx \qquad (22)$$

The graph of x^2 is smooth while its product with cos ($4\pi x$) consists of plus and minus areas (Figure I). As $j \to \infty$ these areas will be of nearly equal size and the resulting cancellation of areas is attended by a loss of significance. Furthermore, as $j \to \infty$, f(x) K(x,j) deviates from a polynomial of low degree, this suggests special methods. Here we will be concerned with the Filon approximation[15] for finite Fourier integrals.

Suppose we can write :

$$f(x) = \sum_{i=1}^{n} a_i \, q_i(x) + \varepsilon(x) \qquad (23)$$
$$. \; x \in [a,b]$$

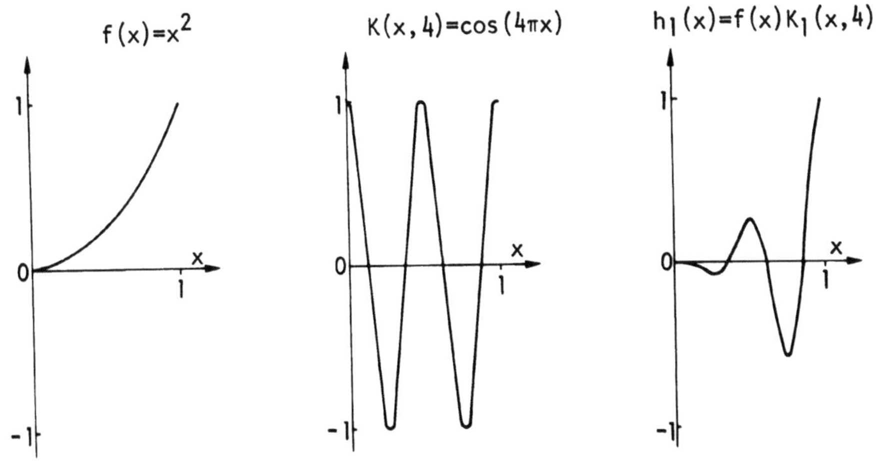

Figure 1

where $\varepsilon(x)$ is small over $[a,b]$ and where the transforms

$$Q_i(j) = \int_a^b q_i(x) \; K(x,j) \; dx \qquad (24)$$
$$. \; i = 1, 2, \ldots n$$

can be computed explicitely in elementary terms. Then

$$I(j) = \int_a^b f(x) \; K(x,j) dx = \sum_{i=1}^{n} a_i \; Q_i(j) + \int_a^b \varepsilon(x) \; K(x,j) \; dx \qquad (25a)$$

$$\simeq \sum_{i=1}^{n} a_i \; Q_i(j)$$

Filon has deduced the formulas when $f(x)$ is approximated by parabolic arcs. Consider the integral

$$I(j) = \int_a^b f(x) \cos (jx)\, dx \qquad (26)$$

with interval [a,b] divided into 2N subintervals of equal length h,

$$h = \frac{b - a}{2N} \qquad (27)$$

over each double subinterval, f(x) is interpolated by a parabola. In such a case, Fourier integrals can be computed explicitly leading to the following rules of approximate integration :

$$S(j)=\int_a^b f(x) \sin(jx)dx=h\{-\alpha[f(b) \cos(jb)-f(a) \cos (ja)] + \beta S_{2p} + \gamma S_{2p-1}\} \qquad (28a)$$

and

$$C(j)= \int_a^b f(x) \cos (jx)dx = h\{\alpha [f(b) \sin(jb) - f(a)\sin (ja)] + \beta C_{2p} + \gamma C_{2p-1}\} \qquad (28b)$$

with

$$C_{2p} = \sum_{i=o}^{N} f(a+i2h) \cos(j(a+i2h))- \frac{1}{2} [f(a) \cos(ja) + f(b) \cos (jb)] \qquad (29a)$$

and

$$C_{2p-1} = \sum_{i=1}^{N} f(a + hi') \cos (j(a + hi')) \qquad (29b)$$

$$. \; i' = 2i - 1$$

S_{2p} and S_{2p-1} are obtained in a similar way by interchanging cosine and sine in the formulas. The expressions of θ, α, β and γ are respectively :

$$\left.\begin{aligned} \theta &= jh \; ; \; \beta = 2 \left(\frac{1 + \cos^2\theta}{\theta^2} - \frac{\sin^2\theta}{\theta^3}\right) \\ \alpha &= \frac{1}{\theta} + \frac{\sin 2\theta}{2\theta^2} - \frac{2 \sin^2\theta}{\theta^3} \; ; \; \gamma = 4 \left(\frac{\sin\theta}{\theta^3} - \frac{\cos\theta}{\theta^2}\right) \end{aligned}\right\} \qquad (30)$$

It should be very easy to introduce this method in polymer programs. However, further investigations in that direction should be undertaken to compute oscillatory integrals in periodic systems. The complete deduction of the above formulas can be found in the original paper[15].

III. FIRST DERIVATIVE OF THE ENERGY BANDS

IIIa. Computation of the First Derivative of the Energy Bands

If F(k) depends on a parameter k and if $\phi_n(k,\underline{r})$ is an eigenvector normalized to unity then[16] :

$$\frac{dE_n(k)}{dk} = (\phi_n(k), \frac{d\underset{\sim}{F}(k)}{dk} \phi_n(k)) \tag{31}$$

It is also possible to obtain the local derivative of $E_n(k)$ when the atomic basis set is not orthogonal. Effectively $E_n(k)$ is a multivalued function of k, solution of the generalized eigenvalue equation :

$$\underset{\sim}{F}(k)\ \underline{C}_n(k) = \underset{\sim}{S}(k)\ \underline{C}_n(k)\ E_n(k) \tag{32}$$

the values of $E_n(k)$ are obtained by the following inner product,

$$E_n(k) = \underline{C}_n^+(k)\ \underset{\sim}{F}(k)\ \underline{C}_n(k) \tag{33}$$

with the normalization condition :

$$\underline{C}_n^+(k)\ \underset{\sim}{S}(k)\ \underline{C}_n(k) = 1 \tag{34}$$

Taking the first derivative of the left and right sides of (33) we have :

$$\frac{dE_n(k)}{dk} = \frac{d\underline{C}_n^+(k)}{dk} \underset{\sim}{F}(k)\ \underline{C}_n(k) + \underline{C}_n^+(k) \frac{d\underset{\sim}{F}(k)}{dk} \underline{C}_n(k) + \underline{C}_n^+(k) \underset{\sim}{F}(k) \frac{d\underline{C}_n(k)}{dk} \tag{35}$$

By noting that

$$\underset{\sim}{F}(k)\ \underline{C}_n(k) = \underset{\sim}{S}(k)\ \underline{C}_n(k)\ E_n(k) \tag{36}$$

and from properties of the hermitian matrices $\underset{\sim}{F}(k)$ and $\underset{\sim}{S}(k)$ we have also :

$$\underline{C}_n^+ (k)\ \underset{\sim}{F}(k) = \underline{C}_n^+ (k)\ \underset{\sim}{S}(k)\ E_n(k) \qquad (37)$$

Introducing these latter relations in (35) we finally end up with

$$\frac{d\underline{E}_n(k)}{dk} = \underline{C}_n^+(k)\ \frac{d\underset{\sim}{F}(k)}{dk}\ \underline{C}_n(k) - \underline{C}_n^+(k)\ \frac{d\underset{\sim}{S}(k)}{dk}$$

$$\underline{C}_n(k)\ E_n(k) \qquad (38)$$

In an orthogonal atomic basis set expression (38) reduces to

$$\frac{dE_n(k)}{dk} = \underline{C}_n^+(k)\ \frac{d\underset{\sim}{F}(k)}{dk}\ \underline{C}_n(k) \qquad (39)$$

Now if we remember the trigonometric expansions of $\underset{\sim}{F}(k)$ and $\underset{\sim}{S}(k)$ it is very easy to apply (38) or (39) to obtain numerical values of $dE_n(k)/dk$.

The derivatives of $\underset{\sim}{F}(k)$ and $\underset{\sim}{S}(k)$ are simply :

$$\frac{d\underset{\sim}{F}(k)}{dk} = ia \sum_j j\ F_{pq}(j)\ \exp\ (ikja) \qquad (40a)$$

and

$$\frac{d\underset{\sim}{S}(k)}{dk} = ia \sum_j j\ S_{pq}(j)\ \exp\ (ikja) \qquad (40b)$$

It is straightforward to introduce a simple scheme of computation at the end of variation loops to evaluate the derivative of $\{E_n(k)\}$.

IIIb. Simple Application to the Infinite Polyene HCO

The first derivative of $\underset{\sim}{F}(x)$ (see table I) is :

$$\frac{d\underset{\sim}{F}(x)}{dx} = \begin{bmatrix} 0 & -i\beta e^{-ix} \\ & \\ i\beta e^{ix} & 0 \end{bmatrix} \qquad (41)$$

Because of the orthogonal atomic basis set of the Hückel method we apply expression (39) to calculate

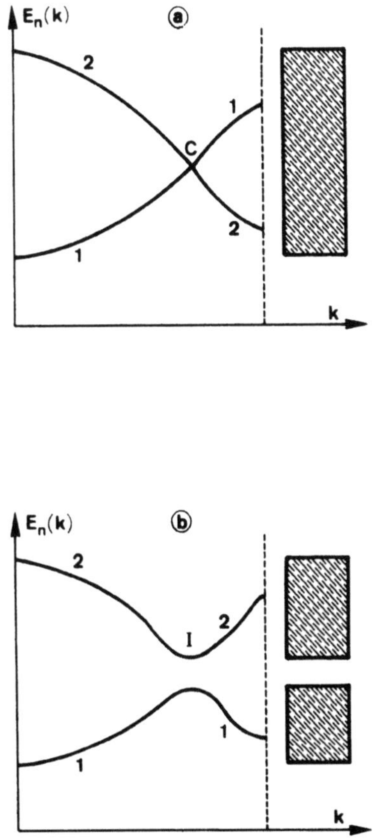

Figure 2

$$\frac{dE_n(x)}{dx} = \left(\frac{1}{\sqrt{2}} \quad \frac{1}{\sqrt{2}} \frac{\omega^*(x)}{|\omega(x)|}\right) \frac{d\underset{\sim}{F}(x)}{dx} \begin{pmatrix} \frac{1}{\sqrt{2}} \\ \frac{1}{\sqrt{2}} \frac{\omega(x)}{|\omega(x)|} \end{pmatrix} \tag{42}$$

Noting that $\omega(x)/|\omega(x)|$ is equal to $\exp(ix/2)$ it is an easy matter to obtain :

$$\frac{dE_n(x)}{dx} = -\beta \sin \frac{x}{2} \tag{43}$$

which is well the first derivative of

$$E_n(x) = \alpha + 2\beta \cos \frac{x}{2} \tag{44}$$

IIIc. On the Use of the First Derivative in Labelling Energy Bands

In the reduced zone scheme $E_n(k)$ is a multivalued function of k and the successive values of n for any given k correspond first to the inner core electrons, then to valence bands (or sometimes to conduction band) and finally to the unoccupied bands. Bands are usually given an index increasing in order of increasing energy (Figure IIb). Difficulties arise because the rule breaks down at degeneracy points, C in Figure IIa. Origin of crossing or noncrossing bands will be discussed in a forthcoming lecture[18]. A solution to this problem may be initiated with the help of group theory but it is also possible to use the simple information of the first derivative of energy bands. This is useful because we only know the values of $E_n(k)$ at certain k points and it is very easy to get in trouble with the assignment of n to a particular value of $E_n(k)$. In figure II, the two possibilities have equal chances if we don't know much about symmetry rules. The resulting properties could be extremely different; in figure IIb there is a band gap and in Figure IIa there is no band gap. The use of the first derivative may give some information to clarify the situation. In figures III and IV[19] we observe from the graph of the first derivative that the solution of the Figure IIIb is the only one possible. A detailed example of the use of the first derivative may also be found in reference (20).

In our applications with the POLYCNDO program we use systematically the first derivative. The program is linked to a BENSON plotter and the variations of $E_n(k)$ are graphically represented by a rotating fixed length vector centered at $(E_n(k),k)$ points (see figures V and VI[21]). An arbitrary angle of 80° (or -80°)

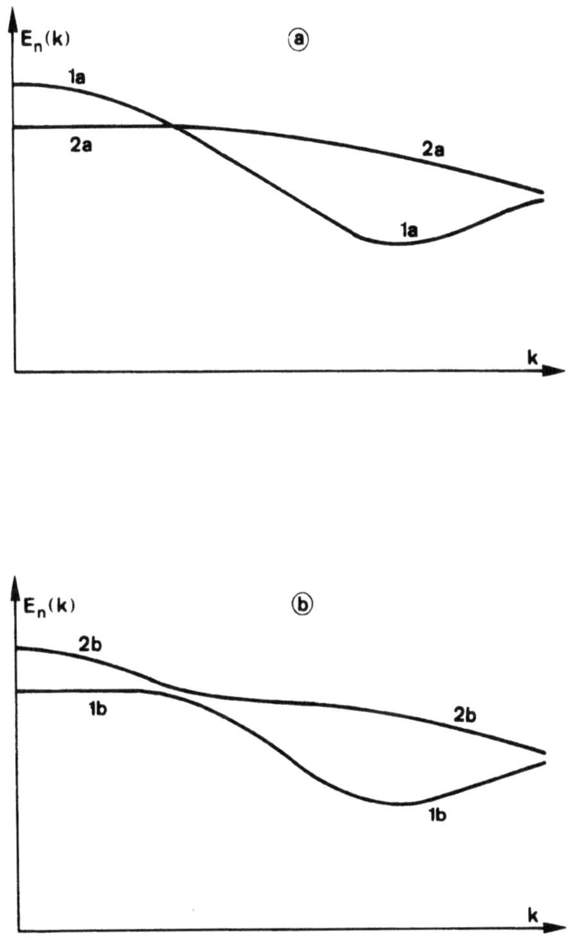

Figure 3

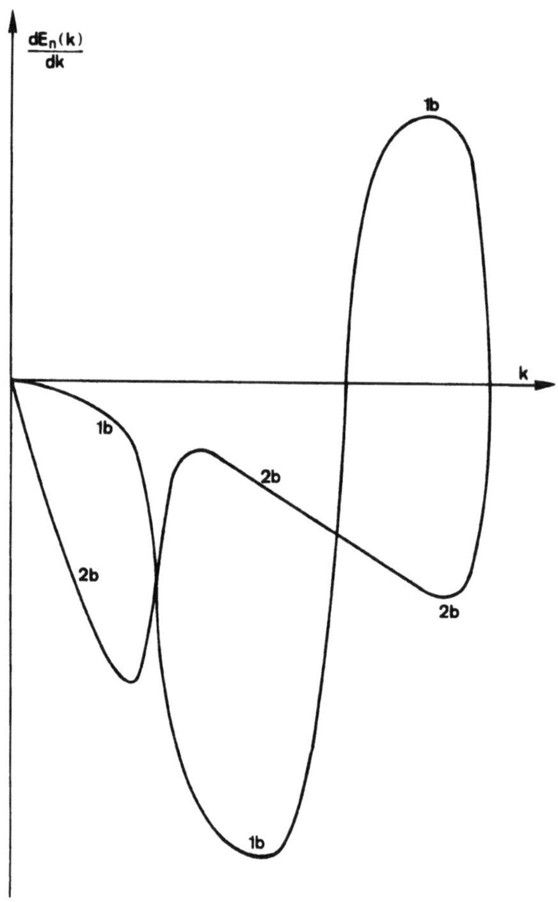

Figure 4

is attributed to the maximum derivative of the whole band structure and the other angles vary within the interval (-80°, 80°). When two or more bands are not enough separated in the figure it is possible to enlarge the graphic. Our experience has shown that nearly all kind of bands may be labelled by this procedure and the time of data manipulation is drastically reduced. Once this work is achieved, the ordered data are stored on a magnetic device and the computation of the density of electronic states can follows.

IV. DENSITY OF ELECTRONIC STATES

IVa. Deduction for one-dimensional systems

In introductory lectures it has been shown that as a consequence of Born Karman conditions the wave function has the following properties :

$$\psi(x + Na) = \psi(x) \tag{45a}$$

$$|\psi(x + ja)|^2 = |\psi(x)|^2 \tag{45b}$$

$$\psi(x + ja) = \exp\left(\frac{2\pi i j}{N}\right) \psi(x) = \exp(ika)\, \psi(x) \tag{45c}$$

we see that $k = \frac{2\pi}{a} \cdot \frac{j}{N}$ and if N, the number of elementary cells, goes up to ∞ k becomes a continuous variable with usual interval: $[-\pi/a, +\pi/a]$. For each nondegenerate band there are two electrons with opposite spin in each primitive cell of the polymer of for each allowed k-vector because there are as many allowed wave vector in the Brillouin zone as there are unit cells in the block of the polymer. The length of that unit cell is a and the length of the corresponding unit cell in the reciprocal space is defined as $2\pi/a$. The length of k-space per allowed k-vector is $2\pi/aN$ and there are $Na/2$ allowed k-vector per unit length of reciprocal space multiplied by 2 due to spin. It follows that, for unit length of the polymer, the number of states in length dk of the reciprocal space is

$$dn = \frac{1}{\pi}\, dk \tag{46}$$

and the number of states per unit energy interval, dn/dE, is called the density of electronic states. In terms of k :

$$D(E) = \frac{1}{\pi} \frac{dk}{dE} \tag{47}$$

In general, for many bands in the Brillouin zone, we have

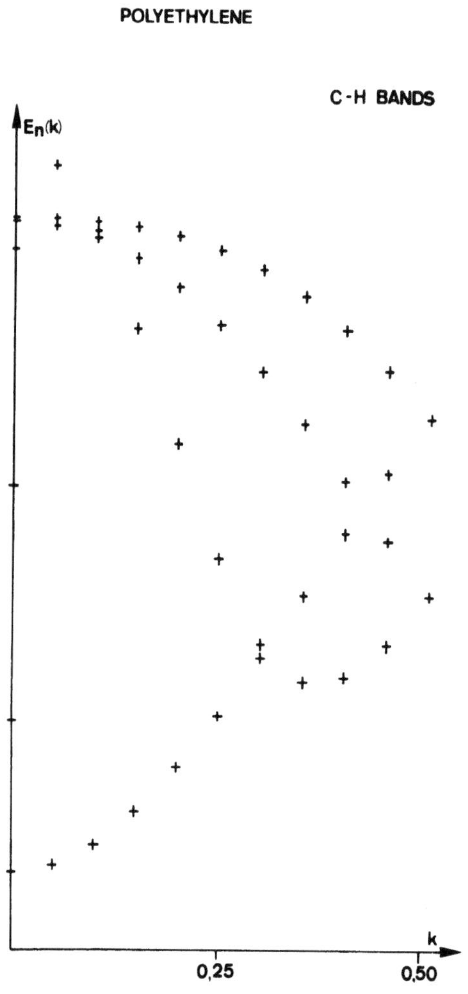

Figure 5

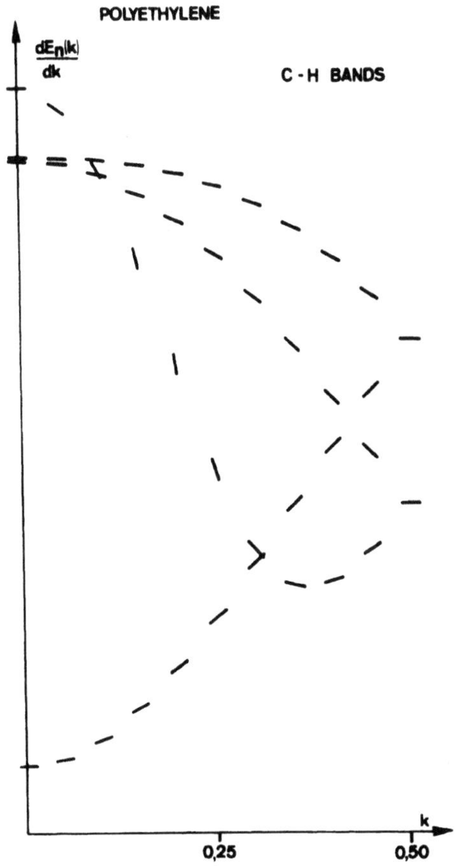

Figure 6

$$D(E) = \frac{1}{\pi} \sum_{n=1}^{n_f} \left(\frac{dk}{dE_n(k)}\right)_E \qquad (48)$$

The number of electronic states with energy less than E, also called the integrated density of electronic states, is defined as

$$I(E) = \int_{-\infty}^{E} D(E)\, dE \qquad (49)$$

There are certain advantage to studying the density of states function instead of the energy band structure itself :

- It corresponds to an entire statistical description of the distribution of the electronic levels
- The density of states function is more closely related to experiment than the energy band structure and particularly to ESCA spectroscopy in the case of polymers[22].
- When the number of energy bands becomes important the density of electronic states appears to be the only one practicable way of comparison between quantum theory and experimental results.

IVb. Application to the Infinite Polyene HCO

The first energy band $E_\pi(k)$ of the infinite polyene has the following analytical expression :

$$E_\pi(k) = \alpha + 2\,\beta \cos\frac{ka}{2}\ ;\ E_\pi(x) = \alpha + 2\,\beta \cos\frac{x}{2} \qquad (49)$$

Applying expression (47) (because there is only one occupied band) it is easy to deduce the density of electronic states in such a case :

$$D(E) = \frac{1}{\pi} \left(\frac{dx}{dE_\pi(x)}\right)_E = \frac{1}{\pi} \left(\frac{1}{-\beta \sin\frac{x}{2}}\right)_E \qquad (50)$$

E(x), dE(x) and $D(E)$ are represented in Figure VII. We observe an infinite discontinuity at $E = \alpha + 2\,\beta$. This is a necessary consequence of the energy bands properties, ($E_n(k) = E_n(-k)$, $dE_n(k)/dk = 0$ at $k = \pm\ \pi/a$ and at those k-points were $dE_n(k)/dk = 0$), leading to characteristic discontinuities in $D(E)$ of one-dimensional systems. However the area under this $D(E)$ curve remains finite and is equal to the number of electronic states in the band, namely 2:

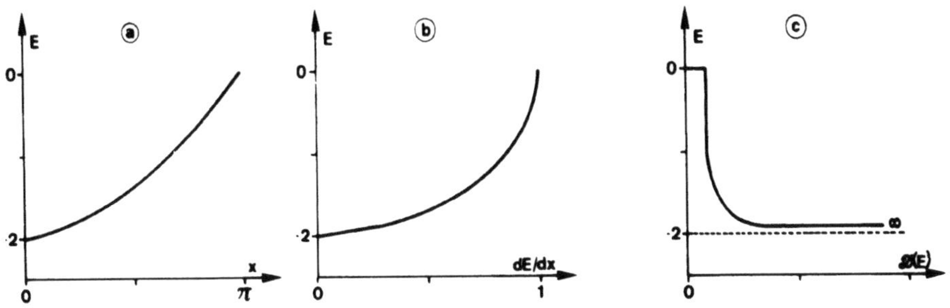

Figure 7

$$\int_{\alpha+2}^{\alpha} - \frac{dE}{\pi \quad \beta\sin[\arccos(\frac{E-\alpha}{2\beta})]} = 2 \tag{51}$$

changing E into $\frac{x}{2}$ we have :

$$\int_{-\frac{\pi}{2}}^{\frac{\pi}{2}} \frac{-2\beta\sin\frac{x}{2}}{-\pi\beta\sin\frac{x}{2}}\, d\,\frac{x}{2} = \frac{2}{\pi}\int_{-\frac{\pi}{2}}^{+\frac{\pi}{2}} du = 2 \tag{52}$$

IVc. Numerical Calculation of D(E)[24]

In real systems the number of bands is greater than in the previous example and the discontinuities occuring when $dE_n(k)/dk=0$ force to compute a new function D(E) which is an average of $\mathcal{D}(E)$ over an energy interval ΔE. To briefly summarize the computation of D(E) we will introduce a new definition of $\mathcal{D}(E)$ equivalent to the classical one :

$$\mathcal{D}(E) = \frac{1}{\pi} \sum_{n=1}^{nf} \int_{B.Z.} \delta[E-E_n(k)]dk \tag{53}$$

Following A. BRUST[23] we define the histogram of the density of states $D(E_i)$ by the relation :

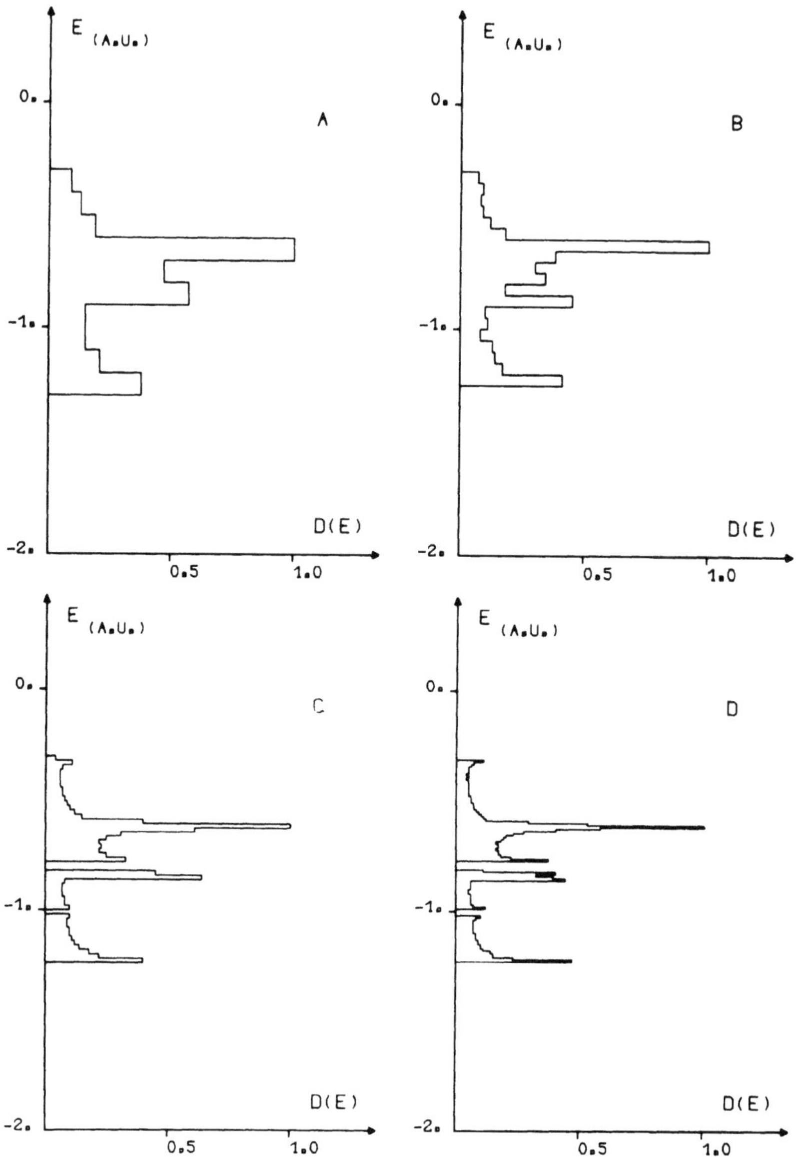
E (A.U.)
0.
A
-1.
-2.
D(E)
0.5
1.0
E (A.U.)
0.
B
-1.
-2.
D(E)
0.5
1.0
E (A.U.)
0.
C
-1.
-2.
D(E)
0.5
1.0
E (A.U.)
0.
D
-1.
-2.
D(E)
0.5
1.0

Figure 8

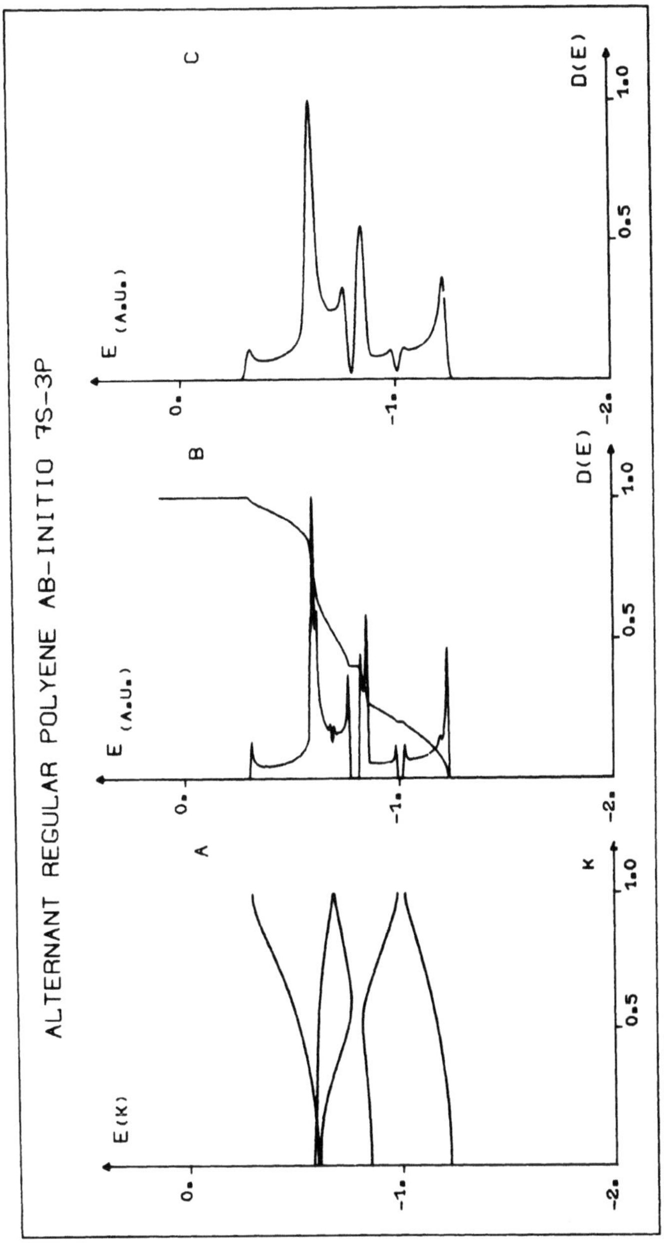
ALTERNANT REGULAR POLYENE AB-INITIO 7S-3P
A
E(K)
K
0.
-1.
-2.
0.5
1.0
B
E (A.U.)
D(E)
C

Figure 9

$$D(E_i) = C \sum_{n=1}^{n_f} \sum_{k}^{n_k} \{\delta^{\Delta E} [E_i - E_n(k)]\} \quad (54)$$

with

- $E_{i+1} = E_i + \Delta E$

- $\delta^{\Delta E} [E_i - E_n(k)] = \begin{cases} 1 \text{ if } |E_n(k) - E_i| \leqslant \Delta E/2 \\ 0 \text{ otherwise} \end{cases} \quad (55)$

- C = scaling factor

We see at once that (54) is a particularization of (53). In our calculations we have chosen $\Delta E = 0.01$ a.u. since the mean resolution of ESCA spectrometer is about 0.02 a.u. The equidistant sampling has been preferred to the random one to achieve the discrete summation over the N_k k-points (expression (54)). From computations on ab initio polyene we found that $N_k = 20000$ is enough to stabilize histograms with 3 significant figures[22]. The histograms of the electronic density of states of ab initio alternant polyene with ΔE respectively equal to A = 0.01 a.u., B = 0.05 a.u., C = 0.02 and D = 0.01 a.u. are shown in figure VIII. The graph of histogram D remains discontinuous and it is worth to smooth it. We adopted the technique of J.T. WABER consisting in a superposition of five histograms ($\Delta E = 0.01$ a.u.) each shifted by a value of $\Delta E/5$ and added together. The result is quite satisfactory as shown in figure IX for alternant ab initio polyene[25]. In our program we have also added the computation of I(E) (figure IVb) and the convolution of D(E) with a gaussian of full width at half-maximum equal to 0.026 a.u. to simulate experimental spectra (figure IXc). The energy bands are interpolated by an Aitken-Lagrange method. Figure IX is a direct output of the program on the BENSON plotter. The program uses 8×10^3 words and needs, on a PDP 11/45, 10 minutes for each band. Two main improvements are to come out very soon : a faster interpolation algorithm and the introduction of the cross section to simulate in a better way the experimental spectra.

REFERENCES

1. J.M. André, "Ab Initio and Semi-Empirical Band Structure Calculations of Polymers" , NATO ASI, Namur (Belgium) 1974.

+ *Chargé de Recherches du "Fonds National Belge de la Recherche Scientifique".*

2. J. Ladik, "Different LCAO Band Structure Calculation Methods for Periodic Polymers and Molecular Crystals", NATO ASI, Namur (Belgium) 1974.

3. R. McWeeny and B.T. Suteliffe, "Methods of Molecular Quantum Mechanics" Academic Press, 1969.

4. J. Delhalle, Internat. J. Quant. Chem., 8, 201, 1974.

5. J.M. André, G. Leroy, Theoret. Chim. Acta (Berlin), 9, 123, 1967.

6. DQSF-routine, IBM System/360, SSP (360A-CM-03X), version III.

7. J. Delhalle, "Etude Théorique de la Structure de Bandes des Systèmes Périodiques en Méthode LCAO", UCL, Louvain, (Belgium) 1970.

8. A.D. McLean, M. Yoshimine, IBM San José, RJ-308, 1964.

9. S.M. Chase, L.D. Fosdick, Comm. ACM, 12, 453, 1969.

10. S.M. Chase, L.D. Fosdick, Comm. ACM, 12, 457, 1969.

11. S.M. Chase, L.D. Fosdick, Comm. ACM, 13, 263, 1970.

12. S.F. O'Shea, D.P. Santry, Chem. Phys. Letters, 25, 164, 1974.

13. A. Karpfen, J. Ladik, P. Russeger, S. Suhai, Theoretica Chim. Acta, 34, 115, 1974.

14. P.J. Davis, P. Rabinowitz, "Numerical Integration", Blaisdell, 1967.

15. K.N.G. Filon, Proc. Roy. Soc. Edinburgh, 49, 38, 1928-1929.

16. G. Del Re, J. Ladik, G. Biczo, Phys. Rev., 155, 997, 1967.

17. J. Delhalle, Annales Soc. Scient. Brux., 86, 277, 1972.

18. W.L. McCubbin, "Effect of Symmetry on the Band Structure of Polymers", NATO ASI, Namur (Belgium), 1974.

19. J. Delhalle, Ph. D. Thesis, Louvain (Belgium), 1974.

20. J.M. André, J. Delhalle, G. Kapsomenos, G. Leroy, Chem. Phys. Letters, 14, 485, 1972.

21. C. Pivont, J.M. André, S. Delhalle, J. Delhalle, to publish.

22. J. Delhalle, J.M. André, S. Delhalle, J.J Pireaux, R. Caudano, J. Verbist, J. Chem. Phys., 60, 595, 1974.

23. J. Brust, "Methods of Computational Physics", 8, 52, Academic Press 1968.

24. J. Delhalle, submitted to Bull. Soc. Chim. Belges.

25. J.M. André, G. Leroy, Int. J. Quant. Chem., 5, 557, 1971.

OPTICAL PROPERTIES OF MOLECULAR CRYSTALS*

G. D. Mahan
Physics Department
Indiana University
Bloomington, Indiana 47401

I. INTRODUCTION

A measurement of the optical properties of electronic systems is usually the best way to obtain information about energy levels. This is particularly true for molecules, since the excitation energies of the molecule are in the range of frequency suitable for optical measurement. There are, of course, many types of optical measurements : one and two photon absorption,[1] reflection, and Raman scattering, to just name some common methods. All of these provide different types of information. These lectures will not be concerned with the different experimental methods. Instead, we will be mainly concerned with the general dielectric properties, which can be measured a number of ways.

Optical measurements can be performed on isolated molecules by doing experiments when the molecule is in a dilute gas, dissolved in a liquid solvent, or introduced as an impurity in another solid. The gas and liquid spectra do not provide any information on the orientation of the molecules, and organic molecules are very asymmetric. So if the optical spectrum shows an intense absorption band, which indicates an electronic transition between two states, one has no information about which electronic transitions are involved. This condition is changed if the molecules are loosely packed into a crystal. With a knowledge of the crystal structure, the asymmetry of the molecule usually helps in determining the symmetry of the electronic states involved in the absorption band. Thus it is necessary to do optical measurements in crystals to learn specific information about

*Research Supported by the National Science Foundation

molecular energy levels. Of course, this raises new problems, since in the crystal the molecules are close together, and begin to interact with each other. In some cases this interaction is weak, and the crystal spectrum is little changed from the solution or gas spectrum. In other cases the interactions are strong, and the changes are large.

These lectures are on the methods of interpreting, and theoretically calculating, the optical properties of molecular crystals. This is not a systematic review, since, fortunately, good ones are available.[2,3] It will be a pedagogical development of the subject. The subject has a number of subtle points, which are easy to get wrong. Many of these are concerned with the method of taking dipole sums, and whether long range and retardation are included correctly. At one point I estimated that fifty per cent of the theoretical articles published in this field were wrong because these points were misunderstood. My goal in these lectures is to discuss these points in sufficient detail that they are understood by everyone.

Of course, numerical examples will be provided which apply to real systems. For weakly interacting systems, the series of ring compounds naphtalene, anthracene, tetracene and phenanthrene were selected. The most experimental and theoretical work has been done on these systems, particularly anthracene. As an example of a strongly interacting molecular solid, in Chapter V we discuss monolayer crystals of trans-bixin methylester.

II. OPTICAL PROPERTIES OF MOLECULES

2.1. Optical Constants

The optical properties of an isolated molecule may be described by the complex polarizability

$$\alpha(\omega) = \frac{e^2}{m} \sum_{\ell} \frac{f_\ell}{\omega_\ell^2 - \omega^2 - i\omega\gamma_\ell} \tag{2.1}$$

This is a summation of terms, where each term is characterized by three numbers: the resonance frequency ω_ℓ, the intensity factor f_ℓ which is called the oscillator strength, and a dissipative component γ_ℓ-this latter term is usually quite small. For example, in a gas of density ρ of these molecules, the complex dielectric function is

$$\varepsilon(\omega) = 1 + \frac{4\pi\rho\alpha(\omega)}{1 - \frac{4\pi}{3}\rho\alpha(\omega)}$$

The complex functions $\alpha(\omega)$ and $\varepsilon(\omega)$ have real and imaginary parts

$$\alpha = \alpha_1 + i\alpha_2$$

$$\varepsilon = \varepsilon_1 + i\varepsilon_2$$

The experimental quantities which are usually measured are the refractive index n and extinction coefficient K, which are given by

$$n + iK = (\varepsilon)^{1/2}$$

If the intensity of a light beam decays in a distance d by an amount

$$I = I_0 e^{-Ad}$$

then the absorption coefficient A is given by $A = 2K\omega/c$.

These results simplify whenever the molecules in the gas are very dilute. Then one has

$$\varepsilon_1 \simeq n^2 = 1$$

$$K = \frac{1}{2}\varepsilon_2 = 2\pi\alpha_2\rho = \frac{2\pi e^2\rho}{m} \sum_\ell \frac{f_\ell \omega \gamma_\ell}{(\omega^2-\omega_\ell^2)^2+\omega^2\gamma_\ell^2}$$

The extinction coefficient K is a sum of lorentzian absorption bands. A measurement of absorption spectra in a gas usually shows a series of sharp absorption lines. The frequency of the line gives ω_ℓ while the intensity gives the oscillator strength f_ℓ. Thus these quantities are easily measured, and are known for many molecules. These experimental values provide the starting point for the calculation of the spectra of molecular solids. The molecular data may also be obtained by measuring the molecule dissolved in liquid solvents, or as an impurity in other solids.

It is also possible to calculate the quantities ω_ℓ and f_ℓ on a purely theoretical basis. That is the subject of the other lectures. These calculations produce a description of the electron energy levels E_i in the solid, and also wave functions Ψ_i. The frequency ω_ℓ is the difference between two energy levels, and the oscillator strength is given in terms of the wave functions

$$\hbar\omega_\alpha = E_i - E_j$$

$$f_\alpha = 2\frac{\omega_\ell m}{\hbar}(\hat{e}_0 \cdot \underline{\mu}_\alpha)^2 \qquad (2.2)$$

$$\underline{\mu}_\alpha = \int d^3r\, \Psi_i(r)^* \underline{r}\, \Psi_f(r) \qquad (2.3)$$

where $\hat{e}_o$ is the polarization direction of the incident electric field of the optical beam. The definition of the oscillator strength is for each molecule, and obeys the Thomas-Kuhn sum rule[4]

$$\sum_{\ell} f_{\ell} = \text{number of electrons in molecule.}$$

Many of the computations of energy levels use as the comparison to experiment the optical data on absorption frequencies and oscillator strength.

2.2 π-Electrons

Electrons in molecules are either in π-bonds or σ-bonds. The σ-bonds are more tightly bound, and require more energy to excite them. They are typically observed in the optical spectra in the far ultraviolet-at photon energies above 10 eV. The excitations of the π-bonds are at lower photon energies. Thus it is the π-electrons which are important in the ordinary optical spectra of visible light, and to which we shall concentrate our attention. This does not mean that the σ-electrons may be ignored entirely. Indeed, they have an important contribution, which shall be discussed in section 5.4.

The molecules with which we shall be concerned are all planar, with the π-electron in p_z orbitals which are perpendicular to the plane of the molecule (Fig. 1). In this case one can prove that the transition moment (2.3) for π to π transitions is always a vector in the plane of the molecule. We shall use LCAO theory to prove this theorem. Let $\phi(\underset{\sim}{r}-\underset{\sim}{R}_m)$ be a p_z wave-function which is centered at the site $\underset{\sim}{R}_m$. The π-electron wavefunctions are linear combinations of these orbitals.

$$\Psi_i(r) = \sum_m a_{m,i} \, \phi(\underset{\sim}{r}-\underset{\sim}{R}_m)$$

$$\Psi_f(r) = \sum_m a_{m,f} \, \phi(r-R_m)$$

If the optical transition causes the electron to change from the state i to f, the transition moment $\underset{\sim}{\mu}$ is

$$\underset{\sim}{\mu} = \sum_{n,m} a^*_{m,f} \, a_{n,i} \int dr \, \phi^*(r-R_m) \, \underset{\sim}{r} \, \phi(\underset{\sim}{r}-R_n)$$

If we assume that

$$\int d^3r \, \phi(r-R_m) \, \phi(r-R_n) = \delta_{nm}$$

$$\int d^3r \, \phi(r-R_m)(\underset{\sim}{r}-\underset{\sim}{R}_m) \, \phi(r-R_n) = 0$$

then we obtain the simple answer

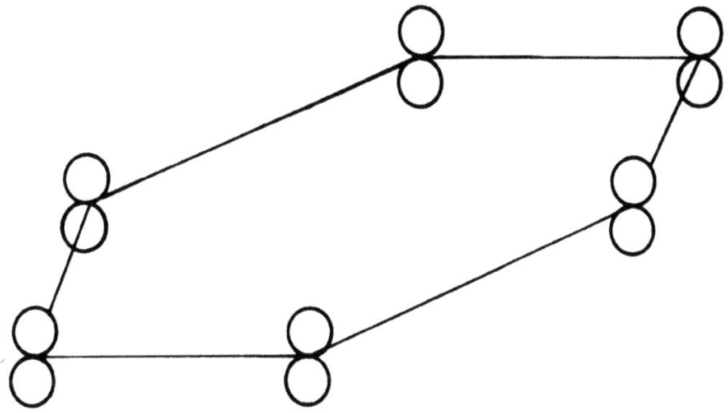

Fig. 1. A schematic picture showing a benzene ring, with the six π-bonds which are perpendicular to the plane of the molecule.

$$\mu = \sum_m a^*_{mf} a_{mi} R_m$$

As illustrated in Fig. 1, if R_m has its origin at the center of the molecule, then all the R_m are in the plane, and so is μ_α. The same result is obtained if the origin is selected someplace else, since the orthogonality of the initial and final wave functions requires that

$$0 = \sum_m a^*_{mf} a_{mi}$$

so that adding a constant vector onto R_m gives zero contribution.

2.3 Molecular Symmetry

The transition moment is a vector, and as such has three components. But we just proved that for π to π transitions in planar molecules, one component is missing, that which is perpendicular to the plane. The symmetry of the molecule, as derived from group theory, often puts further restrictions on the components of the transition moments ; namely, that the transition moment must point along a symmetry direction of the molecule, if one exists.

As an example, consider the case of anthracene. The molecule has the symmetry group D_{2h}, whose character table is given below[5]

	D_{2h}	E	C_{2x}	C_{2y}	C_{2z}	I	σ	σ	σ
	A_{1g}	1	1	1	1	1	1	1	1
	B_{1g}	1	-1	-1	1	1	-1	-1	1
	B_{2g}	1	-1	1	-1	1	-1	1	-1
	B_{3g}	1	1	-1	-1	1	1	-1	-1
	A_{1u}	1	1	1	1	-1	-1	-1	-1
z	B_{1u}	1	-1	-1	1	-1	1	-1	1
y	B_{2u}	1	-1	1	-1	-1	1	-1	1
x	B_{3u}	1	1	-1	-1	-1	-1	1	1

Each of the electronic states belongs to one of these eight representations. Similarly, each optical transition is identified with just one of these representations. Since optical transitions have the symmetry of a vector, they must transform according to x, y, or z. As illustrated in the table, B_{3u} transition transforms according to x, and must have its transition moment in the x direction, while B_{2u} transitions moments are in the y direction. As illustrated in

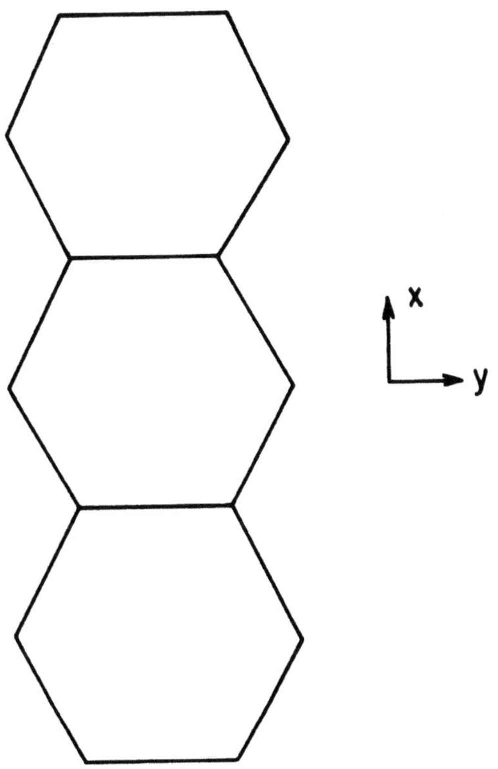

Fig. 2. The (x,y) coordinate system which is used in the analysis of the anthracene molecule.

Fig. 2, x and y are the long and short dimension of the molecular plane. Thus we follow the conventions introduced by Craig and Hobbins,[5] and call "long axis" transitions those belonging to B_{3u} and which have their transition moment along the long direction of the plane, and "short axis" those belonging to B_{2u}. These are sometimes designated L (long axis), M (short axis), and N (normal to the plane).

Thus for π to π optical transitions, each transition moment points in either the x or y direction of the molecular plane. Of course, if one observes an absorption line in the gas spectra, it is not possible to deduce which of these two possibilities applies to that transition. This is quite easy to tell from the spectra of molecular solids.

There are some planar molecules for which the transition moment does not have to be in a single direction. One prominent example is benzene, which is shown in Fig. 1. This result may be deduced from group theory, but it is intuitively obvious that the symmetry is high enough so that there is no reason to differentiate x from y.

Table I shows some optical transition data in planar molecules. We have listed the energy $\hbar\omega_\alpha$ in units of electron volts (1ev = 8066 cm^{-1}), the dimensionless oscillator strength, and the transition moment in units of angstrom. The latter two quantities are related by equation (2.2). Most of our data is taken from Philpott. Phenanthrene does not have the molecular symmetry D_{2h}, and therefore not the representations B_{2u} or B_{3u}. We have listed the transitions according th whether they are long or short axis, putting the long axis under B_{3u} and the short under B_{2u}. Several very weak ($f \sim 0.01$) transitions have been omitted.

Many of the optical transitions listed in Table I are actually each a series of transitions, corresponding to an electronic transition plus the excitation of n molecular vibrations. The different lines correspond to letting n = 0, 1, 2, etc. as different numbers of quantized vibrations are excited. A typical experimental spectrum is shown in Fig. 3 for tetracene. The solution spectrum is also shown (dashed line). The energy unit kK = $10^3 cm^{-1}$.

The lowest energy optical transition of moderate oscillator strength ($f \sim .3$) is short axis polarized. Its transition energy declines with increasing number of rings in the molecule. The next transition is an intense ($f \sim 5$) long axis transition which also declines in energy with increasing length of the molecule. Many higher energy transitions occur in the ultraviolet, and a few of these are also indicated in Table I. Most are short axis (M) polarized, and anthracene has only one significant long axis transition for frequencies less than 10 eV.[6]

Table I : Molecular Transitions[a]

	Naphthalene			Anthracene			Tetracene (Naphthacene)			Phenanthrene		
	$\hbar\omega_\alpha$	f_α	μ_α	$\hbar\omega_\alpha$	f_α	μ_α	$\hbar\omega_\alpha$	f_α	μ_α	$\hbar\omega_\alpha$	f_α	μ_α
B_{2u} M	4.5	0.30	0.53	3.4	0.30	0.61	2.8	0.30	0.60	4.1	0.50	0.70
B_{3u} L	5.8	5.3	1.86	4.9	4.5	1.87	4.8	4.5	1.88	4.9	3.3	1.59
B_{2u} M	7.5	1.8	0.96	5.6	0.6	0.64	4.35	2.4	0.46	5.8	1.8	1.09
B_{2u} M				6.7	1.2	0.83	5.5	1.5	1.03			

a. M.R. Philpott, J. Chem. Phys. 50, 5117 (1969).

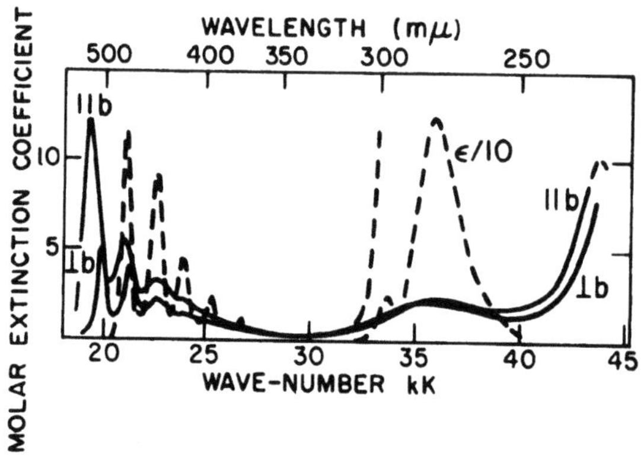

Fig. 3. The absorption spectra of tetracene. The dotted line is the solution spectra, and the solid lines are the crystal spectra for a (001) face with polarizations ⊥b and ǁb. The M axis spectra at 20 kK shows the vibrational splittings in both the solution and crystal spectra. From A. Bree and L.E. Lyons, J. Chem. Society (London) 1960, 5206.

III. DIPOLE THEORY OF CRYSTAL SPECTRA

3.1 Crystal Symmetry

The molecules in Table I from crystals in which the molecules retain their identity and are loosely bound together. The bonding is primarily by van der Waals forces. Almost without exception, the unit cell of the crystal has more than one molecule. All those in Table I have two molecules per unit cell. The group theoretical symmetry of the crystal is different than that of the individual molecule, since the crystal symmetry is determined by the operations which leave the crystal invariant.

As an example, again consider anthracene. The unit cell of the crystal has the shape shown in Fig. 4. The axes are a, b, and c. The edge b is perpendicular to both a and c, but a and c are at angle θ. These parameters are listed in Table II, along with those of naphthalene and phenanthrene which have the same structure. The symmetry of phenanthrene is different because of different symmetry of the molecule. The crystal symmetry of tetracene is though to be a slight distortion of this structure. A mutually orthogonal coordinate system is taken in the directions of a, b, and c', where c' is shown to be perpendicular to the ab plane.

The molecules are located at the corners and the face-center of the ab plane. They stick up in the c direction, although those at the corners point in a different direction than those in the face-center. Thus they form a herringbone structure in the ab plane. These planes of molecules are bound fairly tightly, and they are stacked rather loosely together in the c direction to form solids. The crystals cleave rather easily along the ab plane, and most optical experiments are done with this surface : (001) face. Table II also shows the vector direction of the long axis and short axis of the molecules at the corner and face center. These provide the orientation of the molecule, and will be useful later. The vector coordinate system is (a,b,c').

Our concern is with the optical properties of these crystals. Since they are very anisotropic, it takes more than one optical constant, at any frequency, to describe their properties. The dielectric function $\varepsilon(\omega)$ of a solid is actually a tensor $\varepsilon_{ij}(\omega)$. Since it is symmetric $\varepsilon_{ij} = \varepsilon_{ji}$, then a solid requires at most just six coefficients. This is true in tetracene. The anthracene structure has sufficient symmetry that the number of dielectric constants is reduced to four, and the tensor has the form

Table II: Crystal Structure Data.

	a	b	c	θ	$\hat{\mu}_2$	$\hat{\mu}_M$
Naphthalene[a]	8.23	6.00	8.66	57°05'	(-.438,-.210,.874)	(.321,.872,.370)
					(-.438,+.210,.874)	(.321,-.872,.370)
Anthracene[b]	8.56	6.04	11.16	55°18'	(-.496,-.125,.859)	(.323,.982,.316)
					(-.496,+.125,.859)	(.323,-.892,.316)
Phenanthrene[c]	8.57	6.11	9.47	82°30'	(.255,-.086,.963)	(.441,-.876,.195)
					(.255,+.086,.963)	(.441,+.876,-.195)

a. D.W.J. Cruickshank, Acta Cryst. 10, 504 (1957).
b. D.W.H. Cruickshank, Acta Cryst. 9, 915 (1956).
c. R. Mason, Mol. Phys. 4, 413 (1961).

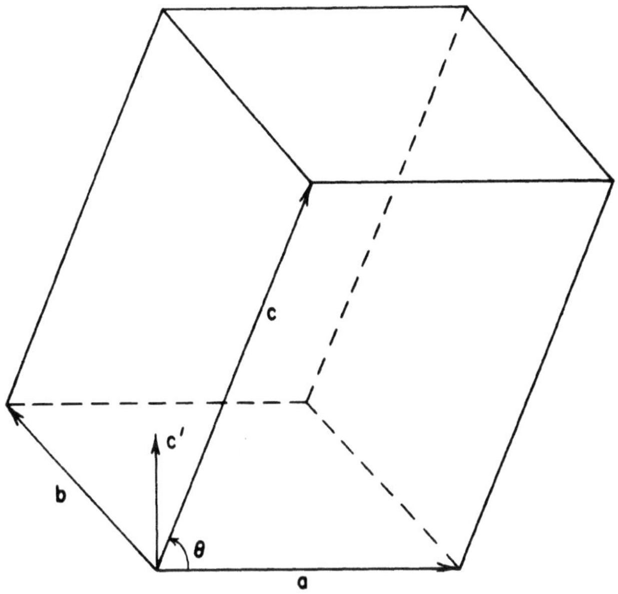

Fig. 4. The unit cell of anthracene. The axis b is perpendicular to a and c, and makes an angle θ with c. The axis c' is perpendicular to the ab plane.

$$\varepsilon_{ij} = \begin{bmatrix} \varepsilon_{11} & 0 & \varepsilon_{13} \\ 0 & \varepsilon_{22} & 0 \\ \varepsilon_{13} & 0 & \varepsilon_{33} \end{bmatrix} \tag{3.1}$$

The off-diagonal term ε_{13} links the a and c directions, as is obvious from the structure in Fig. 4. These have been measured.[7,8]

The optical constants-refractive index and extinction coefficient-are also anisotropic. Their behavior may be deduced from the dielectric function using the standard theories of wave propagation in anisotropic media. We will not discuss this rather old and dull subject, but just quote results whenever needed.

3.2 Oriented Gas Model

The dielectric constants and optical constants of the crystal are defined with regard to the crystal axes a, b, and c'. But the optical transitions occur in the molecules, where the long and short axes have no simple relationship to the crystal axes, as is shown in Table II. We must answer the question - what is the relationship between the optical symmetry of the molecule and the different symmetry of the crystal ?

The question was answered by Davydov.[2] He constructed wavefunctions of the crystal from linear combinations of molecular wavefunctions. Call $\Psi_f^{(1)}$ and $\Psi_f^{(2)}$ the wavefunctions of the molecules at the corner and face-center of the unit cell. Then unit cell wavefunctions $\Psi_f^{(\alpha)}$ and $\Psi_f^{(\beta)}$ are the linear combinations

$$\Psi_f^{(\alpha)} = \frac{1}{\sqrt{2}} (\Psi_f^{(1)} + \Psi_f^{(2)}) \tag{3.2}$$

$$\Psi_f^{(\beta)} = \frac{1}{\sqrt{2}} (\Psi_f^{(1)} - \Psi_f^{(2)}) \tag{3.3}$$

Let us consider an optical transition from molecular state $\underline{i}$ to $\underline{f}$. The initial state i of each molecule at $\underset{\sim}{R}_\ell$ has a wavefunction $\Psi_i(\bar{r},R_\ell)$. The initial state of the crystal is a product of the wavefunctions of each molecule. This is the ground state wavefunction

$$\Psi_G = \prod_{R\ell} \Psi_i(r,R_\ell)$$

The state Ψ_f has the one molecule excited into the state f, while all others are still in their ground state

$$\Psi_f^{(1,2)}(r,R_\ell) = \Psi_f^{(1,2)}(r,R_\ell) \prod_{R_i}{}' \Psi_i(r,R_i)$$

$$\Psi_f^{\alpha,\beta} = \frac{1}{\sqrt{2}} (\Psi_f^{(1)} \pm \Psi_f^{(2)})$$

A calculation of the transition moments from the ground state to a crystal state α or β gives

$$\underset{\sim}{\mu}_\alpha = \int d^3r \ \Psi_f^\alpha(r) \ \underset{\sim}{r} \ \Psi_G = \frac{1}{\sqrt{2}} (\underset{\sim}{\mu}_1 + \underset{\sim}{\mu}_2)$$

$$\underset{\sim}{\mu}_\beta = \int d^3r \ \Psi_f^\alpha(r) \ \underset{\sim}{r} \ \Psi_G = \frac{1}{\sqrt{2}} (\mu_1 - \mu_2)$$

The transition moment of the crystal unit cell is the sum or difference of those of the molecules in the unit cell. In anthracene, according to Table II, a long axis transition has the two possible crystal transition moments

$$\underset{\sim}{\mu}_\alpha = \sqrt{2} \mu_0 (-0.496, 0, 0.859)$$

$$\underset{\sim}{\mu}_\beta = \sqrt{2} \mu_0 (0, -0.125, 0)$$

where μ_0 is the magnitude of the molecular moment. The moment $\underset{\sim}{\mu}_\beta$ points just in the b direction - i.e., along a crystalline axis. Similarly, $\underset{\sim}{\mu}_\alpha$ points in the ac plane. Hence light which is polarized in the b direction will excite the state Ψ_f^β which has no transition moment in the ac plane. This is why ε_{23} and ε_{12} are both zero in (3.1). Similarly, light polarized in the ac plane will excite the Ψ_f^α state, which has no transition moment in the b direction. This is illustrated pictorially in Fig. 5. The transition moments $\underset{\sim}{\mu}_\alpha$ and $\underset{\sim}{\mu}_\beta$ are obtained by adding $\underset{\sim}{\mu}_1$ and $\underset{\sim}{\mu}_2$ with different phases.

The two transition moments μ_α and μ_β are not the same length. Indeed, the ratio of their squares is

$$\left(\frac{\mu_\alpha}{\mu_\beta}\right)^2 = 85$$

Notice that this ratio is independent of the moment for any particular transition, but is only determined by the way the molecule sits in the crystal lattice. Since in (2.2) the oscillator strength is proportional to the square of the moments, we deduce that the ratio of oscillator strengths for a long axis transition in anthracene is

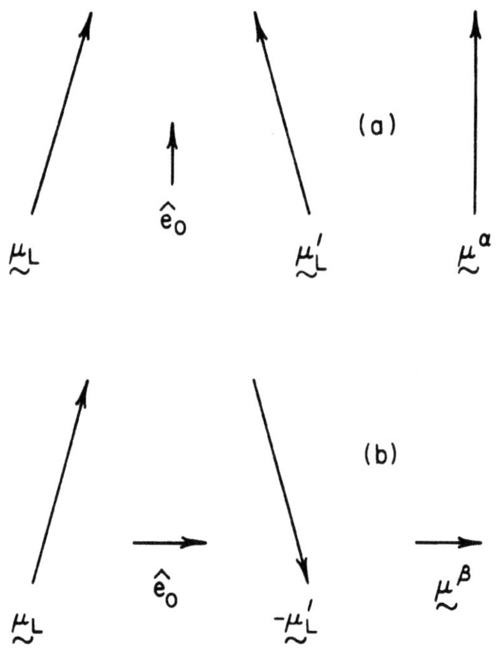

Fig. 5. (a) An electric field in the direction $\hat{e}_0$ makes the two transition moments in the unit cell add up to give μ_α.
(b) Polarization in the other direction, and the two transition moments to give μ_β.

$$\left(\frac{f_{ac}}{f_b}\right)_L = 85$$

This is a very striking prediction, and says that long axis transitions should be very much stronger for ac polarized light than for b polarized light. Fig. 6 shows that it is experimentally verified. The very strong absorption band at 40kK (5eV) was identified as due to a long axis transition, and it is indeed much stronger $\perp$b than $\parallel$ b. The same calculation may be done for short axis transitions, with the result

$$\left(\frac{f_{ac}}{f_b}\right)_M = 0.25$$

So short axis transitions should be stronger in the b direction. This is also verified in Fig. 6, since the absorption band at 25 kK (3.4eV) is short axis, and has stronger absorption for polarization $\parallel$ b.

The ratio of the intensities in the ac and b direction is called the "polarization ratio." A polarization ratio greater than one signifies an L transition , and less than one signifies an M transition. Thus the optical spectrum of crystals provides direct information on the symmetry of the molecular transition - of course it is necessary to know the crystal structure.

This method of calculating polarization ratios is called the "oriented gas model." We are ignoring all interactions between molecules, which is why it is called a "gas " theory. The polarization ratios come only from the orientation of the molecules.

A measurement of polarization ratios usually yields a result which is different than 85 or 0.25. This is evident by inspecting Fig. 6. These deviations result from the interactions between molecules. The rest of these lectures will be spent discussing how the interactions affect the optical spectra. The interactions do affect the polarization ratios, but typically on the order of a factor of two. So there is no change in the qualitative conclusion that a polarization ratio greater than one is L axis, and less than one is M axis. Thus the oriented gas model still serves as a useful guide for interpreting the optical spectra.

3.3 Davydov Splittings

Each molecule is composed of a sum of nuclei and electrons. The coulomb interactions between these electrons and nuclei define a Hamiltonian which is complicated to solve. This problem must be solved completely in order to obtain an accurate theory of crystal spectra. Fortunately there is one shortcut. An approximate theory of crystal spectra can be obtained without any knowledge of molecu-

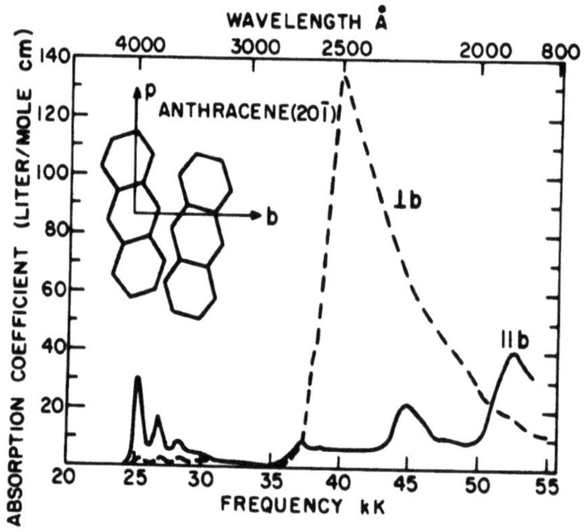

Fig. 6. The anthracene absorption spectra for a (201) face. The absorption band at 38 kK is strongest for polarization ⊥b, which indicates that it is a L axis transition. The transition at 25 kK is stronger for polarization ‖b, which indicates that it is an M axis transition. From L.B Clark and M.R. Philpott, J. Chem. Phys. 53, 3790 (1970).

lar wavefunctions. This is the dipole theory, and requires as imputs only the molecular resonance frequencies ω_ℓ and transition moments $\underset{\sim}{\mu}_\ell$. These, of course, can be obtained experimentally. Thus dipole theory is very expedient, since the only parameters required for the calculation of crystal spectra are the experimental parameters. These are obtained directly from the gas or solution spectra.

In an optical experiment, a light beam of wave vector $\underset{\sim}{k}$ is passing through the solid. Absorption of a light quanta creates an electronic state of wave vector $\underset{\sim}{k}$. The crystal wavefunctions of wave vector $\underset{\sim}{k}$ are linear combinations of the unit cell wavefunctions

$$\Psi^{(\alpha,\beta)}_{\underset{\sim}{k},f}(r) = \frac{1}{\sqrt{N}} \sum_\ell \Psi_f^{(\alpha,\beta)}\, e^{i\underset{\sim}{k}.\underset{\sim}{R}_\ell}$$

where N is the number of unit cells in the crystal. The Hamiltonian of the solid can be written as a sum of Hamiltonians of the individual molecules H_ℓ, plus a term which accounts for the interaction between molecules

$$H = \sum_\ell H_\ell + \sum_{\ell,m} V_{\ell,m} \tag{3.4}$$

where $V_{\ell m}$ describes the coulomb interaction between the electrons and nuclei of one molecule with those of another. In the Hartree approximation, the energy of the excited state (α,f) or (β,f) is

$$E_f^{(\alpha,\beta)}(k) = \int dr\, \Psi^{(\alpha,\beta)*}_{k,f}\, H\, \Psi^{(\alpha,\beta)}_{k,f}$$

with the result

$$E_f^{(\alpha,\beta)}(k) = \hbar\omega_f + \frac{1}{2}\sum_{\ell,m} e^{i\underset{\sim}{k}.\underset{\sim}{R}_{\ell m}} \int dr_\ell \int dr_{\ell m} [\Psi_f^{(1)}(r_\ell,R_\ell)^* \times$$

$$\Psi_i(r_\ell,R_\ell) \pm \Psi_f^{(2)}(r_\ell,R_\ell)^*\Psi_i^{(2)}(r_\ell,R_\ell)]\; V_{\ell,m} \times$$

$$[\Psi f^{(1)}(r_m,R_m)^*\Psi_i^{(1)}(r_m,R_m) \pm \Psi_f^{(2)}(r_m,R_m)^*\Psi_i^{(2)}(r_m,R_m)]$$

The last term must be evaluated. Let us use a multipole expansion. Since the molecules are electrically neutral, there is no monopole term. The first contribution is the dipole-dipole term

$$V_{\ell,m} = -e^2 \underset{\sim}{r}_\ell\, \underset{\approx}{\Phi}(R_{\ell m})\, \underset{\sim}{r}_m$$

$$\phi_{\mu\nu}(R) = \frac{\delta_{\mu\nu}}{R^3} - 3\frac{R_\mu R_\nu}{R^5} \tag{3.6}$$

where r_ℓ and r_m are electron coordinates in molecules ℓ and m. The r integrals can be done, and give the transition moments (2.3)

$$E_f^{(\alpha,\beta)}(K) = \hbar\omega_f + \frac{1}{2} N \sum_{\ell,m} e^{ik.R_{\ell m}} \underset{\sim}{\mu}_\ell^{(\alpha,\beta)} \underset{\approx}{\phi}(R_{\ell m}) \underset{\sim}{\mu}_m^{(\alpha,\beta)} \tag{3.7}$$

Thus the transition moments interact among themselves via the dipole-dipole interaction, and contribute to the energy of the excited molecular state. A more complete analysis would include octopole and higher moments as well. These higher moments cannot be deduced from experiment, and when included have been used as fitting parameters. Only the dipole term will be kept in the present discussion, since it is of interest to see how well it can do alone.

Equation (3.7) may be rewritten as

$$E_f^{(\alpha,\beta)}(k) = \hbar\omega_f + \frac{1}{N} \sum_{\ell,m} e^{ik.R_{\ell m}} [\mu^{(1)} \cdot\phi\cdot \mu^{(1)} \pm \mu^{(1)} \cdot\phi\cdot \mu^{(2)}] \tag{3.8}$$

where the + sign is for α states and the - sign for β states. We have used the result that

$$\sum_{\ell m} e^{ik.R_{\ell m}} \mu^{(1)} \phi\cdot\mu^{(1)} = \sum_{\ell m} e^{ik.R_{\ell m}} \mu^{(2)}\cdot\phi\cdot\mu^{(2)}$$

which is true for the anthracene lattice. The α and β state now have different energies. Again this is obvious from Fig. 5, since the dipole-dipole interactions depend upon which way the dipoles are pointing, at least for interactions between the two different kinds of dipoles. Thus a light beam polarized in the b direction will have a different intensity and resonance energy than one in the ac plane. The energy difference between the two values of resonance frequency is called the "Davydov splitting." This is another quantity which may be calculated theoretically. In the above simple theory, this splitting is given by

$$E^{(\alpha)} - E^{(\beta)} = \frac{2}{N} \sum_{\ell,m} e^{ik.R_{\ell m}} \mu^{(1)}\cdot\phi(R_{\ell m})\cdot\mu^{(2)} \tag{3.9}$$

It is proportional to the square of the transition moment, and hence is proportional to the oscillator strength. So weak transitions (small f) will have small splittings, and strong ones will have large splittings. For the optical transitions shown in Table I, the weak (f = 0.3) M transitions have small splittings, while the intense L transitions have very large splittings.

When one attempts to compute the Davydov splittings using (3.9), the answers do not agree with experiment. Many of the initial efforts, beginning with Craig and Hobbins,[5] attempted to explain the observed Davydov splittings in the 3.4eV M transition in anthracene. Sometimes the predicted splittings had the wrong sign, and were off by an order of magnitude. Although it must be acknowledged that the experimentalists also had difficulty agreeing on the result. The early theoretical efforts attempted to include the interactions with the higher excited states.[9] We have already shown that the optical electric field excites the transition moments of the M state, and these interact with each other to affect the energy of the M state in the crystal. But the optical field also excites the L transition moments, and these can also interact with the M moments to affect the energy of the M state. In fact this should be an important process because the L moments are so large. The initial attempts to account for this process proceeded by perturbation theory. First order perturbation theory was used to calculate how the wavefunctions are altered by mixing the higher energy states j

$$\tilde{\Psi}_f^{(\alpha)}(R_\ell) = \Psi_f^{(\alpha)} + \sum_{\substack{j,m \\ \gamma=0\beta}} \Psi_j^{(\gamma)}(r,R_m) \frac{\mu_j^\gamma . \phi(R_{\ell m}) . \mu_f^\alpha}{E_j - E_f}$$

These new wavefunctions are used to find polarization ratios. Similarly, the energy change is computed by second order perturbation theory

$$\Delta E_f^{(\alpha)} = \frac{1}{N} \sum_{\substack{j,m,\ell \\ \gamma}} \frac{(\mu_j^{(\gamma)} . \phi(R_{\ell m}) . \mu_f^{(\alpha)})^2}{E_j - E_f}$$

These perturbations make a sizable contribution, but in most cases worsen the comparison with experiment.

The higher energy states are obviously important, but they are not being included correctly. The reason is that perturbation theory does not converge in this case. This assertion may be understood by considering coulombs law in an isotropic dielectric

$$\frac{Q_1 Q_2}{\varepsilon R}$$

The interaction between the two charges is reduced by the dielectric constant ε. What physical process is represented by ε? Each charge, say Q_1, polarizes the atoms or molecules in the solid, and this polarization acts to reduce the potential on the other charge. This, of course, is exactly the same process we are trying to calculate for the Davydov splittings. The optical electric field polarizes the molecules (i.e., transition moments) and this polarization affects the energy of the transition. Our perturbation calculation above was merely an attempt to compute the dielectric function, or at least how the dielectric properties of the solid affect the optical spectra. The perturbation series does not converge if ε is greater than two, which is usually always the case. This is because ε equals one plus some function of the molecular polarizability

$$\varepsilon = 1 + f(\alpha)$$

The earlier attempts to do perturbation theory are simply trying to evaluate the dielectric properties by expanding the series

$$\frac{1}{\varepsilon} = 1 - f + f^2 - \ldots$$

which does not converge if $f > 1$.

The dielectric problem requires a strong coupling theory, since perturbation theory does not converge. Fortunately a strong coupling, self consistent, theory is not difficult. In fact, the calculations are no more difficult to perform than those for perturbation theory. These methods will be described in the next sections.

3.4 Classical Dipole Theory

Classical dipole theory has a long history and is discussed in many references.[10-14] Mahan[13] was the first to do a self consistent strong coupling theory for molecular solids. He calculated the Davydov splittings in anthracene, and for the first time obtained agreement with experiments without having to introduce any unknown and varying parameters in the theory. The only ingredients in the theory are the experimentally determined crystal structure, resonance frequencies, and oscillator strengths.

The theoretical equations may be derived from a rigorous quantum mechanical formalism. But the resulting equations have a simple classical interpretation. Here we will present just the classical derivation, which yields the same result.

We begin by examining how an oscillating electric field $E_0 e^{-i\omega t}$ affects a charge e and mass m which is connected to a harmonic spring of resonance frequency ω_0. The classical equation for the charge displacement x is

$$\ddot{x} + \omega_0^2 x = \frac{e}{m} E_0 e^{-i\omega t}$$

and so the induced polarization of the charge is

$$P = ex = \frac{e^2}{m} \frac{1}{\omega_0^2 - \omega^2} E_0 e^{-i\omega t}$$

Since $P = \alpha E$, the polarizability of our spring and charge is simply

$$\alpha = \frac{e^2}{m(\omega_0^2 - \omega^2)}$$

This is very similar to the molecular polarizability (2.1). The molecular polarizability may be derived in an equivalent fashion by assuming the molecule is a sum of springs of different frequencies ω_ℓ, and each spring has an effective charge $e\sqrt{f_\ell}$. The total polarization of the molecule when acted upon by an oscillating electric field is

$$\underset{\sim}{P} = \underset{\approx}{\alpha} \cdot E$$

The polarizability of the molecule is a tensor, and quite anisotropic

$$\underset{\approx}{\alpha} = \frac{e^2}{m} \sum_j f_j \frac{\hat{d}_j \, \hat{d}_j}{\omega_j^2 - \omega^2 - i\gamma_j \omega} \qquad (3.10)$$

The form (2.1) is actually the spherical average of (3.10) which is measured in gas spectra. The unit vectors $\hat{d}_j$ point in the direction of the transition moment

$$\underset{\sim}{\mu} = \mu_j \, \hat{d}_j$$

Now consider an optical wave which generates an oscillating electric field

$$E(r,t) = E_0 \, e^{i(\underset{\sim}{k} \cdot \underset{\sim}{r} - \omega t)}$$

in the crystal. An important quantity is the local electric field $E^{local}(R_m,t)$ which acts upon the molecule at the point R_m. The polarization of the molecule is proportional to the local electric field

$$P_m = \underset{\approx}{\alpha} \cdot E^{local}(R_m) \qquad (3.11)$$

The local electric field is the sum of the optical field E plus the electric field resulting from the dipole-dipole interaction from the polarization on all of the other molecules.

$$E^{local}(R_\ell) = E(R_\ell) - \sum_{m \neq \ell} \underset{\approx}{\phi}(R_{\ell m}) \cdot \underset{\sim}{P}_m \qquad (3.12)$$

Equations (3.11) and (3.12) may be combined to eliminate the local electric field

$$\underset{\sim}{P}_\ell = \underset{\approx}{\alpha} \cdot [\, \underset{\sim}{E} - \sum_{m\neq\ell} \underset{\approx}{\phi}(\underset{\sim}{R}_{\ell m}) \cdot \underset{\sim}{P}_m \,] \tag{3.13}$$

This equation must now be solved to obtain the polarization $\underset{\sim}{P}_m$ on each molecule. If there are two molecules per unit cell, we define a crystalline polarization for each molecule in the unit cell

$$\underset{\sim}{P}_k^{(1)} = \frac{1}{\sqrt{N}} \sum_\ell e^{ik.R_\ell} \underset{\sim}{P}_\ell^{(1)}$$

$$\underset{\sim}{P}_k^{(2)} = \frac{1}{\sqrt{N}} \quad e^{ik.R_\ell} \underset{\sim}{P}_\ell^{(2)}$$

where the sum R_ℓ runs over each unit cell. The polarizations $P^{(1)}$ and $P^{(2)}$ will be different because the two molecules are oriented in different directions. Equation (3.13) may now be written as

$$\underset{\sim}{P}_k^{(1)} = \underset{\approx}{\alpha}^{(1)} \cdot \{\underset{\sim}{E}_0 - \frac{4\pi}{V_0} [\underset{\approx}{T}^e \cdot \underset{\sim}{P}^{(1)} + \underset{\approx}{T}^{(i)} \cdot \underset{\sim}{P}^{(2)}]\} \tag{3.14a}$$

$$\underset{\sim}{P}_K^{(2)} = \underset{\approx}{\alpha}^{(2)} \cdot \{\underset{\sim}{E}_0 - \frac{4\pi}{V_0} [\underset{\approx}{T}^i \cdot \underset{\sim}{P}^{(2)} + \underset{\approx}{T}^e \cdot \underset{\sim}{P}^{(2)}]\} \tag{3.14b}$$

where we have introduced the dipole sums

$$\underset{\approx}{T}^e = \frac{V_0}{4\pi} \sum_\ell e^{ik.R_\ell} \underset{\approx}{\phi}(R_\ell) \tag{3.15a}$$

$$\underset{\approx}{T}^i = \frac{V_0}{4\pi} \sum_\ell e^{ik.(R_\ell+2)} \underset{\approx}{\phi}(R_\ell+\tau) \tag{3.15b}$$

which are also tensors. The vector $\underset{\sim}{\tau}$ goes from the molecule in the corner to the face-center of the unit cell. The detailed properties of these sums will be described in the next section. T^e involves the dipole interactions between translationally equivalent molecules - (1) and (1) or (2) and (2) - while T^i involves the inequivalent molecules - (1) and (2) or <u>vice versa</u>. We also introduce the dimensionless polarizability per unit volume

$$\bar{\underset{\approx}{\alpha}}^{(j)} = \frac{1}{V_0} \underset{\approx}{\alpha}^{(j)}$$

Since the molecules at (1) and (2) are identical, then $\bar{\underset{\approx}{\alpha}}^{(1)}$ and $\bar{\underset{\approx}{\alpha}}^{(2)}$ are also identical except for the differences in the orientation of the molecules. Since the vectors $P^{(1,2)}$ are three dimensional, (3.14a) and (3.14b) define a set of six equations which may be solved for the six components of polarization. Thus we will get a result of the form

$$\underset{\sim}{P}_k^{(1)} = \underset{\approx}{S}^{(1)} \cdot \underset{\sim}{E}_0$$

$$\underset{\sim}{P}_k^{(2)} = \underset{\approx}{S}^{(2)} \cdot \underset{\sim}{E}_0$$

where $S_{ij}^{(1,2)}$ are a complicated function of the molecular polarizability and the dipole sums $T^{e,i}$. From the definition of the dielectric function

$$E_i + \frac{4\pi}{V_0}(P_i^{(1)} + P_i^{(2)}) = \varepsilon_{ij}E_j$$

we derive that the crystal dielectric function is

$$\varepsilon_{ij} = 1 + \frac{4\pi}{V_0}(S_{ij}^{(1)} + S_{ij}^{(2)}) \qquad (3.16)$$

This dielectric function will have resonance absorption lines, and these are the new crystal oscillator frequencies. Thus the singularities in (3.16) will define the new crystal frequencies. These depend upon the polarization of the light, so that different components of the tensor (3.1) will resonate at different frequencies.

Although this sounds like a complicated procedure, the crystal symmetry usually permits a great simplification. Thus in anthracene, the set of equations reduces to two coupled equations rather than six. This reduction is effected by realizing that the polarization for π to π transitions must point in either the long or short axis of the molecule.

$$\underset{\sim}{P}^{(1)} = \hat{d}_L^{(1)} P_L^{(1)} + \hat{d}_M^{(1)} P_M^{(1)}$$

$$\underset{\sim}{P}^{(2)} = \hat{d}_L^{(2)} P_L^{(2)} + \hat{d}_M^{(2)} P_M^{(2)}$$

and the same for the polarizability

$$\underset{\approx}{\bar{\alpha}}^{(1)} = \bar{\alpha}_L \hat{d}_L^{(1)} \hat{d}_L^{(1)} + \bar{\alpha}_M \hat{d}_M^{(1)} \hat{d}_M^{(1)}$$

$$\underset{\approx}{\bar{\alpha}}^{(2)} = \bar{\alpha}_L \hat{d}_L^{(2)} \hat{d}_L^{(2)} + \bar{\alpha}_M \hat{d}_M^{(2)} \hat{d}_M^{(2)}$$

Thus we have reduced the number of equations from six to four

$$P_L^{(1)} = 4\pi\bar{\alpha}_L \left(\frac{V_0}{4\pi} \underset{\sim}{E}_0 \cdot \hat{d}_L^{(1)} - T_{LL}^e P_L^{(1)} - T_{LM}^e P_M^{(1)} - T_{LL}^i P_L^{(2)} - T_{LM}^i P_M^{(2)}\right)$$

$$P_L^{(2)} = 4\pi\bar{\alpha}_L \left(\frac{V_0}{4\pi} \underset{\sim}{E}_0 \cdot \hat{d}_L^{(2)} - T_{LL}^e P_L^{(2)} - T_{LM}^e P_M^{(2)} - T_{LL}^i P_L^{(1)} - T_{LM}^i P_M^{(1)}\right) \qquad \text{etc.}$$

where

$$T^e_{LL} = \hat{d}^{(1)}_L \cdot T^e \cdot \hat{d}^{(1)}_L = \hat{d}^{(2)}_L \cdot T^e \cdot \hat{d}^{(2)}_L$$

$$T^e_{LM} = \hat{d}_L \cdot T^e \cdot \hat{d}_M =$$

etc.

By adding and subtracting these, we find that the α equations decouple from the β equations

$$P^\alpha_L = 4\pi\,\bar{\alpha}_L \left(\frac{V_0}{4\pi} \underset{\sim}{E}_0 \cdot \hat{d}^\alpha_L - T^\alpha_{LL} P^\alpha_L - T^\alpha_{LM} P^\alpha_M\right)$$

$$P^\alpha_M = 4\pi\,\bar{\alpha}_M \left(\frac{V_0}{4\pi} \underset{\sim}{E}_0 \cdot d^\alpha_M - T^\alpha_{MM} P^\alpha_M - T^\alpha_{ML} P^\alpha_L\right)$$

$$P^\beta_L = 4\pi\,\alpha_L \left(\frac{V_0}{4\pi} E_0 \cdot d^\beta_L - T^\beta_{LL} P^\beta_L - T^\beta_{LM} P^\beta_M\right)$$

$$T^{\alpha,\beta}_{LL} = T^e_{LL} \pm T^i_{LL}$$

$$d^{\alpha,\beta}_L = \hat{d}^1_L \pm \hat{d}^2_L$$

Thus we get the solution for the optical dielectric function

$$\hat{e}_0 \cdot \underset{\approx}{\hat{\varepsilon}} - \hat{e}_0 \equiv \varepsilon^{\alpha,\beta} = 1 + \frac{1}{2}\frac{\alpha}{\Delta}\,[4\pi\bar{\alpha}_L\,(1 + 4\pi\bar{\alpha}_M\,T_{MM})(\hat{\varepsilon}_0 \cdot \hat{d}_L)^2 + 4\pi\bar{\alpha}_M \times$$

$$(1 + 4\pi\bar{\alpha}_L\,T^{\alpha,\beta}_{LL})(\hat{e}_0 \cdot \hat{d}^{\alpha,\beta}_M)^2 - 2(4\pi\hat{d}_L)(4\pi\alpha_M)(e_0 \cdot \hat{d}_L)(e_0 \cdot \hat{d}_M)T_{LM}]$$

$$\Delta = (1 + 4\pi\bar{\alpha}_L\,T_{LL})(1 + 4\pi\bar{\alpha}_M\,T_{MM}) - 4\pi\bar{\alpha}_L\,4\pi\bar{\alpha}_M\,T^2_{LM} \qquad (3.17)$$

where ε^β is for b polarized light, and ε^α is for ac polarized light. All quantities T and $\hat{d}$ should have α or β superscripts. This is the most general solution for π to π transitions.

The singularities in the dielectric function occur where $\Delta=0$. Let us solve the problem we formulated at the beginning - the resonance frequency of the M transition at 3.4eV. We separate out its contribution to the conductivity

$$\bar{\alpha}_M = \omega^2_P \frac{f_0}{\omega^2_0 - \omega^2} + \bar{\alpha}'_M \qquad (3.18)$$

$$\omega_P^{\;2} = \frac{4\pi e^2}{mV_0}$$

where $\bar{\alpha}'_M$ is the contribution of the other M states which are at higher energy. Solving the equation $\Delta=0$ for the resonance frequency yields

$$\omega^2 = \omega_0^2 + \omega_P^2 f_0 \frac{[T_{MM}(1 + 4\pi\bar{\alpha}_L T_{LL}) - 4\pi\bar{\alpha}_L T_{LM}^2]}{(1 + 4\pi\bar{\alpha}_L T_{LL})(1 + 4\pi\,\bar{\alpha}'_M T_{MM}) - (4\pi\bar{\alpha}_L)(4\pi\bar{\alpha}'_M)\,T_{LM}^2} \tag{3.19}$$

We wish to compare this with (3.7) which in the same notation is

$$\omega = \omega_0 + \frac{4\pi\mu_M^2}{V_0} T_{MM} \tag{3.7'}$$

To compare (3.19) with (3.7'), we eliminate the interaction of the M dipole with the L ($T_{LM}=0$) which yields

$$\omega^2 = \omega_0^2 + \omega_P^2 f_0 T_{MM}$$

Taking the square root

$$\omega \cong \omega_0 + \frac{1}{2}\frac{\omega_P^2 f_0}{\omega_0} T_{MM} = \omega_0 + \frac{4\pi\mu_M^2}{V_0} T_{MM}$$

Thus the two results are comparable. The interaction with the L states is retained through the term T_{LM}. The specific properties of the L and M states at higher energy enter through the polarizabilities $\bar{\alpha}_L$ and $\bar{\alpha}'_M$. Thus the mixing with higher states is also included.

In making the separation (3.18) it is assumed that $\omega \sim \omega_0$ so that the first term on the right is the only one whose frequency dependence is important, and that $\bar{\alpha}'_M$ is a constant. This assumption is not accurate enough if there are several absorption levels very close together in energy, or if the oscillator strength f_0 is large and the energy shift is large. In that case, rather than seek an analytical solution of the form (3.19), one can just use a computer to plot $\Delta(\omega)$ and look for the zeros, and locate roots in this fashion. The same equation (3.19) is obtained for both α and β polarizations. The values of ω are different because T^{α}_{LL}, T^{α}_{MM} ans T^{α}_{LM} do not have the same values as T^{β}_{LL}, T^{β}_{MM} and T^{β}_{LM}.

3.5 Dipole Sums

In this section we will discuss the method of evaluating the dipole sums of the form

$$\underset{\approx}{T}^{e}(k) = \frac{V_0}{4\pi} \sum_{R_\ell} e^{ik.R_\ell} \underset{\approx}{\phi}(R_\ell)$$

$$\underset{\approx}{T}^{i}(k) = \frac{V_0}{4\pi} \sum_{R_\ell} e^{ik.(R_\ell+\tau)} \underset{\approx}{\phi}(R_\ell+\tau)$$

where $\underset{\sim}{R}_m$ are lattice vectors, while $\underset{\sim}{\tau}$ is not. So $\underset{\approx}{T}^e$ is a summation over dipoles on similar sites of the lattice. Obviously the summation excludes the origin - the dipole acting upon itself. Of course, the dipole interacting with itself has important effects. But this affects the energies of the isolated molecules, and since we are using the experimental frequencies, then this self interaction is already included. The sum T^i involves the interaction of a dipole on one lattice site with the sum of all dipoles on the other type of lattice sites. The superscripts "e" and "i" mean equivalent and inequivalent. If there are L different molecules per unit cell, there would be L! different lattice sums. We shall restrict ourselves to the case L=2.

These summations are tricky because they are conditionally convergent. First discuss the case when k=0

$$T^e(0) = \frac{V_0}{4\pi} \sum_{R_\ell} \phi(R_\ell)$$

At large values of R we do not err much by changing the summation to an integration

$$\int^{R_0} R^2 dR \frac{1}{R^3} \sim \ell n\, R_0$$

which diverges slowly as the upper limit of the distance R_0 increases. This physical phenomenon is well known in classical electrostatics: if a dielectric body is uniformly polarized, the internal electric field depends upon the shape of the body - a different result being obtained for a sphere than a wafer. The T-summation is just the internal electric field if μ is the idpole moment $E = \frac{4\pi}{V_0} \underset{\approx}{T} \cdot \underset{\sim}{\mu}$ on each lattice site. So we can obtain the right answer to the dipole sum by using our knowledge of classical electrostatics.

At ordinary optical frequencies the wave vector of light $\underset{\sim}{k}$ is of the order of 10^5 cm^{-1}. If the experimental sample has dimension $\ell \sim 1$ cm while the atomic unit cell has dimension $a \sim 10^{-8}$ cm, then the wave vector obeys the inequality

$$\frac{1}{a} \gg k \gg \frac{1}{\ell}$$

In this case we will show below that the lattice sums may be written as

$$T^{e,i}_{\mu\nu}(k) = \frac{k_\mu k_\nu}{k^2} - \frac{1}{3}\delta_{\mu\nu} + t^{e,i'}_{\mu\nu} \quad \frac{k_\mu k_\nu}{k^2} + t^{e,i}_{\mu\nu} \qquad (3.20)$$

where the tensors $t^{e'}$ and $t^{i'}$ are independent of wave vector, and all the wave vector dependence is in the first term $\hat{k}\hat{k}$. We emphasize that this is not the limit $\underset{\sim}{k}=0$, since this was discussed previously and shown to depend upon the shape of the experimental sample. The tensors t^e and t^i are symmetric so in general each have six components. However, in monoclinic structures such as anthracene, all second rank tensors have the general form (3.1) wiht only four constants instead of six. The number of independent constants is further reduced by the fact that the trace of these tensors is zero

$$\mathrm{Tr}\; t^{e'} = \mathrm{Tr}\; t^{i'} = 0$$

This follows from the fact that the trace of the dipole-dipole interaction is zero

$$0 = \mathrm{Tr}\;\phi_{\mu\nu}(R) = \phi_{xx}(R) + \phi_{yy}(R) + \phi_{zz}(R)$$

and also that

$$\mathrm{Tr}\left(\frac{k_\mu k_\nu}{k^2} - \frac{1}{3}\delta_{\mu\nu}\right) = 0.$$

Thus we need to find six parameters - three for t^e and three for t^i.

Many optical experiments are done on very thin crystals. In this case it is often not true that $k\ell \gg 1$ as we previously required. Actually the theory is probably valid in this case anyway. Thin samples can only be made if the crystal cleaves easily, and this is only possible if there is a planar geometry as in anthracene. But Mahan and Overmair[15] showed generally that the theory was valid in planar configurations if the sample is more than a few layers thick. This was confirmed for the case molecular crystals by Philpott's calculations.[16]

One way to evaluate the dipole sums is to just start summing the local dipoles near the central one. A systematic procedure could be developed, wheregy shells of dipoles are summed, and each shell is further from the central one. Such a procedure, in fact, just gives $t^{e'}$ or $t^{i'}$. One does not get the term $\hat{k}\hat{k} - \frac{1}{3}\underset{\approx}{I}$, which comes from summing the dipoles which are very far away. Some theorists have summed the dipoles locally, and interpreted the result as T^e or T^i, thereby omitting the long range term $\hat{k}\hat{k} - \frac{1}{3}\underset{\approx}{I}$. This is an incorrect procedure, and leads to incorrect results.

Since the dipole sums are conditionally convergent, how do we know the "correct" answer? The correct answer is found by comparing the result to one which is known and accepted. We shall now do such a calculation in order to establish the correctness of the $\hat{R}\hat{R}$ term. We shall calculate the dielectric constant of an extraordinary ray.

For our proof we shall choose a simple model and a simple lattice structure. Fig. 7 shows a hexagonal lattice wherein the dielectric properties in the x and y directions are identical, but they are different in the z direction. The dielectric tensor has the components

$$\varepsilon_{ij} = \begin{pmatrix} \varepsilon_1 & & 0 \\ & \varepsilon_1 & \\ 0 & & \varepsilon_3 \end{pmatrix}$$

An optical wave is going in the k direction at an angle θ to the z axis, with the electric field polarized in the xz plane. This is called an extraordinary ray. We wish to calculate its effective dielectric function. We start from Maxwells equations

$$\nabla \times E = -\frac{1}{c}\frac{\partial}{\partial t} H$$

$$\nabla \times H = -\frac{1}{c}\frac{\partial}{\partial t} \underset{\sim}{D}$$

and eliminate H in the usual way to obtain

$$\underset{\sim}{k} \times (\underset{\sim}{k} \times \underset{\sim}{E}) = -\frac{\omega^2}{c^2} \underset{\sim}{D}$$

In component form this is

$$k_z^2 E_x - k_x k_z E_z = \frac{\omega^2}{c^2} D_x = \frac{\omega^2}{c^2} \varepsilon_1 E_x$$

$$k_x^2 E_z - k_x k_z E_x = \frac{\omega^2}{c^2} D_z = \frac{\omega^2}{c^2} \varepsilon_3 E_z$$

Solving these equations for the refractive index

$$n = \frac{kc}{\omega}$$

yields the standard result which is derived in many textbooks

$$n^2 = \frac{1}{\dfrac{\cos^2\theta}{\varepsilon_1} + \dfrac{\sin^2\theta}{\varepsilon_3}} \qquad (3.21)$$

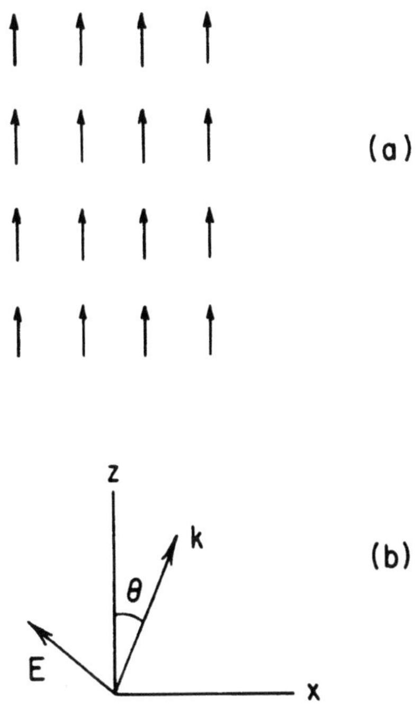

Fig. 7 (a) A model dielectric where the induce polarization moments can only point in the z direction.
(b) An extraordinary ray in this solid, where the $\underset{\sim}{k}$ vector has an angle θ with respect to the z direction, and the electric field is in the xz plane.

We will now derive this same result using the classical dipole methods. Let the classical dipoles be constrained to lie in the z direction, so that polarization only develops in this direction

$$\underset{\sim}{P} = \hat{z}\, P_z$$

$$\underset{\approx}{\alpha} = \hat{z}\hat{z}\, \alpha_0$$

The dielectric components ε_1 and ε_3 have the simple form

$$\varepsilon_1 = 1$$

$$\varepsilon_3 = 1 + \frac{4\pi\alpha_0}{1 + 4\pi\alpha_0\, T_{zz}(k=k_x)} \tag{3.22}$$

The classical dipole theory leads to the self consistent equation

$$\underset{\sim}{P} = \underset{\approx}{\alpha}\cdot(\underset{\sim}{E}_0 - \frac{4\pi T}{V_0} \cdot \underset{\sim}{P})$$

which is easily solved since only the z component enters

$$P_z = \frac{\alpha_0(\hat{z}\cdot E_0)}{1 + 4\pi\bar{\alpha}_0\, T_{zz}}$$

and we derive for the refractive index

$$n^2 = 1 + \frac{4\pi\bar{\alpha}_0\, (\hat{z}.\hat{e}_0)^2}{1 + 4\pi\bar{\alpha}_0\, T_{zz}(\underset{\sim}{k})} \tag{3.23}$$

First consider the possibility that T_{zz} does not depend upon wave vector, but is just a constant obtained by summing neighbours. Combining (3.22) and (3.23) yields

$$n^2 = \varepsilon_1 \cos^2\theta + \varepsilon_3 \sin^2\theta .$$

This is not the same as (3.21) and is wrong. So the assumption that T_{zz} is independent of k direction leads to an incorrect physical result. Now if we assume the form (3.20), we get that

$$T_{zz} = \cos^2\theta + t_3$$

where $t_3 = t_3' - 1/3$ is a constant. This makes

$$\varepsilon_1 = 1$$

$$\varepsilon_3 = 1 + \frac{4\pi\alpha_0}{1 + 4\pi\alpha_0 t_3}$$

when these are put into (3.23), exactly the right result (3.21) is obtained. Thus it is necessary to include a term of the form $\hat{R}\hat{R}$ in order to get the result (3.21) which is known from classical optics. Of course, since the trace of T vanishes, we must only add traceless terms, so we really must add $\hat{R}\hat{R} - \frac{1}{3}\underset{\approx}{I}$ to the sum obtained by local summation.

The original method of summing is called Ewalds Method.[18] It was invented by Ewald for summing coulombic interactions - to obtain Madelung energies - and extended to dipoles by Kornfeld.[19] The version we use was modified by Born and Bradburn.[20] Our discussion follows one we gave earlier.[21]

It is easiest to first fine the coulomb sum

$$V(k,r) = \frac{V_0}{4\pi} \sum_m \frac{e^{ik.R_m}}{|r-R_m|}$$

since

$$T_{\mu\nu}(k) = - \left(\frac{\partial}{\partial x_\mu} \frac{\partial}{\partial x_\nu} V(k,\underset{\sim}{r}) \right)_{r=0} \tag{3.24}$$

First we do the case T^e for summing over translationally equivalent sites, where the R_m=0 term is omitted from the summation. The integral identity

$$\frac{1}{r} = \frac{2}{\sqrt{\pi}} \int_0^\infty dx\, e^{-x^2 r^2}$$

is used to write

$$\frac{1}{|\underset{\sim}{r}-\underset{\sim}{R}_m|} = \frac{2}{\sqrt{\pi}} \left(\int_0^\eta dx\, e^{-x^2(\underset{\sim}{r}-\underset{\sim}{R}_m)^2} + \int_\eta^\infty dx\, e^{-x^2(\underset{\sim}{r}-\underset{\sim}{R}_m)^2} \right)$$

where the constant η is arbitrary. Optimum values of η will be specified later. This is put into the summation for V(k,r)

$$V(\underset{\sim}{k},\underset{\sim}{r}) = \frac{V_0}{4\pi} \frac{2}{\sqrt{\pi}} \left(\int_0^\eta dx \sum_m e^{i\underset{\sim}{k}.\underset{\sim}{R}_m} e^{-x^2(\underset{\sim}{r}-\underset{\sim}{R}_m)^2} - \int_0^\eta dx\, e^{-x^2 r^2} + \int_\eta^\infty dx \sum_m{}' e^{i\underset{\sim}{k}.\underset{\sim}{R}_m} e^{-x^2(\underset{\sim}{r}-\underset{\sim}{R}_m)^2} \right)$$

The term $\underset{\sim}{R}_m$=0 has been included in the first integral, and subtracted out again in the second integral. The first integral is called

$V_1(k,r)$

$$V_1(\underset{\sim}{k},\underset{\sim}{r}) = \frac{V_0}{2\pi^{3/2}} \int_0^\eta dx \sum_m e^{i\underset{\sim}{k}.\underset{\sim}{R}_m} e^{-x^2(\underset{\sim}{r}-\underset{\sim}{R}_m)^2}$$

It is a periodic function of r, and so may be expanded in the reciprocal lattice vectors of the solid.

$$V_1(\underset{\sim}{k},\underset{\sim}{r}) = \sum_{\underset{\sim}{G}} e^{i\underset{\sim}{r}.(\underset{\sim}{k}+\underset{\sim}{G})} U_{\underset{\sim}{G}}$$

where

$$U_G = V_0^{-1} \int_{cell} dr' \; e^{-i\underset{\sim}{r}'.(\underset{\sim}{k}+\underset{\sim}{G})} V_1(\underset{\sim}{k},\underset{\sim}{r}') = \frac{e^{iG.\tau}}{2\pi^{3/2}} \int_0^\eta dx \times$$

$$\int_{crystal} dr' \; e^{-i\underset{\sim}{r}'.(\underset{\sim}{k}+\underset{\sim}{G})} e^{-x^2 r'^2}$$

The integral over the unit cell plus the summation over cells has been combined into an integration over the entire volume of the crystal. The volume V_0 of the unit cell is simply given by the product of the vectors defining the unit cell of the crystal

$$V_0 = \underset{\sim}{a}\cdot(\underset{\sim}{b}\times\underset{\sim}{c})$$

The lattice vector $\underset{\sim}{\tau}$ is from the central site to any one of those being summed over. The factor $\exp(-i\,\underset{\sim}{G}.\underset{\sim}{\tau})$ equals one for sums over translationally equivalent sites but may not for the inequivalent case. The integrals may be done, and

$$V_G = \frac{e^{-i\underset{\sim}{G}.\underset{\sim}{\tau}} e^{-(\underset{\sim}{k}+\underset{\sim}{G})^2/4\eta^2}}{(\underset{\sim}{k}+\underset{\sim}{G})^2}$$

Putting all of these into (3.24) and doing the straightforward gradients, the dipolar sum for translationally equivalent sites is

$$T^e_{\mu\nu}(k) = -\frac{V_0\eta^3}{3\pi^{3/2}}\delta_{\mu\nu} + \sum_{\underset{\sim}{G}} \frac{(\underset{\sim}{k}+\underset{\sim}{G})_\mu(\underset{\sim}{k}+\underset{\sim}{G})_\nu}{(\underset{\sim}{k}+\underset{\sim}{G})^2} e^{-(\underset{\sim}{k}+\underset{\sim}{G})^2/4\eta^2} + \frac{V_0}{4\pi}\sum_m{}'$$

$$e^{i\underset{\sim}{k}.\underset{\sim}{R}_m} \left[\frac{1}{R_m^3}\left(\delta_{\mu\nu} - \frac{3R_{m\mu}R_{m\nu}}{R_m^2}\right)\left(\mathrm{erfc}(\eta R_m) + \frac{2}{\sqrt{\pi}}\eta R_m e^{-(\eta R_m)^2}\right) - \frac{4\eta^3}{\sqrt{\pi}} \frac{R_{m\mu}R_{m\nu}}{R_m^2} e^{-(\eta R_m)^2} \right] \quad (3.25)$$

where erfc is the standard complimentary error function. It is easy to extract the wave vector dependence from this expression. The vectors $\underset{\sim}{G}$ are of magnitude $2\pi/a$ where a is a lattice constant. Thus

$G \gg k$. If we chose η to be of the magnitude of $\eta \sim 1/a$, then both the sums over lattice vector $\underset{\sim}{R}_m$ and the sums over reciprocal lattice vectors $\underset{\sim}{G}$ converge very rapidly. The numerical result is independent of the specific choice of η, but both series converge rapidly - thereby reducing the numerical labor - if η is chosen well. The only term with significant k dependence is the $\underset{\sim}{G}=0$ term in the summation over reciprocal lattice vectors. This term may be rewritten as

$$\frac{k_\mu k_\nu}{k^2} + \sum_G{}' \frac{G_\mu G_\nu}{G^2} e^{-G^2/4\eta^2}$$

where the prime means the $\underset{\sim}{G}=0$ term is now excluded from the sum. Thus we have shown that T^e may be written as

$$T^e_{\mu\nu}(k) = \frac{k_\mu k_\nu}{k^2} + t^e_{\mu\nu}$$

where $t^e_{\mu\nu}$ is constant, independent of k for values of k appropriate for optical frequencies.

Equation (3.25) is valid for all values of k, and numerical values have been found for values of k over the entire Brillouin zone of the solid - up to values of the size $\underset{\sim}{k} \sim \underset{\sim}{G}$.[22] Although these are not necessary for optical calculations, they can be used for other purposes.

The results for translationally inequivalent sums is similar. The first term in (3.25), which came from subtracting out the $R_m=0$ term, is not needed. The factor $\exp(iG.\tau)$ must now be included. The result is

$$T^i_{\mu\nu}(k) = \sum_{\underset{\sim}{G}} \frac{(k+G)_\mu (k+G)_\nu}{(k+G)^2} e^{i\underset{\sim}{G}.\underset{\sim}{\tau}} e^{-(k+G)^2/4\eta^2} + \frac{V_0}{4\pi} \sum_{R_m} e^{ik.r_m}$$
$$\left[\frac{-4\eta^3}{(\pi)^{1/2}} \frac{r_{m_\mu} r_{m_\nu}}{r_m^2} e^{-(\eta r_m)^2} + \frac{1}{r_m^3} (\delta_{\mu\nu} - \frac{3 r_{m_\mu} r_{m_\nu}}{r_m^2}) \right.$$
$$\left. (\mathrm{erfc}(\eta r_m) + \frac{2}{\sqrt{\pi}} \eta r_m e^{-(\eta r_m)^2}) \right]$$

where $\underset{\sim}{r}_m = \underset{\sim}{R}_m + \tau$.

We will now give a short course in solid state physics, for those with a scanty background in this subject. The three dimensional lattice structure is defined by three vectors $\underset{\sim}{R}_1$, $\underset{\sim}{R}_2$, $\underset{\sim}{R}_3$, which are just the same as the $\underset{\sim}{a}$, $\underset{\sim}{b}$, $\underset{\sim}{c}$ we used earlier. Any translationally equivalent molecule can be reached by the vector

$$R_m = \alpha \underset{\sim}{R}_1 + \beta \underset{\sim}{R}_2 + \gamma \underset{\sim}{R}_3$$

where α, β, γ are integers. The three basic reciprocal lattice vectors are

$$\underset{\sim}{G}_1 = \frac{2\pi}{V_0} \underset{\sim}{R}_2 \times \underset{\sim}{R}_3$$

$$\underset{\sim}{G}_2 = \frac{2\pi}{V_0} \underset{\sim}{R}_3 \times \underset{\sim}{R}_1$$

$$\underset{\sim}{G}_3 = \frac{2\pi}{V_0} \underset{\sim}{R}_1 \times \underset{\sim}{R}_2$$

and a general reciprocal lattice vector is formed by

$$\underset{\sim}{G} = r\underset{\sim}{G}_1 + s\underset{\sim}{G}_2 + t\underset{\sim}{G}_3$$

where r, s, t are also integers. With definition of $\underset{\sim}{G}_i$, notice that the vector product

$$\underset{\sim}{G}_i \cdot \underset{\sim}{R}_j = 2\pi\delta_{ij} \qquad (i,j = 1, 2, 3)$$

The three basic vectors $\underset{\sim}{G}_1$, $\underset{\sim}{G}_2$, $\underset{\sim}{G}_3$ are not mutually orthogonal unless $\underset{\sim}{R}_1$, $\underset{\sim}{R}_2$, $\underset{\sim}{R}_3$ are also. The general product

$$\underset{\sim}{G} \cdot \underset{\sim}{R}_m = 2\pi(\alpha r + \beta s + \gamma t) = 2\pi \qquad X(\text{integer})$$

so that

$$e^{i \underset{\sim}{G} \cdot \underset{\sim}{R}_m} = 1$$

Summation over $\underset{\sim}{G}$ mean summations over all possible values of the integers r,s,t.

3.6 Numerical Results

The crystal structure data of Table II can be used with the formulas of the previous section to compute the dipole sums. For example, the result for anthracene is

$$t^e = \begin{pmatrix} -0.15 & 0 & -.016 \\ 0 & -.82 & 0 \\ -.016 & 0 & -.025 \end{pmatrix}$$

$$t^{i} = \begin{pmatrix} -1.24 & 0 & .016 \\ 0 & -.35 & 0 \\ -.016 & 0 & .59 \end{pmatrix}$$

The trace of each matrix is one. These tensors need to be multiplied by the transition moment unit vectors $\hat{\mu}_{L,M}^{(1,2)}$. These results are shown in Table III. The dimensional units are $(\text{cm } \text{Å}^2)^{-1}$, which is obtained by multiplying by $4\pi e^2/V_0$. So multiplying these numbers by the observed transition moment magnitudes in Å (Table I) gives the energies in the spectroscopic units of cm^{-1}. What is actually given in Table III are the quantities

$$t_{LL}^{e} = \frac{4\pi e^2}{V_0} \hat{\mu}_L^{(1)} \cdot t^{e} \cdot \hat{\mu}_L^{(1)}$$

The top half of the Table gives $\underset{\approx}{t}$, while the bottom half gives $\underset{\approx}{T} = \hat{k}\hat{k} + \underset{\approx}{t}$ for the choice of $k = (001)$ - which would be normal incidence on an ab plane. This is a common experimental configuration. This separation is done to show that the $\hat{k}\hat{k}$ term has a sizable influence.

The values of T depend upon the direction of the wave vector of light, and thus the absorption frequencies in the solid also depend upon the direction of the optical wave vector. Of course, this is very well known to be correct, since it follows directly from the classical equation (3.21) for extraordinary rays. This has been explicitly verified by direct measurement in anthracene and other materials.[23,24]

The best measured Davydov splittings are for the M axis transition in anthracene and tetracene. They shall be used as the basis of comparison between theory and experiment. We do not discuss naphthalene or phenanthrene because they each have weak transitions ($f \sim .01$) in the vicinity of the M transitions, which renders difficult the theoretical and experimental analysis. Neither tetracene nor anthracene has this difficulty.

Fig. 6 shows that this transition in anthracene is a series of transitions which are split by the molecular vibrational frequency of 1400 cm^{-1}. The same phenomenon is observed in tetracene, with nearly the same frequency. Thus the polarizability of the M transition should be written as a sum over the vibrational overtones

$$\bar{\alpha}_M = \omega_P^2 f_0 \sum_{n=0}^{5} \frac{\xi_n^2}{(\omega_0 + n\omega_1)^2 - \omega^2} + \bar{\alpha}'_M$$

where ξ_n^2 are the Franck Condon factors which sum to unity, and

Table III Dipole Sums (cm^{-1} angstrom^{-2})[a]

	t^e_{LL}	t^e_{LM}	t^e_{MM}	t^i_{LL}	t^i_{LM}	t^i_{MM}
Anthracene	-172	329	-2077	389	996	660
Naphthalene	-776	320	-2135	116	666	784
Phananthrene	- 97	-209	-1967	1500	-704	196

$\underset{\sim}{k} = (001)$

	T^e_{LL}	T^e_{LM}	T^e_{MM}	T^i_{LL}	T^i_{LM}	T^i_{MM}
Anthracene	2100	1165	-1769	2661	1832	967
Naphthalene	2328	1635	-1578	3219	1981	1341
Phenanthrene	2658	-765	-1855	4255	-1260	309

a. M.R. Philpott, J. Chem. Phys. 50, 5117 (1969).

$\hbar\omega_1 = 1400 cm^{-1}$. These vibrational levels are sufficiently close together that when this form of the polarizability is put into the equation (3.17) $\Delta=0$, the resulting equation to find the roots must be solved numerically.

The multiplicity of vibrational levels requires that the Davydov splitting be calculated - and measured- for each one. Experimental results are shown in Table IV for the lowest level in anthracene and tetracene. The first theoretical calculation of this type, using the self consistent dipole model, was done by Mahan[13] who obtained for anthracene the theoretical Davydov splittings of 228, 105, 59, and 26 cm^{-1}. This calculation only included the first M and first L state in the analysis. Later Philpott redid the calculation and incorporated more excited states in the level analysis.[14] His results are shown in Table IV. These higher states change the theoretical splittings by making them larger. It must be concluded that the dipole model satisfactorily accounts for the Davydov splittings in these materials. Similar good agreement is obtained between the measured and theoretical polarization ratios.

Although this success is gratifying, it is also deceiving. Theorists tend to report their successful results, and let me now confess one of my unsuccessful ones. The most general form for the molecular polarizability

$$\alpha_{\mu\nu} = \alpha_L \hat{\mu}_L \hat{\mu}_L + \alpha_M \hat{\mu}_M \hat{\mu}_M + \alpha_N \hat{\mu}_N \hat{\mu}_N$$

contains just three parameters α_L, α_M, α_N. Thus, if the dipole theory were really valid, the <u>four</u> dielectric constants in (3.1) should be expressible in terms of these <u>three</u> parameters. Numerically it just does not work for anthracene. Furthermore, Philpott showed that the states he included in his anthracene analysis do not explain the (001) refractive index data - in fact they miss badly.[25] It is not surprising, since the dipole theory omits the short range interaction between molecules due to higher moments, and exchange interactions. Indeed, the surprise is that the dipole theory works so well for Davydov splittings. My experience is that this is also the case in inorganic solids, where the splittings are given accurately by dipole theory, but the other quantities are not.[24]

3.7 More Dipole Theory

The point dipole model is a mathematically interesting model of the dielectric properties of a solid. Many exact results in the theory of dielectrics can be obtained using this model. We will mention some of these results. They are not necessarily relevant to the optical properties of solids, but they may be of interest to those concerned with other dielectric properties. The point dipole model

Table IV Davydov Splittings : (cm^{-1})

Anthracene

h	Theory	Experiments				
0	297[a]	220[b]	230[c]	195[d]	190[e]	190[f]
1	112	150	145	140	86	110
2	59	40	80	80	24	60

Tetracene

n	Theory	Experiments		
0	631[a]	630[g]	620[h]	630[i]
1	265	270	275	222
2	138	90	125	153
3	90			

a. M. R. Philpott, J. Chem. Phys. 50, 5117 (1969).

b. H. C. Wolf, A. Naturforsch 13a, 414 (1958).

c. M. S. Brodin and S.V. Morisova, Opt. Spektrosk 10, 473 (1960) (Opt. Spectrosc. 10, 242 (1960)).

d. T.A. Calxton, D.P. Craig T. Thirunamachandran, J. Chem. Phys. 35, 1525 (1961).

e. A. Matsui, J. Phys. Soc. Japan 21, 2212 (1966).

f. L.B. Clark and M. R. Philpott, J. Chem. Phys. 53, 3790 (1970).

g. A. Bree and L.E. Lyons, J. Chem. Soc. 1960, 5206.

h. A. F. Prikhotko and A. F. Skorobogatko, Opt. Spektrosk. 20,65 (1966).

i. J. Tanaka, Bull. Chem. Soc. Japan 38, 86 (1965).

makes the basic assumption that the atoms in the solid are infinitesimal in size. The local electric field acting on an atom produces the polarization

$$\underset{\sim}{P} = \underset{\approx}{\alpha} \cdot \underset{\sim}{E}^{\text{local}}$$

and the local electric field at a site is the sum of the dipole-dipole interactions among all of these polarizations.

One result which can be obtained exactly is the interaction potential between interstitial impurity charges Q_1 and Q_2 which are located at r_1 and r_2. The exact result is[26,27]

$$V = Q_1 Q_2 \left[\frac{1}{R} - \frac{4\pi}{V} \sum_k e^{i\underset{\sim}{k}.\underset{\sim}{R}} \underset{\sim}{W}(\underset{\sim}{k},\underset{\sim}{r}_1) \cdot \underset{\approx}{\alpha} \cdot \underset{\approx}{G}(\underset{\sim}{h}) \cdot \underset{\sim}{W}(-\underset{\sim}{k},\underset{\sim}{r}_2) \right]$$

where

$$\underset{\sim}{W}(\underset{\sim}{k},\underset{\sim}{r}_1) = \frac{V_0}{4\pi} \sum_{R_m} e^{ik.(R_m - r_1)} \frac{(\underset{\sim}{R}_m - \underset{\sim}{r}_1)}{(\underset{\sim}{R}_m - \underset{\sim}{r}_2)^3}$$

$$[\underset{\approx}{I} + 4\pi\, \underset{\approx}{\alpha} \cdot \underset{\approx}{T}] \cdot \underset{\approx}{G} = \underset{\approx}{I}$$

At large separations $\underset{\sim}{R} = \underset{\sim}{r}_1 - \underset{\sim}{r}_2$ this does go into Coulomb's law, but the advantage of this result is that it can be used to calculate interactions at close separations. Similar formulas have been derived for the interaction among dipoles.[26] This latter result has recently been extended to liquids.[28]

Another exact result is the binding energy of solids from van der Waals forces. The usual London's interaction energy is based upon second order perturbation theory, while the exact result is equivalent to summing second, third, fourth, and all orders of perturbation theory. This exact formula is[29]

$$V_i = \frac{1}{4\pi N} \sum_k \int_{-\infty}^{\infty} dy \log [\Delta(\underset{\sim}{k}, iy)]$$

where the determinant is

$$\Delta(k,z) = \det [1 + 4\pi\ \underset{\approx}{\bar{\alpha}}.(iy).\ \underset{\approx}{T}(k)]$$

These formulas have been used by Lucas to calculate the binding energy of rare gas solids.[22]

The dipole theory has been used to calculate surface properties of solids, including image charge theory from a microscopic basis.[15,30]

IV MORE EXACT THEORIES

4.1 Retardation

Some workers in this field have tried to "improve" the theory by including the retardation of the dipole-dipole interaction when evaluating dipole sums.[31,32] The argument seems to be that if the long range terms are important, then retardation must be also. This argument is incorrect. In fact, using retardation in evaluating the dipole sums leads to an incorrect answer. It is rigorously correct to use instantaneous dipole sums and rigorously incorrect to include retardation. Although this conclusion is a bit surprising, it follows directly from Maxwell's equations, and is a simple consequence of classical physics. This section will be devoted to justifying these conclusions.

Maxwell's equations are

$$\nabla \times \mathbf{D} = 4\pi\rho \tag{4.1}$$

$$\nabla \cdot \mathbf{B} = 0 \tag{4.2}$$

$$\nabla \times \mathbf{H} = \frac{4\pi}{c}\mathbf{J} + \frac{1}{c}\frac{\partial}{\partial t}\mathbf{D} \tag{4.3}$$

$$\nabla \times \mathbf{E} + \frac{1}{c}\frac{\partial}{\partial t}\mathbf{B} = 0 \tag{4.4}$$

We shall keep the derivation simple, and assume that no free charges or currents are present

$$\rho = 0$$

$$\mathbf{J} = 0$$

This is a satisfactory approximation in an insulator. The polarization term will be retained, since it is the term of interest.

A fundamental theorem of vector calculus states that any vector function $\mathbf{c}(\mathbf{r})$ may be written as the sum of a rotational and irrotational term, where

$$\mathbf{c} = \mathbf{c}_i + \mathbf{c}_r$$

$$\mathbf{c}_i = \nabla \xi \qquad \nabla \times \mathbf{c}_i = 0$$

$$\mathbf{c}_r = \nabla \times \boldsymbol{\alpha} \qquad \nabla \cdot \mathbf{c}_r = 0$$

If we apply this theorem to (4.2) then $\mathbf{B}$ must be expressible only as a rotational part, which is called the vector potential $\mathbf{A}$

$$\mathbf{B} = \nabla \times \mathbf{A}$$

This is put into (4.4)

$$\nabla \times \left[E + \frac{1}{c} \frac{\partial A}{\partial t} \right] = 0$$

If we use the theorem again, the part in square brackets is irrotational, so must be expressible as the gradient of a scalar potential ϕ

$$E = -\frac{1}{c} \frac{\partial A}{\partial t} - \nabla\phi \qquad (4.5)$$

These two definitions are inserted into the remaining two Maxwell's equations (4.1) and (4.3) to give

$$\nabla^2\phi + \frac{1}{c} \frac{\partial}{\partial t} (\nabla . A) = 4\pi \, \nabla . P \qquad (4.6)$$

$$\nabla^2 A - \frac{1}{c^2} \frac{\partial^2}{\partial t^2} A - \nabla(\nabla . A + \frac{1}{c} \frac{\partial \phi}{\partial t}) = -\frac{4\pi}{c} \frac{\partial}{\partial t} P \qquad (4.7)$$

These are four equations for the four functions ϕ, A_x, A_y, and A_z. Nevertheless, the four equations are not independent. If we multiply the first by $\frac{1}{c} \frac{\partial}{\partial t}$ and the second by ∇, and add the two together, there results

$$0 = 0$$

One of these four equations is not independent, so we have three equations and four unknowns. Thus we have a redundancy, and one more additional condition must be imposed in order to solve the equations. This is called a gauge condition. Each possible gauge condition results in different values for the scalar and vector potentials. It does not change the value of the physically observable quantities E and B, since they turn out to be independent of the gauge.

A. Coulomb Gauge

We shall assume that

$$\nabla \cdot A = 0$$

This is called the "transverse gauge", or sometimes the "Coulomb gauge." Our equations (4.6) and (4.7) simplify to

$$\nabla^2\phi = 4\pi \, \nabla . P \qquad (4.8)$$

$$\nabla^2 A - \frac{1}{c^2} \frac{\partial^2 A}{\partial t^2} = -\frac{4\pi}{c} \frac{\partial}{\partial t} (P - \frac{1}{4\pi} \nabla\phi) \qquad (4.9)$$

The solution to the first equation is easy

$$\phi(r) = -\int \frac{d^3r' \, \underset{\sim}{\nabla}' . P(r')}{|r-r'|} \qquad (4.10)$$

which after an integration by parts may also be written as

$$\phi(r) = \int d^3r' \, \underset{\sim}{P}(r') \, . \underset{\sim}{\nabla}' \left(\frac{1}{|\underset{\sim}{r}-\underset{\sim}{r}'|} \right) \qquad (4.10')$$

The right hand side of (4.9) contains the factor

$$\underset{\sim}{P}_r = \underset{\sim}{P} - \frac{1}{4\pi} \underset{\sim}{\nabla}\phi$$

which is the rotational part of the polarization vector. One can prove it is rotational by simply proving that

$$0 = \underset{\sim}{\nabla}.\underset{\sim}{P}_r = \nabla.P - \frac{1}{4\pi} \nabla^2\phi$$

This follows immediately from (4.10) with the use of the identity

$$\nabla^2 \frac{1}{|r-r'|} = - 4\pi \, \delta\,(\underset{\sim}{r}-\underset{\sim}{r}')$$

Thus (4.9) may be written as

$$\nabla^2\underset{\sim}{A} - \frac{1}{c^2} \frac{\partial^2}{\partial t^2} \underset{\sim}{A} = - \frac{4\pi}{c} \frac{\partial}{\partial t} \underset{\sim}{P}_r$$

An operation by $\underset{\sim}{\nabla}$ on both sides of this equation yields the identity that both sides are zero.

These equations are quite general, and apply in any circumstance. Let us now examine them for an optical wave of frequency ω and wave vector $\underset{\sim}{k}$ travelling through the crystal. The electric field also has rotational and irrotational parts, and the rotational part is

$$\underset{\sim}{E}_r = - \frac{1}{c} \frac{\partial}{\partial t} \underset{\sim}{A}$$

and the irrotational part is $-\nabla\phi$. The rotational part obeys the equation

$$(\nabla^2 + \frac{\omega^2}{c^2}) \, \underset{\sim}{E}_r = - \frac{4\pi\omega^2}{c^2} \underset{\sim}{P}_r \qquad (4.11)$$

Since

$$\nabla . \, \underset{\sim}{E}_r = \underset{\sim}{k}.E_r = 0$$

then $\underset{\sim}{E}_r$ is perpendicular to $\underset{\sim}{k}$, and so is $\underset{\sim}{P}_r$. If $\underset{\sim}{E}_r$ is in the direction $\hat{e}_0$, then so is $\underset{\sim}{P}_r$

$$\underset{\sim}{P}_r = \hat{e}_0(\hat{e}_0 \cdot \underset{\sim}{P})$$

So that (4.11) is simplified to

$$\left(\nabla^2 + \frac{\omega^2}{c^2} \right) E_r = -4\pi \frac{\omega^2}{c^2} (\hat{e}_C \cdot \underset{\sim}{P})$$

Now in the point dipole model, it is assumed that the polarization is of infinitesimal size, and is located at the centers of the atomic positions

$$\underset{\sim}{P}(r) = \sum_{\ell} \underset{\sim}{P}_{\ell} \ \delta(\underset{\sim}{r} - \underset{\sim}{R}_{\ell}) \tag{4.12}$$

The polarization $\underset{\sim}{P}_{\ell}$ at each atomic side $\underset{\sim}{R}_{\ell}$ is proportional to the electric field at that atomic site

$$\underset{\sim}{P}_{\ell} = \underset{\approx}{\alpha}_{\ell} \cdot \underset{\approx}{E}(\underset{\sim}{R}_{\ell}) \tag{4.13}$$

The electric field is given by (4.5) to be

$$\underset{\sim}{E} = \underset{\sim}{E}_r - \underset{\sim}{\nabla} \phi$$

where from (4.10')

$$\nabla\phi(r) = \sum_{\ell} \underset{\approx}{\phi} \ (r - R_{\ell}) \cdot \underset{\sim}{P}_{\ell} .$$

Thus we have derived the set of coupled equations

$$(\nabla^2 + \omega^2/c^2) \ E_r \ e^{i\underset{\sim}{k} \cdot \underset{\sim}{r}} = \frac{\omega^2}{c^2} 4\pi \ \hat{e}_0 \cdot \sum_{\ell} P_{\ell} \ \delta(r - R_{\ell}) \tag{4.14}$$

$$\underset{\sim}{P}_{\ell} = \underset{\approx}{\alpha}_{\ell} \cdot [\underset{\sim}{E}_r - \sum_m \phi(R_{\ell} - R_m) \cdot \underset{\sim}{P}_m] \tag{4.15}$$

The solution to (4.14) proceeds as follows. Assume the optical wave is going perpendicular to planes which are separated by the distance a. So the lattice vector $R_{\ell} = (\ell a, \rho_j)$ where ρ_j is a two dimensional lattice vector in the plane.

$$\delta(r - R_{\ell}) = \delta(Z - \ell a) \ \delta(\underset{\sim}{\rho} - \underset{\sim}{\rho}_j)$$

If all polarizations in the plane have the same value, then the two dimensional sum

$$\sum_{\rho j} \delta(\underset{\sim}{\rho} - \underset{\sim}{\rho}_j)$$

is periodic in $\underset{\sim}{\rho}$, so may be expanded in the two dimensional lattice vectors $\underset{\sim}{g}$ of the plane

$$\sum_{\rho_j} \delta(\underset{\sim}{\rho}-\underset{\sim}{\rho}_i) = \frac{1}{A_0} \sum_g e^{ig\cdot\rho}$$

when $A_0 = V_0/a$ is the area of a unit cell in the plane. The electric field may be similarly expanded

$$E_r(\underset{\sim}{r}) = \sum_g E_r(g,z)\, e^{ig\cdot\rho}$$

In (4.14), the equation for each $\underset{\sim}{g}$ component is independent

$$\left(\frac{\partial^2}{\partial z^2} - g^2 + \omega^2/c^2\right) E_r(g,z) = -\frac{\omega^2}{c^2}\frac{4\pi}{A_0}\sum_\ell (\hat{e}_0\cdot\underset{\sim}{P}_\ell)\,\delta(z-\ell a)$$

which has the solution

$$E_r(g=0,z) = -\frac{2\pi i\omega}{cA_0}\sum_\ell \hat{e}_0 \cdot \hat{\underset{\sim}{P}}_\ell\, e^{i\frac{\omega}{c}|z-\ell a|} \tag{4.16}$$

$$E_r(g,z) = -\frac{2\pi\omega^2}{c^2 A_0\sqrt{g^2-\omega^2/c^2}}\sum_\ell \hat{e}_0\cdot P_\ell\, e^{-|z-\ell a|\sqrt{g^2-\omega^2/c^2}}$$

Since $gz \sim 2\pi$ is large, the terms with $g\neq 0$ may be neglected. Thus E_r is essentially given by (4.16). Now if we assume that $E_r(\ell)$ and P_ℓ botn are propagating modes with a spatial dependence

$$P_\ell = P_0\, e^{in\frac{\omega}{c}\ell a}$$

$$E_r = E_{r0}\, e^{in\omega\ell a/c}$$

where n is the refractive index, then inserting this into (4.16) gives

$$(n^2-1)E_{r0} = \frac{4\pi}{V_0}\hat{e}_0\cdot\hat{P}_0 \tag{4.17}$$

which is the desired result. Equations (4.15) and (4.17) were the equations we solved in Section 3.4, and called classical dipole theory. There we solved the case of several molecules per unit cell, while here we just did one molecule per cell.

In Section 3.4 we called E_r the applied field. That previous nomenclature is incorrect. An oscillating dipole created both a short range dipole field, and a long range radiation field, the latter is E_r. The set of coupled equations (4.14) and (4.15) define how these short and long range fields cooperate to describe electromagnetic mode propagation in crystals. A solution of these equations yields the normal modes of the solid. In an experiment, an electric field is applied to the crystal, for example, by sending a light beam at the sample. This beam strikes the surface of the sample, and excites the normal modes of the solid, which propagate on through the crystal. Thus E_r is not the applied field, but is instead the normal mode field which is obtained by the self consistent solution of the interacting dipole problem. This point of view is called "Polariton theory," and was first described correctly in The Dynamical Theory of Crystal Lattices by Born and Huang.[10-12] This derivation has been carried through in great detail. Nothing has been omitted, and no retardation phenomenon has been dropped. Yet (4.15) shows that the instantaneous dipole sum is to be taken. The retardation is in the equations - in equation (4.14) for E_r - and has been included. But the dipole summation part (4.13) is summed instantaneously, since the retardation is elsewhere in the coupled equations. If we were to assume, in an ad hoc fashion as some are wont to do, that the dipole interaction $\phi_{\mu\nu}$ in (3.6) is also retarded, then we would be putting retardation in the equations twice - we would be overcounting retardation.

These results were derived using the coulomb guage. But the final result is independent of the choice of gauge, since it involves an observable quantity - the electric field. This may be a little hard to believe because the derivation seemed to rely heavily upon the condition that $\nabla . A = 0$. The best way to demonstrate that the results are independent of gauge is to derive them again in another gauge. We shall do so in the Lorenz gauge. This also provides another opportunity to elucidate the fact that the dipole sums should be taken in an instantaneous fashion

B. Lorentz Gauge

The Lorentz gauge assumes that

$$\nabla . A + \frac{1}{c}\frac{\partial \phi}{\partial t} = 0 \qquad (4.18)$$

so that (4.6) and (4.7) become

$$\left(\nabla^2 - \frac{1}{c^2}\frac{\partial^2}{\partial t^2}\right) \phi = 4\pi\, \underset{\sim}{\nabla} . \underset{\sim}{P} \qquad (4.19)$$

$$\left(\nabla^2 - \frac{1}{c^2}\frac{\partial^2}{\partial t^2}\right) \underset{\sim}{A} = - \frac{4\pi}{c}\frac{\partial}{\partial t} \underset{\sim}{P} \qquad (4.20)$$

The easiest way to solve this equation is by the use of Hertz potentials.[33] The Hertz potential is the vector function $\underset{\sim}{\pi}(\underset{\sim}{r})$ which obeys the equation

$$\left(\nabla^2 - \frac{1}{c^2} \frac{\partial^2}{\partial t^2} \right) \underset{\sim}{\pi}(r) = 4\pi \underset{\sim}{P}(r) \qquad (4.21)$$

The vector and scalar potentials are

$$\underset{\sim}{A} = -\frac{1}{c} \frac{\partial \underset{\sim}{\pi}}{\partial t}$$

$$\phi = \underset{\sim}{\nabla} \cdot \underset{\sim}{\pi}$$

So they satisfy the equations (4.19) and (4.20), and also the gauge condition (4.18). From (4.5) the electric field is given by

$$E = \left[\frac{1}{c^2} \frac{\partial^2}{\partial t^2} - \underset{\sim}{\nabla} \underset{\sim}{\nabla} \right] \cdot \underset{\sim}{\pi}$$

The general solution to (4.21) is quite complicated. The result is much simpler if a simple oscillatory time dependence $\exp(-i\omega t)$ is assumed for P and E. Then the result is

$$\underset{\sim}{\pi}(r,t) = e^{-i\omega t} \int d^3r' \frac{\underset{\sim}{P}(r')}{|r-r'|} e^{i\frac{\omega}{c}r'}$$

So in the point dipole model (4.12) we get an expression for the local electric field

$$E(r,t) = -e^{i\omega t} \left(\frac{\omega^2}{c^2} \underset{\approx}{I} + \underset{\sim}{\nabla} \underset{\sim}{\nabla} \right) \cdot \sum_{R_\ell} \underset{\sim}{P}_\ell \frac{e^{i\frac{\omega}{c}R_\ell}}{|r-R_\ell|} \qquad (4.22)$$

The operator

$$\left(\frac{\omega^2}{c^2} \underset{\approx}{I} + \underset{\sim}{\nabla} \underset{\sim}{\nabla} \right) \frac{e^{i\frac{\omega}{c}r}}{r} = e^{i\frac{\omega}{c}r} \left(\frac{1}{R} \left(\frac{\omega}{c}\right)^2 [\underset{\approx}{I} - \hat{R}\hat{R}] - \underset{\approx}{\phi}(R) \times \left(1 - i\frac{r\omega}{c}\right) \right)$$

generates the retarded dipole-dipole interaction function. Thus (4.22) states that the electric field at any point is obtained by adding up the retarded dipole-dipole interaction from all of the oscillating dipoles. The polarization equation (4.13) completes a pair of coupled equations which completely describe the normal modes of the dielectric solid. At this point the question is naturally asked: why do we now have retardation in the dipole interaction, whereas we did not before? The answer is simple: we had retardation before. We will show that the single equation (4.22) is identical to the pair (4.14) and (4.15) of coupled equations we solved previously. In particular, the local electric field E is identical to

$$E(r) = E_r - \sum_{\ell} \phi(r-R_\ell) \cdot P_\ell$$

In the previous treatment, retardation was included through the radiation field term E_r, which was the irrotational term in the electric field.

We will prove this in the fashion of the prior proof. A planewise geometry will be assumed, with lattice vectors $R_\ell = (\ell z, \rho j)$, and the values of P_ℓ is the same in each plane (only true for one molecule per unit cell). Now the sum over the plane is

$$F(z,\rho) = \sum_{\rho j} \frac{e^{i\frac{\omega}{c}|r-R_\ell|}}{|r-R_\ell|} = \sum_g F_g \, e^{i\rho\cdot g}$$

which is also expanded in reciprocal lattice vectors $\underset{\sim}{g}$ of the plane, with the coefficients

$$F_{g=0} = \frac{2\pi}{A_0} \frac{ic}{\omega} \, e^{i\frac{\omega}{c}|z-\ell a|}$$

$$F_g = \frac{2}{A_0} \frac{1}{\sqrt{g^2-\omega^2/c^2}} \, e^{-|z-\ell a|\sqrt{g^2-\omega^2/c^2}}$$

The electric field E is also expanded in lattice vectors $\underset{\sim}{g}$, and fourier coefficients set equal on the two sides of the equation

$$\underset{\sim}{E}_g = -\left(\frac{\omega^2}{c^2} + \underset{\sim}{\nabla}\,\underset{\sim}{\nabla}\right) \sum_\ell \underset{\sim}{P}_\ell \sum_g e^{i\rho\cdot g} F_g$$

In the $\underset{\sim}{g} = 0$ term the gradient operators give zero since $\underset{\sim}{P}$ is assumed to be perpendicular to z, the wave vector direction. So this becomes

$$E_{g=0} = -\frac{2\pi i\omega}{A_0 c} \sum_\ell \underset{\sim}{P}_\ell \, e^{i\frac{\omega}{c}|z-\ell a|}$$

and comparing with (4.16) shows that it is identical to E_r. For the terms with $g \neq 0$, one can ignore the factors $\frac{\omega^2}{c^2}$ compared to g^2, so that our electric field expression becomes

$$E = E_r - \frac{2\pi}{A_0} \underset{\sim}{\nabla}\,\underset{\sim}{\nabla} \cdot \sum_\ell \underset{\sim}{P}_\ell \sum_{g\neq 0} \frac{e^{ig\cdot\rho}}{g} e^{-g|z-\ell a|}$$

It can be shown that the second term is just the static dipole-dipole interaction which is needed to make this therm be the required form (4.15)

$$E = E_r - \sum_\ell \phi(r-R_\ell) \cdot \underset{\sim}{P}_\ell$$

In fact, the sum over planar reciprocal lattice vectors is not only the dipole-dipole interaction, it is a rather convenient way to sum the dipolar interaction. It tends to be numerically faster than the Ewald method described earlier. This plane-wise formula was introduced by Nijboer and de Wette.[34,35]

4.2 Agranovich Theory

Agranovich appears to be the first one to derive the proper equations without making the dipole approximation.[36] He neglects exchange interactions, but his formalism is correct in the Hartree approximation. We will not derive his results, but merely write them down and make them look plausible. A proper derivation uses the quantum exciton theory. However, we shall arrive at the same results by analogy with classical dipole theory.

In classical dipole theory, the molecular polarizability $\underset{\approx}{\alpha}$ arises from a set of oscillators, and each oscillator has a frequency ω_α and an oscillator strength f_α. Let $D_{\alpha,\ell}$ be the amplitude of the α excitation on the molecule located at the position $\underset{\sim}{R}_\ell$, so that the total polarization of the molecule at R_ℓ is

$$\underset{\sim}{P}_\ell = \sum_\alpha D_{\alpha,\ell}\, \underset{\sim}{\mu}_\alpha \tag{4.23}$$

Next we must describe how an excitation α at molecule R_ℓ interacts with a different excitation β at another site $R_{\ell'}$. This interaction is

$$M_{\alpha\ell,\beta\ell'} = \int d^3r_1 \int d^3r^2\ \psi_\alpha^*(r_1-R_\ell)\ \psi_i\ (r_1-R_\ell)\ \frac{e^2}{(\underset{\sim}{r}_1-\underset{\sim}{r}_2)}\ \psi_\beta^*\ (r_2-R_{\ell'}) \times$$
$$\psi_i(r_2-R_{\ell'})$$

where the interaction is through the coulomb interaction. In analogy with the classical dipole theory of Section 3.4, the equation governing the amplitudes is

$$(\omega_\alpha^2-\omega^2)D_{\alpha\ell} = -2\omega_\alpha \sum_{m,\beta} M_{\alpha\ell,\beta m}\ D_{\beta m} + \frac{e^2}{m}\frac{f_\alpha}{\mu_\alpha}\ \hat{\mu}_{\alpha\ell}\cdot\underset{\sim}{E} \tag{4.24}$$

This result is appropriate for a wave of frequency ω, and the term $\omega^2 D$ arose from the classical acceleration $\frac{\partial^2}{\partial t^2}$ D. Although we will not derive this expression here, we will show that is does exactly give our previous equations when the dipole approximation is employed. In the dipole approximation, the interaction M is approximated by

$$M_{\alpha\ell,\beta m} = e^2 \underset{\sim}{\mu}_{\alpha\ell} \cdot \underset{\approx}{\phi}(R_{\ell m}) \cdot \underset{\sim}{\mu}_{\beta m}$$

so that (4.24) becomes

$$(\omega_\alpha^2 - \omega^2) D_{\alpha\ell} = -2\omega_\alpha e^2 \underset{\sim}{\mu}_{\alpha\ell} \cdot \sum_m \underset{\approx}{\phi}(R_{\ell m}) \cdot \underset{\sim}{P}_m + \frac{e^2}{m} \frac{f_\alpha}{\mu_\alpha} \hat{\mu}_{\alpha\ell} \cdot \underset{\sim}{E}_\ell$$

If we multiply this by $\underset{\sim}{\mu}_{\alpha\ell}(\omega_\alpha^2 - \omega^2)^{-1}$ and sum over α, we obtain

$$\underset{\sim}{P}_\ell = \underset{\approx}{\alpha}_\ell(\omega) \cdot \{ \underset{\sim}{E}_\ell - \sum_m \underset{\approx}{\phi}(R_{\ell m}) \cdot \underset{\sim}{P}_m \}$$

where (3.10) has been used for the molecular polarizability. The above equation is just the form for the dipole approximation (4.15) which was derived in the prior section. The form (4.24) is the right form to use whenever the dipole approximation is inadequate, and it is necessary to include a more accurate value of the coulomb matrix elements.

These equations can be further simplified by changing to crystalline wave vector states k. Thus define

$$D_{\alpha k} = \frac{1}{\sqrt{N}} \sum_\ell e^{ik.R_\ell} D_{\alpha\ell}$$

$$M_{\alpha,\beta}(k) = \sum_\ell e^{ik.(R_\ell - R_m)} M_{\alpha\ell,\beta m}$$

where in the latter case we used the fact that $M_{\alpha\ell,\beta m}$ is actually a spatial function only of the difference $\underset{\sim}{R}_\ell - \underset{\sim}{R}_m$. The simplified equation is

$$(\omega_\alpha^2 - \omega^2) D_{\alpha k} = -2\omega_\alpha \sum_\beta M_{\alpha,\beta}(\underset{\sim}{k}) D_{\beta k} + \frac{e^2}{m} \frac{f_\alpha}{\mu_\alpha} \hat{\mu}_\alpha \cdot \underset{\sim}{E}_k \qquad (4.25)$$

This form of the equation is only appropriate for one molecule per unit cell of the crystal. For more than one molecule per unit cell, an additional index (superscript) should be added to account for which molecule. The interaction M has to be evaluated for each pair of molecules in the cell, and the sum over β on the right hand side must also include a sum over all of the molecules in the unit cell.

There are two aspects to (4.25) which make it difficult to do numerical calculations. First, to evaluate the matrix element M requires a knowledge of the molecular wave functions. These are dif-

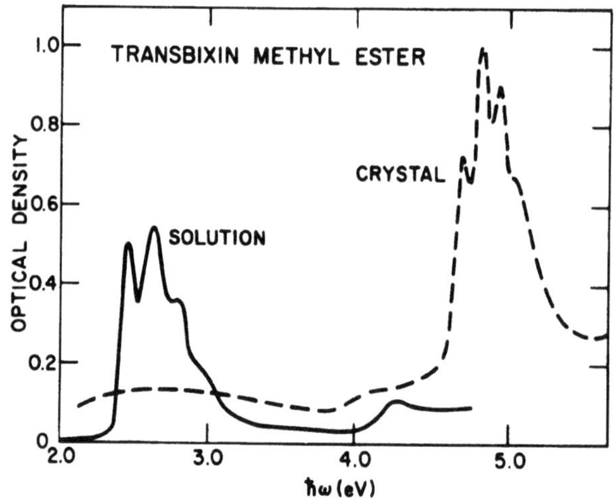

Fig. 8 The optical absorption of transbixin methyl ester molecules in solution (solid line) and in monolayer single crystals according to Bergeron and Slusarczuk.

ficult to obtain, particularly for excited states! But that difficulty is gradually being overcome since molecular wave functions are being computed with increasing accuracy.

The second aspect to these equations is that they are infinite in dimensionality. Each molecule has literally an infinite number of excited states which must be summed in the β summation. Of course, in a practical calculation, the sum over excited states might be limited to a few states which are low in energy or high in oscillator strength, although it is not obvious that this approximation is accurate. Nevertheless, the dimensionality is still high. For example, if one wanted to include five excited states in a crystal with two molecules per unit cell, the dimensionality of the equations is ten. Solving the similar problem in anthracene in the dipole approximation, the dimensionality was only four, and was down to two after application of symmetry.

The first application of these equations to molecular solids was by Silbey, Jortner, and Rice for anthracene.[37] They did actually calculate the matrix elements M for nearby molecules and used the dipole approximation for the interaction of distant molecules. This is an excellent procedure. Unfortunately, they made one mistake which marred their computations, and made their results less accurate. Instead of diagonalizing (4.25) they used the less accurate form

$$(\omega_\alpha-\omega)\, D_\alpha = - \sum_\beta M_{\alpha\beta}\, D_\beta + \frac{e^2 f\alpha}{2m\omega_\alpha \mu_\alpha}\ \hat{\mu}_\alpha \cdot \underset{\sim}{E}$$

Philpott[38] has recalculated the crystal spectra using the matrix elements of Silbey, Jortner, and Rice, and the correct formula. The n = 0 Davydov splitting agrees better with the experiment, although the $n \geqslant 1$ values are unchanged by the use of better matrix elements. One problem with these types of calculations is that molecular orbital theory does a very poor job of predicting oscillator strengths, and presumedly an equally poor job of calculating optical matrix elements.

In the next chapter we report our results of calculations on the monolayer crystals of trans bixin methylester.

V. MONOLAYER CRYSTALS OF TRANS BIXIN METHYLESTER

5.1 Experimental

Bergeron and Slusarczuk[39] have measured the optical properties of transbixin methylester molecules. The solid line in Fig.8 shows the absorption spectra of isolated molecules measured in solution spectra. The main band is centered at 2.6 eV. These molecules were made into monolayer single crystals, whose absorption spectrum is also shown by the dashed line. The main absorption band is now moved to 4.9eV. This is a very large crystal field splitting, and shows that we have a strongly coupled system.

The crystal structure was measured through electron diffraction studies. The molecules are long zig-zag chains of about twenty carbon atoms, with alternate double bonds.[40] They line up parallel in the crystal, and form a monolayer which is one chain length thick, and which extends endlessly in two dimensions. The two dimensional structure is a slightly distorted hexagonal close packed (hcp) with an average spacing of 5.3A. The structure is very much like one gets stacking long, rond, fence posts on a pile which extends endlessly in height and width.

The point dipole model is obviously inadequate for the present problem. The molecules are about 28 Å long. In the crystal, they are oriented parallel to each other and separated by about 5.3 Å. The electronic states involved in the transition are distributed over the length of the molecule - they are linear combinations of π orbitals, one on each carbon site. A point-dipole calculation was performed for the present system, but it predicted results which were nonsense. There seems to be no advantage on trying to patch up the dipole calculation by adding octipole or higher moments. Instead, it is just as easy and far more accurate to use the full electron-electron interactions, as we have done here.

We noted earlier that one important effect of higher states was to screen the electron-electron interactions. In our calculation, we exactly diagonalize the equation (4.24) for the π-electron system. This includes this screening in a rigorous way. The σ-electrons also act to screen the interactions.[41-42] For a bulk, three dimensional, solid, this may be satisfactorily included by adding a dielectric constant to the electron-electron interaction. It is not immediately obvious that this is a correct procedure in our problem, since we have a two dimensional monolayer of finite thickness. However, the detailed calculations in Section 5.4 show, somewhat surprisingly, that a single dielectric constant describes the screening adequately. Although this screening is not large in our loosely packed system - we estimate $\varepsilon^{\sigma} = 1.5$ - it affects the results significantly.

5.2 Molecular Properties

In this section we summarize the theory of molecular properties which are described in Ref. 43. This theory was for a chain of 2N carbon atoms which formed 1N conjugated double bonds. The Trans-Bixin Methylester molecule has 20 carbon atoms, but the methylester end groups cooperate to give a total of 11 conjugated double bonds. So one might choose N to be either 10 or 11. We have performed the crystal field calculation for both cases, and find that they predict very similar results. Here we will report the results for N = 11. This choice was selected, partly because the number of double bonds seems to be the important factor, and partly because the spectrum of the isolated molecule agrees very well with the N = 11 predictions of Ref. 43.

The 2N system (N = 11) of π-electrons has 2N orbital states, and each is spin degenerate. So N orbital states are occupied and N are empty. We adopt the notation of Ref. 43, and call the occupied states ψ_m^+ (m = 1,2,N) and the empty states ψ_m^-(m = 1,2,3,...N). They have energies $\varepsilon_m^\pm$. The ordering of these levels is indicated in Fig. 9. The highest occupied level is ε_N^+ and the lowest empty level is ε_N^-. If we choose the zero of the energy to be midway between ε_N^+ and ε_N^-, then

$$\varepsilon_m^\pm = \mp \varepsilon_m = \mp [\beta_1^2 + \beta_2^2 + 2\beta_1\beta_2 \cos\theta_m]^{1/2}$$

$$\beta_1 = -4.28165 \text{ eV}$$

$$\beta_2 = -3.23271 \text{ eV}$$

and θ_m is an angle which characterizes state m. They are the m roots of the equation

$$\beta_2 \sin(N\theta) + \beta_1 \sin(N+1)\theta = 0$$

arranged in the order $\pi > \theta_N > \theta_{N-1} > \ldots\theta_1 > 0$

The π electron on each carbon site s(s=1,2,3,...,2N) are described by a localized orbital $\Phi_s(\underset{\sim}{r}-\underset{\sim}{R}_s)$ centered about atomic position $\underset{\sim}{R}_s$. These are assumed to be strictly orthogonal and localized, so that for any function $g(\underset{\sim}{R})$

$$\int d^3R\, g(R)\, \Phi_s(\underset{\sim}{r}-\underset{\sim}{R}_s)\, \Phi_r(\underset{\sim}{r}-\underset{\sim}{R}_r) = \delta_{rs}\, g(\underset{\sim}{R}_s)$$

The wave functions of the molecules are LCAO's

$$\psi_m^\pm(r) = \sum_{s=1}^{2N} C_{m,s}^\pm \Phi_s \tag{5.1}$$

Because of the alternate double bonds, the even $C_{m,2s}^\pm$ and odd $C_{2s-1}^\pm$ coefficients differ

$$C_{m,2s}^\pm = K_m \sin(s\theta_m)$$

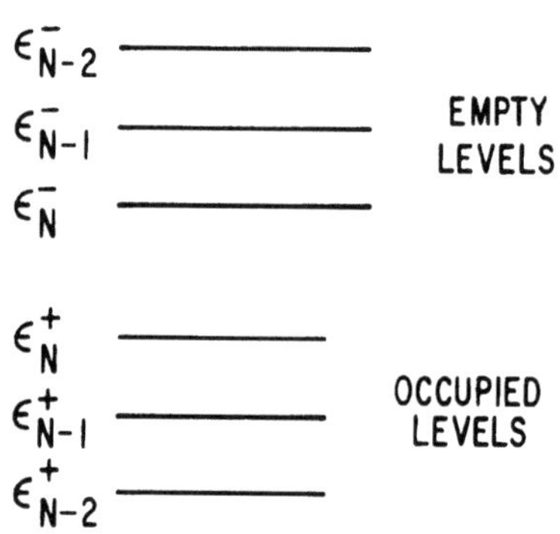

Fig. 9 The scheme of labelling electronic states in the molecule. The highest occupied level is ε^+_N, the lowest empty levels is ε^-_N.

$$C^{\pm}_{m,2s-1} = \mp (-)^m K_m \sin(N+1-s)\,\theta_m$$

$$K_m = [\,N - \cos(N+1)\theta_m \sin(N\theta_m)/\sin\theta_m\,]^{-1/2} \tag{5.2}$$

The coefficients $C^{\mu}_{m,s}$ for an orthonormal set

$$\sum_{s=1}^{2N} C^{\mu}_{m,s} C^{\nu}_{m,s} = \delta_{\mu,\nu}\,\delta_{m,n}$$

$$\sum_{m=1}^{N} \sum_{\nu=1}^{2} C^{\nu}_{m,r} C^{\nu}_{m,s} = \delta_{r,s}$$

This completely describes the wave functions and energies of the molecular states.

In optical transition, the electron goes from the state ψ^+_n to ψ^-_m. We designate this transition as (n,m). The allowed optical transitions must have n minus m an even integer or zero, the forbidden transitions have n - m an odd integer. There are N^2/z allowed optical transitions if N is even, and $(N^2 + 1)/2$ if N is odd - there are 61 for N=11. We designate the allowed transition by the subscript $\alpha(\alpha = 1,2,\ldots,61)$ where we order the transitions by increasing energy. The excitation energy is

$$\hbar\omega_\alpha = \varepsilon_n + \varepsilon_m$$

The dipole transition moment has a length

$$\underset{\sim}{p}_\alpha = \int d^3r\, \psi^+_n(\underset{\sim}{r})\, \underset{\sim}{r}\, \psi^-_m(\underset{\sim}{r}) = \sum_{s=1}^{2N} C^+_{n,s} \underset{\sim}{R}_s C^-_{m,s} \tag{5.3}$$

By using (5.2) this can be rearranged into the form

$$\underset{\sim}{p}_\alpha = 2K_n K_m \sum_{s=1}^{N} \underset{\sim}{R}_{2s} \sin(s\theta n)\sin(s\theta m)$$

And the oscillator strength is given by

$$f_\alpha = \frac{2}{3} m\, p_\alpha^{\,2}\, \omega_\alpha/\hbar = .08749\,(\hbar\omega_\alpha/\text{eV})(p_\alpha/\text{Å})^2 \tag{5.4}$$

In table V are listed the first 16 allowed transitions.

Table V

MOLECULE				CRYSTAL (NO SCREENING)										CRYSTAL (SCREENING)			
				$N_s = 1$		$N_s = 4$		$N_s = 9$		$N_s = 16$		$N_s = 61$		$N_s = 9$		$N_s = 16$	
α	(n,m)	$h\omega_\alpha$	f_α	$h\Omega_\alpha$	F_α	$h\Omega_\alpha$	F_α	$h\Omega_\alpha$	F_α	$h\Omega_\alpha$	F_α	$h\Omega_\alpha$	F_α	$h\Omega_\alpha$	F_α	$h\Omega_\alpha$	F_α
1	(11,11)	2.640	2.066	4.693	2.066	3.725	.477	3.716	.491	3.714	.491	3.713	.486	3.624	.776	3.622	.774
2	(10,10)	3.969	.715			5.166	-.459	4.985	.328	4.971	.338	4.964	.313	4.606	.785	4.599	.757
3	(9, 11)	4.124	.009			5.293	2.781	5.137	.972	4.126	.912	5.122	.852	4.859	.803	4.853	.766
4	(11, 9)	4.124	.009			4.124	.0	4.124	.0	4.124	.0	4.124	.0	4.124	.0	4.124	.0
5	(9 , 9)	5.608	.260					5.870	-.174	5.859	-.104	5.855	-.104	5.728	.074	5.726	.067
6	(8, 10)	5.635	.007					5.917	.625	5.906	.505	5.906	.425	5.803	.211	5.801	.191
7	(10, 8)	5.635	.007					5.635	.0	5.635	.0	5.635	.0	5.635	.0	5.635	.0
8	(7, 11)	5.786	.003					6.237	.838	6.222	.694	6.128	.637	6.065	.431	6.060	.390
9	(11, 7)	5.786	.003					5.786	.0	5.786	.0	5.786	.0	5.786	.0	5.786	.0
10	(6, 10)	7.204	.003							7.233	10^{-4}	7.233	10^{-4}			7.222	10^{-4}
11	(10, 6)	7.204	.003							7.204	.0	7.204	.0			7.204	.0
12	(5, 11)	7.209	.001							7.266	-.037	7.266	-.037			7.250	-.034
13	(11, 5)	7.209	.001							7.209	.0	7.209	.0			7.209	.0
14	(7, 9)	7.271	.003							7.339	-.176	7.338	-.171			7.316	-.142
15	(9, 7)	7.271	.003							7.271	.0	7.271	.0			7.271	.0
16	(8, 8)	7.301	.112							7.481	.599	7.477	.547			7.408	.439

The first four columns contain the value of α, the values of (n,m), the values of $\hbar\omega$ in eV, and f_α. The first transition (11,11), at 2.640 eV has the largest oscillator strength, and only the transitions with n = m have any appreciable oscillator strength. In fact, for N = 11 we have found that the transitions with n = m contain 98% of the π-to-π oscillator strength. One can also see that all transition for which n + m equals the same constant occur at nearly the same energy. For example, when n + m = 18 the transitions occur at 5.6 - 5.8 eV, while for n + m = 16 they occur at 7.2 - 7.3 eV.

The oscillator strengths f_α listed in Table V are slightly different from those given in Ref. 43. This is because we have made several approximations in our calculation. The first of these is the neglect of the transverse component of the transition moment $\underset{\sim}{p}_\alpha$ in (2.3). The carbon atoms lie in a zig-zag line, with a bond angle of 120°. So $\underset{\sim}{p}_\alpha$ has its major component along the long axis, the z direction, and has a much smaller component perpendicular to this. We neglect the transverse part. This is because in the crystal we do not know the orientation of the molecule, and hence the direction of the transverse component. Rather than guess a direction, it is easier to omit this small term. The other approximation is to ignore the slight (< 1%) change in bond length at the edges of the molecule. So we set all double bond lengths a = 1.366 Å, and all single bond lengths b = 1.444 Å. With these two approximations, the position of each π electron becomes just a vector in a Z direction.

$$\underset{\sim}{R}_{2s} = Z_{2s} = \hat{Z}\ \{(s-N/2)(a+b) - b/2\} \cos(30°) \qquad (5.5a)$$

$$\underset{\sim}{R}_{2s-1} = Z_{2s-1} = \hat{Z}\{(s-1-N/2)(a+b) + b/2\} \cos(30°) \qquad (5.5b)$$

This puts Z = 0 at the center of the molecule, so that $Z_{2N+1-2s} = -Z_{2s}$.

The oscillator strengths calculated from (5.4), and listed in Table V, are for a single electron. But because of spin degeneracy, each of the orbital levels in Fig. 9 are doubly degenerate, and occupied orbitals contain two electrons. So the total theoretical oscillator strength is twice that given in Table V. This predicts that the lowest molecular transition (11,11) or α = 1 has an oscillator strength of 4.1. The observed value is about 2. So the Huckel wavefunctions we are using predict too much oscillator strength. This is a common difficulty which is characteristic of Huckel wavefunctions.[44] We will therefore have to employ the common but deplorable procedure of reducing our matrix elements by a factor of $\sqrt{2}$ in order that they predict the correct oscillator strength. This will be done in the crystal field calculation also, as is common.

We have also calculated all of the f_α for π-to-π transition in the molecule. The sum of these values exhausts the f-sum rule.

5.3 Crystal Field Theory

For our calculations, we have assumed that the crystal structure is a two-dimensional hexagonal-close-packed (hcp) lattice. The measured lattice structure is a slight distortion of this, but this difference is small. The calculations were done assuming a separation between molecules of d = 5.3 Å. Let $\underset{\sim}{\rho}_j$ be a vector in the (x,y) plane which denotes the position of the jth molecule, and its wavefunction can be written $\psi_m^{\pm}(\underset{\sim}{r}-\underset{\sim}{\rho}_j)$.

In the crystal there are electron-electron interactions between states on different molecules. As an electron changes its state during an optical transition, the electron-electron interactions are altered. Thus we are interested in evaluating terms like

$$M_{k\ell,mn} = \sum_{\underset{\sim}{\rho}_j}' \int d^3r_1 d^3r_2\, \psi_k^-(\underset{\sim}{r}_1)\; \psi_\ell^+(\underset{\sim}{r}_1) \frac{e^2}{|\underset{\sim}{r}_1-\underset{\sim}{r}_2+\underset{\sim}{\rho}_j|} \times \psi_m^-(\underset{\sim}{r}_2)\; \psi_n^+(\underset{\sim}{r}_2) \tag{5.6}$$

$$N_{mn} = \sum_{\underset{\sim}{\rho}_j}' \int d^3r_1 d^3r_2\, \psi_m^-(\underset{\sim}{r}_1)\; \psi_n^+(\underset{\sim}{r}_1) \frac{e^2}{|\underset{\sim}{r}_1-\underset{\sim}{r}_2+\underset{\sim}{\rho}_j|} \rho(\underset{\sim}{r}_2) \tag{5.7}$$

$$\rho(\underset{\sim}{r}) = 2\sum_{m=1}^{N} \psi_m^-(\underset{\sim}{r})^2 - \rho_{ion}(\underset{\sim}{r})$$

The term (5.6) is the interaction between transitions in two different molecules - i.e., between two excitons - and is summed over all possible pairs of molecules. The second term (5.7) is the interaction between a transition on one molecule and the ground state of another molecule; ρ(r) is the charge density, which must average to zero for the neutral molecules. The term (5.7) is zero in our model, and it is small even in more realistic models. We shall neglect it, and only discuss (5.6).

The integral in (5.6) can be evaluated using the wavefunctions and other relationships described in the prior section.

$$M_{k\ell,mn} = \sum_{s,r=1}^{2N} C_{k,s}^- \; C_{\ell,s}^+ \; F(Z_s - Z_r)\; C_{m,r}^- \; C_{n,r}^+ \tag{5.8}$$

Where the unscreened coulomb interaction in the crystal is

$$F(Z) = e^2 \sum_{\ell j}{}' \; [\rho_j^2 + Z^2]^{-1/2} \qquad (5.9)$$

The prime on the summation means that the $\rho_j=0$ term is omitted. The screening by the σ-electrons will be introduced **in a later** section. Now let us be concerned only with the π-electron system. One can further simplify (5.8) by using (5.2).

$$M_{k\ell,mn} = [\,1 + (-)^{k+\ell+m+n}\,]\, K_k K_\ell K_m K_n \sum_{r,s=1}^{N} \sin(s\theta_k)\, \sin(s\theta_\ell)\, \sin(r\theta_m) \times$$

$$\sin(r\theta_n)\, [\, F(Z_{2r} - Z_{2s}) - (-)^{k+\ell} F(Z_{2r} + Z_{2s})] \qquad (5.10)$$

Note that for the allowed optical transition, n+m and k+ℓ are even integers, so the terms $(-1)^{k+\cdots}$ may be replaced by 1. The matrix element is also finite for interactions between two optically forbidden levels where n + m and k + ℓ are both odd. But M is zero when evaluated between an allowed and a forbidden state, so we only need to consider optically allowed states in our calculation.

Let us now examine the form of the effective interaction F(Z). The first thing one notices is that the sum over ρ_j in (5.9) does not converge at large ρ_j values. So, instead define F(Z) by

$$F(Z) = e^2 \sum_{\rho_j}{}' \; [\, (\rho_j^2 + Z^2)^{-1/2} - \rho_j^{-1}\,] \qquad (5.11)$$

This now converges. The second term in (5.11), which has been included somewhat arbitrarily, is a constant, independent of Z. Hence it contributes nothing to the matrix element M. This is obvious from (5.6) because of the orthogonality of the wavefunctions. Define a dimensionless interaction V by

$$F(Z) = \frac{e^2}{d} V(Z/d)$$

where d is the molecular separation. We have calculated V(Z/d) by Ewald's Method, and the details are provided in Section 5.6. The results are shown in Fig. 10 for a hcp lattice. For the present calculation we need $0 \leq Z/d \leq 5.1$, which is about the range of values shown in the Fig. 10.

For small values of $Z \ll d$, we see from Fig. 10 that $V(Z/d) \sim Z^2$ so $F(Z) \sim Z^2$. We note that the approximation

$$F(Z) = \text{CONSTANT} \times Z^2$$

corresponds to the dipole approximation. This approximation would certainly be very poor in the present calculation. Fig. 10 shows that, over the range of values we need for Z, F(Z) is more closely approximated by a linear function of Z than a quadratic one. In any case, we have used the true potential V(Z/d) and F(Z).

If we denote $\alpha = (m,n)$ and $\beta = (k,\ell)$, then the matrix element in (3.5) couples the α and β optical transitions. We must now employ these matrix elements in a calculation of the crystal energy levels and oscillator strengths. From (4.17), (4.23), and (4.25) we need to solve

$$(\omega_\alpha^2-\omega^2)\,D_\alpha + 2\,\omega_\alpha \sum_\beta M_{\alpha\beta}D_\beta = \frac{3e^2}{mn}\,\frac{f_\alpha}{\mu_\alpha}(\mu_\alpha \cdot \underset{\sim}{E}) \qquad (5.12a)$$

$$(\varepsilon-1)\,E_r = \frac{4\pi}{V_0}\sum_\alpha D_\alpha(\hat{\mu}_\alpha \cdot \hat{e}_0) \qquad (5.12b)$$

where $\varepsilon = n^2 = (kc/\omega)^2$. Equation (5.12) must be modified to account for the spin degeneracy of the orbital states. Denote the classical dipole for the spin up and down states as $D_{\alpha\uparrow}$ and $D_{\alpha\downarrow}$. If α and β refer only to the orbital part of the wavefunction, then

$$(\omega^2-\omega^2)D_{\alpha\uparrow} + 2\omega_\alpha \sum_\beta M_{\alpha\beta}(D_{\beta\uparrow}+D_{\beta\downarrow}) = 3\frac{e^2}{m}\,\frac{f_\alpha}{\mu_\alpha}\,(\hat{\mu}_\alpha \cdot \underset{\sim}{E})$$

$$(\omega_\alpha^2-\omega^2)D_{\alpha\downarrow} + 2\omega_\alpha \sum_\beta M_{\alpha\beta}\,(D_{\beta\uparrow}+D_{\beta\downarrow}) = 3\,\frac{e^2}{m}\,\frac{f_\alpha}{\mu_\alpha}(\mu_\alpha \cdot \underset{\sim}{E})$$

$$(\varepsilon-1)\,E_r = \frac{4\pi}{V_0}\sum_\alpha (D_{\alpha\uparrow}+D_{\alpha\downarrow})\,(\hat{\mu}_\alpha \cdot \hat{e}_0)$$

The electron-electron interactions only affect the combination $D_{\alpha\uparrow}+D_{\alpha\downarrow}$. So if we define

$$D_{\alpha\pm} = D_{\alpha\uparrow} \pm D_{\alpha\downarrow}$$

we get that

$$D_{\alpha-} = 0$$

and

$$(\omega_\alpha^2-\omega^2)D_{\alpha+} + 4\omega_\alpha \sum_\beta M_{\alpha\beta}\,D_{\beta+} = 6\,\frac{e^2}{m}\,\frac{f_\alpha}{\mu_\alpha}\,(\hat{\mu}_\alpha \cdot \underset{\sim}{E}) \qquad (5.12a')$$

$$(\varepsilon-1)\,E_r = \frac{4\pi}{V_0}\sum_\alpha D_{\alpha+}(\hat{\mu}_\alpha \cdot \hat{e}_0) \qquad (5.12b')$$

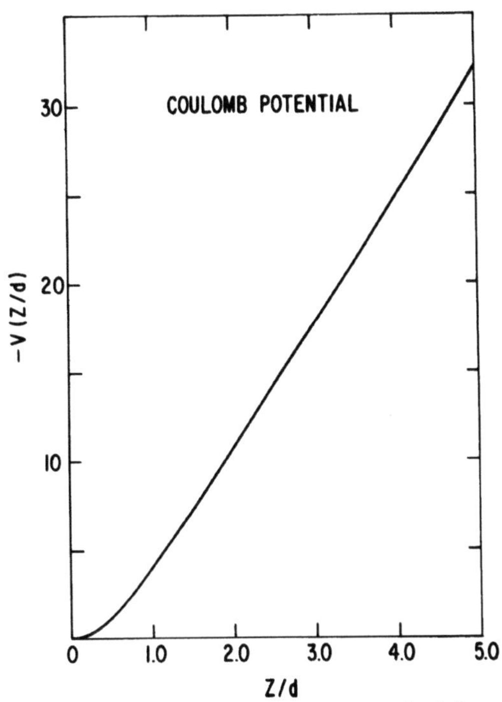

Fig. 10. The effective coulomb potential V(z/d), whose calculation is described in the Appendix. Note that V is negative, so here we have plotted -V.

The oscillator $D_{\alpha+}$ acts as if it has twice the oscillator strength and twice the interaction strength as any single spin oscillator.

These equations may be simplified in the present problem by recognizing that $\hat{\mu}_\alpha$ points only in the z direction, and then performing a spherical average over all possible directions of $\hat{e}_0$. Furthermore, by defining

$$\omega_p^2 = 4\pi e^2/mV_0$$

$$m_{\alpha\beta} = 4(\omega_\alpha\omega_\beta)^{1/2} M_{\alpha\beta}$$

$$C_\alpha = D_\alpha [e^2E(\omega_\alpha)^{1/2}/mc]^{-1} \tag{5.13}$$

the equations (5.12) simplify to

$$(\omega_\alpha^2-\omega^2) C_\alpha + \sum_\beta m_{\alpha\beta} C_\beta = 2f_\alpha(\omega_\alpha)^{-1/2} \tag{5.14a}$$

$$\varepsilon(\omega) = (kc/\omega)^2 = 1 + \omega_p^2 \sum_\alpha C_\alpha(\omega_\alpha)^{1/2} \tag{5.14b}$$

The matrix $m_{\alpha\beta}$ is now symmetric

$$m_{\alpha\beta} = m_{\beta\alpha}$$

The optical properties are obtained by solving (5.14a) for the coefficients C_α, and inserting these into the dielectric function (5.15b). In the molecular case, where $m_{\alpha\beta} = 0$, we get the standard result

$$\varepsilon(\omega) = 1 + 2\omega_p^2 \sum_\alpha \frac{f_\alpha}{\omega_\alpha^2 - \omega^2}$$

In the crystal case, we anticipate that there should be a new set of eigenfrequencies Ω_α and oscillator strengths F_α, so that we should be able to write

$$\varepsilon(\omega) = 1 + \omega_p^2 \sum_\alpha \frac{F_\alpha}{\Omega_\alpha^2 - \omega^2} \tag{5.15}$$

The crystal spectrum is obtained by solving (5.14). The left hand side of (5.14a) defines a matrix

$$A_{\alpha\beta} = m_{\alpha\beta} + \omega_\alpha^2 \delta_{\alpha,\beta} \tag{5.16}$$

The eigenvalues of this matrix provide the values of Ω_α^2. Furthermore, the eigenvectors define a unitary matrix $S_{\alpha\beta}$ such that

$$\sum_\beta S_{\alpha\beta} (S^{-1})_{\beta\gamma} = \sum_\beta (S^{-1})_{\alpha\beta} S_{\beta\gamma} = \delta_{\alpha\gamma}$$

$$\sum_{\beta\gamma} S_{\alpha\beta} A_{\beta\gamma} (S^{-1})_{\gamma\lambda} = \Omega_\alpha^2 \delta_{\alpha\lambda}$$

So if we start from (5.14a), which we write as

$$\sum_\beta (\Omega_\lambda^2 - \omega^2) S_{\lambda\beta} C_\beta = 2 \sum_\alpha S_{\lambda\alpha} f_\alpha (\omega_\alpha)^{-1/2}$$

whence

$$C_\beta = 2 \sum_{\lambda\beta} (S^{-1})_{\beta\lambda} S_{\lambda\alpha} f_\alpha (\omega_\alpha)^{-1/2} (\Omega_\lambda^2 - \omega^2)^{-1}$$

so from (5.14b) we get

$$\varepsilon(\omega) = 1 + 2\omega_p^2 \sum_{\alpha\beta\lambda} \frac{(S^{-1})_{\beta\lambda} S_{\lambda\alpha}}{\Omega_\lambda^2 - \omega^2} f_\alpha (\omega_\beta/\omega_\alpha)^{1/2}$$

This has the form of (5.15), with

$$F_\lambda = 2 \sum_{\alpha\beta} (S^{-1})_{\beta\lambda} S_{\lambda\alpha} f_\alpha (\omega_\beta/\omega_\alpha)^{1/2} \tag{5.17}$$

One can also see immediately that, with the definition (5.15), the f-sum rule is automatically obeyed.

$$\sum_\lambda F_\lambda = 2 \sum_\alpha f_\alpha$$

The numerical calculations proceeded in the following order. Using the energies ε_m listed in Ref.43, we calculated the angles θ_m. These were used to calculate the transition moments from (5.3) and oscillator strength from (5.4). These later results are listed in Table V. Then we calculated the matrix components M from (5.10) and (5.13). $m_{\alpha\beta}$ and f_α were divided by 2.0, in order to give agreement with the molecular spectra. Next we diagonalized the matrix components $A_{\alpha\beta}$ in (5.16), thereby obtaining it eigenenergies and eigenvectors. The eigenvectors were used to calculate the oscillator strength F_α from (5.17).

Some of the components from the matrix $m_{\alpha\beta}$ are listed in Table

VI. These components have the dimensions of $(\text{energy})^2$, and the units are $(\text{eV})^2$. The largest component is m_{11}, which is the diagonal element for the lowest optical transition. One might expect this to be the largest interaction component because this level has by far the largest oscillator strength. On the other hand, Table VI shows that the size of the oscillator strength has little relation to the size of the interaction $m_{\alpha\beta}$. For example, levels $\alpha = 3$ and 4 have an oscillator strength < 0.01, yet their interaction terms $m_{3\alpha}$ and $m_{4\alpha}$ are not significantly smaller than the others. This is quite different than the results of a point dipole model, in which $m_{\alpha\beta}$ is proportional to the transition moment, and shows the necessity of calculating the true electron-electron interactions. This molecular crystal is clearly a strongly coupled system.

The molecular levels $\alpha = 3$ and 4 are degenerate, $\hbar\omega_3 = \hbar\omega_4$. Their interaction components are also equal,

$$m_{3\alpha} = m_{4\alpha}$$

This can be seen in Table VI, and proven as a general rule from (5.10) and (5.13). This is true for any degenerate pair (n,m) and (m,n), such as $\alpha = 6,7$ and 8,9 etc.

The large size of the m_{11} interaction term qualitatively explains the large frequency shifts observed between the molecular and crystal spectra. For example, if we were to ignore the interactions of the $\alpha = 1$ level with other levels, then we predict

$$\hbar\Omega_1 = [(\hbar\omega_1)^2 + m_{11}]^{1/2} = 4.69 \text{ eV}$$

This is a sizable shift $\hbar\omega_1 = 2.64$ eV. In fact, it predicts that this level is now at higher energies than the levels $\alpha = 2,3,$ and 4. But because all of these levels are strongly coupled, this prediction is premature, and we must really include a larger number of levels in the diagonalization procedure. So we have adopted the procedure of successively increasing the number of states N_s in the diagonalization scheme. The steps $N_s = 1,4,9,16,$ and 61 are shown in Table I. Each step except the last includes a new set of levels which occur at nearly the same energy. We were hopeful that, as the size of N_s was increased, the calculated properties of the lowest-energy crystal states would converge to reasonable values. Table V shows that this indeed happens. The results for $N_s = 16$ do not look significantly different than those for $N_s = 61$. This occurs because the higher states have negligible oscillator strength, and only small interactions with the states at lower energy.

Table V shows that some crystal transitions have zero oscillator strength while others have negative oscillator strength. Those

Table VI : $M_{\alpha\beta}$ $(ev)^2$

β/α	1	2	3	4
1	15.06	5.63	-2.13	-2.13
2	5.63	5.91	2.93	2.93
3	-2.13	2.93	3.94	3.94
4	-2.13	2.93	3.94	3.94

The assumption that the π- and σ-electrons have separate electronic properties implies that the polarizability from each component is additive

$$\alpha_{\mu\nu} = \alpha^{\pi}_{\mu\nu} + \alpha^{\sigma}_{\mu\nu} \qquad (5.19)$$

Now the calculation of the optical transitions of the molecular solid is simply a crystal field calculation. We are just trying to self-consistently determine how the electric fields on one molecule affect the energy levels on other molecules. The calculations of the previous section, which only included the π-electrons, therefore only included the contributions of $\alpha^{\pi}_{\mu\nu}$. Our treatment of the π-electrons did not assume that the π-electrons could be represented by a polarizability as in (5.18). Such an assumption is equivalent to the dipole approximation, as has been discussed in the Introduction.

Our treatment of the σ-electrons is going to be based on the premise that their influence can be described by a local polarizability $\alpha^{\sigma}_{\mu\nu}$. This polarizability applies on an atom-to-atom basis, so of E^r_μ is the electric field at atom r, then the σ-electron polarization at atom r' is

$$P^{r'}_{\mu} = \delta_{rr'}\, \alpha^{\sigma}_{\mu\nu}\, E^{r}_{\nu}$$

So we are assuming that polarizing our σ-electron has no influence on other σ-electrons in the same molecule, and only the original σ-electron is polarized.

From a quantum mechanical viewpoint, the polarizability arises from virtual transitions involving the σ-electrons. These are with zero oscillator strength are, in fact, forbidden. These additional forbidden transitions arise because of the degeneracies in the molecular transitions. For example, take the degenerate states $\alpha = 3$ and 4, and recombine their amplitude factors C_3 and C_4 in (5.14a)

$$C_3^{\pm} = C_3 \pm C_4$$

Then one finds in (5.14a) that the C_3^- state has zero oscillator strength and zero interactions with other levels. This cancellation occurs because 3 and 4 have the same matrix elements with other states, so C_3^- becomes a forbidden transition. The state C_3^+ has an oscillator strength $2f_3$, and still has the interactions with other levels. Thus there is one forbidden transition for each degenerate pair (n,m) and (m,n) of allowed molecular transitions.

The occurrence of some negative oscillator strengths in the calculated results is a greater concern. Since the actual oscillator strengths must be positive, this shows that our results are, in some way, in error. We believe that this error is from omission of the screening caused by the σ-electrons. This subject is discussed in the next section.

In Fig. 11, we show the behavior of the three lowest levels as the molecular spacing d of the hcp lattice is changed from 5.0 Å to 10.0 Å. The three energies $\hbar\Omega_\alpha$ change smoothly, with no crossings. The three oscillator strengths change a great deal. At d = 10 Å they have their molecular ordering $F_1 > F_2 > F_3$, although F_3 is still much larger than its molecular value of .018.

5.4. Screening by σ-Electrons

Only a limited set of states was included in the calculations of the previous section. The discussion was restricted to transition where the initial and final state were both bound π-electrons. There are two other types of electron states for these molecules : the bound states and the positive energy states in the continuum. These states are often ignored because they are only reached by optical transition at high energy. However, in spite of the fact that these transitions are not excited directly in the experiment, they still play an important role.

The influence of these other transitions is best understood from a classical point of view. If one puts a d.c. electric field on an atom, the atom responds by polarizing with a polarizability $\alpha_{\mu\nu}$

$$P_\mu = \alpha_{\mu\nu} E_\nu \qquad (5.18)$$

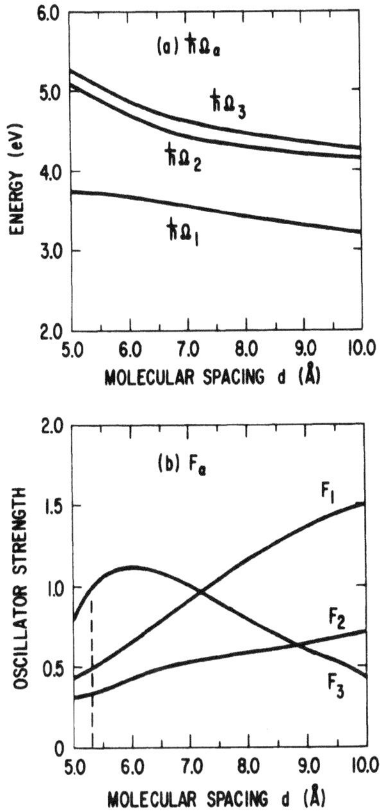

Fig. 11. The theoretical changes in the properties of the first three allowed transitions as the molecular spacing of the hcp lattice is increased.

(a) The absorption energies $\hbar\Omega_\alpha$ change in a smooth way with no crossings.

(b) The oscillator strengths change dramatically, with the very weak molecular level f_3 becoming the strongest at the crystal spacing of 5.3 Å (dashed line). A diagonalization matrix of $N_s = 9$ was used in calculating these curves. This calculation omits σ-electron screening.

in the high energy transitions mentioned at the beginning of this section. So the influence of these transitions is included by using the polarizatility $\alpha^{\sigma}_{\mu\nu}$. This does not include the transitions involving π-electrons going to continuum states. However, we assume that these effects can also be lumped into this core polarizability function $\alpha^{\sigma}_{\mu\nu}$.

The simplest approach is to divide the interaction F(z) by ε^{σ},[14,15] so that (5.12a) now reads

$$(\omega_\alpha^2-\omega^2)D_{\alpha+} + \frac{4\omega_\alpha}{\varepsilon^\sigma}\sum_\beta m_{\alpha\beta}D_{\beta+} = 6\frac{e^2 f_\alpha}{m\mu_\alpha}\hat{\mu}_\alpha\cdot\underset{\sim}{E} \qquad (5.20)$$

The dielectric constant ε^{σ} is not very large because it only includes the contribution of the σ-electrons. ε^{σ} is actually frequency dependent, but that should be unimportant for the low energies at which we calculate.

Equation (5.20) is certainly valid in a bulk, three dimensional solid.[41,42] In the present problem, with a thin two dimensional monolayer, one might initially question using a bulk dielectric procedure. So some calculations were done to test this approach. The exact screening function is given by a series in α whose leading terms are[26]

$$V(\underset{\sim}{R}_1,\underset{\sim}{R}_2)_{\text{Total}} = e^2\left[\frac{1}{(\underset{\sim}{R}_1-\underset{\sim}{R}_2)} - \sum_j{}' \frac{(\underset{\sim}{R}_1-\underset{\sim}{R}_j)_\mu\,\alpha_{\mu\nu}(\underset{\sim}{R}_2-\underset{\sim}{R}_j)_\nu}{(\underset{\sim}{R}_1-\underset{\sim}{R}_j)^3\,(\underset{\sim}{R}_2-\underset{\sim}{R}_j)^3} + \ldots\right] \qquad (5.21)$$

Let us simplify the present problem by calling $\alpha_{\mu\nu}$ an isotropic atomic polarizability $\alpha_0\delta_{\mu\nu}$. In this case (5.21) becomes

$$V(Z_r,Z_{r'})_{\text{Total}} = \frac{e^2}{d}\left[V(z_r-z_{r'}) - \tilde{V}(z_r,z_{r'}) + \ldots\right] \qquad (5.22)$$

$$\tilde{V}(z_r,z_{r'}) = \frac{\alpha_0}{d^3}\sum_s U(z_r-z_s)\,U(z_{r'}-z_s) \qquad (5.23)$$

$$z_r = Z_r/d$$

where V(z) is the dimensionless interaction discussed in Section 5.6 and

$$U(z) = \frac{d}{dz}V(z)$$

Note that V(z) is an even function of z, and U(z) is an odd function, and at large z

$$\lim_{|z| \to \omega} U(z) = - \frac{2\pi}{A} \operatorname{sgn}(z)$$

The function $\tilde{V}(z_r, z_{r'})$ is plotted in Fig.12. One sees that to a good approximation it can be described as a function of $(z_r - z_{r'})$ so that edge effects are small. In this plot we have used $\alpha_0 = 1.0 \times 10^{-24}$ cm^3 whcih we estimate from graphite.[45] Another important feature of $\tilde{V}(z)$ is that it has nearly the same shape as V(z). So it is a good approximation to write

$$\tilde{V}(z) = \gamma V(z) + \delta \tag{5.24}$$

where γ and δ are constants. The constant δ is unimportant since we are evaluating our potential between orthogonal wavefunctions. The condition (5.24) needs to be satisfied in order to express the σ-electron contribution as a screening function. For if (5.24) is true we may write

$$V(z)_{\text{Total}} = \frac{e^2}{d\varepsilon^\sigma} V(z) + \text{Constant} \tag{5.25}$$

where

$$(\varepsilon^\sigma)^{-1} = 1 - \gamma + O(\alpha^2)$$

We will use (5.25) and estimate ε^σ by

$$\varepsilon^\sigma = 1 + 4\pi\, \bar{\alpha}/(1 - 4\pi\, \bar{\alpha}/3) \tag{5.26}$$

$$\bar{\alpha} = 4\pi_0/A(a + b)\sqrt{3}$$

The effects of this screening is evident in the crystal energy levels and oscillator strengths listed in the last columns for $N_s = 9$ and 16. One systematic effect of screening is to lower the energy of all of the interacting levels from their counterpart values in the unscreened case. This simply occurs because screening lowers all matrix elements, and reduces all the shifts in energy levels. On the other hand, the oscillator strengths are completely jumbled by screening, and bear little relationship to the unscreened values. Fig.13 shows how the first six levels and oscillator strengths change with lattice separation d. In computing this curve, note that ε^σ also changes with d because of the change in area A. The claculated oscillator strengths in this figure are significantly different than in Fig.11, which shows that even a little screening is important. At d = 5.3 we get $\varepsilon^\sigma = 1.5$.

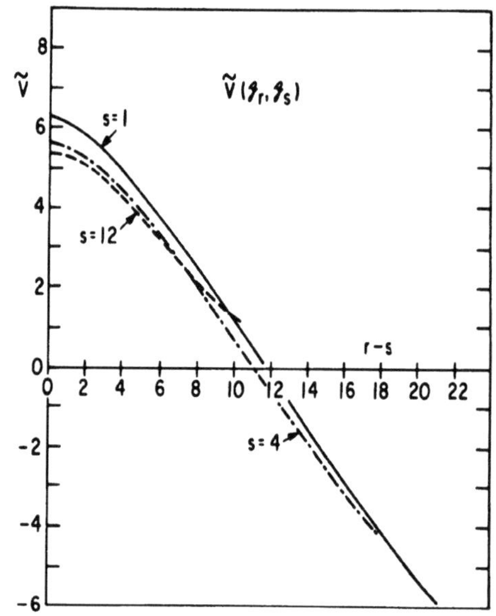

Fig. 12. The effective potential $\tilde{V}$ which arises because of linear polarization, as given in Eq. (5.23). The potential depends upon the two sites r and s, but is apparently nearly a function of $Z_r - Z_s$. The function is symmetric $\tilde{V}(Z_r, Z_s) = \tilde{V}(Z_s, Z_r)$.

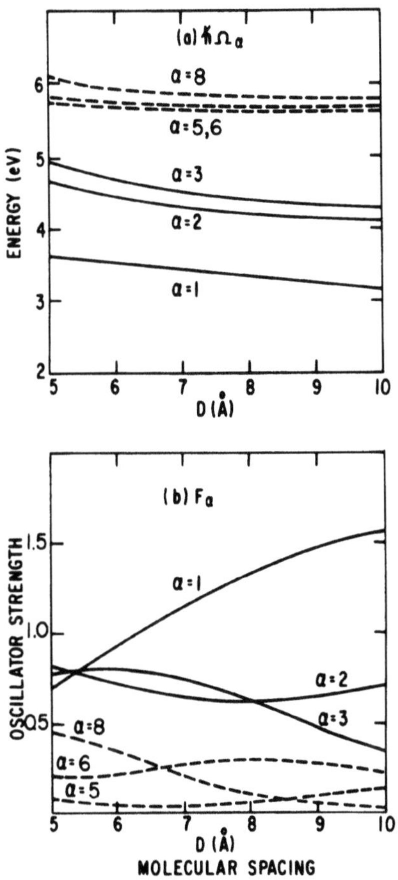

Fig. 13. The theoretical changes in the properties of the first six allowed transitions as the spacing of the hcp lattice is changed. The screening of the σ-electrons has been included. This screening also depends upon d. Otherwise this is the same as Fig. 10, with which it should be compared.

5.5 Comparison to Experiment

We will now compare our theoretical results to the experiments of Bergeron and Slusarczak. In making this comparison, we bear in mind that Huckel wavefunctions give good energy levels but poor oscillator strengths. So we expect that we are able to predict absorption levels accurately, but oscillator strengths less ably.

The experiments are limited to the frequency range below 5.5 eV. This includes only the three lowest states in the spectrum, and we shall restrict our discussion to these.

The experimental spectrum of the isolated molecule shows a large absorption band at 2.60 eV which is the transition $\alpha = 1$. A weaker band at 4.26 eV we associate with the $\alpha = 2$ transition. The predicted level $\alpha = 3, 4$ is too weak to show up experimentally.

For the monolayer crystal, the calculations predict three main features :

(1) all absorption energies increase as the electron-electron interactions are increased. As in Fig.13, the spacing between levels changes little, and all levels appear to increase together;
(2) the oscillator strength F_1 of the first level decreases steadily as the interaction is increased. This oscillator strength gets distributed among the other levels.
(3) the oscillator strength F_3, unobservable on the isolated molecule, increases dramatically in the solid, and can even become the strongest level.

These results suggest the following interpretation of the observed crystal spectra. The low energy shoulder on the absorption spectra, which begins about 3.8 eV and is centered about 4.2 eV, we interpret as the level $\alpha = 1$. This absorption band occurs at a higher frequency, and with less oscillator strength, then we predict. The main band, centered at 4.9 eV and with a 0.5 eV width, we interpret as the sum of the two bands $\alpha = 2$ and $\alpha = 3$. The energy difference between $\hbar\Omega_2$ and $\hbar\Omega_3$ is much less than the width of this band. The observed absorption band has a series of three or four maximum separated by $\Delta = 0.12 \pm .01$ eV. This latter structure may be caused by the main stretching mode vibration of the molecule.

So we conclude that the main theoretical predictions do explain the data : all absorption energies increase in energy, but the oscillator strength of the molecular band f_1 gets transferred in the crystal to the band F_3. The calculations show these trends quite clearly. The experiments show that the transfer of oscillator strength from f_1 to F_3 takes place to a larger extent than predicted by the calculations. This no doubt reflects the inadequacies of the Huckel wavefunctions used in the calculations. We only regret that better wavefunctions are not available.

5.6. Coulomb Potential

The effective coulomb potential has been evaluated by Ewald's Method with the adaptation to two dimensions suggested by Nijboer and de Wette. We wish to evaluate

$$F(Z) = e^2 \sum_{\underset{\sim}{\rho}_j}{}' \left\{ (\rho_j^2 + Z^2)^{-1/2} - \rho_j^{-1} \right\} \tag{5.27}$$

where the sum $\underset{\sim}{\rho}_j$ runs over the two dimensional crystal lattice in the (x,y) plane. The term $\underset{\sim}{\rho}_j = 0$ is not included in the summation. Following Ewald's Method, we insert an integral and break up the regions of integration

$$F(Z) = \frac{2e^2}{\pi^{1/2}} \sum_{\underset{\sim}{\rho}_j}{}' \int_0^\infty dt\, e^{-t^2\rho_j^2} [e^{-t^2Z^2} - 1] \tag{5.28}$$

$$F(Z) = \frac{2e^2}{\pi^{1/2}} \left\{ \sum_{\underset{\sim}{\rho}_j} \int_0^S dt\, e^{-t^2\rho_j^2} [e^{-t^2Z^2} - 1] - \int_0^S dt (e^{-t^2Z^2} - 1) + \right.$$

$$\left. \sum_{\underset{\sim}{\rho}_j}{}' \int_S^\infty dt\, e^{-t^2\rho_j^2} [e^{-t^2Z^2} - 1] \right\}$$

The optimum value of $S \sim (\text{lattice constant})^{-1}$ will be specified below. The last term now converges rapidly as one takes the summation over $\underset{\sim}{\rho}_j$. In the first term we have added in the $\underset{\sim}{\rho}_j = 0$ term, and subtracted it out in the second term. The first term in (5.27) may be written as

$$\lim_{\rho\to 0} B(Z,\underset{\sim}{\rho}) = \lim_{\underset{\sim}{\rho}\to 0} \frac{2e^2}{\pi^{1/2}} \sum_{\underset{\sim}{\rho}_j} \int_0^s dt\; e^{-t^2(\underset{\sim}{\rho}_j-\underset{\sim}{\rho})^2} [e^{-t^2Z^2}-1] \tag{5.29}$$

The vector $\underset{\sim}{\rho}$ is in the (x,y) plane. Since $B(\underset{\sim}{Z},\underset{\sim}{\rho})$ is a periodic function of $\underset{\sim}{\rho}$ in the two-dimensional lattice, it can be expanded in reciprocal lattice vectors

$$B(Z,\underset{\sim}{\rho}) = \frac{2e^2}{\pi^{1/2}} \int_0^s dt\; (e^{-t^2Z^2}-1) \sum_{\underset{\sim}{G}} e^{i\underset{\sim}{\rho}\cdot\underset{\sim}{G}} V_{\underset{\sim}{G}}$$

We get

$$V_{\underset{\sim}{G}} = \frac{\pi}{At^2}\; e^{-G^2/4t^2}$$

where $A = \sqrt{3}\, d^2/2$ for hcp is the area of the unit cell of the lattice. The t integral may be performed to provide the result for B(Z,0).

$$B(Z,0) = \frac{e^2\pi}{A} \Bigg\{ \frac{2}{s\pi^{1/2}} (1-e^{-Z^2s^2}) - 2Z\, \mathrm{erf}(sZ) + \sum_{\underset{\sim}{G}}{}' \Big[e^{GZ} \mathrm{erfc}\Big(\frac{G}{2s} + Zs\Big) + e^{-GZ} \mathrm{erfc}\Big(\frac{G}{2s} - Zs\Big) - 2\mathrm{erfc}(G/2s)\Big]\; G^{-1} \Bigg\} \tag{5.30}$$

The first term in (5.30) is the $\underset{\sim}{G} = 0$ contribution. Thus our final result is

$$F(Z) = e^2 \Bigg\{ \frac{2s}{\pi^{1/2}} - \Big(\frac{1}{Z} + \frac{2\pi Z}{A}\Big)\mathrm{erf}(sZ) + \frac{2\pi^{1/2}}{A} (1 - e^{-Z^2s^2}) + \frac{\pi}{A} \sum_{\underset{\sim}{G}}{}' [e^{GZ} \mathrm{erfc}(G/2s+Zs) + e^{-GZ}\mathrm{erfc}(G/2s-Zs) - Z\mathrm{erfc}(G/2s)\,]\,G^{-1} + \sum_{\underset{\sim}{\rho}_j}{}' \,[\,(\rho_j^2 + Z^2)^{-1/2} \mathrm{erfc}(s(\rho_j^2+Z^2)^{1/2}) - \rho_j^{-1}\, \mathrm{erfc}(s\rho_j)] \Bigg\} \tag{5.31}$$

This expression for F(Z) converges rapidly for all values of Z.

This is an important feature, because Z ranges in value from zero up to five times the lattice spacing.

The present calculation used the two dimensional hcp lattice. If we denote the molecular separation by d, then the optimum value for s is

$$s = (G_0/2d)^{1/2} = \frac{1}{d}(2\pi/\sqrt{3})^{1/2}$$

Where G_0 is the lowest reciprocal lattice vector $G_0 = 4\pi/d\sqrt{3}$. With this choice of s, both the lattice and reciprocal lattice sums in (5.31) converge so rapidly that one need only take the first term.

REFERENCES

1. D. Frohlich and H. Mahr, Phys. Rev. Letters 16, 895 (1966).
2. A.S. Davydov, Theory of Molecular Excitations, (Plenum, New York, 1971).
3. M.R. Philpott, Advances in Chemical Physics, edited by I. Prigogine and S.A. Rice (John Wiley & Sons, Inc., New York, 1973) pp. 228-341.
4. A.S. Davydov, Quantum Mechanics, (Addison Wesley, Reading, 1965) p. 318.
5. D.P. Craig and P.C. Hobbins, J. Chem. Soc. 1955, 539, 2302, 2309.
6. E.E. Koch and A. Otto, Phys. Stat. Sol. (b) 51, 69 (1972).
7. K.S. Sundararajan, A. Krist, 93, 238 (1936).
8. I. Nakada, J. Phys. Soc. Japan 17, 113 (1962).
9. D. Fox and S. Yatsiv, Phys. Rev. 108, 938 (1957).
10. M. Born and K. Huang, Dynamical Theory of Crystal Lattices, (Oxford University Press, 1954).
11. U. Fano, Phys. Rev. 103, 1202 (1956) ; 118, 451 (1960).
12. J.J. Hopfield, Phys. Rev. 112, 1555 (1958).
13. G.D. Mahan, J. Chem. Phys. 41, 2930 (1964).

14. M.R. Philpott, J. Chem. Phys. 50, 5117 (1969).

15. G.D. Mahan and G. Overmair, Phys. Rev. 183, 834 (1969).

16. M.R. Philpott , J. Chem. Phys. 58, 588 (1973).

17. L.D. Landau and E.M. Lifshitz, Electrodynamics of Continuous Media, (Pergamon, New York, 1960) p.321.

18. P.P. Ewald, Ann. Physik 64, 253 (1921).

19. H. Kornfeld, Z. Physik 22, (1924).

20. M, Born and M. Bradburn, Proc. Comb. Phil. Soc. 39, 104 (1942).

21. G.D. Mahan, Ph. D. Thesis, University of California, Berkeley, 1964 (unpublished).

22. A.A. Lucas, Physica 35, 353 (1967) ; 39, 5 (1968).

23. L.B. Clark and M.R. Philpott , J. Chem. Phys. 53, 3790 (1970).

24. L.C. Kravitz, J.D. Kingsley, and E.L. Elkins, J. Chem. Phys. 49, 4600 (1068) ; J.D. Kingsley, G.D. Mahan, and L.C. Kravitz, J. Chem. Phys. 49, 4611 (1968).

25. M.R. Philpott, J. Chem. Phys. 52, 1984 (1970).

26. G.D. Mahan, Phys. Rev. 153, 983 (1967).

27. G.D. Mahan and R.M. Mazo, Phys. Rev. 175, 1191 (1968).

28. G. Nienhuis and J.M. Deutch, J. Chem. Phys. 56, 235, 1819, 5511 (1972).

29. G.D. Mahan, J. Chem. Phys. 43, 1569 (1965).

30. G.D. Mahan, Phys. Rev. B5, 739 (1972).

31. D.P. Craig and P.D. Dacre, Proc. Roy. Soc. A310, 297 (1969); D.P. Craig and L.A. Dissado, Proc. Roy. Soc. A310, 313 (1969); M. Tanaka and J. Tanaka, Mol. Phys. 16, 1 (1969).

32. J.S. Avery, Proc. Phys. Soc. (London) 89, 6771 (1966).

33. W.K.H. Panofsky and M. Phillips, Classical Electricity and Magnetism, (Addison-Wesley, Reading, Mass. 1956) p. 219.

34. B.R.A. Nijboer and F.E. deWette, Physics 23, 309 (1957); 24, 442 (1958).

35. F.E. deWette and G.E. Schacher, Phys. Rev. 137, A78 (1965).

36. V.M. Agranovich, Zhur Eksp. i Theor. Fiz. 37, 430 (1959) (Soviet Physics - JETP 10, 307 (1960) ; Fiz Tverd. Tela. 3, 811 (1961) (Soviet Physics - Solid State 3, 592 (1961)).

37. R. Silbey, J. Jortner, and S.A. Rice, J. Chem. Phys. 42, 1515 (1965).

38. M.R. Philpott, Chem. Phys. Letters 17, 57 (1972).

39. J.A. Bergeron and G.M. Slusarczuk, General Electric Research and Development Center (unpublished).

40. J.N. Murrell, The Theory of the Electronic Spectra of Organic Molecules, (John Wiley & Sons, New York, 1963).

41. H. Herzenberg, D. Sherrington, and M. Suveges, Proc. Phys. Soc. (London) 84, 465 (1964).

42. R.A. Harris, J. Chem. Phys. 48, 3600 (1968).

43. H. Suzuki and S. Mizuhashi, J. Phys. Soc. Japan 19, 724 (1964).

44. R.G. Parr, Quantum Teory of Molecular Electronic Structure, (W.A. Benjamin, New York, 1963).

45. E.A. Taft and H.R. Phillip, Phys. Rev. 138, A197 (1965).

ELECTRONIC AND STRUCTURAL CONSIDERATIONS IN THE QUEST FOR EXCITONIC POLYMERIC SUPERCONDUCTORS

W. A. Little

Physics Department

Stanford University

Stanford, CA 94305

BACKGROUND : SUPERCONDUCTIVITY AND THE EXCITON MECHANISM [1]

The electrical resistivity of a metal decreases as one lowers the temperature following a behavior illustrated in Fig. 1.

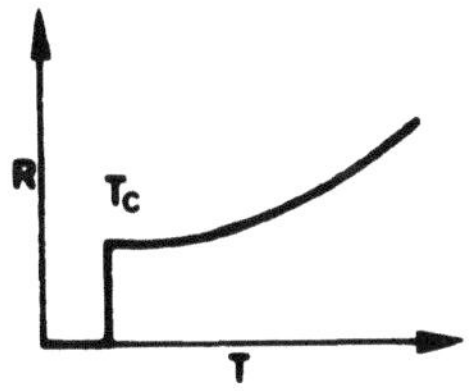

Fig. 1

In many metals and alloys an abrupt transition is observed at a temperature T_c of the order of a few degrees Kelvin below which the resistivity drops to zero. In this superconducting state the metal can carry a current with no dissipation of energy. This remarkable property has been used in recent years for powerful electromagnets which require refrigeration to maintain them in the superconducting state but which do not dissipate energy in operation. Fields up to 150 kG have been produced and magnets many meters in diameter have been built. It is anticipated that in the next decade massive use will be made of superconductivity in electrical generators, ship propulsion and high speed ground transportation.

In addition to having zero resisstance a superconductor also excludes magnetic field from its interior, where B = 0. Actually two kinds of superconductors are known - a type I superconductor which excludes fields up to a critical value H_c and above this reverts to the normal state, and a type II superconductor which excludes fields up to H_{c1}, then the field penetrates but the material remains superconducting up to a higher critical field H_{c2}. The type II superconductors are important in technological applications for while H_c is typically less than 1000 gauss, H_{c2} can exceed 200kG.

The principle objection to the use of superconductivity in large scale application is the need for sophisticated refrigeration. Most superconductors have transition temperatures of order 5 K. The highest known transition temperature (for the alloy Nb_3Ge) is only 23.8 K - just above the boiling point of liquid H_2. From systematic studies of many thousands of alloys and from the development of powerful theoretical methods the reasons for the limitations in the achievable transition temperatures have emerged. The theory has also revealed a possible alternative route to obtaining dramatically increased transition temperatures and this would involve certain types of polymeric conductors. We need to understand some of this theory to determine the structural and electronic requirements for these polymers.

The essential difference between a superconductor and a normal conductor is that the elctrons in the former are not free to move independently but are bound as pairs and the pairs are constrained to be in the same quantum state. This was pointed out in a classic paper by Bardeen, Cooper and Schoeffer - the BCS theory - in 1957. In a conventional superconductor the pairing interaction results from the interaction between the electrons and the lattice. The strength of this phonon induced electron-electron attraction depends upon the mass of the ions of the lattice. One finds that

$$T_c \cong \hbar\omega_D \exp \frac{-1}{N(0)\lambda} \tag{1}$$

where ω_D is the Debye frequency of the lattice, N(0) is the density of states at the Fermi surface, and λ the effective electron-electron attraction. From (1) one finds $T_c \approx M^{-1/2}$. Metals having different isotopic compositions exhibit a dependence of T_c upon isotopic mass of this form.

From (1) we can understand why some limit exists on the attainable value of T_c. Suppose we model the ionic lattice as a system of uncoupled ions of mass M on springs, of spring constant, k. Then $\omega_D = \sqrt{k/M}$. If k is very large ω_D will be large but λ will be small. The reason is that the attraction results from the distortion of the lattice by the electrons. If the lattice is too stiff the distortion will be small and the net attraction small. The small value of λ in the exponential results in a very small value for T_c. If k is allowed to become weaker, λ increases and T_c rises, however, as the lattice becomes very soft the decrease in the pre-exponential term, $\hbar\omega_D$ begins to dominate and T_c reaches a maximum and then falls. Thus we expect that T_c will be limited for each type of lattice. A further limitation results from the, so called, renormalization of the phonon frequencies. This results from the influence of the electron-phonon coupling on the spring constant itself. If λ gets too large the lattice becomes unstable and the resultant new structure will have a lower λ and smaller T_c.

The exciton mechanism[2] is based on using an electron-exciton mechanism instead of the electron phonon mechanism. Consider a model illustrated in Fig. 2.

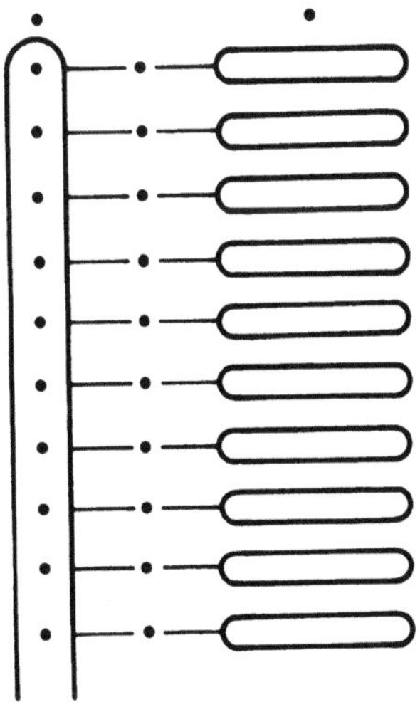

Fig. 2

An electron moving along the polymeric spine will polarize the adjacent side chain dyes. A second electron some distance behind the first will be attracted to this induced charge. This results is an effective electron-electron attraction. Applying the conventional BCS theory yields a $T_c \approx \hbar\omega_{ex} \exp \frac{-1}{N(0)\lambda}$ but where, ω_{ex} is the characteristic frequency of the excitations of the side chains. Provided λ can be made adequately large then T_c will be scaled up by a factor of

order $\left(\omega_{ex}/\omega_D \approx M_{ion}/m_e \right)^{1/2}$, which is of order 300.

This then is the basis for the hope of achievong superconductivity even at room temperature.

This has been known for over a decade yet no "excitionic superconductor" is presently known to exist. The problem is that the criteria which the structures must satisfy are extremely demanding. Nevertheless, as we shall show later certain types of structures appear to be capable of satisfying the theoretical criteria and if these can be prepared one could test the basic problem.

MODEL ONE-DIMENTIONAL SYSTEMS

The first system[2] which was suggested as a possible excito nic superconductor was that shown in Fig. 3. This presented the chemists with a formidable synthetic problem. While it might be possible at this time to prepare such materials by solid state polymerization there are reasons to believe that it still might not satisfy the theoretical criteria for superconductivity.

In general, the highly unsaturated nature of both the proposed carbon chain "spine" and the link to the side-chain of the model put harsh constraints on the conditions of any planned synthesis. Alternative models have been considered in which bonds of a variety of different strength would couple together the constituent pieces. These include [3,4] various TCNQ charge transfer complexes which form stacked arrays or chains such that the conductivity along the stack is large. These are not true polymers but rather are bound together largely by crystal forces. Such chains have been considered as possible conductive spines for our model. However, the diffuse distribution of the conduction electrons over the TCNQ molecule makes it difficult to conceive of a suitable polarizable counter ion which could couple strongly enough to the electrons in such a spine. Tetrathiofulvalene (TTF) perhaps comes closest to satisfying this.

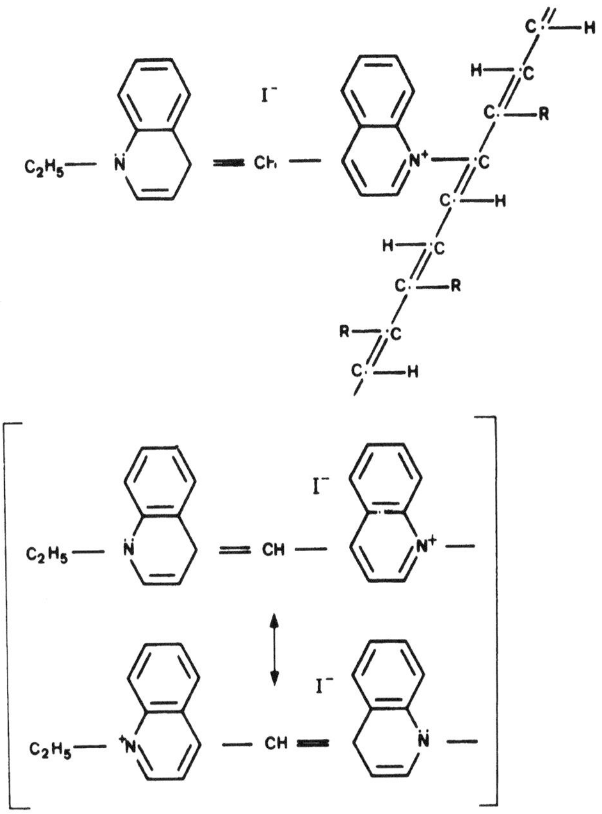

Fig. 3

A second possibility is the class of stacked square planar complexes of platinum studied by Krogmann[5]. In these the d_z^2 orbital of the platinum atoms in the complex overlap with those of the adjacent platinum atoms in the stack as illustrated in Fig. 4. These materials are excellent conductors in the direction of the chain and have a very much reduced conductivity in the transverse direction. Here the conduction electrons are well localized on the platinum atoms and it would appear that a polarizable substituent could couple strongly to these electrons. The synthesis of such a model system appears to be significantly easier than the original, fully covalently coupled structure.

All the above materials are, so called, one dimensional conductors, i.e. the electronic transport properties are determined by the one dimensional chain of atoms or molecules. Such systems have certain properties not found in systems which are more isotropic. Some of these properties can interfere with any incipient tendency towards superconductivity.

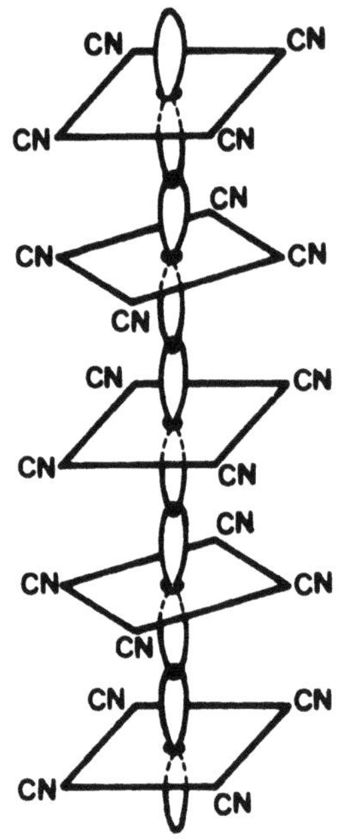

$K_2Pt(CN)_4Br_{0.3}.nH_2O$

Fig. 4

The first problem is the Peierls instability. Consider a linear chain of equally spaced atoms. One may describe the electrons in the bands by a simple Hückle calculation. In a large system the levels will form a near continuum. These states will be filled up to the Fermi energy. Let us suppose the band is half filled. Now if one introduces a periodic distortion in the lattice with a period double that of the atomic spacing then the band will be split into two bands and the gap will fall at the Fermi surface. This would leave an insulating ground state.

From Fig. 5 we see that such a distortion lowers the energy of the occupied states and raises the energy of the unoccupied states. Then at T = 0°K the system will spontaneously distort to lower its energy and thus become insulating. At finite temperatures the thermal occupation of the states in the upper part of the band inhibits this Peierls distortion and above a characteristic Peierls' temperature no distortion occurs. The above model neglects any pairing interaction resulting from the exciton coupling. This interaction mixes states from the lower with the upper part of the band. This is somewhat similar to the effects of temperature and likewise suppresses the Peierls distortion. Thus the superconducting and Peierls tendencies compete with one another.

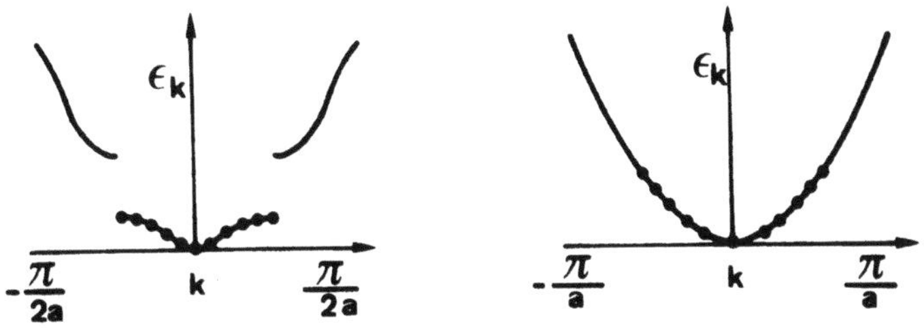

Fig. 5

A second problem is related to the effects of disorder. In a simple Hückle calculation of a chain of identical atoms with the same Coulomb, α and overlap integrals β one obtains the familiar band-like states. However, if α or β have a small random component added to them such as might result from random disorder in the chain, then upon diagonalizing the Hückle matrix one finds all the states are now localized. This is unique to one dimension and has a market effect upon the transport properties. Conduction must occur through some form of hopping. It is believed that this effect would have a less serious effect upon superconductivity for tunnelling between localized regions could still be expected to yield bulk superconductivity.

A third problem is the effect of fluctuations in these linear chains. No true phase transition can occur in a truly one dimensional systems so one finds always a gradual transition from one phase to another. This is expected to smear both the Peierls transition and any superconducting transition over a wide temperature range. This smearing, being a consequence of the limited dimensionality of the system, is very sensitive to interchain coupling. With sufficient interchain coupling the filamentary systems can develop enough three dimensional character to suppress the Peierls transition and yet leave many of the properties of a filamentary superconductor unchanged.

In the final lecture we will discuss an attempt to calculate from first principles the transition temperature of a proposed excitonic superconductor band on the platinum square planar complexes.

A PROPOSED EXCITONIC SUPERCONDUCTOR

In this lecture we will discuss the elements of a calculation of the transition temperature of a particular system which we have proposed as a model excitonic superconductor. We will see from it what are the key electronic and structural requirements of the model.

The elementary discussion given earlier showed that the type of system one would need would be one with a higly conductive spine to which would be attached a series of highly polarizable side-chains. Consider the molecule illustrated in Fig. 6.

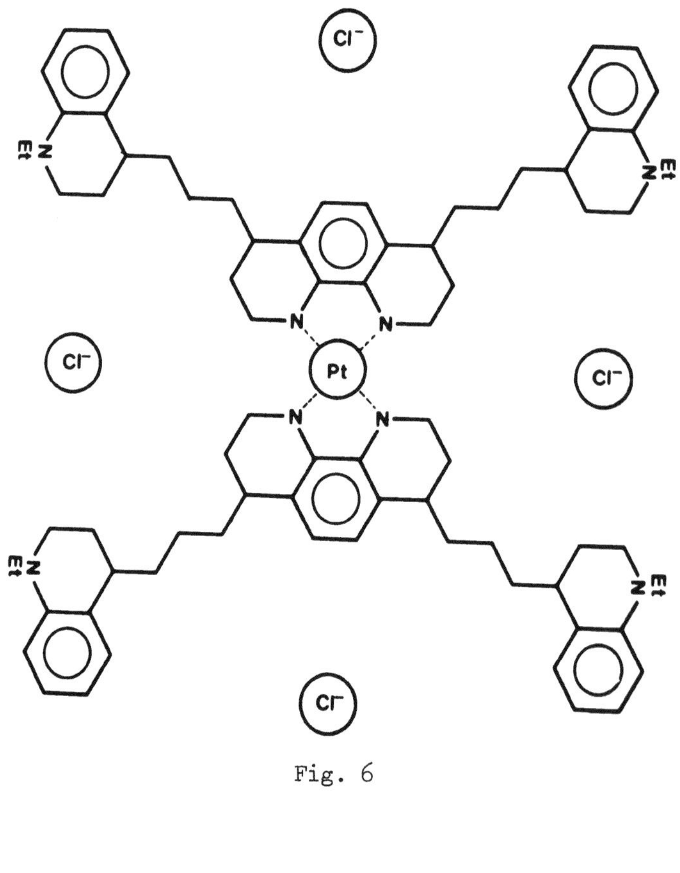

Fig. 6

The phenanthroline ligands can complex directly with platinum. In the 4 and 7 positions the phenanthroline is coupled to form a cyanine like dye. Prior to complexation each half of the phenanthroline acts like a cyanine-dye base. Upon complexation the base is converted to the dye. In such a dye the transition charge moves between the quinoline nitrogen far from the platinum to the nitrogen adjacent to the platinum.

Earlier studies had shown that certain phenanthroline complexes of platinum form a stacked array with each platinum atom stacked above the other. This geometric arrangement would provide the conductive "spine". The ligand structure of Fig. 6 recently has been synthesized but the complex has yet to be prepared. Given the complex it is still not known whether the presence of the large substituents would prevent the formation of the stacked chain. However, we shall assume here that the chain can be formed by some means and will then calculate its properties from the geometry of the proposed structure. We can learn much from so doing.

The main calculations which have to be made are the following. First, we must calculate the polarizability of the dye-like substituents. Next we must take into account the interaction between each of the polarizable substituents on the square planer complex and between those on different complexes in the stack. This interaction causes each excited state of the individual dye to form an exciton band and the calculation gives us the dispersion relationship of these bands.

Next we must calculate the band structure of the platinum chain spine ignoring both long range Coulomb effects and the exciton interaction. As we do not know the exact metal-metal spacing, we calculate this band structure for a range of such spacings.

With this band structure we can determine the screening of the Coulomb field from the movement of electrons in the spine and from those in adjacent spines. We must also determine the effective dielectric constant of the neighborhood of the spine and then find the net Coulomb repulsion of the electrons.

The attractive component which results from the interaction between the electrons and the excitons is obtained from the exciton wave functions and exciton band structure. The sum of this term and the Coulomb term gives the net interaction.

This then is all the information which is needed to determine T_c and other properties of the superconducting state. The transition temperature is given by an integral equation whose kernel contains the net interaction and the density of the states from the electron band structure. It may be solved by iteration.

A calculation of the above form has recently been completed. We start with a LCAO calculation of the polarizability of the phenanthroline structure using a modified Pariser-Parr-Pople technique using certain elements of the random phase approximation[6]. A configuration interaction (C.I.) calculation is made to determine the low lying excited states and their transition densities. Next we use these results to calculate the exciton band and exciton wave fuctions using a second C.I. calculation.

In a separate calculation we determine the structure of the 5-d and 6-p bands of the platinum chain. This is done using a multiple scattering procedure[7]. Reasonable agreement is obtained with experimental measurements on $K_2Pt(CN)_4\ Br_{0.3}\ 3\ H_2O$.

Previously[8] we had solved the problems of the screening of the Coulomb field in these filamentary type materials and we thus find the net Coulomb interaction between electrons in the spine and between electrons in the spine and on the side chains. The latter allows us to calculate the attractive component resulting from the interaction with the exciton system.

Using all the above information we can set up and solve the integral equations for the superconducting gap parameter and T_c. The result of this calculation was that, assuming no Peierls distortion, superconducting transition temperatures of order 3000K should result[9]. While one should not take the numerical value too seriously it does indicate that strong coupling effects can be expected in such structures. The most striking features of the calculations though were the effects of small changes in the ligand system. If the polarizable ligands were moved 2Å further from the Pt atom, T_c was driven to zero. Also if a ligand structure with only two polarizable side chains (instead of four) was used, again superconductivity was not found. Variations of the metal-metal distance and the Fermi energy had a weak effect upon T_c.

We conclude then that high temperature superconductivity may be a possibility in these systems but that a high density, close packing of the polarizable substituents is absolutely essential for its attainment.

REFERENCES

1. Several excellent books describe the superconducting state, C. Kittel, "Intro to Solid State Physics", J. Willey,New York, 1966 ; P. G. de Gennes, "Superconductivity of Metals and Alloys" Benjamin, 1966 ; J. R. Schrieffer, "Theory of Superconductivity" Benjamin, 1964.
2. W. A. Little, Phys. Rev. A134, 1416 (1964).
3. I. F. Shchegolev, Phys. Stat. Sol (A) 12, 9 (1972).
4. E. B. Yagubskii and M. L. Khidekel, Russian Chem. Reviews 41, 1011 (1972).
5. K. Krogmann, Angew. Chem. Inst. Ed. 8, 35 (1969).
6. H. Gutfreund and W. A. Little, J. Chem. Phys. 50, 4468 (1969).
7. K.H Johnson, Advances in Quantum Chemistry 7, ed. P. O. Löwdin.
8. D.Davis, Phys. Rev. B7, 129 (1973).
9. D.Davis, Ph. D. thesis, Stanford University (1974).

EFFECT OF SYMMETRY ON THE BAND STRUCTURE OF POLYMERS

W.L. McCubbin

School of Mathematics and Physics

University of East Anglia

Norwich U.K.

I. INTRODUCTION REMARKS

It may appear surprising that any lectures on the elementary symmetry properties of polymers would be considered necessary in an Advanced Study Institute of this kind. However, if one examines no more than the literature on polyethylene band structures, one finds papers by Imamura[1], Morokuma[2] and Clementi[3] in which some fundamental consequences of the molecular symmetry were not appreciated. It is therefore quite appropriate that we use the regular polyethylene chain to illustrate in some detail how the line group representations are generated and how they are used.

In fact, the dependence of the band structure on molecular (crystal) symmetry is an elementary problem only if one restricts oneself to an exceedingly idealised model of the real system. Almost all work on the electronic states of polymers of quantum mechanical nature has been done on ideal chains[4], i.e. on single molecules which are assumed quasi-infinite, extended, rigid and chemically homogeneous. Where tentative extension to 3D polymer crystals has been attempted, the crystal has been assumed to be an ordered stack of such ideal chains.

The real polymer chain is frequently not in an extended conformation; the most famous example is polyethylene in which the chains are folded with a fold length of approximately 120 Å. There exists, therefore, a super-lattice (wholly ignored by the naive theory), the effects of which on the band structure need to be examined.

The real chain is, also, never rigid. However, for the majority of systems, in which the thermal vibrations involve only small oscillations about the lattice points of a given structure, the effect on the symmetry of electronic wave functions can be ignored[5]. (The other possibility is that transitions occur between near-minimal-energy configurations of different symmetry: for that one would require a group theory for non-rigid molecules, which is still in the process of development[6].

The assumption that the chain is quasi-infinite may appear relatively harmless; after all, the ends are so far apart that a carrier moving in the "bulk" of the chain will be unaffected by the existence of the ends. However, the transport process in macroscopic specimens necessarily involves carrier transfert from one chain to another, just as in molecular solids of low molecular weight. Thus bulk conductivity will not be found to possess a simple relation to the carrier effective mass along the chain.

The actual situation is even worse than this, at least in polymers with a negative electron affinity (i.e. lowest unoccupied states above the zero of energy). In such cases an unpaired "conduction" electron (i.e. one injected into the system or arising from the dissociation of a hole-electron pair formed in some excitation process) will show no tendency at all to wander along a particular chain. Consequently, the translational symmetry of the chain will be largely irrelevant to the motion of the electron. (It will be the crystal symmetry, if any, that dominates the motion). This brings us to the realisation that if one follows the by now conventional approach of starting with the single chain, one must discuss the relevant symmetry properties in a self-consistent manner. That is to say, suppose one assumes the charge carriers (conduction electrons) to exist in states that belong to one or other irreducible representation of the symmetry group of the Hamitonian of the single chain and then uses this to set up a band calculation. If the resulting dispersion relations E(K) show that the virtual states of this Hamiltonian are not bound states, then the assumption regarding the states occupied by the conduction electron is open to criticism.

Finally, the existence of inhomogeneities in high concentration (either chemical impurities, which we shall not discuss, or topological disorder, which we shall) provides the coup-de-grâce to the applicability of the ideal band structure. This will be the topic to which we address ourselves in the second part of this Course. Perhaps enough has been said to indicate that the symmetry problem in polymers is not a trivial one-and also that a better title for the series might have been "The Effects of Symmetry Loss on the Band Structure of Polymers".

II. LINE GROUPS OF IDEAL CHAINS

The Line Group is the space group of a system possessing translational symmetry in only one dimension. Sometimes the term Band Group is to be found in the literature[7], but this refers to one or other of the seven possible frieze patterns, i.e. to systems that are essentially planar. We shall not wish to limit our discussion to planar systems, however, and the term line group is therefore preferable.

Consider an infinite periodic chain with repeat units containing a fixed number of atoms. It will in general contain two categories of atom : those comprising the main chain, i.e. the skeleton of the molecule in question, and the atoms (or groups) bonded to main chain atoms. Pragmatically, we may define a main chain atom as an atom the removal of which leads to the formation of two chain ends. (In other than covalent systems, the identity of main-chain atoms is not always obvious). Whenever the geometry of the side-atoms displays the same symmetry as the main chain (e.g. the H atoms in polyethylene) it is convenient to ignore the side-atoms in working out the properties of the relevant line group. In order to fix our ideas, then, we shall take the planar zig-zag chain as a model for what we have to say about line groups. The notation and method will, however, apply to more general chains.

2.1 Enumeration of symmetry elements

(a) Infinite line group G_∞

It will be recalled that for <u>finite</u> systems all mirror planes and axes of rotation must intersect at a point and that this point is to be taken as the origin for all symmetry operations. This limitation does not apply to infinite systems and one must, in the first instance, list all symmetry elements.

The symmetry elements of the infinite planar zig-zag chain are as follows. The operations generated by these elements will, in general, be combined with a translation nt and, where indicated, with a sub-primitive translation τ .

1. Two mirror planes per cell, $\sigma(x)$, $\sigma'(x)$.
2. Two two-fold rotation axes per cell, $C_2(y)$, $C'_2(y)$.
3. Two txo-fold rotation axes per cell, $C_2(z)$, $C'_2(z)$ (τ)
4. Two inversion centres per cell, i_n, i'_n.
5. One glide plane $\sigma(y)$. (τ)
6. One mirror plane $\sigma(z)$.

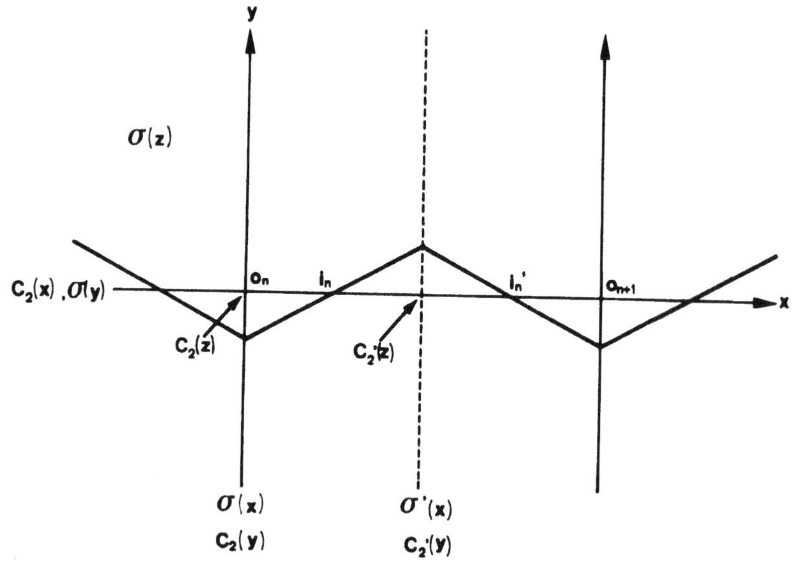

Figure 1. Symmetry elements of the infinite zig-zag chain.

7. One two-fold screw axis $C_2(x)$ (τ)

The dashed elements are equivalent to the undashed because in each case another element can be found, the operation of which carries the dashed element into the undashed. The remaining elements generate an infinite number of operations. This means that the line group is infinite and the number of irreducible representations is infinite. In order to avail oneself of the theory of finite groups, one introduces cyclic boundary conditions.

(b) Finite line group G

Impose cyclic boundary conditions over N cells. Then operations of the finite space group that differ only by a translation of Nt are taken to be the same operation and the number of symmetry operations, etc. becomes finite.

As usual, this condition is not strictly meaningful in so far as the periodicity it imposes is not present in the physical system. While it is found that the numerical effects* of this (say on total energies) are usually negligible, it should be realised that the class structure and character table of the finite group are slightly different for odd and even N. This difference really disappears in the limit of infinite N.

Let us suppose that all operations are performed with respect to the origin O_n in figure 1. (It is clear that the identity of the mixed operations, i.e. these involving both point and translation operations, depends somewhat on the choice of origin). The symmetry operations of the polyethylene chain are shown in Table 1.

2.2. Translation subgroup T and factor group F.

We require a methodical way of separating the primitive translational part of a general symmetry operation from the part involving point operations and sub-primitive translations. This is conveniently done by means of decomposition of G into left or right cosets.

Let b be an element of H, a subgroup of G. Let a_1 be an element of G that does not belong to H, then the collection $\{a_1 b\}$, for all $b \in H$, is a left coset of H. It can be written symbolically as a_1H. By choosing a second element a_2 of G that does not belong to H or a_1H, one generates the left coset a_2H. Continuing in this way, one exhausts all elements of the finite group G, and the decom-

*These arise from the need to interpolate the band structure between the N values of k.

Table 1

Symmetry operations of the polyethylene chain

Symmetry elements	Typical operations	Basic operations
Screw axis $C_2(x)$	$\{C_2(x)\|\tau\}$	
Axis $C_2(y)$	$\{C_2(y)\|0\}$	
Axis $C_2(z)$	$\{C_2(z)\|\tau\}$	
Inversion centre	$\{i\|\tau\}$	
Plane $\sigma(x)$	$\{\sigma(x)\|0\}$	$\{\sigma(x)\|0\}$
Glide plane $\sigma(y)$	$\{\sigma(y)\|\tau\}$	$\{\sigma(y)\|\tau\}$
Plane $\sigma(z)$	$\{\sigma(z)\|0\}$	$\{\sigma(z)\|0\}$
Primitive translations T^n	$\{\varepsilon\|nt\}$	$\{\varepsilon\|nt\}$

position into cosets is expressed thus :

$$G = H + a_1H + a_2H + \ldots\ldots a_mH$$

In the present case, the subgroup in question is the translation subgroup $T = \{T^1, T^2 \ \ldots, T^N = E\}$, where the pure translation T^n may be given the Koster symbol $\{\varepsilon| \ nt\}$. It is readily shown that T is an invariant abelian subgroup of G. One consequence of the invariance property is that left and right cosets are identical, and we shall proceed in terms of left cosets only. Since the elements $a_1 \ldots a_m$ will be pure point operations, possibly accompanied by a sub-primitive translation , we may write

$$G = \{\alpha_1 \ |\tau_1\}T + \{\alpha_2 \ |\tau_2\}T + \ldots\{\alpha_f| \ \tau_f\}T$$

in which $\{\alpha_1| \ \tau_1\}$ is the identify $\{\varepsilon|0\}$ and (for the zig-zag chain) τ_i is either 0 or τ. As ordinarily defined, the collection of cosets $\{H, a_1H, \ldots a_mH\}$ constitutes a subgroup of G, called the factor group F, in which H is the unit element. If the unit element of F is denoted f_1 then

$$F = \{\{ \alpha_1| \ \tau_1\}f_1 \ , \{ \alpha_2| \ \tau_2\}f_1 \ , \{\alpha_3| \ \tau_3\}f_1\ldots\ldots$$

$$\{ \alpha_f| \ \tau_f\}f_1\}.$$

For practical purposes, this is equivalent to the collection of coset representatives themselves, i.e. we determine the properties of the factor group as if it were the collection $\{\{\alpha_1| \ \tau_1\}$, $\{\alpha_2| \ \tau_2\} \ldots\ldots \{\alpha_f| \ \tau_f\}\}$. In the case of the zig-zag chain, we make the identifications $\{\alpha_1| \ \tau_1\}= \{\varepsilon|0\}$,$\{ \alpha_2| \ \tau_2\} = \{C_2(x)|\tau\}$, ... $\{\alpha_8| \ \tau_8\} = \{\sigma(z)| \ 0\}$. Hence H is of order 8 and is isomorphous with ($\approx$) the point group D_{2h}.

The irreducible representation of T and F are easily found. As usual, one makes use of the pair of equations

$$\sum_i d_i^2 = \text{order of group} \qquad (1)$$

$$\text{no. of irr.resp} = \text{no. of classes} \qquad (2)$$

where d_i = dimension of ith representation.
Since $T^1$is abelian, each element belongs to a separate class, and since there are thus N classes, there must be N irreducible representations. Consequently, (1) is satisfied only if $d_1 = d_2 = \ldots$ $d_N = 1$. Thus all N irreducible representations of T are one- dimensional.

The character table is found by noting that any eigenfunction of the Hamiltonian is also an eigenfunction of the translation operator T^n, i.e. $T^n \psi = \lambda_n \psi$. Since $\lambda_N = 1$, by the cyclic boundary condition, and given the algebraic properties of T^n, λ_n must be one of the Nth roots of unity. In physical notation,

$$\lambda_n = e^{iknt}, \text{ where } k = \frac{2\pi m}{Nt}$$

$$\text{and} \quad m = 0, \pm 1, \pm 2, \ldots, \frac{N}{2} \text{ (N even)}$$

$$= 0, \pm 1, \ldots\ldots \pm\frac{N-1}{2} \text{ (N odd)}.$$

Thus, once k is given, the representation of T^n is simply e^{iknt}.

In writing down the character table for $F \approx D_{2h}$, and later on for G, it is unnecessary to retain all the symmatry operations. From the multiplication table (Table 2), it may be confirmed that those operations not given a Koster symbol in the final column of Table 1 can be written

$$\{C_2(x)|\tau\} = \{\sigma(y)\ |0\}\ \{\sigma(z)\ |0\}$$

$$\{C_2(y)|0\} = \{\sigma(x)\ |0\}\ \{\sigma(x)\ |0\}$$

$$\{C_2(z)|0\} = \{\sigma(x)\ |0\}\ \{\sigma(y)\ |\tau\}$$

$$\{i|\ \tau\ \} = \{\sigma(x)\ |0\}\ \{\sigma(y)\ |\tau\}\{\sigma(z)|0\}.$$

We are therefore at liberty to consider the reflections as basic operations and to carry only these into the following discussion. (There is nothing unique, of course, about the reflections; other operations could equally well have been chosen).

2.3. Class structure of G.

By use of the multiplication tabel, all elements, of form $\{\alpha|nt+m\tau\}$, can be grouped in conjugate classes. Since this cannot be demonstrated in general terms, independent of the value of N, it is instructive to look at a particular example.

Example 1. A zig-zag chain with cyclic boundary conditions over N=6 unit cells. It can be shown that there are 18 classes. They are :

1. $\{\varepsilon|0\}$
2. $\{\varepsilon|1t\}$, $\{\varepsilon|5t\}$
3. $\{\varepsilon|2t\}$, $\{\varepsilon|4t\}$
4. $\{\varepsilon|3t\}$
5. $\{\sigma(z)|0\}$
6. $\{\sigma(z)|1t\}$, $\{\sigma(z)|5t\}$
7. $\{\sigma(z)|2t\}$, $\{\sigma(z)|4t\}$

8. $\{\sigma(z)|3t\}$
9. $\{C_2(z)|\tau\}$, $\{C_2(x)|11\tau\}$
10. $\{C_2(x)|3\tau\}$, $\{C_2(x)|9\tau\}$
11. $\{C_2(x)|5\tau\}$, $\{C_2(x)|7\tau\}$
12. $\{\sigma(y)|\tau\}$, $\{\sigma(y)|11\tau\}$
13. $\{\sigma(y)|3\tau\}$, $\{\sigma(y)|9\tau\}$
14. $\{\sigma(y)|5\tau\}$, $\{\sigma(y)|7\tau\}$
15. $\{\sigma(x)|0\}$, $\{\sigma(x)|1t\}$, $\{\sigma(x)|2t\}$, $\{\sigma(x)|3t\}$, $\{\sigma(x)|4t\}$,$\{\sigma(x)|5t\}$
16. $\{i|\tau\}$, $\{i|3\tau\}$, $\{i|5\tau\}$, $\{i|7\tau\}$, $\{i|9\tau\}$, $\{i|11\tau\}$
17. $\{C_2(y)|\tau\}$, $\{C_2(y)|3\tau\}$, $\{C_2(y)|5\tau\}$, $\{C_2(y)|7\tau\}$, $\{C_2(y)|9\tau\}$, $\{C_2(y)|11\tau\}$
18. $\{C_2(z)|0\}$, $\{C_2(z)|1t\}$, $\{C_2(z)|2t\}$, $\{C_2(z)|3t\}$, $\{C_2(z)|4t\}$, $\{C_2(z)|5t\}$

The purpose of enumerating the classes is to obtain at least a preliminary indication of the number of irreductible representations of each dimension.

Suppose G has x 1-dimensional and y 2-dimensional representations. Then using (1) and (2) and the fact that $g = fN$ (where f and N are the orders of the factor and translation subgroups, respectively),

$$x(1)^2 + y(2)^2 = 8 \times 6 = 48$$

and $$x + y = 18$$

Whence $x = 8$ and $y = 10$. This suggests that there are 8 1-dimensional and 10 2-dimensional representations; it does not establish that there are no representations of higher dimensionality. Completeness can only be established after the representations of G have been found.

We observe in passing that the allowed values of k for this case are $k = 0, \pm\frac{\pi}{2t}, \pm\frac{2}{3}\frac{\pi}{t}, \frac{\pi}{t}$. It will be shown later that the 18 representations are associated in quite a specific way with the various values of k.

2.4. Irreducible representations of G.

The three main steps mentioned above were :

1. listing of symmetry elements and operations,
2. investigation of the subgroup structure of G,
3. investigation of the class structure of G.

They were in a sense preliminary to the fourth and central task of finding the irreducible representations of G. By this we mean

Table 2

Multiplication table for $D_{2h} \approx G(k = 0)$

The operations are denoted by the name of the corresponding element.

D_{2h}	ε	$C_2(x)$	$C_2(y)$	$C_2(z)$	i	$\sigma(x)$	$\sigma(y)$	$\sigma(z)$
E	ε	$C_2(x)$	$C_2(y)$	$C_2(z)$	i	$\sigma(x)$	$\sigma(y)$	$\sigma(z)$
$C_2(x)$	$C_2(x)$	ε	$C_2(z)$	$C_2(y)$	$\sigma(x)$	i	$\sigma(z)$	$\sigma(y)$
$C_2(y)$	$C_2(y)$	$C_2(z)$	ε	$C_2(x)$	$\sigma(y)$	$\sigma(z)$	i	$\sigma(x)$
$C_2(z)$	$C_2(z)$	$C_2(y)$	$C_2(x)$	ε	$\sigma(z)$	$\sigma(y)$	$\sigma(x)$	i
i	i	$\sigma(x)$	$\sigma(y)$	$\sigma(z)$	ε	$C_2(x)$	$C_2(y)$	$C_2(z)$
$\sigma(x)$	$\sigma(x)$	i	$\sigma(z)$	$\sigma(y)$	$C_2(x)$	ε	$C_2(z)$	$C_2(y)$
$\sigma(y)$	$\sigma(y)$	$\sigma(z)$	i	$\sigma(x)$	$C_2(y)$	$C_2(z)$	ε	$C_2(x)$
$\sigma(z)$	$\sigma(z)$	$\sigma(y)$	$\sigma(x)$	i	$C_2(z)$	$C_2(y)$	$C_2(x)$	ε

deriving the representation matrices for each irreducible representation of dimension 1, 2 etc.

Particular solutions of the Schrödinger equation corresponding to energy E_k are Bloch functions of the form

$$\psi_k(r) = u_k(r)\, e^{ikr} .$$

From the time-reversal symmetry of the Schrödinger equation, $\psi_{-k}(r)$ is also a solution of energy E_k and hence for $k > 0$ the general solution has the form

$$\chi(r) = A\psi_k(r) + B\psi_{-k}(r) .$$

In general some operations of G will transform $k \rightarrow -k$ and we therefore need to work with the general solution. We say that the pair ψ_k, ψ_{-k} constitute a basis for the 2-dimensional irreducible representations of G. It is covenient to write this in vector form

$$\underline{\psi}_{-k} = \begin{pmatrix} \psi_k \\ \psi_{-k} \end{pmatrix}$$

Then, any operator $A \in G$ is represented, in the qth 2-dimensional representation, by the matrix D_q of order 2 given by

$$A\underline{\psi}_k^{(q)} = D_q\, \underline{\psi}_k^{(q)}$$

When $k = 0$, $\psi_k = \psi_{-k} = \psi_0$ and the representations are one-dimensional. The elements of the translational group, $\{e^{iknt}\}$, all reduce to unity and the representations of G and the 8 1-dimensional representations of $F \approx D_{2h}$. The truncated character table of D_{2h} is given in Table 3.

(An interesting question, which does not seem to have been asked, concerns the identity of the second solution of the Schrödinger equation at $k = 0$, which we know must exist. By differentiation of the Schrödinger equation with respect to k, it can be verified that the second solution is

$$\phi_o(r) = \left| \frac{\partial \psi_k(r)}{\partial k} \right|_{k=0}$$

Since $$\phi_o = \left| \frac{\partial u_k}{\partial k} + ixu_k \right|_{k=0} \rightarrow \infty \quad \text{as } |x| \rightarrow \infty$$

it is not a physical solution.

Table 3

Truncated character table for $D_{2h} \approx G(k = 0)$ showing only the basic operations.

D_{2h}	$\{\varepsilon\|0\}$	$\{\sigma(x)\|0\}$	$\{\sigma(y)\|\tau\}$	$\{\sigma(z)\|0\}$
A_g	1	1	1	1
A_u	1	-1	-1	-1
B_{1g}	1	1	-1	-1
B_{1u}	1	-1	1	1
B_{2g}	1	-1	-1	1
B_{2u}	1	1	1	-1
B_{3g}	1	-1	1	-1
B_{3u}	1	1	-1	1

To be mathematically correct, therefore, we should think of two distinct one-dimensional representations at k = o, for one of which the basis function is unphysical.

It can be shown that the ensemble of operations which leave k invariant to within an r.l.v., i.e. for which $Ak = k + K_e$, itself constitutes a group K, the importance of which lies in the fact that the representations of K can be built up into representations of G. Look at the basic operations : only $\{\sigma(y)|\tau\}$ and $\{\sigma(z)|0\} \in K$. Thus the elements of K are all powers and products of $\{\sigma(y)|\tau\}$ and $\{\sigma(z)|0\}$ viz., $\{\varepsilon|nt\}$, $\{\sigma(y)|\tau\}$, $\{\sigma(z)|0\}$ and $\{\sigma(y)|\tau\}\{\sigma(z)|0\} = \{C_2(x)|\tau\}$. K is thus isomorphous with C_{2v}; the character table is shown in the upper part of Table 4. Here, as usual, A(B) refers to symmetry (antisymmetry) with regard to a rotation of $\frac{2\pi}{n}$ about axis C_n and suffix 1 (or 2) indicates symmetry (antisymmetry) with regard to the plane of the chain (i.e. σ or π symmetry). Since $\{\sigma(y)|\tau\}$ and $\{C_2(x)|\tau\}$ have the effect of multiplying any function by the phase term $\pm e^{ik\tau}$ these are the required characters.

The 2D representations of G can now be constructed by supplying a second term identical to the entries of the first part of Table 4, except that k is replaced by -k. In doing this we are simply implementing the earlier statement that $\psi(k,r)$ and $\psi(-k,r)$ form a basis for the 2D representations of G.

It will be observed that one of the basic operations, namely $\{\sigma(x)|0\}$ is still not accounted for. By writting

$$\sigma\psi = \begin{pmatrix} \sigma_1 & \sigma_2 \\ \sigma_3 & \sigma_4 \end{pmatrix} \begin{pmatrix} \psi_k \\ \psi_{-k} \end{pmatrix}$$

and observing that the result of this operation must be $\pm\begin{pmatrix} \psi_{-k} \\ \psi_k \end{pmatrix}$, we find that $\sigma_1 = \sigma_4 = 0$ and $\sigma_2 = \sigma_3 = \pm 1$. Since $\begin{pmatrix} 0 & 1 \\ 1 & 0 \end{pmatrix}$ and $\begin{pmatrix} 0 & -1 \\ -1 & 0 \end{pmatrix}$ have the same character, it is immaterial from a group theoretical point of view which matrix one uses, and we chose the former. It should be emphasised that this choice does imply a choice of the sign of the p_x-type symmetry orbitals (see below).

These are not the only 2D irreducible representations of G. By putting $k = \frac{\pi}{t}$ in the first four rows, and recognising the equivalence of representations with the same character, one finds that ${}^2\Gamma_1$, ${}^2\Gamma_3 \rightarrow {}^2\Gamma_5$ and ${}^2\Gamma_2$, ${}^2\Gamma_4 \rightarrow {}^2\Gamma_6$.

The first four 2D representations of G occur for each value of k or pairs (k,-k) other than 0, π/t. For N even, there are $\frac{N-2}{2}$ such values of $|k|$ and therefore 2(N-1) representations.

Table 4

Matrices for the basic operations in the two-dimensional representations of G and the related one-dimensional representations of the star group K.

$K \approx C_{2v}$	$\{\varepsilon\|nt\}$	$\{C_2(x)\|\tau\}$	$\{\sigma(y)\|\tau\}$	$\{\sigma(z)\|0\}$
A_1	e^{iknt}	$e^{ik\tau}$	$e^{ik\tau}$	1
A_2	e^{iknt}	$e^{ik\tau}$	$-e^{ik\tau}$	-1
B_1	e^{iknt}	$-e^{ik\tau}$	$-e^{ik\tau}$	1
B_2	e^{iknt}	$-e^{ik\tau}$	$e^{ik\tau}$	-1

G	$\{\varepsilon\|nt\}$	$\{\sigma(x)\|0\}$	$\{\sigma(y)\|\tau\}$	$\{\sigma(z)\|0\}$
${}^2\Gamma_1$ (A_1)	$\begin{pmatrix} e^{iknt} & 0 \\ 0 & e^{-iknt} \end{pmatrix}$	$\begin{pmatrix} 0 & 1 \\ 1 & 0 \end{pmatrix}$	$\begin{pmatrix} e^{ik\tau} & 0 \\ 0 & e^{-ik\tau} \end{pmatrix}$	$\begin{pmatrix} 1 & 0 \\ 0 & 1 \end{pmatrix}$
${}^2\Gamma_2$ (A_2)	$\begin{pmatrix} e^{iknt} & 0 \\ 0 & e^{-iknt} \end{pmatrix}$	$\begin{pmatrix} 0 & 1 \\ 1 & 0 \end{pmatrix}$	$\begin{pmatrix} -e^{ik\tau} & 0 \\ 0 & -e^{-ik\tau} \end{pmatrix}$	$\begin{pmatrix} -1 & 0 \\ 0 & -1 \end{pmatrix}$
${}^2\Gamma_3$ (B_1)	$\begin{pmatrix} e^{iknt} & 0 \\ 0 & e^{-iknt} \end{pmatrix}$	$\begin{pmatrix} 0 & 1 \\ 1 & 0 \end{pmatrix}$	$\begin{pmatrix} -e^{ik\tau} & 0 \\ 0 & -e^{-ik\tau} \end{pmatrix}$	$\begin{pmatrix} 1 & 0 \\ 0 & 1 \end{pmatrix}$
${}^2\Gamma_4$ (B_2)	$\begin{pmatrix} e^{iknt} & 0 \\ 0 & e^{-iknt} \end{pmatrix}$	$\begin{pmatrix} 0 & 1 \\ 1 & 0 \end{pmatrix}$	$\begin{pmatrix} -e^{ik\tau} & 0 \\ 0 & e^{-ik\tau} \end{pmatrix}$	$\begin{pmatrix} -1 & 0 \\ 0 & -1 \end{pmatrix}$
${}^2\Gamma_5$ (A)	$\begin{pmatrix} e^{in\pi} & 0 \\ 0 & e^{-in\pi} \end{pmatrix}$	$\begin{pmatrix} 0 & 1 \\ 1 & 0 \end{pmatrix}$	$\begin{pmatrix} i & 0 \\ 0 & -1 \end{pmatrix}$	$\begin{pmatrix} 1 & 0 \\ 0 & 1 \end{pmatrix}$
${}^2\Gamma_6$ (B)	$\begin{pmatrix} e^{in\pi} & 0 \\ 0 & e^{-in\pi} \end{pmatrix}$	$\begin{pmatrix} 0 & 1 \\ 1 & 0 \end{pmatrix}$	$\begin{pmatrix} i & 0 \\ 0 & -1 \end{pmatrix}$	$\begin{pmatrix} -1 & 0 \\ 0 & -1 \end{pmatrix}$

Including those at $k = \pi/t$ (or $k = -\pi/t$) there are in all $2(N-1)$ irreducible representations of order 2.

For N odd, i.e. $k = 0, \pm 1, \pm 2, \ldots, \frac{N-1}{2}$ there are no representations corresponding to points on the edge of the Brillouin zone. Thus excluding $k = 0$, there are $\frac{N-1}{2}$ values of $|k|$, and again $2(N-1)$ 2D representations in all.

Example 1 continued: No. of 2D reps $= 2(N-1) = 2 \times 5 = 10$

Recall, for $N = 6$, $k = 0, \pm \frac{\pi}{2t}, \pm\frac{2}{3}\frac{\pi}{t}, \frac{\pi}{t}$

$k = 0$	$\pm \frac{\pi}{2t}$	$\pm\frac{2}{3}\frac{\pi}{t}$	$\frac{\pi}{t}$
↑	↑	↑	↑
8	4	4	2
1D reps	2D reps	2D reps	2D reps

(2D reps total: 10 2D reps)

Completeness can now be established by use of the results

$$\sum_{j}^{\text{no. of classes}} n_j \chi_j^{px} \chi_j^{q} = g\, \delta_{pq}$$

where n_j = no. of symmetry operations in the jth class

χ_j^p = character in the jth class for the pth operations.

This is left as an exercise

III. SYMMETRY ORBITALS FOR IDEAL CHAINS

Assume that all the electrons of the system are described by wave functions that possess the full symmetry of the ideal chain; i.e. belong to one or other of the irreducible representations of G for the chain. These wavefunctions will in general be linear combinations of Bloch-type symmetry orbitals based on atomic orbitals of given quantum number. For example, out of the jth atomic orbital ϕ_j, one may construct the unit cell functions

$$u_{jn} = \phi_j(\bar{r} - \bar{R}_n) + \sum_{\ell} e^{ik\tau_\ell} \phi_j(\bar{r} - \bar{R}_n - \bar{a}_\ell)$$

where $R_n = nt$ locates the atom at the origin of the nth cell.

The associated symmetry orbitals may then be written

$$\psi_{kj}(\phi_j) = \sum_n e^{ikR_n} u_{jn}$$

$$\psi_{-kj}(\phi_j) = \sum_n e^{-ikR_n} u_{jn}$$

The electron wavefunction χ_{kf}, where f is the band index, will be of form

$$\chi_{kf} = \sum_{j=1}^{M} c_j^{kf} \psi_{kj} \qquad (3)$$

where M is the total number of orbitals per cell.

The allocation of symmetry orbitals to the irreducible representations of G tells one immediately which of the coefficients c_j^{kf} must be zero in (3).

3.1. Symmatry orbitals for the ideal polyethylene chain

In writing down the unit cell functions for the polyethylene chain it is convenient to distinguish bonding and antibonding combinations, i.e. to write

$$u_{jn}^{\pm} = \phi_j(\bar{r} - \bar{R}_n) \pm e^{ik\tau} \phi_j(\bar{r} - \bar{R}_n - \bar{a})$$

where ϕ_j refers either to an orbital of carbon or to the symmetric or antisymmetric combination of the hydrogen orbitals relating to a given carbon.
(In the latter case, the symmetry referred to is with respect to reflection in the plane of the molecule).

One now proceeds to apply the basic operations of G to each symmetry orbital in turn. In the present problem some time-saving is possible by noting that only two operators, $\{\sigma(y)|\tau\}$ and $\{\sigma(z)|0\}$, need be used; the characters of the other operations are identical in each representation.

Example 2. Under $\{\sigma(y)|\tau\}$

$$s(\bar{r},\bar{R}_n) \rightarrow s(r,R_n + \tau) = \begin{cases} e^{ik\tau} s(r,R_n), & 0 < k < \pi/t \\ e^{-ik\tau} s(r,R_n), & 0 > k > -\pi/t \end{cases}$$

and similarly for the orbital $s(r,R_n+a)$

$$\therefore \sum e^{ikR_n} \{s(r,R_n) + e^{ik\tau} s(r,R_n + a)\} \rightarrow$$

$$e^{ik\tau} \sum e^{ikR_n} \{s(r,R_n) + e^{ik\tau} s(r,R_n+a)\}$$

and similarly for $\psi_{-k}^{+}(s)$. Therefore $\{\sigma(y)|\tau\}\ \psi_{\pm k}^{+}(s) = e^{\pm ik\tau} \psi_{\pm k}^{+}(s)$.

And by inspection $\{\sigma(z)|0\}\ \psi^{+}_{\pm k}(s) = \psi^{+}_{\pm k}(s)$. Hence the matrices for $\{\sigma(y)|\tau\}$ and $\{\sigma(z)|0\}$ are, respectively,

$$\begin{pmatrix} e^{ik\tau} & 0 \\ 0 & e^{-ik\tau} \end{pmatrix} \quad \text{and} \quad \begin{pmatrix} 1 & 0 \\ 0 & 1 \end{pmatrix}.$$

Consequently $\psi^{+}_{k}(s)$ and $\psi^{+}_{-k}(s)$ form a basis for the representation of A_1 of G.

By continuing this process for the other pairs of basic functions, and repeating it for the special values $k = 0$ and $k = \pi/t$, one obtains Table 5. Anyone going through this in detail and taking trouble to check on the matrix representation of $\{\sigma(x)|0\}$ will encounter an apparent difficulty with $\psi_k(p_x)$. This is connected with the choice of sign referred to above. Since s, p_y and p_z orbitals do not change sign on reflection, the matrix

$\begin{pmatrix} 0 & 1 \\ 1 & 0 \end{pmatrix}$ is correct for them. However,

$$\{\sigma(x)|0\} \text{ maps } p_x(r - R_n) \to - p_x(r - R_n)$$

$$p_x(r - R_n - a) \to - p_x(r - R_n + a)$$

$$= - e^{-2ik\tau}\, p_x(r - R_n - a)$$

Hence

$$\{\sigma(x)|0\}\psi^{+}_{k}(p_x) = \{\sigma(x)|0\}\ \Sigma\ e^{ikR_n}\{p_x(r - R_n) + e^{-ik\tau}\, p_x(r - R_n - a)\}$$

$$= - \Sigma\ e^{-ikR_n}\{p_x(r - R_n) + e^{-ik\tau}\, p_x(r-R_n-a)\}$$

$$= - \psi^{+}_{-k}(p_x)$$

Similarly, $\{\sigma(x)|0\}\ \psi^{+}_{-k}(p_x) = - \psi^{+}_{k}(p_x)$.

Hence, when applied to a function of p_x symmetry, the matrix should $\begin{pmatrix} 0 & -1 \\ -1 & 0 \end{pmatrix}$. It is, however, desirable that the matrix for a given operation should not depend on the nature of the basis function, even if the two forms are of the same character.

This question is readily resolved by defining the symmetry orbital $\psi^{+}_{-k}(p_x)$ as

$$\psi^{+}_{-k}(p_x) = - \Sigma\ e^{-ikR_n}\{p_x(r - R_n) + e^{-ik\tau}\, p_x(r - R_n - a)\}$$

and similarly for $\bar{\psi}_{-k}(p_x)$. With these definitions, the matrix $\begin{pmatrix} 0 & 1 \\ 1 & 0 \end{pmatrix}$ holds for all functions.

IV. EFFECT OF SYMMETRY ON THE BAND STRUCTURE.

The effect of symmetry is principally the determination of degeneracties (or absence of degeneracies) in the band structure. Therefore we discuss briefly some of the main concepts and results.

4.1. Degeneracies.

Consider the spin-free Hamiltonian

$$H = -\nabla^2 + V - A$$

in which V (the Coulomb operator) and A (the exchange operator) are real and possess the symmetry of the lattice. Degenracy in the solutions of the Schrödinger equation may arise from 3 distinct properties of the Hamiltonian, two formal and one arithmetical.

1. The reality of H. If H is real, the spin-free Schrödinger equation is invariant under complex conjugation and time reversal:

   i.e. if $\left(H - i\hbar \frac{\partial}{\partial t}\right) \psi\,(r,t) = 0$

   then $\left(H - i\hbar \frac{\partial}{\partial t}\right) \psi^*(r,-t) = 0.$

   In that sense complex conjugation and time-reversal are equivalent (one "undoes" the other). Note that the time-reversal operation is itself equivalent to the reversal of linear and angular momenta.
2. The spatial symmetry of H, through V.
3. The numerical content of H, i.e. the magnitude of V and A.

We note that degeneracy arising from (3), i.e. one that can be shifted in k-space (or removed altogether) by a small change in the magnitude of V, is called "accidental". The other causes produce "essential" degeneracies. While property (1) alone is sufficient to produce essential degeneracy, e.g. E(k) = E(-k), it is the degeneracies arising from property (2) that most usually engages our attention.

Type 1. Degeneracy of ψ_k and ψ_{-k}. As noted, it is this degeneracy (arising from property(1)) that accounts for the two-dimensionality of the irreducible representatives of G for k on Δ.

Types 2. Degeneracy of states belonging to different representations of the space group G.

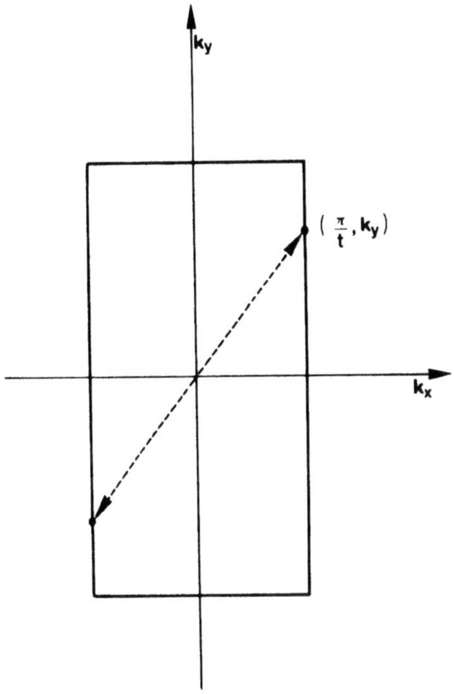

Figure 2. To illustrate Heine's proof of Herring's Theorem.

(a) Accidental degeneracy.

For most k in the B.Z., states belonging to different representations will have different energies, but through numerical chance two bands may tend to the same energy as k tends to some value k_1. Since a periodic potential does not couple states belonging to different representations, a crossing of bands at k_1 is allowed.

(b) Essential degeneracy.

Herrings Theorem[8] : If a boundary plane of the B.Z. is perpendicular to a two-fold screw axis, no element of whose coset is a pure two-fold notation, then the energy bands must stick together in pairs at points in this plane.

Heine's restricted proof for a 2D lattice with a glide plane, a mirror plane and an inversion centre.

Let $\psi(x,y)$ be an eigensolution at $k = \left(\frac{\pi}{t}, k_y\right)$. Applying the time-reversal operator I,

$I\psi = \psi^*$ has the wave vector $-k = \left(-\frac{\pi}{t}, k_y\right) = \left(\frac{\pi}{t}, -k_y\right)$.

Applying the glide operation $g = \{\sigma(y)|\tau\}$ to $I\psi$ one obtains $g\,I\psi$ which once more has the wave vector $k = \{\frac{\pi}{t}, k_y\}$. (Fig. 2 may assist in visualising these operations).

Now suppose that $g\,I\,\psi = \alpha\psi$, α = complex constant (4)

Then $g\,I\,g\,I\,\psi = g\,I(\alpha\psi) = \alpha^* g\,I\psi = \alpha^*\alpha\psi$ (5)

However $g\,I\,g\,I\,\psi = \psi(x+t,y) = e^{i\frac{\pi}{t}.t}\,\psi(x,y) = -\psi$ (6)

Since $\alpha^*\alpha > 0$ (5) and (6) are contradictory, therefore (4) is false. Therefore, ψ and g ψ must be linearly independent. Since g and I are symmetry elements, g I must commute with the Hamiltonian. Therefore, ψ and $g\,I\,\psi$ have the same energy. Therefore the irreducible representations on the Brillouin boundary are at least two-dimensional and the bands must stick together there.

The perceptive reader may criticise this proof on the grounds that it does no more than prove that the eibenstates are doubly degenerate on the Brillouin boundary. Did we really need to prove this, since we know that the Schrödinger equation must possess two linearly independent solutions for all k and since we recognise that we cannot now invoke the degeneracy k and -k eigensolutions to generate our two-dimensional representation owing to the equivalence of $-\frac{\pi}{t}$ and $\frac{\pi}{t}$?

If one recalls the discussion regarding the second solution at Γ , one will agree that there is some obligation to demonstrate that the second solution is physical. Therefore the finding of an acceptable linearly independent solution, namely $g\,I\,\psi$, is an im-

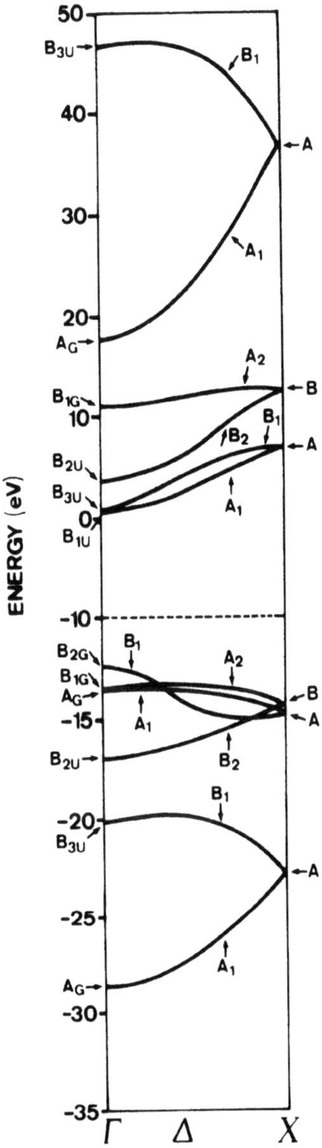

Figure 3. Polyethylene band structure.

portant part of establishing the sticking together of bands. Only by such a sticking together can the double-degeneracy be supplied.

Type 3. Degeneracy of states belonging to the same irreducible representation of G.

As we have seen, the periodic crystal potential will, in general, couple states belonging to the same representation of the space group. The perturbation will, in general, push apart the bands that would otherwise cross at some value of k.

However, it is mathematically possible for the perturbation to be vanishingly small, in which case a crossing or touching would occur. It is thought that degeneracies of this kind are very rare in actual systems.

4.2. Effect of symmetry on the band structure of polyethylene

First of all, let us look briefly at how the symmetry assignments of fig.3 were made. It will be realised that the computer output consists of a string of eigenvalues giving a distribution of points on the E(k) diagram. How is one to connect them up ? If one consults the accompanying table of eigenfunctions, i.e. the table of LCAO coefficients C_j^{kf} mentioned previously, one can see immediately for which orbitals j the $C_j^{kf} \neq 0$. For example, the uppermost bound state $\sim$ -12.5 eV at Γ, will show only the -combination of carbon p_x orbitals to be non-zero. On consulting Table 5 one immediately concludes the state must be of B_{2g} symmetry. And so on for the other points.

It is always advisable to check that no errors have been made by use of the Compatibility Relations. It is clear from Table 5 that there is a contraction in the ratio 2:1 in the number of irreducible representations as k leaves the point Γ and again as it arrives at X. Thus an irreducible representation becomes a reducible representation in a k-domain to its left. By the use of Table 3 and 4 it is a simple matter to confirm the following scheme:

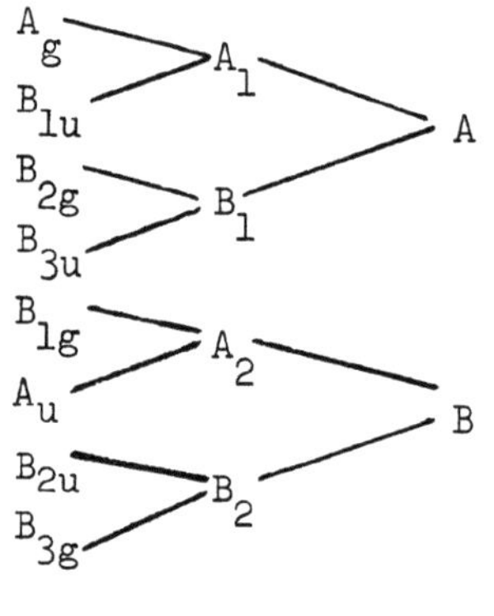

Table 5. Symmetry of the function $\psi_k^{\pm}$ (ϕ) at k = 0, 0 < |k| < π/t and k = π/t. The first four orbitals refer to carbon ; the final two refer to the symmetric (S_H) and antisymmetric (A_H) combinations of hydrogen 1s function bonded to the same carbon atom.

	Representation	S	P_x	P_y	P_z	S_H	A_H
Γ	A_g	+	0	-	0	+	0
	Au	0	0	0	0	0	0
	B_{1g}	0	0	0	-	0	-
	B_{1u}	0	+	0	0	0	0
	B_{2g}	0	-	0	0	0	0
	B_{2u}	0	0	0	+	0	+
	B_{3g}	0	0	0	0	0	0
	B_{3u}	-	0	+	0	-	0
Δ	A_1	+	+	-	0	+	0
	A_2	0	0	0	-	0	-
	B_1	-	-	+	0	-	0
	B_2	0	0	0	+	0	+
X	A	±	±	±	0	±	0
	B	0	0	0	±	0	±

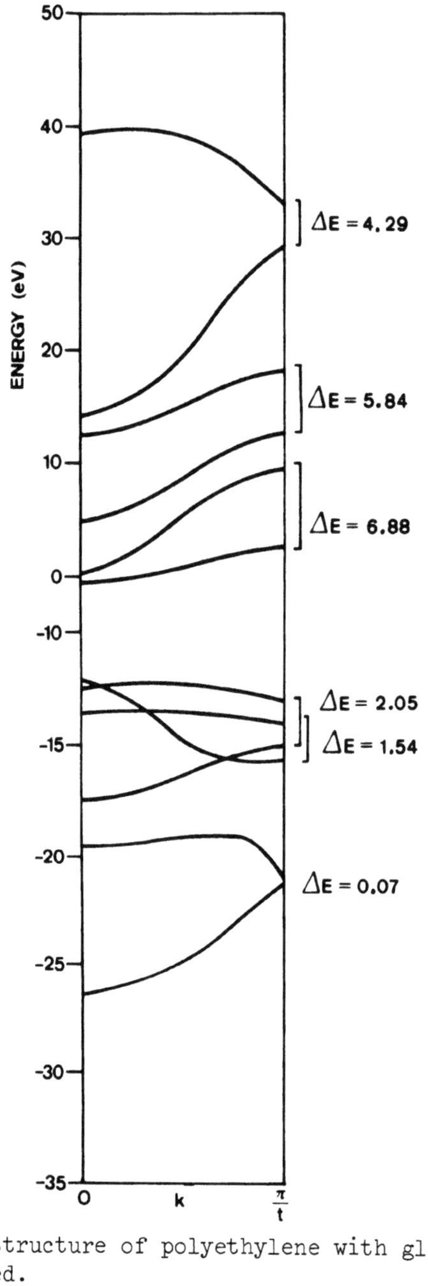

Figure 4. Band structure of polyethylene with glide symmetry removed.

It is easily seen that the connections made in fig.3, for example $B_{2g} \rightarrow B_1 \rightarrow A \rightarrow A_1 \rightarrow A_g$, are compatible with this scheme.

Having made and verified the symmetry assignments, one may proceed to note the existence of Type 2 degeneracies of both essential and accidental kind. As k moves away from X along Δ, the bands symmetry-separate due to the presence of the glide operation which, as Table 4 shows, provides the only difference in character between the representations A_1 and B_1 (or A_2 and B_2). At Γ a further summetry-separation occurs between states composed of orbitals of p_x -symmetry and those composed of s and p_y orbitals.

Confirmation, if that be needed, that the essential degeneracy at X is indeed due to glide symmetry is provided by fig.4. Here we calculated the band structure of polyethylene with the glide removed. Successive HCH angles were increased and decreased by 40°, keeping all other band angles and lengths constant.

V. GENERALISATION TO MORE COMPLEX CHAINS

5.1. Ideal chains.

The simplest extensions of the above theory are as follows:

1. Chains of planar zig-zag structure but having side-atoms other than hydrogen. The only difference from the polyethylene case is the greater number of symmetry orbitals and the extra effort in allocation them to the irreducible representation of G. An example would be PTFE $(CF_2 - CF_2 -)_n$ which to a high degree of accuracy may be considered planar.
2. Chains with a symmetry differing from ideal polyethylene only in the absence of the glide plane. Examples would be the mono-substituted paraffins $(CH_2 - CHR -)_n$ or the long polyenes in which a long-short alteration is thought to exist in the main chain.
3. Helical structures, which PTFE really is. For polymers with a well-defined main chain it is quite profitable to set up a band calculation using only the screw symmetry. Planar systems then appear as a special case.

The above procedures carry over without elaboration and these cases may be left as exercises.

5.2. Quasi-ideal chains

By a quasi-ideal chain we mean one which does not possess <u>one</u>

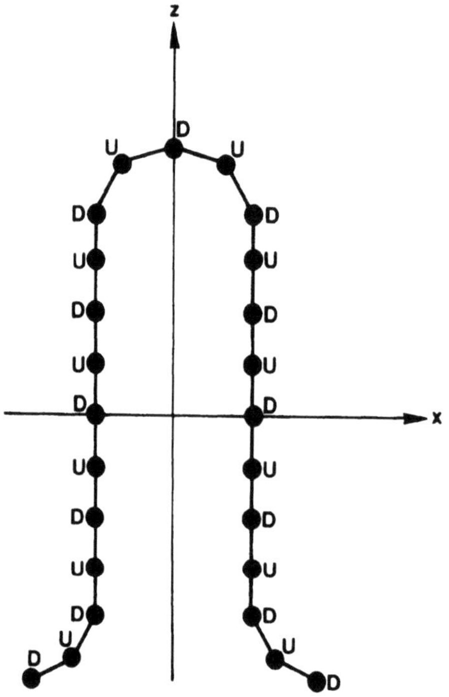

Figure 5. A highly simplified model of a chain fold.
D and U label atoms below and above the Y-plane.

of the four defining properties of an ideal chain. Perhaps the most realistic from a chemical point of view is that concerning chemical impurities. If the mean spacing between impurities > 100 Å, then they are essentially non-interacting and the treatment of such "isolated" impurities follows, in principle, much the same lines as in semiconductor theory. Qualitatively, one expects the band structure of the best lattice to remain, but with the addition of impurity levels. Important though such calculations may appear, none applicable to polymers are known to the author. Hopefully this lack will be remedied in the near future.

Possibly more interesting to the physicist are the chemically pure chains that are not in the extended conformation. What happens to the line group when the chain is folded ? Before that question may be answered, one must know the geometry in the fold region. The simplest possible case for the zig-zag chain is shown in fig.5. The highest symmetry group of this chain (occurring at $k = 0$ and $k = \pi/a$) is isomorphous with C_{2v}. One can thus immediately deduce that there is no essential degenaracy at $k = \pi/a$ in this case.

However, one must ask oneself what the merit of a band calculation on the single isolated chain would be. With such a large number of atoms per unit cell, the eigenvalue calculation would certainly be expensive and to scientific value ought therefore to be ascertained beforehand. It should first of all be remembered that the folded conformation is characteristic of the condensed state. Therefore, one may not obtain quantitative results by considering the crystalline lamella as a bundle of non-interacting folded chains. However, even if one were willing to settle for qualitative results of the kind just mentioned, one encounters the difficulty that the fold regions are poorly defined experimentally. On the one hand, there is the knowledge that rotations involving only four bonds will suffice (and a corresponding temptation to build models incorporating this feature) and on the other, some comparative melting-point studies(9) that suggest the material in the fold may contain $\sim$ 40 main-chain bonds.

The models of charge transport that emerge from these extreme models of folding are quite distinct, but to discuss them further is outside the scope of this lecture. It is sufficient to point out that even in a lamella that we think of as crystalline, there may be significant lengths of chain (the fold regions) that are rotationally disordered. This is true a fortiori of the material existing inbetween lamellae and it provided the motivation to study disordered polymer chains.

ACKNOWLEDGMENTS

I have benefited considerably from conversations on group theory with various colleagues over the years. I should especially wish to record my indebtedness to Dr. M. Boon, formerly of the Battelle Institute, Geneva, Dr. I. D. C. Gurney of the University of Newcastle upon Tyne, and Dr. A. A. Cottey of the University of East Anglia. I have also made considerable use of the Sandia Corporation Research Report SC-4655(RR) by K. E. Lawson and G. A. Crosby.

REFERENCES.

1. A. Imamura, J. Chem. Phys. 52, 3168 (1970) ;
H. Fujita and A. Imamura, ibid. 53, 4555 (1970) .

2. K. Morokuma, Chem. Phys. Letters 6, 186 (1970);
J. Chem. Phys. 54, 962 (1971)

3. E. Clementi, J. Chem. Phys. 54, 2492 (1971).

4. The underlying assumptions are discussed by W. L. McCubbin Chem. Phys. Letters, 8, 507 (1971), and in more detail in Quantum Chemistry Group Report 207, University of Uppsala, 1967.

5. J. T. Hougen, J. Chem. Phys. 37, 1433 (1962);
ibid. 39, 358 (1963).

6. See, for example, S. L. Altman, Mol. Phys. 21, 587 (1971).

7. A. Pabst, Acta. Cryst. 10, 851 (1957).

8. C. Herring, Phys. Rev. 52, 361, 365 (1937).

9. J. R. Knox, J. Polym. Sci. C No. 18 p.69 (1967).

X-RAY STRUCTURE DETERMINATION OF POLYMERS

Edward Atkins

H.H. Wills Physics Laboratory
University of Bristol
BRISTOL BS8 1TL

England

THEORETICAL ASPECTS OF X-RAY DIFFRACTION

1. Scattering Amplitude

The basic criteria for diffraction is that the wavelength of the incident electromagnetic radiation should be of the same order of magnitude as the distance between scattering particles. This for structure on an atomic scale X-rays are usually used since they have a wavelength $\simeq$ 0.1mm (10^{-9}m) ; although electrons and neutrons also fall in this wavelength range. For matter with a regular periodicity the pattern of scattered radiation consists of a set of discrete beams and the rules governing them can be conveniently expressed in the form of Bragg's equation

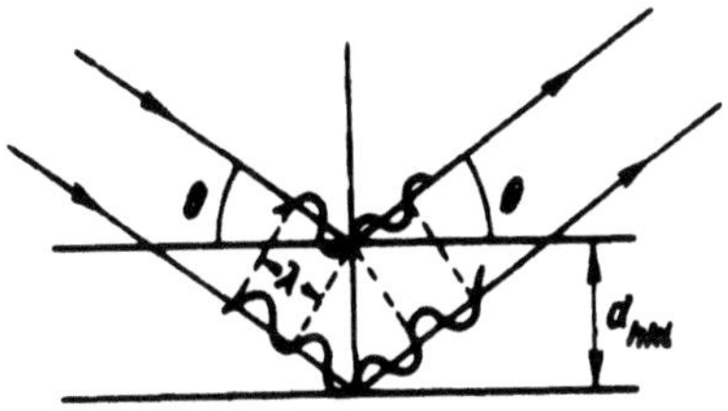

fig. 1. derivation of Braggs law

$$n\lambda = d_{hk\ell} \sin\theta \text{ , where n is an integer} \qquad (1)$$

Note the diffracted angle = 2θ

2. Reciprocal Space

Consider the scattering from two centres ; one centre being of the origin, the other specified by a vector $\underline{r}$ (fig. 2.)

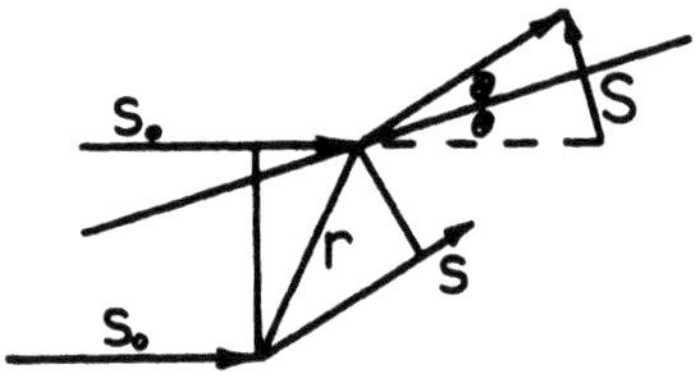

fig. 2. scattering at two centres.

Let $\underline{s}_0$ be the incident wave vector of modulus $1/\lambda$ and $\underline{s}$ be the emergent wave vector again of modulus $1/\lambda$

$$\text{i.e } |\underline{s}_0| = |\underline{s}| = 1/\lambda$$

path difference between waves is given by

$$\underline{s}.\underline{r} - \underline{s}_0.\underline{r} = (\underline{s} - \underline{s}_0).\underline{r}$$

It is convenient to use

$\underline{S} = (\underline{s} - \underline{s}_0)$ and from fig.2 it can be seen that

$$|\underline{S}| = 2 \sin\theta / \lambda \qquad (2)$$

The dimensions of $|\underline{S}|$ are reciprocal of length and it is called the reciprocal vector. If $\underline{s}_0$ is fixed then $\underline{S}$ varies in length and direction as $\underline{s}$ rotates. The ends of the vector $\underline{S}$ pass through a region called reciprocal space and, as we shall see later it turns out that this space is where diffraction pattern is found, i.e. where the Fourier transform is plotted.

3. Scattering Amplitude and Fourier Transforms

If the object has n scattering centres the resultant wave is found by

$$\sum_{j=1}^{n} f_j \, e^{2\pi i(\underline{S}.\underline{r}_j)} = F(\underline{S}) \qquad (3a)$$

where $F(\underline{S})$ is called the scattering amplitude and f_j is the scattering power, which can vary, for the jth centre. If the object is continuous the set of discrete points in positions $\underline{r}_j$ may be replaced by a density distribution $\rho(\underline{r})$, such that

$$F(\underline{S}) = \int \rho(\underline{r}) \, e^{2\pi i(\underline{S}.\underline{r})} dv_{\underline{r}} \qquad (3b)$$

where $dv_{\underline{r}}$ is an element of scattering volume.

In Cartesian coordiantes

$$F(XYZ) = \iiint_{xyz=-\infty}^{+\infty} \rho(x,y,z) e^{2\pi i(xX + yY + zZ)} dxdydz \qquad (3c)$$

The integral representations of (3b) and (3c) are most general and all types of scattering from objects of different shapes and distribution of scattering material can be discussed by reference to (3).

Only certain values of the scattering amplitude are realised in the diffraction pattern, since $\underline{S} = \underline{s} - \underline{s}_o$, the ends of this reciprocal vector for a fixed orientation is constant $\underline{s}_o$, lie on a sphere described by the vector $\underline{s}$.

By changing the orientation of the incident beam a whole series of spheres may be generated the outer envelope of which is called the limiting sphere - and no diffraction information can be obtained outside this sphere which has a radius = $2/\lambda$ and is therefore a function of λ. The intensity observed along $\underline{s}$ is proportional to $|F|^2$.

For a crystal consisting of groups of atoms repeating periodically in three dimensions, the electron density $\rho(xyz)$ is a periodic function in all three coordinates and it can be shown for integer values of h,k, and unit cell sides a, b, c that

$$F_{hk\ell} = \frac{1}{abc} \int_0^a \int_0^b \int_0^c \rho(xyz) \, e^{2\pi i(hx/a + ky/b + \ell z/c)} dxdydz \qquad (4)$$

Thus the scattering by crystals only occurs in certain directions. The point at which F differs from zero and takes the value $F_{hk\ell}$ form a periodic distribution in reciprocal space, known as the reciprocal lattice (Fig. 3).

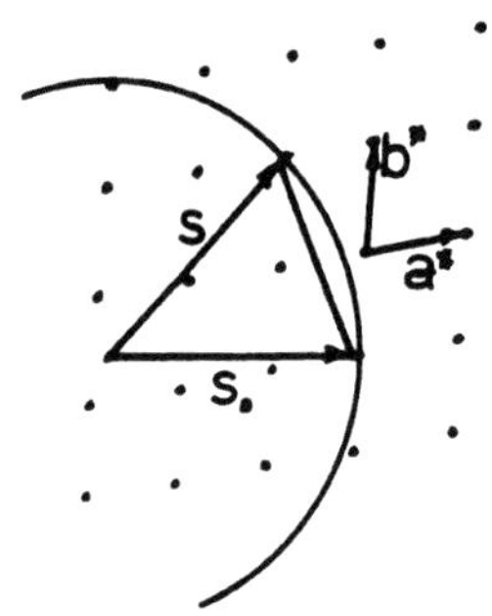

fig.3. Reciprocal lattice and the Ewald sphere.

Each point is characterised by a reciprocal lattice vector

$$\underline{r}^* = h\underline{a}^* + k\underline{b}^* + \ell\underline{c}^*$$

Consider a chain molecule which is periodic in one dimension, if $\underline{c}$ defines this periodicity then

$$F(X,Y,\ell/c) = \Sigma f_j \, e^{2\pi i(x_j.X + y_jY + \ell z_j/c)} \qquad (5)$$

which means that F is continuous in XY planes but discrete at intervals of ℓ/c in the Z direction.

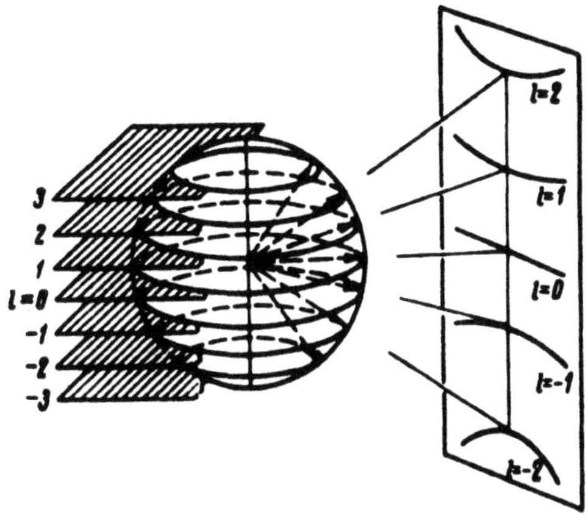

Fig. 4. Layer lines in reciprocal space.

4. Intensity

The scattering amplitude is governed by the electron density $\rho(\underline{r})$, which is equal to the mean number of electrons in an element of volume around a point $\underline{r}$ divided by the volume of this element. The absolute value of the amplitude, and hence the intensity, are deduced from the scattering characteristics of a single electron and Classical electrodynamics enables us to derive an expression of the form

$$I(\underline{S}) = \frac{1}{r^2} I_o \left(\frac{e^2}{mc^2}\right)^2 \frac{1 + \cos^2 2\theta}{2} |F(\underline{S})|^2 \qquad (6)$$

where $I(\underline{S})$ and I_o are the intensities diffracted and incident respectively, r is the distance from the object. The term (e^2/mc^2) is the scattering factor for a single electron and the term $(1+\cos^2 2\theta)/2$ is the polarization factor.

In practise we usually only measure relative intensities and write

$$i(\underline{S}) = F(\underline{S})^2 = F(\underline{S})\, F^* (\underline{S}) \qquad (7)$$

The Fourier integrals given in equations (3) have the property of reciprocity, i.e. we may equally well write

$$\rho(xyz) = \iiint F(XYZ)e^{-2\pi i(xX + yY + zZ)}dXdYdZ \tag{8a}$$

or

$$\rho(\underline{r}) = \int F(\underline{S})e^{-2\pi i(\underline{r}.\underline{S})}dv_{\underline{S}} \tag{8b}$$

This property is useful in diffraction theory since it enables the electron density distribution $\rho(\underline{r})$ to be obtained from the observed diffraction pattern. There is however a basic difficulty since the observed intensity gives only the moduli of the structure amplitude $F(\underline{S}) = \sqrt{I(\underline{S})}$ whereas $\underline{F}(\underline{S}) = |F| \exp(i\alpha)$, where α is the phase. This difficulty is known as the phase problem.

The result of application of the Fourier integral to a function is called the Fourier transform of the function, i.e. $F(\underline{S})$ is the Fourier transform of $\rho(\underline{r})$ and vice versa.

The operation is represented symbolically by the Fourier operator $\mathcal{F}$

$$F(\underline{S}) = \mathcal{F}\ [\rho(\underline{r})\ ]$$

or

$$\rho(\underline{r}) = \mathcal{F}^{-1}\ [F(\underline{S})]$$

where $\mathcal{F}$ or $\mathcal{F}^{-1}$ denotes integration of (3) or (8) over the whole volume of real or reciprocal space and including the exponential terms. Thus $\mathcal{F}\mathcal{F}^{-1} = 1$ and

$$\mathcal{F}^{-1}\ [\mathcal{F}\ [\ \rho(\underline{r})]] = \rho(\underline{r})$$

represented symbolically as

$$F(\underline{S}) \leftarrow \mathcal{F} \rightarrow \rho(\underline{r})$$

Since $\mathcal{F}$ is a complex operator we may express it in real and imaginary parts

$$\mathcal{F} = \mathcal{F}_c + i\ \mathcal{F}_s$$

where the subscripts c and s refer to cosine and sine terms respectively which may be written as :

$$F_c\,[\rho(\underline{r})] = A(\underline{S}) \quad ; \quad F_s\,[\rho(\underline{r})] = B(\underline{S})$$

so

$$F(\underline{S}) = F\,[\rho(\underline{r})] = \int \rho(\underline{r}) \cos 2\pi\, \underline{S}\underline{r} dv_{\underline{r}} + i \int \rho(\underline{r}) \sin 2\pi \underline{S}\underline{r} dv_{\underline{r}}$$

$$= A(\underline{S}) + i\, B(\underline{S})$$

in which $A(\underline{S})$ and $B(\underline{S})$ are respectively the real and imaginary components of $F(\underline{S})$. It can also be represented in terms of modulus $|F|$ and phase α:

$$F = |F| \exp(i\alpha) = |F| \cos\alpha + i\, |F| \sin\alpha$$

so

$$|F|^2 = A^2 + B^2 \; ; \; \tan\alpha = B/A$$

Example

Consider a simple 'top-hat' distribution where $\rho(x) = 1$ for $-a/2 < x < a/2$

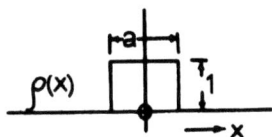

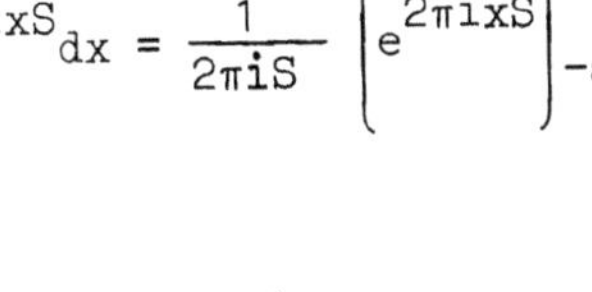

$$F(\underline{S}) = \int_{-a/2}^{+a/2} e^{2\pi i x S} dx = \frac{1}{2\pi i S} \left[e^{2\pi i x S} \right]_{-a/2}^{+a/2}$$

$$\therefore F(S) = a \left(\frac{\sin(\pi a S)}{\pi a S} \right)$$

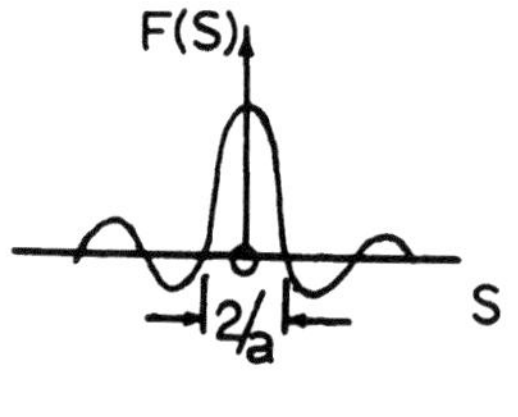

As $a \to \infty$ the diffraction patterns narrows and becomes a δ-function.

5. Convolution Theorem

Suppose we have two functions $f_1(\underline{r})$ and $f_2(\underline{r})$ in real space and that their transforms in reciprocal space are $F_1(\underline{S})$ and $F_2(\underline{S})$ respectively. The mechanism of convolution results in the distribution of function $f_1(\underline{r})$ in accordance with the law specified by the other function $f_2(\underline{r})$.

To explain this process let us multiply their transforms in reciprocal space.

The Fourier transform of the product $F_1(\underline{S})$ and $F_2(\underline{S})$ is

$$\int F_1(\underline{S})\, F_2(\underline{S})\, e^{2\pi i \underline{r}\cdot\underline{S}} dv_{\underline{S}} \tag{9}$$

but

$$F_1(\underline{S}) = \int f_1(\underline{r}')\, e^{-2\pi i \underline{r}'\cdot\underline{S}} dv_{\underline{r}'}$$

where $\underline{r}'$ is a vector in real space which is independent of $\underline{r}$.

Thus we may write for (9)

$$\int F_2(\underline{S}) \left[\int f_1(\underline{r}')\, e^{-2\pi i \underline{r}'\cdot\underline{S}} dv_{\underline{r}'} \right] e^{2\pi i \underline{r}\cdot\underline{S}} dv_{\underline{S}}$$

$$= \int f_1(\underline{r}') \left(\int F_2(\underline{S})\, e^{2\pi i (\underline{r}-\underline{r}')\cdot\underline{S}} dv_{\underline{S}} \right) dv_{\underline{r}'}$$

$$= \int f_1(\underline{r}')\, f_2(\underline{r}-\underline{r}')\, dv_{\underline{r}'} \tag{10}$$

This operation is designated by the notation

$$\widehat{f_1 f_2}(\underline{r}) = \int f_1(r') \, f_2(\underline{r}-r') \, dv_{\underline{r}'} \tag{11}$$

Thus the Fourier transform of the product of $F_1(\underline{S})$ and $F_2(\underline{S})$ is the convolution of the two functions $f_1(\underline{r})$ and $f_2(\underline{r})$.

ELUCIDATION OF MOLECULAR STRUCTURE

The molecular structure is determined first of all by the chemical constitution of the monomeric units which build the polymer chain and secondly by the mechanism of addition of successive monomeric units. Usually the chemical composition of the monomer is known in advance and commonly there is also information concerning the mode of addition ; for example, whether the monomer units are linked head to tail, head to head, tail to tail, or a random mixture of both. To describe the structure of a chain molecule both its configuration and its conformation must be specified.

Configuration - by this we mean the spatial arrangement of bonds in a molecule of given chemical constitution without regard to the multiplicity of spatial arrangements that may arise by rotation about single bonds, e.g.

these two chain segments have different configurations.

Conformations :- are the various spatial arrangements of the atoms in a molecule of established chemical constitution and configuration that may arise through rotations about single bonds, e.g. helices of opposite chirality would have different conformations.

1. Diffraction by Fibres

Many synthetic and naturally occurring polymers are of a fibrous nature with the molecular chains oriented parallel to the fibre axis. X-ray diffraction patterns are usually obtained by placing the fibres at right angles to the X-ray beam. The direction parallel to the fibre axis and through the center of the fibre diffraction pattern is referred to as the meridian, and the direction perpendicular to this is called the equator

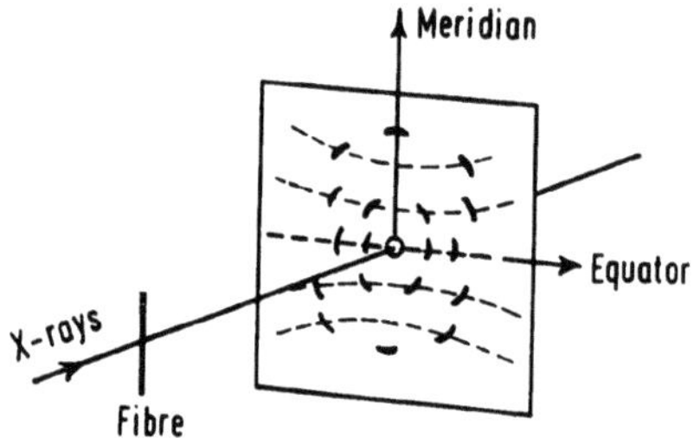

In some fibres the molecules are regularly arranged so as to form crystalline regions, but the different crystalline regions within a fibre are randomly oriented about the fibre axis. The diffraction patterns from such fibres are similar to single crystal rotation photographs, although the spots may be drawn out into arcs due to disorientation, and may be rather broad if the crystalline regions are small.

2. Helical Notation

Very often linear polymeric chains assume regular helical conformations in the condensed phase. A helix may be defined by the number of mononumeric units in the axial identity repeat distance c. If P is the pitch of the helix, and p the projected axial rise per monomer we may write :

$$P/p = u/t = \text{number of monomers per turn},$$

where u and t represent the number of monomers and turns respectively in the identity period c.

3. Fourier Transform of a Helical Molecule

(i) Transform of a Continuous Helix

Consider a continuous helix of negligible thickness, infinite length, with a radius r and pitch P.

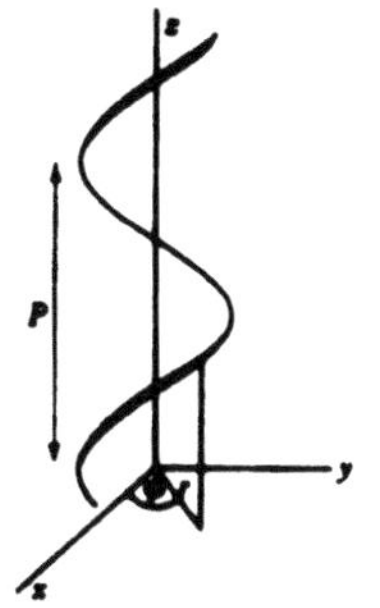

real space

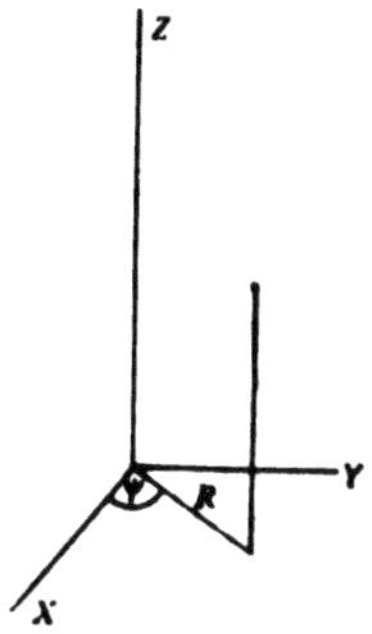

reciprocal space

The parametric equations of a helix are :

$$\left.\begin{aligned} x &= r \cos (2\pi z/P) \\ y &= r \sin (2\pi z/P) \\ z &= z \end{aligned}\right\} \qquad (12)$$

The Fourier transform expression in Cartesian coordinates may be written in this case as (see equation 3c) :

$$T(X,Y,Z) = \int e^{2\pi i\ (xX + yY + zZ)} dv \qquad (13)$$

where X,Y,Z and x,y,z are the reciprocal space and real space coordinates respectively.

For this simple helix dv is proportional to dz and by substituting equations (12) into equation (13) we obtain

$$T(X,Y,Z) = \int_0^P e^{2\pi i\ \{rX \cos(2\pi z/P) + r\,Y \sin(2\pi z/P) + zZ\}} dz$$

If we neglect constants of proportionality and put $R^2 = X^2 + Y^2$ and $\tan \psi = Y/X$ the expression becomes

$$T(R, \psi, Z) = \int_0^P e^{2\pi i\{ Rr \cos(2\pi z/P-\psi) + zZ\}} dz \qquad (14)$$

Now the X-ray scattering from a continuous helix which repeats exactly a vertical distance P is confined to layer lines at heights Z = n/P in reciprocal space (where n is an integer). Thus the integral in equation (14) only has non-zero value when Z = n/P, or

$$T(R,\ \psi,\ n/P) = \int_{0}^{P} e^{2\pi i\ \{Rr\ \cos(2\pi z/P-\psi)\ +\ zn/P\}} dz$$

The integral cannot be solved by ordinary methods of integration but we are able to simplify this expression by utilizing the identity :

$$2\pi i^{n}\ \mathcal{J}(X) \equiv \int_{0}^{2\pi} e^{iX\ \cos\phi}\ e^{in\phi}\ d\phi$$

if we put $X = 2\pi Rr$ and $\phi = 2\pi z/P$ we obtain :

$$T(R,\psi,n/P) = J_{n}(2\pi Rr)\ e^{in(\psi\ +\ \pi/2)} \qquad (15)$$

Now $J_n(X)$ is a Bessel functinn of the first kind, or order n and argument X.

(ii) Properties of Bessel Functions

$$J_{-n}(X) = J_{n}(X)e^{in\pi}$$

That is, when n is even $J_{-n}(X) = J_{n}(X)$

and when n is odd $J_{-n}(X) = -J_{n}(X)$

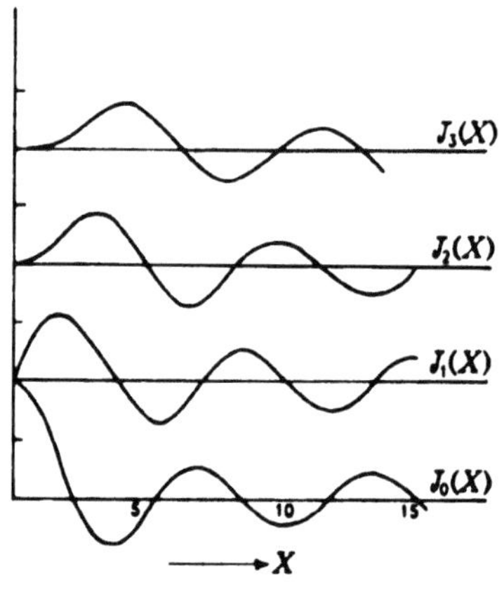

The important points to note are that only the zero order Bessel functions $J_0(X)$ has non-zero value at $X = 0$. Also the value of the first maximum of $J_n(X)$ decreases as n increases and then maximum occurs at larger values of X with increasing value of n.

The intensity distribution on the nth layer line is :

$$|T(R, \psi, n/P)|^2 = |J_n(2\pi Rr)|^2$$

which is independent of ψ.
Since the first maxima of the Bessel functions move further from the origin with increasing order, the centre of the diffraction pattern has a cross-shaped appearance.

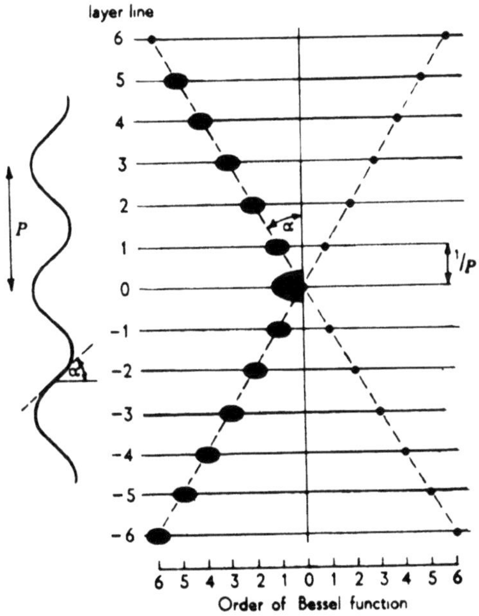

It is often said that the pitch angle α of the helix may be obtained directly from the diffraction pattern by measuring the angle θ, between the meridian and the line drawn through successive maxima. Actually $\tan\theta \simeq 1.1 \tan\alpha$
N.B., there is no intensity on the meridian.

(iii) Transform of a Discontinuous Helix

A discontinuous helix is defined by a set of points occurring with a spacing, parallel to the helix axis, of p, say, but all lying on the continuous helix. The X-rays are imagined to be scattered by these points only. The axial projection of the structure repeats with a periodicity p which gives rise to meridional reflexions at spacings $\pm$ 1/p, $\pm$ 2/p etc. Thus the cross-shaped pattern typical of the continuous helix occurs at every point defined by $\pm$ m/p (where n is an integer) along the meridian.

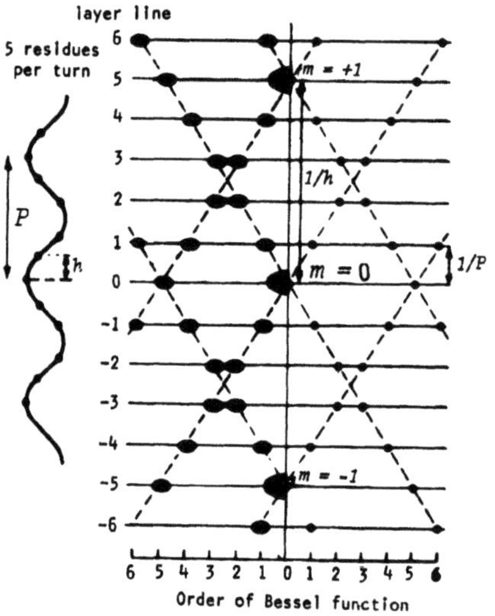

This process is an example of convolution. The multiplication of a continuous helix by a set of planes p apart in real space. Thus the resulting Fourier transform is the Fourier transform of a continuous helix being laid down at points $\pm$ m/p apart in reciprocal space.

The number of monomers per turn of helix is given by P/p which may take any value. Suppose the discontinuous helix repeats exactly in a distance c and let the number of scattering centres in this distance be M, then

$$c = Mp, \text{ or if } N = \text{number of turns of helix, } c = NP$$

The layer line spacing is ℓ/c and since c is the structural repeat along the helix axis intensity occurs at distances given by $Z = \ell/c = n/P + m/p$, where ℓ is the layer line numner. Thus the Fourier transform only has non-zero value governed by the selection rule :

$$\ell = nN + mM \qquad \ldots \quad \ldots \quad \ldots \tag{16}$$

For example in the case of a 8/3 helix :

$$\ell = 3n + 8m$$

ℓ	n	m
0	0	0
1	3	-1
2	-2	1
3	1	0
4	4	-1
5	-1	1
6	2	0
7	-3	2
8	0	1

4. Structure Factor Calculations

In practise a helical structure consists of groups of atoms lying at different radii, angles and heigths. In general the first atom will not start at $x = r$, $y = 0$, $z = 0$ but at $x = r \cos \phi$, $y = r \sin \phi$, $z = z$. To allow for this movement a phase factor $e^{2\pi\ell\ z/c}$ must be introduced. Rotation of the helix through an angle ϕ to bring this point in line with the coordinate system results in the Fourier transform also being rotated in the same direction and by the same amount. Thus a point at (R, ψ, Z) must have originally been at $(R, \psi-\phi, Z)$ and the transform becomes :

$$J_n(2\pi Rr) = e^{i\,[\,n(\psi + \pi/2 - \phi) + 2\pi\ell z/c]}$$

If there are j atoms in the set and the atomic scattering factor of the jth atom is f_j the general expression is

$$T(R, \psi, \ell/c) = \sum_n \sum_j f_j\, J_n(2\pi Rr) e^{i\{n(\psi + \pi/2 - \phi_j) + 2\pi\ell z_j/i\}}$$

subject to the selection rule

$$\ell = Nn + Mm$$

CRYSTAL STRUCTURE DETERMINATION

The analysis of the crystal structure of a polymer necessarily includes both the determination of the relative positions of the atoms which comprise one molecule (chain segment) and the relative positions and orientations of neighbouring molecules in the crystalline lattice. The periodic packing arrangement of molecular chains in a specific manner gives rise to the crystal lattice concept. The basic repeating unit of structure is convoluted with the lattice parameter in real space. The result in Fourier space is to multiply the molecular transform of the unit cell contents by the reciprocal lattice net, i.e. the molecular or unit cell transform is "sampled" only at reciprocal lattice points. Usually the first step in the determination of crystal structure involves the indexing of the X-ray pattern in order to deduce the unit cell shape and size.

1. Determination of Unit Cell Parameters

In general the polymeric material will be in the form of a fibre and we will concern ourselves with indexing fibre diagrams (single crystal rotation photographs) only. The method may also be applied to randomly oriented preparations or chain-fold single crystal mats after appropriate modifications. Suppose that the unit cell is orthorhombic i.e. all the angles are 90° and $a \neq b \neq c$. Let c be the fibre axis (rotation axis). The geometric situation in reciprocal space is shown in figure 5.

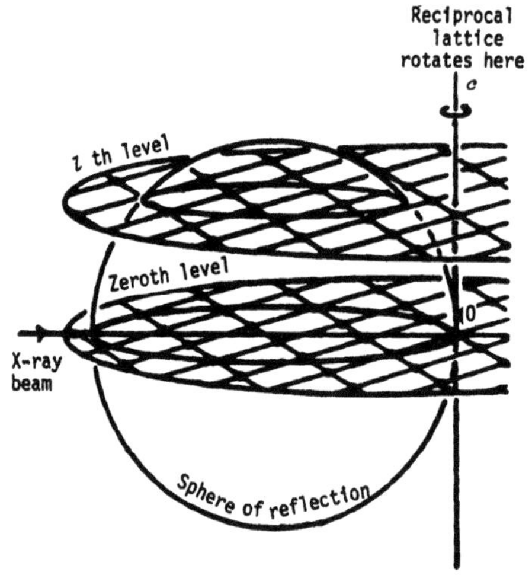

fig.5. reciprocal lattice passing through Ewald sphere as it rotates.

It is a relatively straightforward to calculate the value of c from the layer line spacing. In addition we know that all reflexions on the equator have 1 = 0, for those on the first layer line 1 = 1, for those on the second 1 = 2, and so on. It remains to determine the indices of h and k for each reflexion. Consider first the equatorial reflexions, which have indices of the type (hk0). For an orthorhombic system the equation for the interplanar spacing is

$$d(hk0) = (h^2/a^2 + k^2/b^2)^{-1/2}$$

$$= a/\ [h^2 + k^2/(b/a)^2]^{1/2}$$

which may be rewritten in the form :

$$\log d = \log a - \frac{1}{2} \log [h^2 + k^2/(b/a)^2]$$

A chart for indexing the reflexions of a rectangular reciprocal lattice net can be constructed (Hull-Davey charts) by plotting for each index point (h,k) a curve of b/a versus log d. By marking the logarithm of the observed equatorial spacings on a piece of paper and moving it parallel to the log d axis indices can be selected.

Equations relating the film coordinates (x,y) of a diffraction spot (either flat film or cylindrical film) to the cylindrical polar coordinates of the diffracted beam (ξ, ζ) may be derived. For a flat film they are as follows :

$$x = D \tan \cos^{-1} (2- \zeta^2 - \xi^2/2\sqrt{1-\xi^2})$$

and $y = D.2\ /(2 - \xi^2 - \xi^2)$, where D is the specimen to film distance (see for example : M.J. Buerger, X-ray Crystallography, 1942, J. Wiley). In addition Bernal charts are available to aid the computation.

Careful consideration of figure 7 shows that for ideally oriented fibres in which c^* coincides with the fibre axis, reciprocal lattice points with indices (001) cannot interest the Ewald sphere, so that reflexions cannot appear on the meridian of the diffraction pattern unless the fibre is somewhat inclined with respect to the direct beam. Nevertheless, actual fibres usually possess a sufficiently wide spread of orientations about the ideal fibre axis to generate meridional reflexions when perpendicular to the X-ray beam. Alternatively the fibre can be tilted at specific angles to the X-ray beam to bring particular reflexions into the diffracting condition.

Thus by indexing the reflexions and finding the reciprocal cell parameters allows the real cell parameters to be computed, taking into account the relationship between the two lattices. The data enables the "molecular box" to be established. The contents of the "box" must be determined by the length of time a reciprocal lattice point actually exists in the Ewald sphere. Therefore it is the intensity of the diffraction spots which provide the molecular information concerning the molecules in the unit cell.

2. Intensity of the Diffraction Spots

The ideal intensity of a reflexion (hkl) is derivable for the structure factor, i.e.

$$I(hkl) = |F(hkl)|^2$$

where

$$F(hkl) = \sum_{n=1}^{N} f_n e^{2\pi i(hx_n + ky_n + lz_n)} \qquad (17)$$

In practise a number of experimental factors modify the ideal intensity. The actual intensity observed is given by

$$I(hkl) = P.L.J.A.\ |F|^2 \qquad (18)$$

(i) Polarisation Factor P

The X-rays emitted by an X-ray tube target are unpolarised, but after being scattered by the electrons of an atom they became plane polarised to a degree that is dependent on the Bragg angle θ:

$$P = \frac{1}{2}(1 + \cos^2 2\theta) \qquad (19)$$

(ii) Lorentz Factor L

This term is dependent on the time a reciprocal lattice point spends in the Ewald sphere, i.e. in the diffracting position.

(iii) Multiplicity Factor j

This is the number of different sets of planes, with different (hkl)'s that contribute to a single observed reflexion.

(iv) Absorption Factor A

The absorption factor for X-rays of a particular crystal or other specimen depends not only on its elemental composition and the wave-length of the X-rays but also on the shape and size of the specimen.

(v) Temperature Factor

Because of thermal vibrations of the atoms in a crystal, the ideal intensity is reduced by a factor of the form :

$$e^{-B(\sin^2\theta)/\lambda^2}$$

where B is related to the mean-square amplitude ($\bar{u}^2$) of atomic vibrations by

$$B = 8\pi^2\bar{u}^2$$

3. Crystal Structure

The diffraction from a fibre, although having the appearance of a single crystal rotation diagram, usually has diffraction arcs overlapping along the layer lines which reduced the information available. The probability of overlap occurring is increased by the diffraction broadening caused by the small crystallite size. These effects increase with the diffraction angle and tend to smear the diffracted intensity over so large an area that it is below the threshold for observation. The paucity of diffraction data at large diffraction angles ahs the net effect of imposing an ultimate limit on the accuracy with which atoms can be located - the so called re-solution of the structure. In attempting to solve the crystal structure of a polymer we must therefore recognize at the onset that the limitations affecting the intensity data at our disposal will in general deprive us of the use of conventional methods of crystal structure determination; such as Fourier difference synthesis of full-matrix least squares refinement. In the past it has been usual to compute the Fourier transform of promising molecular structures and to evaluate their relative merits as well as gauge the degree of refinement accomplished by the quality of agreement between observed and calculated structure factors. The reliability index

$$R = \frac{\Sigma|F_o - F_c|}{\Sigma|F_o|}$$

is used to obtain an indication of the agreement between observed and calculated data.

In practise the investigator examines the general distribution of intensities in the diffraction pattern for indications of direct help; for example, (a) whether the main chain possesses a planar zigzag or helical conformation, (b) if helical, what the particular helix is, (c) the number of monomeric units in the unit cell, (d) probable orientation of any planar groups (amide, aromatic etc.) (e) indication of symmetric elements such as screw axes etc. In addition full use should be made of, (a) the stereochemistry of the molecule, (b) any infra-red or other spectroscopic data, (c) analogies with the crystal structure of similar polymers, and (d) compilation of standard bond lengths and angles for single crystal structures of related monomers.

A more recent advance towards bias-free and automatic refinement of polymer structures by constrained least-squares-matrix methods has been reported by Arnott (Arnott, S, 1967, Symp. Fibrous Proteins, p.26, edit. W.G. Crowther, Butterworths, London).

4. Example - Crystal Structure op Poly(ethylene adipate)

$$\left[-(CH_2)_2 - O - \overset{O}{\overset{\|}{C}} - (CH_2)_4 - \overset{O}{\overset{\|}{C}} - O -\right]_n$$

Fibre patterns were recorded in a cylindrical camera with Cukα radiation (λ = 0.154 nm)
(See A. Turner-Jones and C.W. Bunn, Acta Cryst. 15, 105, 1962)

Unit cell

$\underline{a}$ = 0.547 ± 0.003 nm

$\underline{b}$ = 0.723 ± 0.002 nm

$\underline{c}$ = 1.172 ± 0.004 nm (fibre axis)

β = 113.5°

Volume of cell ; V = abc sinβ = 0.4254 nm^3

Molecular weight of monomer, M = 172.18

From the unit cell volume and molecular weight the calculated density for monomers in the cell (N = 2) is

$$d = \frac{NM}{Vx\ (\text{Avogadro's Number})} = \frac{2 \times 172.18}{0.4254 \times 10^{-21} \times 6.023 \times 10^{23}}$$

= 1.34 g/cm^3, which may be compared with the experimental density of the drawn fibre of 1.26 g/cm^3.

Examination of the intensity distribution on the equator favours an arrangement with one chain in the centre of the unit cell and one at each corner : the strong (110) and (020). The absence of all reflexions of the type (h01) for odd 1 indicated a space group $P2_1/a$ (see International Tables of Crystallography). The layer line periodicity of 1.17nm may be identified with the extended monomer repeat of 1.22nm, some slight perturbation being required to reduce this length by 0.05 nm. After refinement the reliability index, R, was found to be 0.19.

DIFFRACTION PATTERNS OF ORIENTED POLYMERS

When linear polymers are subjected to mechanical deformation by cold drawing, rolling, or stretching, the molecular chains tend to align themselves in the general direction of the stress direction or at some angle to it. The material is often termed preferentially oriented. Good orientation gives rise to a well defined fibre pattern with little angular spread of the diffraction arcs. No preferred orientation is equivalent to a power pattern. Many polymers give patterns intermediate between the two types. X-ray diffraction can be used to establish the relative spatial orientations of particular reflexions and also to estimate the angle spread about the mean.

1. Poles

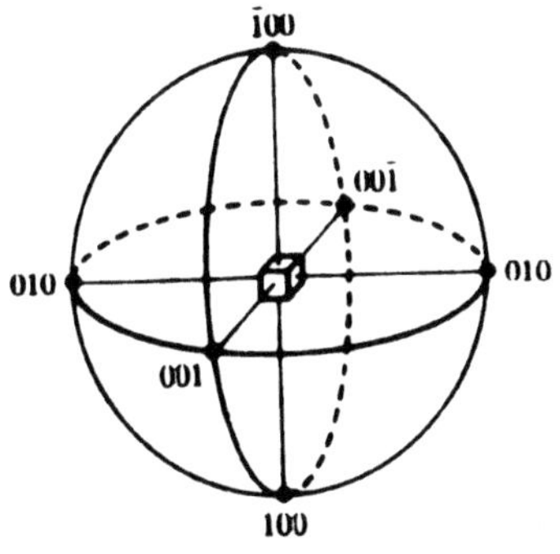

A point is taken in a crystalline substance and an array of normals to the lattice planes are constructed. Intersections of these normals emanating outwards with a sphere centred on the reference point produces a set of poles on the surface of the sphere.

The distribution of such poles is known as the "pole figure".

2. Stereographic Projection

Since the pole figure is a three-dimensional network to display such poles in two-dimensions a stereographic projection is

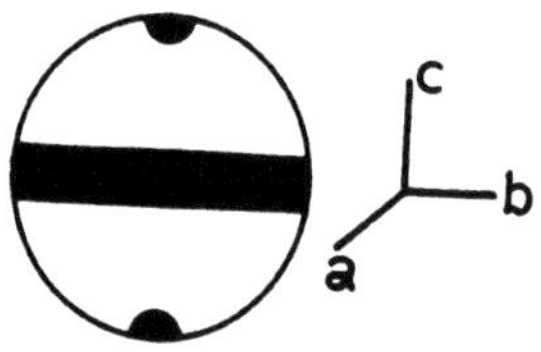

Fig. 6. Illustrates a crystalline unit which is randomly rotated about the c axis.

used. Poles in the northen hemisphere and southern hemispheres are symbolised by solid and open circles respectively. In practise a Wulff net is used to help locate and define the positions of poles on the stereographic projection.

LINE BROADENONG DUE TO FINITE CRYSTALLITE SIZE

The width of a diffraction spot increases as the thickness of the crystal decreases. The width W is usually measured, in radians, at the intensity equal to half the maximum intensity (see fig. 7). As an approximate measure of W we can take half the difference between the two extreme angle at which the intensity is zero, or

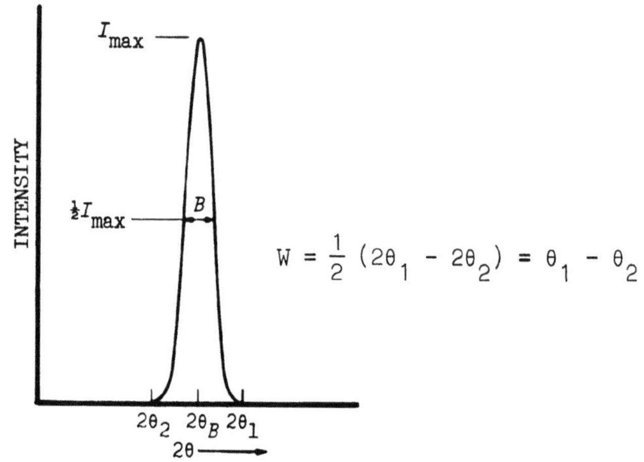

Fig. 7.

The path difference equations for these two angles are

$$2t \sin \theta_1 = (n + 1)\lambda$$

$$2t \sin \theta_2 = (n - 1)\lambda$$

By subtraction we obtain

$$t(\sin \theta_1 - \sin \theta_2) = \lambda$$

or

$$2t \cos \left(\frac{\theta_1 + \theta_2}{2}\right) \sin \left(\frac{\theta_1 - \theta_2}{2}\right) = \lambda$$

But θ_1 and θ_2 are very nearly equal to θ, so that

$$\theta_1 + \theta_2 \simeq 2\theta, \text{ and } \sin \left(\frac{\theta_1 - \theta_2}{2}\right) \simeq \left(\frac{\theta_1 - \theta_2}{2}\right)$$

Therefore $2t\left(\frac{\theta_1 - \theta_2}{2}\right) \cos\theta = \lambda$

$$t = \frac{\lambda}{W \cos\theta} \qquad (20)$$

A more exact treatment to the problem gives

$$t = \frac{0.9\ \lambda}{W \cos \theta}$$

This is known as the Scherrer formula.

Example

Suppose $\lambda = 0.15$ nm, $d = 0.1$ nm, and $\theta = 49°$. Then for a crystalline specimen 1 mm in diameter the line width W, due to the small crystal effect alone, would be about 2×10^{-7} radian, or too small to be observable. Such a crystal would contain some 10^7 lattice planes of the spacing assumed above. However if the crystal were only 50 nm thick, it would contain only 500 planes, and the diffraction peak would be about 4×10^{-3} radian, noticeably broad.

Variance method

This measure of breadth is based on the moment of inertia of the line about its centre of gravity (centroid). If the practical range of integration is from θ_1 to θ_2, the centroid is defined by :

$$\langle Q \rangle \quad \frac{\int_{\theta_1}^{\theta_2} \theta I(\theta) d\theta}{\int_{\theta_1}^{\theta_2} I(\theta) d\theta}$$

If the line is symmetrical the centroid will coincide with the peak. The variance is defined as the mean-square deviation of from its mean value

$$W_\theta = \langle(\theta - \langle\theta\rangle)\rangle^2 = \frac{\int_{\theta_1}^{\theta_2} (\theta - \langle\theta\rangle)^2 I(\theta) d\theta}{\int_{\theta_1}^{\theta_2} I(\theta) d\theta}$$

and the relation between apparent size and variance is :

$$\Delta\theta = \frac{\lambda(\theta_2 - \theta_1)}{4\pi^2 W \cos\theta}$$

LATTICE DISTORTIONS AND PARACRYSTALLINITY

The structure of an ideal crystal is characterized by the dimensions of the unit cell, the electron density distribution within the cell, and the crystal size. We may introduce the concept of the paracrystal as a crystal possessing these same characteristics except that now the dimensions of the unit cell and the electondensity distribution vary statistically from cell to cell. These irregularities in the paracrystalline lattice cause the diffraction spots to be broader and more diffuse than would be expected on the basis of classical crystallite-size effect.

(i) Distortions of the First and Second Kinds

Hosemann has postulated that real crystal structures are subject to two kinds of distortion. In lattices affected only by distortions of the first kind the long range periodicity is preserved, that it, the distortions are displacements of structural elements, such as atoms or molecules.

In a lattice possessing distortions of the second kind the long range-order is lost and each lattice point varies in position only in relation to its nearest neighbours rather than to ideal lattice points. Thus the absolute magnitudes of the displacements from the positions corresponding to a strictly periodic arrangement increase as a function of the square of the distance from any arbitrarily chosen reference point.

(ii) One-dimensional distorted lattice

Consider points distributed along a straight line as shown in fig. 8

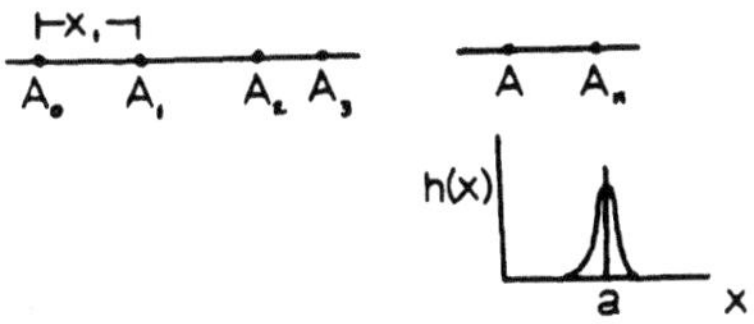

fig. 8. (a) irregular one-dimensional lattice,
(b) typical distribution function h(x).

Let the distance between any two successive points vary about a mean value a, and the distance A_{n-1} to A_n is not related to the next segment, A_n to A_{n+1}, that is, the distances between near neighbours are independent. Furthermore let them all follow the same statistical distribution h(x), where h(x)dx is the probability that distance x_1 between two near neighbours lies between x and x + dx. The quantity h(x) is normalised,

$$\int_0^{\infty} h(x)dx = 1$$

and its average value is

$$a = \int_0^{\infty} xh(x)dx$$

Let us now calculate the distribution function $h_2(x)$ for the distance x_2 between second neighbours, x_2 being the sum of two distances between close neighbours y and x-y. The probability for the occurrence of both these events is the product of the two individual probabilities of having a distance x_1 equal to y and x-y respectively, or h(y) and h(x-y).

Thus total probability is

$$h(x) = \int_0^\infty h(y)\, h(x-y)dy = h(x) * h(x)$$

which is, of course, the convolution of h(x) with itself. Similarly

$$h_3(x) = h_2(x) * h(x)$$

and in general

$$h_n(x) = \underbrace{h(x) * h(x) * h(x) \ldots\ldots\ldots\ldots * h(x)}_{n \text{ times}}$$

All the functions $h_n(x)$ are normalised :

$\int_0^\infty h_n(x)\, dx = 1$, and the average value of x_n is na. The same is true for second, third etc. neighbours.

The points are scattered more and more about the mean lattice positions situated at a, 2a, 3a, etc. At large distances the nth region is so diffuse that it overlaps with its neighbours and the distribution becomes uniform. Since $h_{-n}(x) = h_n(-x)$ the overall distribution is

$$z(x) = \delta(x) + \sum_1^\infty h_n(x) + \sum_1^\infty h_n(-x)$$

(iii) Fourier transform of z(x)

The transform of h(x) is as follows :

$$H(s) = \int h(x) \exp(2\pi isx)dx \qquad (21)$$

and, using the colvolution theorem we have

$$F\, h_n(x) = F\,\underbrace{\left[h(x) * \cdots\cdots\cdots * h(x)\right]}_{n\ \text{times}} = \left[H(s)\right]^n \qquad (22)$$

but since, $h_{-n}(x) = h_n(-x)$

$$F\, h_{-n}(x) = \left[H^*(s)\right]^n$$

Using equation (21) and (22) and remembering that

$F\,\delta(x) = 1$, we obtain

$$F\, z(x) = Z(s) = 1 + H(s) + H^2(s) + \ldots\ldots + H(s) + \left[H(s)\right]^2 + \ldots$$

rewriting with $H(s) = Ae^{i2\pi u}$, gives

$$Z(s) = 1 + 2 \sum_1^\infty A^m \cos(2\pi mu)$$

which can be rewritten as

$$Z(s) = \frac{1 - A^2}{1 + A^2 - 2A\cos(2\pi u)}$$

It turns out that $A = \exp(-2\pi^2 s^2 \Delta^2)$, where Δ is the characteristic width of the distribution $h(x)$ and $u = sa$. The maximum and minimum values of this function occur as the cosine terms fluctuates from +1 to −1

$$Z_{max} = \frac{1+A}{1-A}$$

$$Z_{min} = \frac{1-A}{1+A}$$

and the shape of Z(s) is shown in fig.9

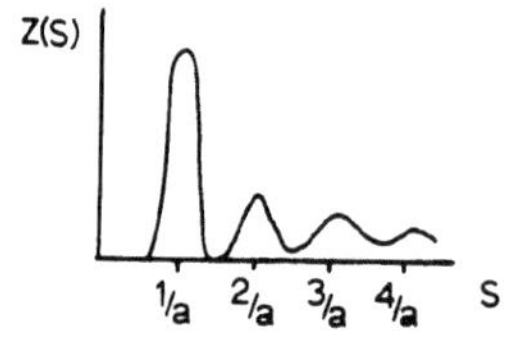

fig. 9. transform of Z(s)

Thus we see a series of maxima for integral values of sa. These maxima decrease in intensity and spread out as a function of the order of the reflexion.

SUGGESTION FOR FURTHER READING

Diffraction of X-rays by Chain Molecules
by B.K. Vainshtein, 1966, Elsevier, Amsterdam.

X-ray Diffraction Methods in Polymer Science
by L. Alexander, 1969, Wiley-Interscience, New York.

X-ray Diffraction by Polymers
by M. Kakudo and N. Kasai, 1972, Elsevier Pub. Co. Amsterdam.

X-ray Diffraction in Crystals, Imperfect Crystals abd Amorphous Bodies
by A. Guinier, 1963, Freeman & Co. San Francisco.

Elements of X-ray Diffraction
by B.D. Cullity, 1956, Addison-Wesley, London.

Optical Transforms
by C.A. Taylor and H. Lipson, 1964, Bell and Sons Ltd., London.

Diffraction of X-rays by Proteins, Nuclei Acids and Viruses
by H.R. Wilson, 1966, Edward Arnold, London.

X-ray Crystallography
by M.J. Buerger, 1962, John Wiley, London.

Chemical Crystallography
by C.W. Bunn, 1945, Oxford Press.

Holmes, K.C. and Blow, D.M., in Methods of Biochemical Analysis, edit. D. Glick, vol. 13, 1965, p. 113, John Wiley, New York.

Natta, G. in Nuovo Cimento, 1960
vol. 15, Series 10, No. 1. p. 1.

EXPERIMENTAL CHARGE DENSITIES IN SOLIDS AND THEIR SIGNIFICANCE FOR THEORETICAL CALCULATIONS

Ph. Coppens

Department of Chemistry

State University of New York at Buffalo

Buffalo, New York 14214

I. INTRODUCTION

The rapid development of X-ray Crystallography in the past decades has provided a practically routine tool for the determination of crystal structures of moderate complexity. In such determinations the atomic positions are deduced from the experimental electron distribution, which scatters X-rays and gives rise to the X-ray diffraction pattern. The elastic amplitude of scattering of the contents of one unit cell or structure factor F, is given by

$$F(\underset{\sim}{S}) = \int_{\substack{\text{unit}\\ \text{cell}}} \rho\,(\underset{\sim}{r}) \exp \frac{2\pi i}{\lambda}\, \underset{\sim}{S}\cdot\underset{\sim}{r}\; d\tau \qquad (1)$$

where ρ is the time-averaged one electron density at the point defined by $\underset{\sim}{r}$, and $\underset{\sim}{S}$ is a reciprocal lattice vector of magnitude $2 \sin \theta/\lambda$.

The elastic scattering defined by (1) is a one-electron property, even though the total scattering, including inelastic processes, is a two-electron property, as given by the quantum mechanical Waller-Hartee theory :[1,2]

$$I_{total}(\underset{\sim}{S}) = \int \psi^{*} \left| \sum_i \exp \frac{2\pi i}{\lambda}\, \underset{\sim}{S}\cdot\underset{\sim}{r} \right|^2 \psi \; d\tau = \sum_i \sum_j \int \psi^{*} \left| \exp \frac{2\pi i}{\lambda} \underset{\sim}{S}\cdot\underset{\sim}{r}_{ij} \right| \psi \; d\tau \qquad (2)$$

where ψ is the electronic wave function, the sum is over all electrons, and $\underset{\sim}{r}_{ij} \equiv \underset{\sim}{r}_i - \underset{\sim}{r}_j$.

The elastic component of the total intensity is given by

$$I_{\text{elastic}}(\underset{\sim}{S}) = \left|\int \psi^* \left| \sum_i \exp \frac{2\pi i}{\lambda} \underset{\sim}{S}\cdot\underset{\sim}{r}_i \right| \psi \; d\tau\right|^2 \quad (3)$$

where the integration of the wave function is over all coordinates but those of the ith electron, leading to the one-electron density structure factor defined in (1).

As described by (1) the structure factors are the Fourier transform of the electron density ρ. According to the Fourier transform theorem ρ is therefore the inverse Fourier transform of the structure factors, the magnitudes of which are experimentally accessible. The following lectures will be concerned with the experimental electron density, complications introduced by the inability to measure the phase of the scattering amplitudes, results obtained in recent studies and their application as a test of the adequacy of theoretical calculations, and the derivation of physical properties from the experimental functions.

II. THE COMPUTATION OF EXPERIMENTAL DENSITIES

2.1 Definition of difference density functions

For a periodic lattice the structure factor F is obtained at values of $\underset{\sim}{S}$ satisfying the Bragg diffraction condition. The electron density ρ is then a summation over all values of F, rather than an integration over a continuous function :

$$\rho(\underset{\sim}{r}) = \frac{1}{V} \sum_{\underset{\sim}{H}} F(\underset{\sim}{H}) \exp -2\pi i \; \underset{\sim}{H}.\underset{\sim}{r} \quad (4)$$

in which $\underset{\sim}{H}$ represents vectors $\underset{\sim}{S}$ fulfilling the diffraction condition, defined by the indices h, k and l :

$$\underset{\sim}{H} = h\underset{\sim}{a}^* + k\,\underset{\sim}{b}^* + l\,\underset{\sim}{c}^* \quad (5)$$

where $\underset{\sim}{a}^*$, $\underset{\sim}{b}^*$ and $\underset{\sim}{c}^*$ are the reciprocal lattice dimensions.

An objection to the use of (4) in the analysis of charge distributions is that it contains a large contribution from the core electrons which are unperturbed by bonding within the accuracy of the diffraction experiment.[3,4] Furthermore, errors are introduced because the accessible range of $\underset{\sim}{H}$ is limited by the wavelength used and restrictions inherent in the experimental equipment.

For both reasons, it is desirable to substrat the inner electrons (the scattering of which persists to large values of $\underset{\sim}{S}$) from the electron distribution. The following two functions may be defined:

$$\rho_{valence}(\underset{\sim}{r}) = \rho(\underset{\sim}{r}) - \sum_{atoms} \rho_{i,core}(\underset{\sim}{r}) =$$

$$\frac{1}{V} \sum_{\underset{\sim}{H}} (F(\underset{\sim}{H}) - F_{core}(\underset{\sim}{H}) \exp(-2\pi i\, \underset{\sim}{H}.\underset{\sim}{r}) \qquad (6)$$

and

$$\rho_{deformation}(\underset{\sim}{r}) = \rho(\underset{\sim}{r}) - \sum_{atoms} \rho_{i,spherical\ atom}(\underset{\sim}{r})$$

$$= \frac{1}{V} \sum_{\underset{\sim}{H}} (F(\underset{\sim}{H}) - F_{spherical\ atom}(\underset{\sim}{H}))\exp(-2\pi i\, \underset{\sim}{H}.\underset{\sim}{r}) \qquad (7)$$

The deformation density has been used extensively in experimental studies, and is very sensitive to the detailed nature of chemical bonding. Unfortunately, it is not routinely calculated in theoretical work. Below we will describe applications of both functions and argue that both should be calculated in theoretical and experimental studies.

2.2 The simple crystal paradox

For a periodic density, the elastic scattering amplitude $F(\underset{\sim}{S})$ is only available at distinct points for which $\underset{\sim}{S} = \underset{\sim}{H}$. Much of the early experimental work was done on simple crystal structures with a small unit cell such as NaCl,[5] and diamond.[6] Such crystals have few or no reflections in the low angle region, in which the scattering of the valence electrons is concentrated. Aluminium, for example, has only five reflections with $\underset{\sim}{S} < 1Å^{-1}$. The situation is only slightly more favorable for silicon which has nine reflections in the low angle region.[7] To base any conclusions on such a small number of measurements extreme accuracy is required ($\sigma(F) \sim$ 1/2% or better).

The amount of information on the valence electrons increases when the unit cell dimensions become larger. Thus, paradoxically, more complicated crystals are more suitable for electron density studies. This is a result of the reciprocal relationship between the crystal and its diffraction pattern, which leads to a very dense diffraction pattern (small distance between reflections) for a large crystallographic unit cell (complicated structure).

Terms in the summation (4) with small $\underset{\sim}{H}$ correspond to density waves with a large wavelength. When $H = 0.2Å^{-1}$, for example, the corresponding contribution to $\rho(\underset{\sim}{r})$ has a periodicity of 5Å. Very few such waves are possible when the periodicity of the lattice is of the same magnitude.

2.3 The assumptions of X-ray crystallography ; the phase problem and the spherical atom and harmonic motion approximations.

The structure factor $F(\underset{\sim}{H})$ is in general a complex quantity, which may be written as

$$F(\underset{\sim}{H}) = A(\underset{\sim}{H}) + i\ B(\underset{\sim}{H}) \qquad (8)$$

or as a vector in the complex plane :

$$\underset{\sim}{F}(\underset{\sim}{H}) = |\underset{\sim}{F}(\underset{\sim}{H})| \ \exp i\phi \qquad (9)$$

where ϕ is the phase of the wave ($\phi = \tan^{-1} B/A$), which is not experimentally accessible. Using Friedel's law ($F(\underset{\sim}{H}) = F^{*}(-\underset{\sim}{H})$ and applying Euler's relation, (4) can be written as :

$$\rho(\underset{\sim}{r}) = \frac{2}{V} \sum_{0}^{+\underset{\sim}{H}} |F(\underset{\sim}{H})| \cos 2\pi(\underset{\sim}{H}.\underset{\sim}{r} - \phi) \qquad (10)$$

where the summation is over half the vectors in reciprocal ($\underset{\sim}{H}$) space. Expression (10) implies that the phase ϕ must be known for the calculation of ρ.

This is of course, the well-known phase problem of X-ray crystallography. In the centro-symmetric case $B(\underset{\sim}{H})$ is zero, because $\rho(\underset{\sim}{r}) = \rho(-\underset{\sim}{r})$, leading to cancellation of the sin terms in (1) and values ϕ equalling either zero or π. The phases can be determined approximately by an impressively successful array of mathematical techniques, developed by X-ray crystallographers over the past decades.[8]

Once the approximate phases are known, further refinement is done by least-squares minimization of the discrepancies Δ between the observed magnitudes of F and those calculated in the spherical-atom approximation :

$$F(\underset{\sim}{H}, \text{calc}) = \sum_{\substack{\text{all} \\ \text{atoms}}} f_n \exp(2\pi i\ \underset{\sim}{H}.\underset{\sim}{r}_n)\ T_n \qquad (11)$$

Here, f_n is the theoretical (free-space atom) scattering factor for the atom at $\underset{\sim}{r}_n$ (= $x_n\underset{\sim}{a} + y_n\underset{\sim}{b} + z_n\underset{\sim}{c}$, where $\underset{\sim}{a}$, $\underset{\sim}{b}$ and $\underset{\sim}{c}$ are the unit cell edges), with a temperature factor T_n, which is the Fourier transform of the therma-vibration smearing function.

Equation (11) contains two assumptions. In the harmonic approximation T_n depends on the mean square amplitude of atomic displacement, which may be a scalar U_n if the displacements are isotropic, or a second rank tensor $U_{ij,n}$ in the case of anisotropic harmonic motion. The second approximation is that f_n can be derived from the wave function of the isolated spherical atom, which is true only if the electron distribution in the crystal can be represented by a superposition of spherical atoms, an assumption challenged in the electron density work. Experience with different scattering factor models (see below) has shown that for crysals with a center of symmetry approximation (11) is sufficiently good to distinguish between the two possible phases; and the electron density summation (10) can therefore be performed with the *calculated* phases and the *observed* amplitudes. For non-centric crystals, however, phase errors of the order of 2-3° introduce an additional term in the average standard deviation of the experimental density[9], which can be written as :

$$< \sigma^2(\rho) > = \frac{2}{V^2} [\sum_{0}^{\underset{\sim}{H}} \sigma^2 F(\underset{\sim}{H}) + f^2(\underset{\sim}{H}) \sigma^2(\phi)] \qquad (12)$$

Non-centric crystal structures are therefore less suitable for charge-density studies and should be avoided if similar information can be obtained from a centrosymmetric solid.

2.4 The use of neutron and high-order X-ray data in the calculation of difference densities

2.4.1. Bias in atomic parameters obtained with the spherical atom approximation

For the difference summations (6) and (7) knowledge of the atomic parameters $\underset{\sim}{r}_n$ and T_n is required. The spherical atom approximation may introduce significant errors in these parameters, as becomes evident on comparison with the results of a neutron structure determination. The structure factor applicable to elastic neutron scattering is identical to (11), but for the replacement of f_n by the nuclear neutron scattering length b_n. Since the dimension of the atomic nuclei are smaller by a factor 10^{-5} than the wavelength of the thermal neutrons used, the nuclei are essentially point scatterers which obey the assumption of spherical symmetry. Furthermore, unlike f_n, b_n is not a function of $\underset{\sim}{S}$ and no further assumptions on the radial distribution of the atomic scattering matter are required.

Discrepancies between X-ray and neutron thermal parameters are often as high as 10% for first row atoms, but typically less for heavier atoms, where the valence electrons constitute a smaller fraction of the total electron density. The anisotropy of the discrepancies in oxalic acid dihydrate and s-triazine is such that in the X-ray analysis the atom appears to be smeared out in directions corresponding to the chemical bonds and lone pair electrons.[10] On the other hand, X-ray parameters describing vibrations perpendicular to the molecular plane in aromatic molecules are often too small,[11,12] suggesting a contraction of electron density into the molecular plane.

A similar, but less pronounced effect is observed for X-ray positional parameters of atoms which are in an asymmetric bonding environment.[13,14,15] Hydrogen atoms with their single (valence) electron are typically displaced by 0.1-0.2Å into the bond to the adjacent atom,[16,17] the displacement being larger for OH than for CH,[18] in agreement with the difference in electronegativity of the oxygen and carbon atoms. For first-row atoms, the asphericity shift is smaller by a factor ten or more,[9] and is typically 0.005-0.01 Å. Atoms with lone-pair electrons may be displaced away from the bonds, towards the lone-pair region, while carbon atoms are shifted into triple bonds[19,20] and towards the center of the ring in aromatic compounds such as p-terphenyl,[21] pyrene[22] and phenanthrene.[23]

The thermal and positional parameter has seriously affects the difference densities according to (6) and (7). It is therefore essential to obtain unbiased values from an independent neutron experiment or from a modified refinement of the X-ray data.

2.4.2 X-N difference densities

The X-N difference densities are defined by (6) and (7) with

$$F_{\text{core(spherical atom)}} = \sum_{\substack{\text{all} \\ \text{atoms}}} f_{n,\text{core(spherical atom)}} \exp(2\pi i \, \underset{\sim}{H} \cdot \underset{\sim}{r}_{n,N}) T_{n,N} \qquad (13)$$

where $\underset{\sim}{r}_{n,N}$ and $T_{n,N}$ are determined in the neutron experiment.

Some representative deformation densities are shown in figs 1 and 2. Their interpretation will be discussed below.

2.4.3. X-X difference densities.

In the absence of neutron results parameter bias may be reduced by modification of the X-ray refinement. Of the different techniques described[24] the one most commonly applied is the high order X-ray refinement, the validity of which is based on the assumption that the scattering of the diffuse valence electrons is limited to the low order region, (which is usually defined as containing all reflections accessible with CuKα radiation (λ = 1.54 A, $S < 1.3$ $Å^{-1}$). This assumption breaks down to some extent because of bonding effects. Evidence for this breakdown is given by

a) the calculated scattering factors of a prepared state oxygen atom.[25]

b) an analysis of the scattering of the features in the theoretical deformation density of dicyanogen[9]

c) comparison of high order refinement parameters with neutron results, which shows that atoms with lone-pair electrons are shifted towards the lone-pair in the high-order refinement.[26]

As the residual high-order valence scattering is associated with the lone-pair electrons, X-X (high angle) deformation maps will give a reasonable representation of the bond density, but be much less reliable in the lone-pair regions. This is confirmed by comparison of X-X and X-N maps on glycylglycine[27] and p-nitropyridine N-oxide[12] and by examination of the ethyleneimine deformation densities in the bonding and carbonyl lone-pair regions.[35,28]

2.5 Experimental considerations

Great care should be taken in collection of experimental data for charge density studies. Generally, several sets of symmetry equivalent reflections are to be measured, and their intensities should agree to 2% or better.

Crystal quality and homogeneity of the X-ray or neutron beam should be carefully analyzed. Data must be corrected for absorption while extinction and multiple reflection effects should be taken into account. Measurement of the experimental scale[29] and collection of data at reduced temperatures are highly desirable if a quantitative analysis is to be made.

III. SURVEY OF RESULTS AND COMPARISON WITH THEORY

3.1 Molecular Crystals

3.1.1. Discussion of the experimental results

Compounds studied by combined X-ray and neutron diffraction

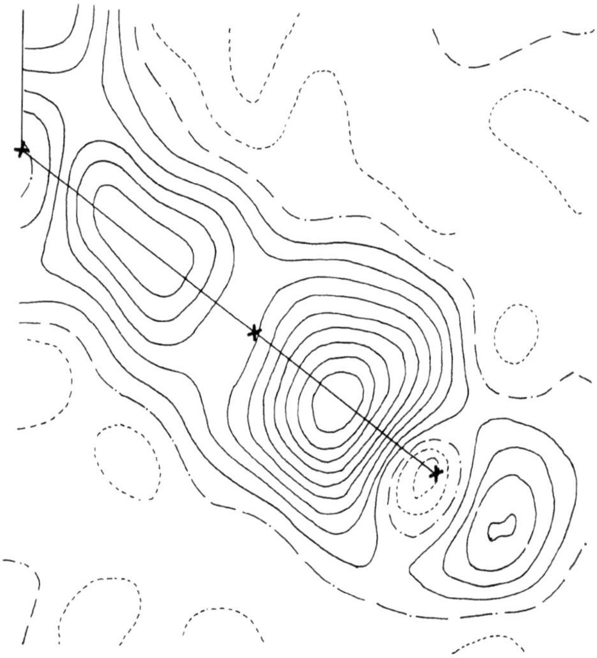

Figure 1. Section of the function ΔQ_{X-N} through the molecular plane of tetracyanoethylene. Contours at 0.10 eÅ^{-3}. Negative contours ---, Zero contour --- The cross to the right indicates the position of the nitrogen atom. The carbon atoms are to the left (Becker, Coppens & Ross, 1974).

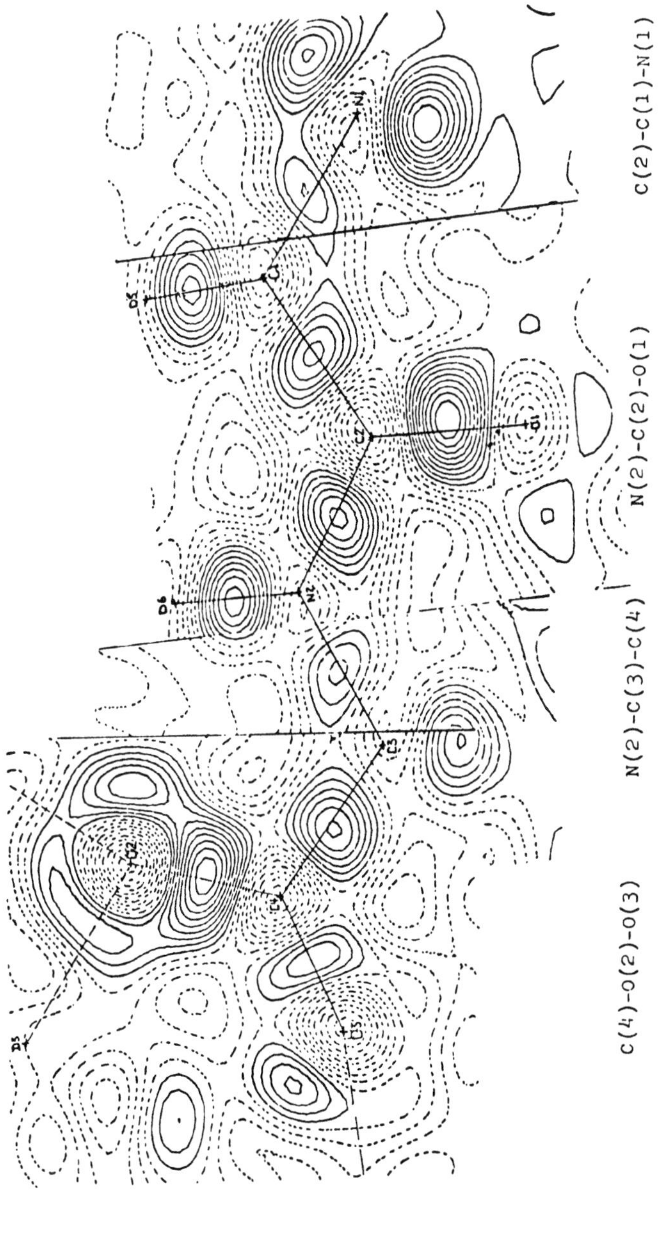

Figure 2. Deformation density in section through the glycylglycine molecule. Contours each 0.05 $e\AA^{-3}$. Negative contours broken.

are listed in table 1. Typical difference maps (figs 1,2) show excess density in the bond and lone-pair regions, and an electron deficiency near the nuclear positions. We will consider the lone-pair and bond features separately :

a) lone-pair peaks. CO oxygen atoms in conjugated systems show a broad "half moon" shaped maximum extended in the molecular plane (cyanuric acid,[26] oxalic acid dihydrate,[30] glycine[31] ammonium oxalate[32]) or two separated maxima (the second CO in cyanuric acid, glycine and ammonium oxalate), suggesting two lone-pairs with approximate sp^2 hybridization. A single maximum is found one the hydroxyl oxygen in oxalic acid. Very sharp, single peaks also appear along the extension of the C≡N bond in cyano-compounds[20,33] and at the back of the carbonyl oxygen in benzenechromium tricarbonyl,[34] indicating sp hybridization of the nitrogen and oxygen lone-pairs. A very pronounced peak extending perpendicular to the molecular plane occurs at the oxygen atom in tetracyanoethyleneoxide.[33] It is similar in shape but more intense than the lone-pair peaks on the water molecules in oxalic acid dihydrate and ammonium oxalate monohydrate, and the general appearance is consistent with approximate sp^3 hybridization of the two lone-pair orbitals.

b) bond density maxima. Distinct maxima of varying height are found on the internuclear axes in CC, CO, CN, BH, CH and NH bonds. The maxima are clearly shifted away from the bond axis, however, in strained three-membered ring systems (ethyleneoxide,[33] and also in X-ray maps of ethyleneimine[35] and cyclopropane,[36]) and in the three-center BHB bond,[49] the bond peaks being displaced outwards in the former and inwards in the latter case.The consistent observation of outward displacements in small ring compounds constitutes a direct proof of the bending of the bonds in strained rings, as proposed first by Coulson and Moffit.[41] Cross sections perpendicular to the bond at bond midpoint show the cylindrical symmetry of the single and triple bonds, and the extension of double or partially double bonds in the direction associated with the π-orbital axis.[37,22,24] Sections perpendicular to the molecular plane containing the bond axis further indicate a polarization of the density towards the more electronegative atom of a heteroatomic bond (glycylglycine,[27] s-triazine[37]). It is of interest that the deformation sections through the bonds in orthorhombic sulfur, show two rather broad maxima rather than a single peak in the bond.[51] This observation may be related to the relative dimensions of the S-S bond length and the radius of the outer maximum of the 3p electrons (2.05 and 0.81 Å respectively). For comparison, the corresponding values for Si, which shows a single maximum are 2.35 and 1.07 Å. A similar effect may then be expected in the O-O bond in hydrogenperoxide (r(O-O) = 1.484 Å ; r(2s) = 0.45 Å).

TABLE I

Combined X-ray and neutron charge density studies

	centric/acentric-temperature (°K)	reference
s-triazine	c-300	37
oxalic acid. 2 H_2O	c-300	30
tetracyanoethyleneoxide	c-300	33
cubic tetracyanoethylene	c-300	20
ammoniumtetroxalate	c-300	42
1.3.5. acetylbenzene	c-300	11
alpha glycine	c-300	31
alpha-glycylglycine	c-300	27
orthorhombic sulfur	c-300	51
cyanuric acid	c-100	26
decaborane	c-113	49
benzenechromiumtricarbonyl	c- 78	34
p-nitropyridine N-oxide	c- 30	12
sucrose	a-300	18
ammoniumoxalate. H_2O	a-300	32
hexamethylenetetramine	a-300	42,52

c) Other features. A large residual peak is observed near the chromium atom in benzenechromium tricarbonyl. Assessment of its physical significance requires additional studies of metallo-organic complexes. A complication arises from the low neutron scattering power of Cr, compared with its dominance in the X-ray scattering. In such a situation errors in the neutron measurement of the Cr parameters are magnified in the X-N map.

The densities on sucrose[18] and glycine[31] suggest that the polarizing influence of hydrogen bonding on the molecular density is measureable, in agreement with theoretical studies of the formamide dimer[38] and the HF..H_2O complex,[39] which predict density changes of 0.1-0.2 eA^{-3}.

A survey of peak heights reveals considerable fluctuations among chemically comparable features.[9] For example, C-C bond peak heights range from 0.25 $e\AA^{-3}$ in oxalic acid to 0.7 $e\AA^{-3}$ in p-nitropyridine N-oxide, and lone-pair densities from 0.15 to 0.65 $e\AA^{-3}$ in the same two compounds. The chemical significance of these differences is largely obscured at present by the effects of thermal averaging, and by an unnecessary lack of knowledge of the absolute scale of the X-ray data.

But, the prospects for more quantitative work seem good as further advances in experiment and interpretation can be expected. In the results obtained so far density peaks and hollows appear where expected from advanced, but not from approximate theoretical calculations.

3.1.2. The effect of thermal averaging

To compare the experimental evidence with theoretical results the effect of thermal smearing is to be considered. The main contribution to thermal vibrations is by low frequency lattice modes. For frequencies lower than 50 cm^{-1} deexcitation is incomplete even below 10°K. Furthermore, the zero-point motion of all modes can not be frozen out, so that the density at rest is a function which cannot be measured even if the resolution of very small wavelengths were available.

It follows that it is necessary for a quantitative comparison to vibrationally average theoretical densities. Extrapolation of experimental densities to zero thermal motion implies a resolution not present in the experimental data.

The convolution of the electron density with a thermal smearing function is relatively straightforward for modes in which the molecule behaves as a rigid body. Using the Fourier transform operator F and a smearing function $T(\underset{\sim}{r})$

$$< \rho(\underset{\sim}{r}) > = \rho(\underset{\sim}{r}) * T(\underset{\sim}{r}) = F^{-1}\, [F(\rho) \quad F(T)] \qquad (14)$$

For internal modes the electron density will in general vary with the nuclear coordinates and a statistical average over all conformations must be taken. In an approximate treatment proposed by Coulson and Thomas[53] the thermal average is calculated by assuming each of the atomic basis functions to vibrate with its nucleus, while the coefficients of the molecular orbitals remain the same during the vibration.

This "convolution" approximation agrees well with more exact calculations for the H_2 molecule and has been applied to acetylene,[54] though it has not been tested for larger molecules.

The work on acetylene indicates that coupling between atoms such as occurs in a molecule is not a major influence, so that mean square amplitudes from neutron refinements or modified X-ray treatments may be used in approximate studies. Experimental studies on the temperature dependence of the deformation scattering in silicon[55] (by monitoring the 222 reflection as a function of temperature) indicate a bond density motion, slightly smaller or equal to the motion of the core electrons.

Approximate models, bases on such relations are readily applied if a Gaussian fit to the overlap and lone-pair peaks is found.[9] They will be applied below in a comparison of theoretical and experimental results on the cyano group.

3.2 Difference densities for some metallic, ionic and covalently bonded crystals

Ionic or "network" crystals have high Debye temperatures, so the thermal smearing of the density is smaller than in molecular crystals. Furthermore, as the atomic positions are often fixed by symmetry considerations, positional parameters are not biased by assumptions inherent in the selection of a refinement model. A major drawback is the small unit cell size of many of these crystals (section 2.2). In aluminium, discrepancies between the structure factors calculated with isolated atom form factors and those from a pseudopotential calculation by Ascarelli and Raccah exist only for the low order (111) and (200) reflections.[43] They both indicate that the valence electrons are more diffuse in the metal than in the isolated atom, as may be expected. The observed values (on an absolute scale) are within the experimental error identical to the results of the Ascarelli and Raccah calculation, but in the case of the (220) reflection disagree with the second pseudo-potential calculation by Walter, Fong and Cohen.[44] The valence charge density in aluminium[43,44] and also in vanadium metal[45] shows its largest maxima between nearest neighbor atoms. Experimental evidence indicates that this is not the case for metallic beryllium

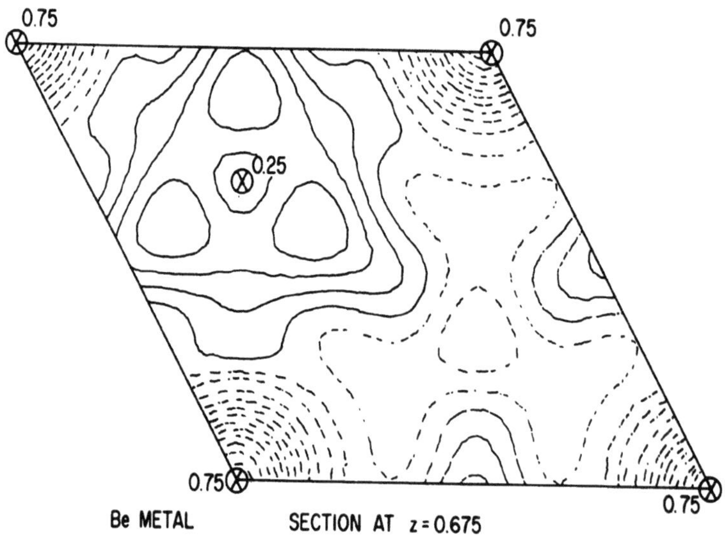

Figure 3. Deformation density in Be metal in a section at z = 0.675 containing the density maximum. Contour 0.025 eÅ^{-3}. Negative contours broken. Crosses indicate projection of Be atoms and their z parameters (Yand and Coppens, to be published).

for which accurate data were collected by Brown.[46] The difference between the OPW (orthogonalized plane wave) scattering factors and the experimental measurements in the region below 0.45 $Å^{-1}$ suggest tight binding i.e. covalent bond formation. Further analysis by Stewart[47] indicates that the bonding is of the multi-center type. This is confirmed by the deformation density map which shows a diffuse distribution with distinct maxima of about 0.08 $eÅ^{-3}$ in the center of a tetrahedron formed by four beryllium atoms (fig. 3).[48]

Density maps of silicon, based on very accurate structure factors from Pendellösung-fringe measurements[56] are shown in Figs. 4 and 5. The cross sections perpendicular to the bonds show almost perfect cylindrical symmetry. The valence density peaks in the midpoint of the bond, unlike valence densities in bonds between first row atoms which tend to show maxima at the nuclear positions. This is compatable with the difference between the radial distributions of the L and M valence electrons.

3.3 Comparison between theory and experiment

3.3.1 Qualitative aspects

How well do the experimental results support theoretical calculations ? Qualitatively agreement is often satisfactory. Thus, calculations on diborane[57] show accumulation of density within the three-center bond, in agreement with experimental densities in the BHB triangles in decaborane.[49] Calculations on cyclopropane[58] show density outside the bonds as observed in the experimental maps. Generally, good quality calculations, with large basis sets, reproduce the overlap and lone-pair densities in molecular crystals. The same is true for pseudopotential calculations on silicon and III-V compounds.[64]

Two further questions occur: a) what sophistication is required for the calculation to give reliable valence and deformation maps. b) how good is the quantitative agreement between experiment and theory.

3.3.2 Experimental results as a test of theoretical adequacy

Calculations on nitrogen and diborane[57,59] and on HF and O_2[60] with basis functions of increasing sophistication show that deformation densities are especially inadequate in the bond regions when minimal basis sets, double zeta or even saturated sp basis sets are used. In particular the addition of polarization functions is required to build up the density in the bonding regions.

The inadequacy of the semi-empirical INDO method and of mini-

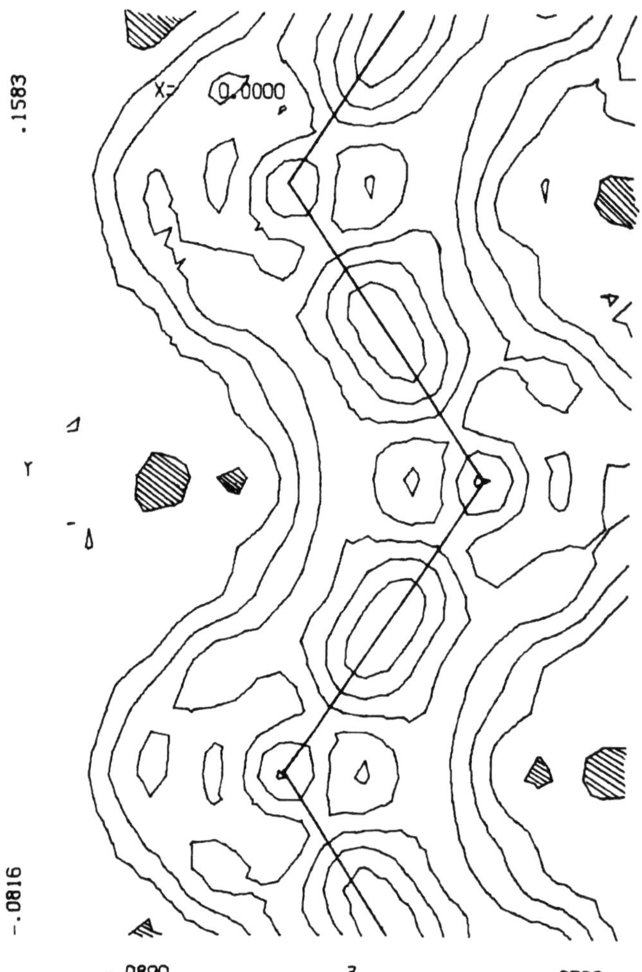

Figure 4a. The valence density in the plane of the Si-Si bonds in silicon. Contours at 0.1 eÅ^{-3}. Negative areas cross-hatched.

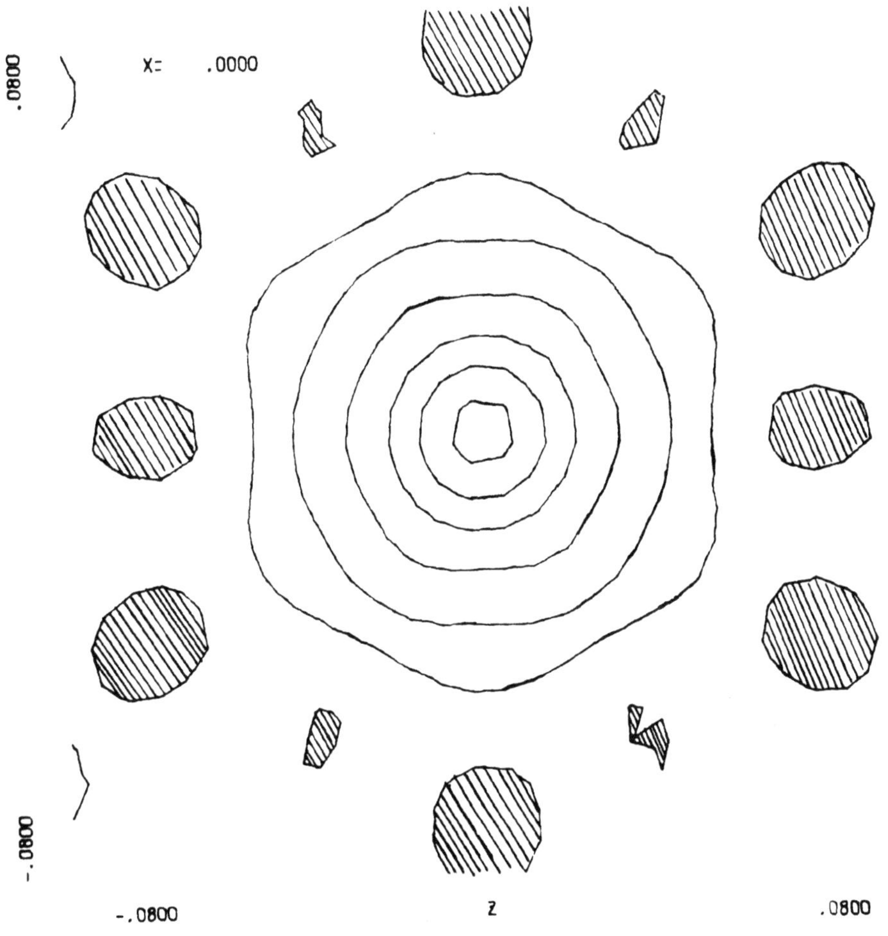

Figure 4b. As 4a, section perpendicular Si-Si bond.

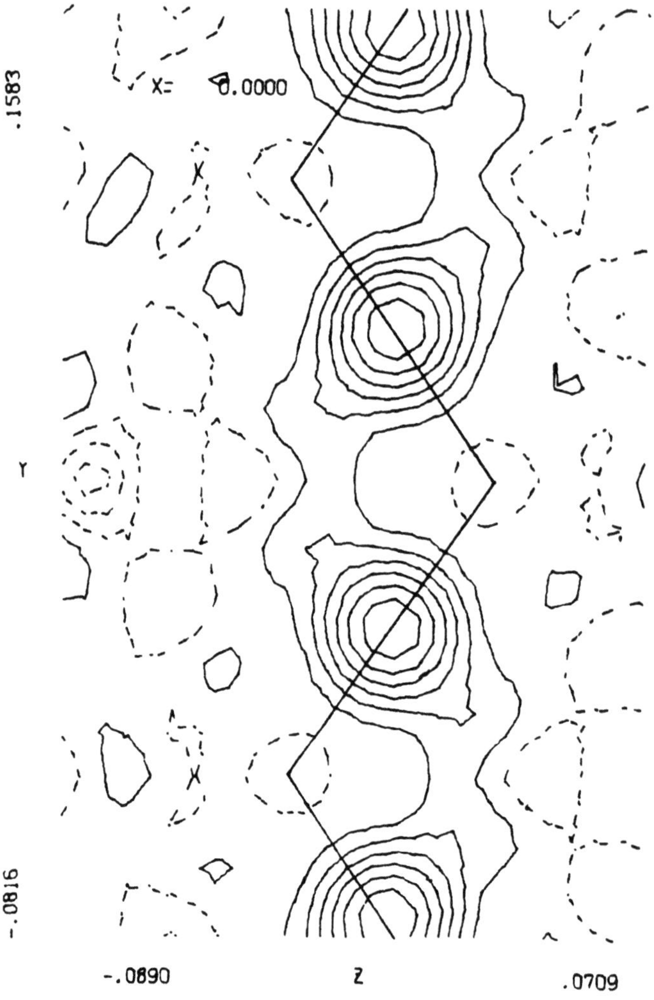

Figure 5. The deformation density in the plane of the Si-Si bonds in silicon. Contours at 0.05 $eÅ^{-3}$ Negative contours broken.

mal-basis set calculations is also apparent from comparison with the experimental difference densities on cyanuric acid.[61] In the theoretical maps the overlap density is almost completely absent, while the lone-pair maxima are several times higher than indicated by experiment, even with allowance for thermal averaging. A study by Bats and Feil[62] in which orbital coefficients and basis functions are varied to give the best fit with the RHF (Restricted Hartree-Fock) electron density in the CN^- ion and LiCN and LiNC molecules, shows that the MBS basis set is inherently too restricted to represent the deformation density, even though the gross features of the total charge distribution may be properly represented.

Even RHF calculations may not always give adequate charge densities, as evidenced by the general decrease in theoretical dipole moment (by about 0.2D) when CI (Configuration Interaction) is included. Calculations on CO by Becker[63] indicate differences of up to 0.2 $e\text{Å}^{-3}$ inthe deformation densities, i.e. about three times the standard deviation of a reasonably accurate experimental determination.

The experimental results on silicon are an unusually good test for calculations, because of their uncommon accuracy. Theoretical calculations have been described with the pseudopotential[64] and a self-consistent orthoganalized plane wave method (SCOPW),[65] but only for the former were valence density maps reported. The peak height in the experimental and theoretical maps are almost exactly equal (0.69 and 0.65 $e\text{Å}^{-3}$ respectively), but the valence density in the pseudopotential map is elongated perpendicular rather than parallel to the bond-axis. The effect is far outside the experimental errors and must be attributed to a deficiency in the calculation. A number of structure factors[56] have been calculated from the SCOPW results. Especially those obtained with the Kohn-Sham-Caspar rather than the Slater exchange term show good agreement with the measurements. A comparison of densities would be highly desirable.

3.3.3 A quantitative comparison

A calculation of a series of deformation densities in linear molecules such as HCN, NCCN and HCCH by Hirshfeld,[66] based on wave-functions of Yoshimine and McLean, shows similar features in comparable molecular fragments, such as the C≡N groups in HCN and NCCN. Small molecules are not readily accessible experimentally, because they tend to be in the gaseous state at room temperature. But approximate molecular transferability in part, justifies comparison of these densities with the experimental results on tetracyanoethylene, shown in fig. 2. Without allowance for thermal motion the most striking discrepancy is the peak height of the lone-pair electrons, the theoretical value being considerably larger than the measurement. Allowance for thermal motion in the harmonic approximation, describing the bond peak by a single Gaussian,[9] shows the lone-pair density to be much more affected by thermal smearing. The thermally corrected values agree in relative height with the experimental results (table 2), but the latter are 10-50% higher, indicating a possible scale factor error in the experimental analysis.

Further studies on the measurement of the experimental scale factor are presently being undertaken. Development of thermal convolution formalisms and theoretical calculations on larger molecules are needed to establish experimental densities as a quantitative test of calculations.

TABLE 2

Peaks heights ($e\mathring{A}^{-3}$) in C≡N, C-C and the lone pair region in TCNE (exp) and NCCN (theoretical)

Theoretical	C≡N	C-C	lone-pair
at rest	1.0	0.6	1.1
thermally averaged	0.65	0.4	0.35
Experiment	0.9	0.6	0.4

IV. DERIVATION OF ONE-ELECTRON PROPERTIES

To what extend can the experimental density be used to derive other functions such as net atomic charges, charge transfer in mixed crystals, dipole and higher moments, field gradients, electrostatic potentials, binding energy and others ? A modest start in this direction has been made. Two distinct approaches are 1) analysis in electron density (direct) space and 2) direct analysis of the observed structure factors i.e. in scattering (reciprocal) space.

4.1 Direct analysis of the electron density

Derivation of one-electron properties by integration of the charge density requires a choice of well defined regions in electron density space . If O is the appropriate operator

$$< \hat{O} > = \int \hat{O} \rho(\underset{\sim}{r})\, d\tau \qquad (15)$$

where the integration is over the region of interest, which may be the immediate surrounding of an ion in an ionic crystal or the volume of the v.d. Waals envelope in a crystal with distinct molecular entities. Integration over the atomic volume in a molecular crystal is far more ambiguous, as no generally accepted definition of what constitues an atom in a molecule exists.[24] At present, use of (15) has been limited to a number of isolated cases discussed below. But it may be anticipated that no reliable values will be derived for properties which depend strongly on the gradient of the density, since steep gradients generally observed near the nucleus in theoretical deformation maps are smeared out by thermal vibrations. The (correlationless) binding energy can, for example be calculated from the one-electron density function and the unbiased nuclear position.[67,68,2] Fair qualitative agreement with spectroscopic values of the binding energy has indeed been obtained from electron diffraction data on gaseous hydrogen, nitrogen and oxygen.[68] But gas-phase electron diffraction is only affected by internal vibrations, which have smaller amplitudes than lattice modes in crystals. Vibrations should affect net ionic and molecular charges less than higher multipoles of the charge distribution, which are increasingly dependent on the position vector $\underset{\sim}{r}$.

4.1.1. Net ionic and molecular charges

Because of the proximity of adjacent ions in a crystal, the charge integration is not as straightforward as may be expected. In the work of kurki-Suonio and collaborators, the separation into component densities is achieved through an analysis of the radial density $4\pi r^2\ \rho\ (r)$.[69] This function, calculated with a theoretical correction for series termination effects, shows a clear minimum at a value of r defined as the radius of best separation. Its value at the minimum, which is different from zero, is a measure of conceptual indeterminacy, and is much higher for metal oxides than for alkali halides. The results indicate approximately doubly and singly ionized states for these two groups of solids. It is interesting that such an analysis leads to non-zero charges when spherical free-space atoms are placed at the ionic positions in the crystal. But, it has been noted that such a hypothetical model bears no relation to reality and should not be used as a basis for an estimate of net charge.[69]

An example of a molecular charge integration is the organic charge transfer complex of tetracyanoethylene (TCNE) with deuteropyrene[70]. The shape of TCNE molecule is such that its volume can be described by a parallelipiped. This facilitates analytical integration of the charge density.[71] The curves of integrated intensity against dimensions of the molecular box show nearly horizontal sections, indicating clear separation of the molecules. The charge transfer in the ground state is found to be zero within experimental error.

Among the most interesting molecular salts, are the mixed crystals of tetracyanoquinodimethanide (TCNQ) and tetrathiofulvalinum (TTF) which show quite unusual electrical properties. Inference based on bond lengths suggest the transfer of about 0.5 electron from TTF to TCNQ.[72] Unfortunately, the molecular volumes are more irregularly shaped and attempts at analytical charge integration as applied to TCNE-pyrene had to be abandoned. Other numerical methods of more general applicability are now being developed.

4.1.2 Dipole and higher moments

The calculation of molecular moments according to (15) seems a fruitful field which is largely unexplored. Studies are now being undertaken in which the integration volume is defined by the v.d Waals radii of the outer atoms. Progress will be reported.

4.2 Reciprocal space analysis

By recasting the structure factor formalism in terms of general density function rather than free-space spherically-averaged atoms, new parameters D may be introduced, which can be optimized in a least squares refinement. Thus, if the structure factor is written as

$$F\ (D_k,\ x_m,\ y_m,\ r_m,\ T_n) \qquad (16)$$

for a unit cell with m nuclei and a density described by k charge and n thermal parameters, least squares refinement can be initiated from an approximate set of starting parameters.

The density parameters D_k may be occupancy factors of spherical valence shells or non-spherical deformation functions, coefficients of radial exponents or elements of the first order density matrix. One-electron density properties may be calculated from the D_k and the basis density functions. The advantage of (16) over analysis in direct space is that thermal motion is explicitly included and that, in principle at least, the results apply to the density at rest. A disadvantage is the necessity to introduce predefined density functions, which are inherently restricted and may therefore projudice the results.

4.2.1 Net atomic charges in molecular crystals

Overlap between atoms is allowed for by projecting the valence electron density into interpenetrating spherical density functions centered at the atoms, assuming again negligible perturbation of the core electrons. Expression 16 becomes :

$$F(\underset{\sim}{H}) = \sum_{\substack{\text{all}\\ \text{atoms}}} [f_{core,i} \exp 2 i\pi \underset{\sim}{H}.\underset{\sim}{r}_i + a_{valence,i} \times f_{valence,i} \exp 2 i\pi \underset{\sim}{H}.\underset{\sim}{r}_i] \, T_i \qquad (17)$$

where $f_{core,i}$ and $f_{valence,i}$ are the Fourier transforms of the core and valence density respectively and $a_{valence,i}$ is the valence shell occupancy parameter. Both Slater-type atomic orbital and isolated atom functions have been used. The best agreement between calculated and observed structure amplitudes is generally obtained with somewhat contracted Slater type orbitals (STO)[73,74] as derived by Hehre et al[75] by a basis-set optimization on a number of small molecular fragments. The experimental value of the scale factor of the mineral kernite also supports the STO functions.[76]

The results qualitatively agree with Mulliken population analyses of theoretical wave functions, though the difference in definition would seem to preclude close agreement. The experimental charges from different determinations generally agree in sign and approximate magnitude when comparable atoms are involved (such as oxygens in water molecules in different crystalline hydrates).[74] The conclusions may be summarized as follows :

1. net atomic charges in molecular crystals are fractions of electron units, even in a mineral such as kernite,[76] a Ni complex[74] or a Zwitter ion like glycylglycine[27].

2. oxygen atoms are negatively charged by 0.2-0.5 electrons, while carbon atoms are positive in groups like carboxyl and carbonyl, but negative in an aliphatic environment. Boron atoms in kernite and borazine[77] are positive by about 0.4 and 0.8 e respectively.

3. atomic charges correlate with atomic electronegativity rather than with number of bonds, no significant differences being observed between trigonal and tetrahedral boron atoms and between bridging and hydroxylic oxygen atoms in the borate polyanion in kernite.

The studies on glycylglycine[27] indicate that refinement of thermal and occupancy parameters only may lead to an underestimate of the net charge. Somewhat larger charges are obtained when the orbital exponents of the atomic density functions are also adjusted. In this way, values of −0.37 and +0.39e are obtained for the biologically important net charges of the carboxylate and ammonio groups.

4.2.2 Orbital exponents

An adjustment of the shape of the density functions (spherical deformation) may be effected by variation of a parameter k which defines the radial dependence of the spherical density function.[78] In the case of an STO orbital :

$$X = X_o \exp{-K\alpha r} \qquad (18)$$

In both glycylglycine and oxalic acid, this leads to slight expansion of the oxygen atoms and a contraction of the carbon atoms relative to isolated atom functions. The most illustrative results are obtained in the analysis of orthorhombic sulfur.[51] The results for this homoatomic solid are intermediate between the isolated atom and the optimized Slater-type functions and correspond to a slight contraction in comparison with the former. However, a combined analysis of the X-ray and neutron data indicates that the one parameter function (18) has inadequate detail and underestimates the expansion near the nuclear positions and the contraction that occurs at 0.6-0.8 Å from the nuclei.

4.2.3 Nonspherical atomic deformation functions

A consistent scheme of atomic deformation functions has been applied by a number of authors.[79,80] In the most general formulation by Stewart[80] surface harmonic deformation functions centered on each of the atoms to represent the atomic as well as the overlap density, apply to the atoms at rest rather than to the averaged density. The highest deformation functions proposed are of hexadecapole symmetry, which can represent the product of two d-type atomic orbitals. Thus, for atom j

$$P_j(\underset{\sim}{r}_j)=P_{j,M}(\underset{\sim}{r}_j) + P_{j,D}(\underset{\sim}{r}_j) + P_{j,O}(\underset{\sim}{r}_j) + P_{j,H}(\underset{\sim}{r}_j) \qquad (19)$$

where the subscripts indicate monopole, dipole, octapole and hexadecapole terms. The radial functions in this analysis are products of Slater-type functions, with orbital exponents being considered as variables in the least-squares treatment, together with the population coefficients of the density terms in (19). From the powder data on diamond the exponent value 1.68(3) bohr^{-1} is obtained, when only spherical (monopole) terms are included. This value agrees well with the standard molecular value (1.72) for the L shell orbital exponent.[75] But when non-spherical deformation terms are added the atom becomes more diffuse, indicating the projection of the diffuse two-center terms in the one-center density function.

One of the aims of such an analysis is the derivation of functions such as local (atomic) dipole and higher moments, molecular moments, field gradients and coupling tensors and their comparison with values from other techniques.

A related formalism, applied by Hirshfeld and collaborators has been used to reconstruct the molecular deformation density at rest from the least-squares population parameters.[81] Considerable similarity with theoretical maps on cyano compounds was obtained in the analysis of tetracyanocyclobutane.

4.2.4 Prospects for measurement of elements of the first-order density matrix

In a formalism more closely related to theoretical formulations the density is represented as a sum products of orbitals centered on atoms μ and ν :

$$\rho(\underset{\sim}{r}) = \sum_{\mu}\sum_{\nu} P_{\mu\nu}\, \phi_\mu\, \phi_\nu \tag{20}$$

or

$$F(\underset{\sim}{H}) = \sum_{\mu}\sum_{\nu} P_{\mu\nu}\, F(\phi_\mu\, \phi_\nu)\, T_{\mu\nu} \tag{21}$$

where F indicates the X-ray scattering Fourier transform operator.

Expression (20) explicitely contains bond scattering terms. But even in a minimal basis set formulation with frozen core densities, the number of parameters $P_{\mu\nu}$ in (21) is very large as each atomic center contains 10 and each bond 16 independent atomic orbital products. Further restrictions therefore have to be introduced to limit the number of parameters and the correlation between parameters in the calculational process. In the application of (21) to experimental data on cyanuric acid and tetracyanoethylene oxide all two-center terms in "long" bonds between non-adjacent atoms were neglected.[61] Electron deformation maps reconstructed from the experimental values of $P_{\mu\nu}$ indicate the formalism to be adequately space-filling, as the features in X-N and X-X maps can be well reproduced. Massa and Clinton[82] have pointed out that the idempotency condition which assures N- representability should be used as a constraint in the fit to the coherent diffraction data.

Extension beyond the minimal basis set level would correspond to an additional number of parameters not warranted by the amount and quality of the experimental information. Analysis at this level is obviously of limited value because the diffraction data have been shown to be superior to minimal basis set densities. Other types of basis function in (20) have to be explored for further work.

V. CONCLUDING REMARKS

Where conclusions about theoretical densities have been possible, more sphisticated calculations are supported and several approximate treatments such as INDO, MBS and pseudopotential methods have been shown to give an inadequate representation of the difference densities. Experimental methods are yet to be developed as a fully quantitative test of theoretical densities. Intermolecular effects like hydrogen bonding and charge transfer in mixed molecular crystals are further areas of potential interest, while only a modest start has been made in the derivation of physical properties from the coherent diffraction data.

REFERENCES

1. I. Waller and D.R. Hartree, Proc. Roy. Soc., A124, 119 (1929).

2. L.S. Bartell and R.M. Gavin, Jr., J. Am. Chem. Soc., 86, 3493 (1964). L.S. Bartell and R.M. Gavin, Jr., J. Chem. Phys., 43, 856 (1965).

3. P.P.M. Groenewegen, J. Zeevalkink and D. Feil, Acta Crystallogr., A27, 487 (1971).

4. R.F. Stewart, private communication.

5. S. Gottlicher and W. Wölfel, Z. Electrochem., 63, 891 (1959).

6. R. Brill, Solid State Physics, 20, 1, (1967).

7. Y.W. Yang and P. Coppens, to be published.

8. Crystallographic Computing, Copenhagen : Munksgaard, 1969, and references given.

9. P. Coppens, Acta Crystallogr., B30, 255 (1974).

10. P. Coppens, Acta Crystallogr., B24, 1272 (1968).

11. B.H. O'Connor and F.H. Moore, Acta Crystallogr., B29, 1903, (1973); B.H. O'Connor, Acta Crystallogr., B29, 1893 (1973).

12. F.K. Ross and P. Coppens, Abstracts ACA Winter Meeting, Paper D3 (1972); Y. Wang, R.H. Blessing, F.K. Ross, J. Williams and P. Coppens, to be published.

13. B. Dawson, Austr. J. Chem., 18, 595 (1965).

14. P. Coppens and C.A. Coulson, Acta Crystallogr., 23, 718.

15. P. Coppens and F.L. Hirshfeld, Isr. J. Chem., 2, 117 (1964).

16. L.H. Jensen and M. Sundaralingam, Science, 145, 1185 (1964).

17. R.F. Stewart, E.R. Davidson and W.T. Simpson, J. Chem. Phys., 42, 3175 (1965).

18. J.C. Hanson, L.C. Sieker and L.H. Jensen, Acta Crystallogr. B29, 797 (1973); J.C. Hanson, L.C. Sieker and L.H. Jensen.

19. H. Irngartinger, L. Leiserowitz and G.M.J. Schmidt, J. Chem. Soc., (B), 497 (1970).

20. P. Becker, P. Coppens and F.K. Ross, J. Am. Chem. Soc., 95, 7604 (1973).

21. H.M. Rietveld, E.N. Maslen and C.J.B. Clews, Acta Crystallogr. B26, 693 (1970).

22. A.C. Hazel, F.K. Larsen and M.S. Lehman, Acta Crystallogr., B28, 2977 (1972).

23. M.I. Day, Y. Okaya and D.E. Cox, Acta Crystallogr., B27, 26 (1971).

24. P. Coppens, Measurement of Electron Densities in Solids by X-ray diffraction in International Reviews of Science. Lancaster : MTP, to be published.

25. P. Coppens, Acta Crystallogr., A25, 180 (1969).

26. P. Coppens and A. Vos, Acta Crystallogr., B27, 146 (1971); G.C. Verschoor and E. Keulen, Acta Crystallogr., B27, 134, (1971).

27. J.F. Griffin and P. Coppens, (1974) to be published.

28. P. Coppens and L. Baine, unpublished results.

29. P. Coppens, Isr. J. Chem., 10, 85 (1972).

30. P. Coppens, T.M. Sabine, R.G. Delaplane and J.A. Ibers, Acta Crystallogr., B25, 2451 (1969).

31. J. Almlöf, Å. Kvick and J.O. Thomas, J. Chem. Phys. (1973), in press; Abstracts First European Crystallography Conf., Bordeaux, Group B3.

32. J.C. Taylor and T.M. Sabine, Acta Crystallogr., B28, 3340, (1972).

33. D.A. Matthews and G.D. Stucky, J. Am. Chem. Soc., 93, 5954 (1971); D.A. Matthews and G.D. Stucky, Trans. Am. Crystallogr., Assoc., 8, 113 (1972); D.A. Matthews, J. Swanson, and G.D. Stucky, J. Am. Chem. Soc., 93,5945 (1971).

34. B. Rees and P. Coppens, Acta Crystallogr., B29, 2515 (1973).

35. T. Ito and T. Sakurai, Acta Crystallogr., B29, 1594 (1974).

36. A. Hartman and F.L. Hirshfeld, Acta Crystallogr., 20, 80 (1966).

37. P. Coppens, Science, 158, 1577 (1967).

38. M. Dreyfus, B. Maigret and A. Pullman, Theor. Chim. Acta, 17, 109 (1970); M. Dreyfus and A. Pullman, ibid., 19, 30 (1970).

39. P.A. Kollman and L.C. Allen, J. Chem. Phys., 52, 5085 (1970).

40. P. Coppens and L. Baine, unpublished results.

41. C.A. Coulson and W.A. Moffit, Phil. Mag., 40, 1 (1949).

42. E.D. Stevens, Experimental Determination of Electron Density Distributions by X-ray Diffraction, Thesis University of California at Davis, (1973).

43. R.J. Temkin, V.E. Henrich and P.M. Raccah, Solid State Commun., 13, 811 (1973).

44. J.P. Walter, C.Y. Fong, and M.L. Cohen, Solid State Commun., 12, 303 (1973).

45. M.V. Linkoaho, Physica Scripta, 5, 271 (1972).

46. P.J. Brown, Phil. Mag., 26, 1377 (1972).

47. R.F. Stewart, ACA Conference Program and Abstracts Series 2, 1, 200 (1973).

48. Y.W. Yang and P. Coppens, unpublished results.

49. R. Brill, H. Dietrich and H. Dierks, Acta Crystallogr., B27, 2003 (1971).

50. J.C. Slater, Quantum Theory of Molecules and Solids, Vol. 2, p. 103, New York : McGraw Hill, 1965.

51. R.H. Blessing, W.F. Cooper, Y.W. Yang, and P. Coppens, ACA Conference Program and Abstracts Series 2, 1, 200 (1973).

52. J.A.K. Duckworth, B.T.M. Willis and G.S. Pawley, Acta Crystallogr., A26, 263.

53. C.A. Coulson and M.W. Thomas, Acta Crystallogr., B27, 1354 (1971); M.W. Thomas, Acta Crystallogr., B27, 1760 (1971); M.W. Thomas, Chem. Phys. Lett., 20, 303 (1973).

54. A.F.J. Ruysink, Electron Distribution in Crystals, Thesis, University of Groningen, The Netherlands, (1973).

55. J.B. Roberto, B.W. Batterman and D. Keating, Phys. Rev. B., 9, 2590 (1974).

56. S. Tanemura and N. Kato, Acta Crystallogr., A28, 69 (1972); P.J.E. Aldred and M. Hart, Proc. Roy. Soc., A332, 223, (1973); P.J.E. Aldred and M. Hart, Proc. Roy. Soc., A332, 239 (1973).

57. E.A. Laws, R.M. Stevens and W.N. Lipscomb, J. Am. Chem. Soc., 94, 4467 (1972); E.A. Laws and W.N. Lipscomb, Isr. J. Chem., 10, 77 (1972).

58. R.M. Stevens, E. Switkes, E.A. Laws and W.N. Lipscomb, J. Am. Chem. Soc., 93, 2603 (1971).

59. P.R. Smith and J.W. Richardson, J. Phys. Chem., 69, 3346, (1965); P.R. Smith and J.W. Richardson, J. Phys. Chem., 71, 924 (1967).

60. P.E. Cade, Trans. Am. Crystallogr. Ass., 8, 1 (1972).

61. D.S. Jones, D. Pautler and P. Coppens, Acta Crystallogr., A28, 635, (1972); J.W. McIver, P. Coppens, and D. Nowak, Chem. Phys. Lett., 11, 82, (1971).

62. J.W. Bats, A. Theoretical Study of the Electron-density in CN^-, LiCN and LiNC. Thesis, Technische Hogeschool Twente, The Netherlands, (1973).

63. P. Becker, private communication.

64. J.P. Walter and M.L. Cohen, Phys. Rev., B4, 1877 (1971).

65. D.J. Stukel and R.N. Euwema, Phys. Rev., B1, 1635, (1970).

66. F.L. Hirshfeld, Acta Cryst., B27, 769 (1971).

67. C. Tavard and M. Roux, Compt. Rend., 260, 4460, 4933 (1965); C. Tavard, M. Rouault and M. Roux, J. Chim. Phys., 62, 1410 (1965).

68. D.A. Kohl and R.A. Bonham, J. Chem. Phys., 47, 1634 (1967).

69. K. Kurki Suonio, Analysis of Crystal Atoms on the Basis of X-ray Diffraction, Italian Crystallographic Association Meeting, Bari, Italy (1971); K. Kurki Suonio and P. Salmo, Ann. Ac. Sci. Fenn., Ser. A VI, 369, (1971).

70. F.K. Larsen and P. Coppens, Acta Crystallogr., in press.

71. P. Coppens and W.C. Hamilton, Acta Crystallogr., B24, 925 (1968).

72. R.H. Blessing and P. Coppens, Solid State Commun., in press.

73. R.F. Stewart, J. Chem. Phys., 51, 4569 (1969); R.F. Stewart, J. Chem. Phys., 58, 4430 (1973).

74. P. Coppens, D. Pautler and J.F. Griffin, J. Am. Chem. Soc., 93, 1051, (1971); P. Coppens, Acta Crystallogr., B27, 1931, (1971).

75. W.J. Hehre, R.F. Stewart and J.A. Pople, J. Chem. Phys., 52, 2769 (1970).

76. W.F. Cooper, F.K. Larsen, P. Coppens, and R.F. Giese, Am. Miner., 58, 21 (1973).

77. P.W.R. Corfield and S.G. Shore, J. Am. Chem. Soc., 95, 1480, (1973).

78. P. Coppens, et al., unpublished results.

79. R.J. Weiss, X-ray Determination of Electron Distributions, (Amsterdam : North Holland Publishing Co.), (1966); B. Dawson, Proc. Roy. Soc., A298, 255 (1967).

80. R.F. Stewart, J. Chem. Phys., 58, 1668 (1973).

81. M. Harel and F.L. Hirshfeld, Abstracts First European Cryst., Conference, Bordeaux, Group B3.

82. L.J. Massa, W.L. Clinton, Trans. Am. Crystallogr., Ass., 8, 149 (1972); W.L. Clinton and L.J. Massa, Phys. Rev. Letters, 20, 1363, (1972).

STRUCTURE AND BONDING IN POLYMERS AS REVEALED BY ELECTRON SPECTROSCOPY FOR CHEMICAL APPLICATIONS (ESCA)

D.T. Clark

Department of Chemistry

University of Durham

South Road, Durham City

I. FOREWORD

The material presented in this review is a fairly literal transcript (with minor modifications) of the lectures to be given at the NATO Summer School. The objective of the course as far as these particular lectures are concerned may be summarised as follows. Firstly, to introduce to a heterogeneous group of experimentalists and theoreticians the powerful experimental technique of ESCA for studies of structure and bonding in general. After reviewing the fundamentals of the technique in terms of the measurements actually made and basic instrumentation, consideration is given to the types of information available from ESCA studies. This is illustrated by reference to simple monomeric systems. Secondly, quantitative theoretical models for the interpretation of experimental data are reviewed. Thirdly, detailed consideration is given to application of ESCA to studies of structure and bonding across a broad front in the polymer field. Finally, some consideration is given to the likely growth areas of the technique in terms of instrumentation, theoretical interpretation of data and areas of application with particular references to the polymer field. As should become apparent the technique provides a most interesting focal point for the mutually beneficial interaction of experimentalists and theoreticians.

The lecture material is conveniently divided into three major sections :

a) A general introduction including a detailed discussion of the fundamentals of ESCA.

b) Theoretical models for the interpretation of ESCA data pertaining to both core and valence photoionizations.

c) A detailed review of applications of ESCA to polymers together with a prognosis for the future.

In the majority of cases the examples chosen to illustrate points arising in sections (a) and (b) derive from the research program in my own laboratories. This may readily be justified on the grounds that it is representative of much of the research work being carried out in many laboratories around the world with the additional and over-riding merit that it is readily to hand for preparation of lecture material and slides. As far as section (c) is concerned our research program on ESCA applied to polymers is currently the most comprehensive and well developed and it seems natural therefore to present our own results.

Since comprehensive reviews have appeared in print very recently on instrumental design[1] and on theoretical aspects of ESCA[2] these aspects are considered only briefly here and these notes consist of a basic introduction to ESCA followed by a comprehensive review of applications to studies of structure and bonding in polymers. The lectures themselves will however, cover the important aspects of both instrumentation and theory.

II. GENERAL INTRODUCTION

Only five or six years ago Electron Spectroscopy for Chemical Applications (ESCA) was a comparatively unknown technique being explored only by workers with the courage and technical expertise necessary to build their own spectrometers.[3,4] However, the published work of these pioneers in the field had indicated the enormous potential of the technique and within a space of three years or so at least ten instrument manufacturers were advertising their wares. The availability of instruments was rapidly capitalized on by the scientific community and a review of the field for the period 1968 to 1971[5] included 166 references. Since that date publications relating to ESCA have continued to proliferate and the technique has proved valuable across a wide range of investigations.

In Durham we have been applying ESCA to studies of structure and bonding encompassing organic, inorganic and polymeric systems.[6] These studies, together with complimentary theoretical analysis, have demonstrated that ESCA is an extremely powerful tool for investigation of structure an bonding with an information content per spectrum unsurpassed by any other spectroscopic method. A further valuable feature of ESCA is the surface sensitivity of the technique and its flexibility as to the physical form in which the sample under investigation may be examined. These aspects are of particular value in studies of molecular solids and crystals, and in particular polymeric materials, since it is often desirable to be able to study the surface as well as the bulk of a solid sample if an understanding of its behaviour is to be gained. Surface sensitive techniques suitable for such studies have been conspicuous by their absence despite the considerable effort and ingenuity which has been expended on surface studies during the last few years. In these particular series of lectures we will be predominantly interested in the application of ESCA to the study of the electronic structures and properties of polymeric materials.

The application of ESCA to structure and bonding in polymers has been largely pioneered at Durham, and it is already clear that the technique has much to offer particularly in investigations of otherwise intractable materials and in studies of specific surface effects. In its application to more routine investigations the technique is able to provide answers to a range of questions rapidly. For example, the detailed interpretation of the core level spectra of an ethylene-tetrafluoroethylene copolymer gives access to the copolymer composition, the relative proportions of various structural features (alternation and blocks), the proportion of methylene groups oxidized to carbonyl, and the amount of hydrogen bonded water associated with the carbonyls. In addition ESCA provides a very convenient means of investigating the valence energy levels of such polymer systems.

The material set out below taken in conjunction with the lectures themselves, should provide a sound basis for introducing ESCA as a powerful tool for studies of structure and bonding in molecular solids and crystals. After a fairly detailed introduction to ESCA as a technique, a comprehensive review of ESCA applied to polymers is given.

III. FUNDAMENTALS OF ESCA

3.1 Introduction

In common with most other spectroscopic methods ESCA is a technique originally developed by physicists and now gradually being taken over by chemists to be developed to its full potential as a tool for investigating structure and bonding. The technique has largely been developed by Professor Kai Siegbahn[3,4] and his collaborators at the University of Uppsala over the past 20 years or so and much of the early work has been extensively documented in the 'first ESCA book'.[3] It is only within the last decade however that the potential of the technique has been revealed with the development of spectrometers of sufficient resolution and sensitivity. In addition to the aesthetically pleasing designation as 'Electron Spectroscopy for Chemical Applications'+ originally coined by Siegbahn, the technique is also variously known as X-ray Photoelectron Spectroscopy (XPS), High Energy Photoelectron Spectroscopy (HEPS), Induced Electron Emission Spectroscopy (IEES), and Photoelectron Spectroscopy of the Inner Shell (PESIS).

The principle advantages of the technique may be summarised as follows :

a) The sample may be solid, liquid or gas (it is as easy to study a high molecular weight polymer as it is to study a gas) and the technique is essentially non-destructive. One notable exception to this generalization is poly (thiocarbonyl fluoride) which depolymerizes rather rapidly under X irradiation. (It should be noted that ESCA is capable as a technique of studying samples directly in solid, liquid or gaseous states).

b) The sample requirement is modest, in favourable cases 1 mg. of solid, 0.1 µl of liquid or 0.5 cc. of gas (at STP). These sample requirements are based on the minimum readily handled with conventional techniques. As will become apparent the sensitivity of the technique is such that a fraction of a monolayer coverage may be detected.

c) The technique has high sensitivity and is independent of the spin properties of any nucleus and is applicable in principle to any element of the periodic table. H and He are exceptions being the only elements for which the core levels are also the valence levels.

d) The information it gives is directly related to the electronic structure of a molecule and the theoretical interpretation is relatively straightforward.

e) Information can be obtained on both the core and valence energy levels of molecules.

These particular advantages of ESCA as a technique make it eminently suitable for the study of polymers.

+ Originally designated'... for Chemical Analysis'

3.2 Properties of Core Orbtials

A clearer understanding of ESCA as a technique is obtained with some knowledge of the properties of core orbitals. The material presented below provides a convenient introduction and is particularly apposite since it was from a research program involving non-empirical quantum mechanical calculations of cross sections through potential energy surfaces for simple reactions that this author proceeded to an experimental interest in ESCA as a technique.

Traditionally chemists have discussed the electronic structure of molecules, dealing only with the valence electrons and neglecting inner shell or core electrons. The reason for this being:

a) core electrons are not explicitly involved in bonding (although most of the total energy of a molecule resides in the core electrons);

b) it is only in the past ten years that sufficient computing capability has become available to allow non-empirical quantum mechanical calculations on molecules in which the core electrons are explicitly considered.

It has become clear however, that although core electrons are not involved in bonding, the core energy levels of a molecule encode a considerable amount of information concerning structure and bonding. This is illustrated by examples of work carried out at Durham over the past few years.

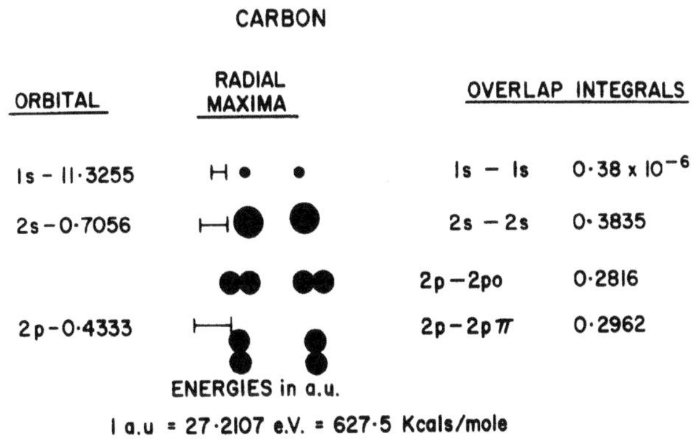

Fig. 1. Radial maxima, energies and overlap integrals for carbon orbitals.

Fig. 1 shows the orbital energies, total energy and radial maxima for the carbon atom calculated non-empirically in a Gaussian basis set. Considering first the orbital energies, it is clear that the 1s (core) level is very much lower in energy than the 2s and 2p (valence) levels. From the radial maxima the 1s orbital is confined to a region in the immediate vicinity of the nucleus whereas the valence orbitals are much "larger". Since the core orbital is essentially localised around the nucleus, overlaps with orbitals on adjacent atoms are negligible. Shown in Fig. 1 are overlap integrals between orbitals on two adjacent carbon atoms with a bond length of 1.39 Å. The negligible value for the overlap integral involving the core orbitals is one reason why they are not involved in bonding.

It has tacitly been assumed by chemists in the past that in discussing the transformation of one molecule into another the energies of the inner shell or core electrons could be taken as constant and effectively ignored. A particularly interesting transformation which illustrates that this is not the case is the transformation of cyclopropyl to allyl cation which occurs in a disrotary fashion. The relevant energy levels are shown in Fig. 2. As far as the carbon atoms are concerned there are fairly drastic charge migrations involved in this transformation. Thus C1 which carries a substantial positive charge in cyclopropyl cation becomes C2 with a substantial negative charge in allyl cation. As a result the C_{1s} orbital energy changes from -11.7122 a.u. to -11.5613 a.u.

The changes in energy of this particular core level is in fact almost three times the total energy change in the transformation.[+] The almost degenerate pair of C_{1s} orbitals for C2 and C3 in cyclopropyl cation change in energy by 0.044 a.u. in the transformation to allyl cation. Inspection of the charge distributions and core energy levels reveals that a more negative energy (i.e. increased binding energy)[++] is associated with an increased positive charge on an atom. The charge on a given atom

[+] It should be remembered that the total energy of a molecule may be expressed as

$$E = \sum_{r=a}^{k} \varepsilon_r - \sum_{\text{pairs } rs} (2J_{rs} - K_{rs}) + V_{nn}$$

where the ε_r are the occupied orbital energie, J_{rs} and K_{rs} are coulomb and exchange repulsion integrals and V_{nn} is the nuclear repulsion energy.

[++] Binding energy is defined here as the energy required to remove the electron in a given orbital to infinity and may be equated to the -ve of the orbital energy (Koopmans' Theorem). The approximations involved in this are discussed elsewhere.

is determined by the valence electron distribution, and the core
is determined by the valence electron distribution, and the core
levels of a molecule reflect this. Clearly although the core levels are not involved in bonding they are a sensitive function of the electronic environment about a given atom.

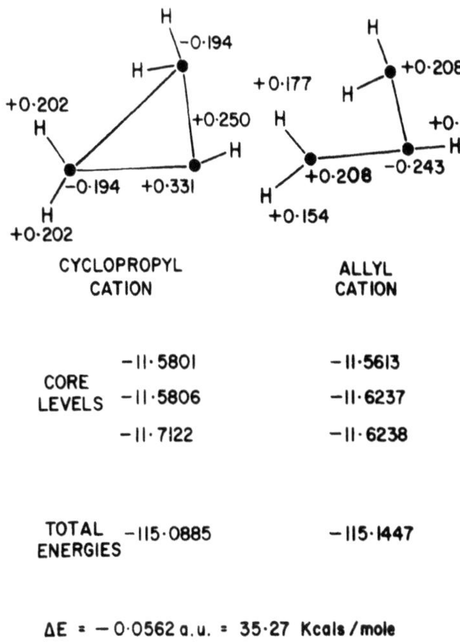

Fig. 2. Computed charge distributions and orbital energies for cyclopropyl and allyl cations.

The different electronic environments about C1 and C2 (C3) in cyclopropyl cation is therefore reflected in the "shift" in C_{1s} binding energies of ∿4.1 eV.

Core electrons are localized in space close to a nucleus and this is reflected in the fact thaat the binding energies are characteristic of a given element. This is emphasized on consideration of the approximate core binding energies for first and second row elements given in Table 1. Clearly on the basis of their core binding energies it is an easy matter to distinguish say sulphur from chlorine.

To summarize :

Core orbitals are essentially localized on atoms, their energies are characteristic for a given element and are sensitive to the electronic environment of an atom. Thus for a given core level of a given element whilst the absolute binding energy for that level is characteristic for the element, differences in electronic environment of a given atom in a molecule give rise to a small range of binding energies (i.e. "shifts" in binding energies are apparent) often characteristic of a particular structural feature.

Table 1. Approximate binding energies for core levels of first and second row elements (in eV)

	Li	Be	B	C	N	O	F	Ne
1s	55	111	188	284	399	532	686	867

	Na	Mg	Al	Si	P	S	Cl	Ar
1s	1072	1305	1560	1839	2149	2472	2823	3203
2s	63	89	118	149	189	229	270	320
$2p_{1/2}$	31	52	74	100	136	165	202	247
$2p_{3/2}$			73	99	135	164	200	245

3.3 The ESCA Experiment

ESCA involves the measurement of binding energies of electrons ejected by interactions of a molecule with a monoenergetic beam of soft X-rays. For a variety of reasons the most commonly employed X-ray sources are $AlK\alpha_{1,2}$ and $MgK\alpha_{1,2}$ with corresponding photon energies of 1486.6 eV and 1253.7eV respectively. In principle all electrons, from the core to the valence levels can be studied and in this respect the technique differs from u.v. photoelectron spectroscopy (UPS) in which only the lower energy valence levels can be studied. The basic processes involved in ESCA are shown in Fig. 3.

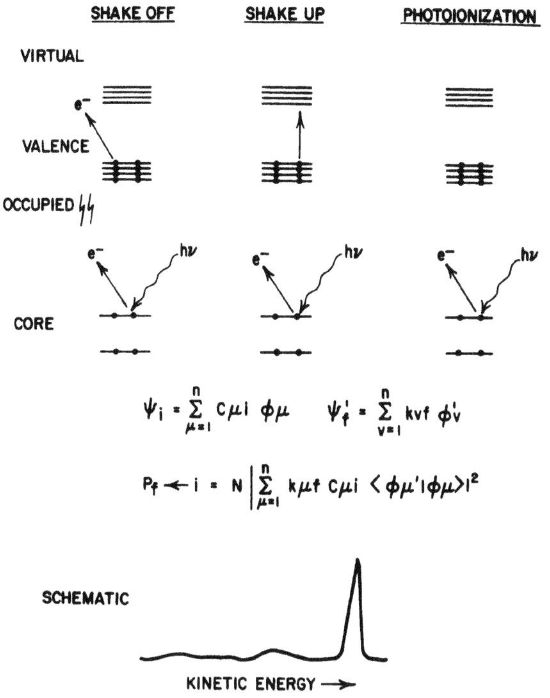

Fig. 3. Schematic illustrating photoionization, shake up and shake off processes.[+]

The removal of a core electron (which is almost completely shielding as far as the valence electrons are concerned) is accompanied by reorganization of the valence electrons in response to the effective increase in nuclear charge. This perturbation gives rise to a finite probability for photoionization to be accompanied by simultaneous excitation of a valence electron from an occupied to an unoccupied level (shake up) or ionization of a

---

[+] There is a close analogy between shake up processes for hole states and the more familiar electronic transitions for the neutral system as, for example, conventionally observed with a U.V./Visible spectrometer. However, whilst the former follow monopole selection rules (the computation of intensities therefore depending on overlap as indicated in Fig. 3), the latter follow dipole selection rules.

valence electron (shake off). These processes giving rise to satellites to the low kinetic energy side of the main photoionization peak, follow monopole selection rules and may well be of considerable importance in the future in elucidating particular aspects of structure and bonding in polymer systems. +

De-excitation of the hole state can occur (Fig. 4) via both fluorescence and Auger processes, for elements of low atom number the latter in fact being the more porbable.

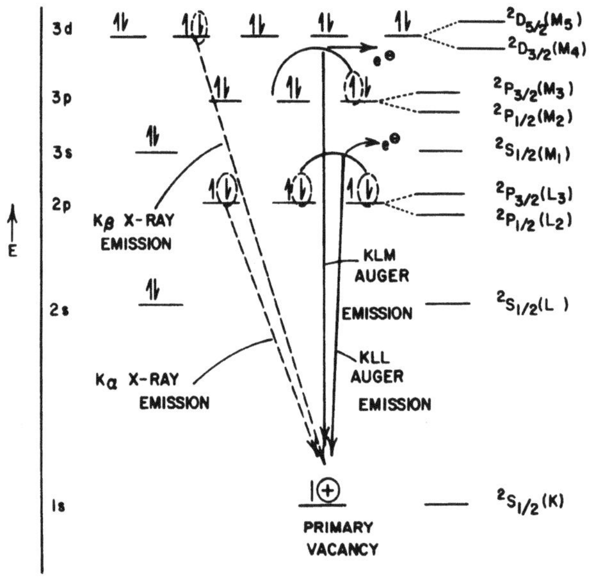

Fig. 4. De-excitation of core hole states.

It should be perhaps emphasized that the lifetimes of the hole states involved in ESCA are typically in the range 10^{-13} - 10^{-17} secs. The basic experimental set up for ESCA is shown in Fig. 5.

The components are :

X-ray generator. The most commonly used X-ray sources are $MgK\alpha_{1,2}$ and $AlK\alpha_{1,2}$ with photon energies (and linewidths) of 1253.7 eV (∿ 0.7 eV) and 1486.6 eV (∿ 1.0 eV) respectively. Harder X-ray sources e.g. $CrK\alpha_1$ 5414.7 eV and $CuK\alpha_1$ 8047.8 eV have much larger inherent linewidths and thus give lower resolution. If this is not important however, the use of sources with widely differing photon energies can be very useful for studying

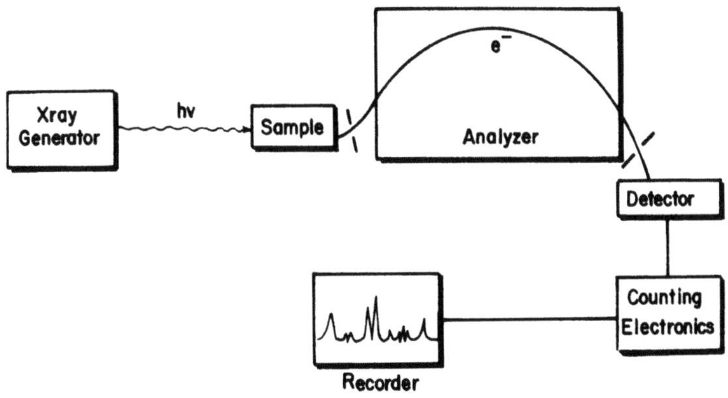

Fig. 5. Schematic of basic experimental set up for ESCA.

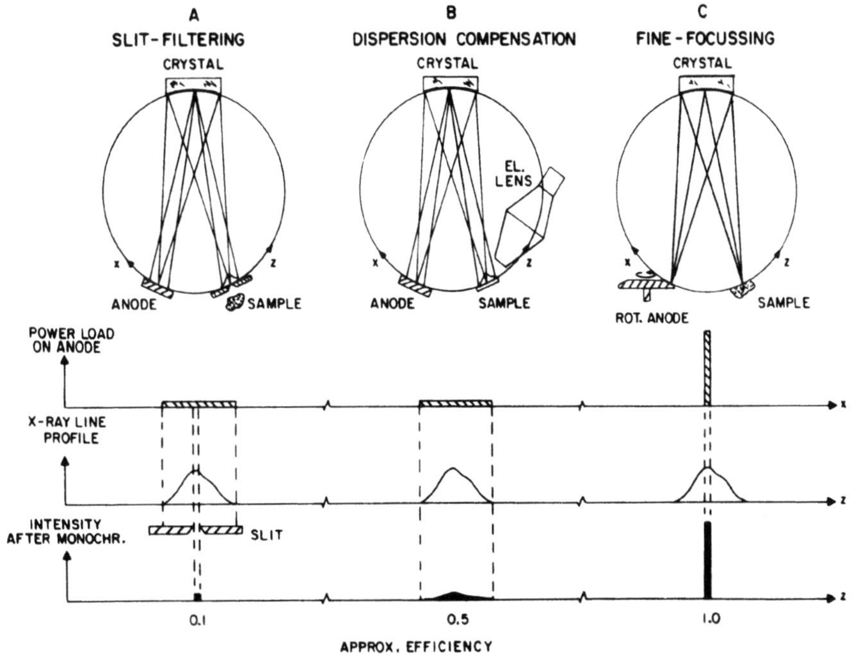

Fig. 6. X-ray monochromatization schemes.

escape depth dependence (see later). In many cases as will become apparent, the inherent width of the excitating radiation provides the dominant contribution to overall linewidth.

For $AlK\alpha_{1,2}$ radiation, the lattice spacing for quartz is such that monochromatization may be achieved by dispersion from suitably bent quartz discs. The three variants which have been proposed[1] are shown in Fig. 6. The first two namely slit filtering and dispersion compensation have been commercially exploited. The most recently developed however, is the elegant fine focussing X-ray monochromatization scheme of Siegbahn and Coworkers.[1] As will become apparent composite linewidths for individual C_{1s} levels obtained with an unmonochromatized $MgK\alpha_{1,2}$ source are typically $\sim$ 1.0 eV[+] whereas with a monochromatized X-ray source linewidths can typically be $\sim$ 0.5 eV with concomitant improvement in resolution. A further advantage occurring from X-ray monochromatization is the removal of satellites($K\alpha_{3,4}$ $K\beta$ etc..) and bremsstrahlung (white radiation) leading to cleaner cut spectra with much improved signal/background. (A development to look forward to in the future is the harnessing of synchrotron radiation (see later))[7].

With $MgK\alpha_{1,2}$ and $AlK\alpha_{1,2}$ photon sources there is sufficient energy to study the 1s and valence levels of first row elements, the 2s, 2p and valence levels of second row elements and so on. There is no particular virtue in studying the most tightly bound core levels of a given element (e.g. say the 1s level of gold), since this may well have a very large natural linewidth (see later) and the required higher energy photon source inevitably has a larger linewidth. Information concerning the valence electron distribution is encoded in all core levels (the information content may differ from level to level, see section on multiplet splittings), so that the core level selected for study depends on the cross section for photoionization, inherent linewidth, binding energy, photon energy and photon linewidth and the information required.

Sample region. The sample region of the spectrometer is usually separated from the X-ray tube by a thin metal window (typically 3/10 thou Al or better, Be) which ensures that electrons from the gun, used to excite the characteristic X irradiation, do not enter the sample region. Polymer samples are conveniently studied as films or powders mounted on a sample probe which may be taken into the spectrometer from atmosphere by means of insertion locks. Samples may thus be readily mounted and inserted into the spectrometer greatly facilitating routine analyses. Provision is usually made to enable samples to be heated or cooled in situ and an ancil-

---

+ Defined as Full Width at Half Maximum (FWHM)

lary sample preparation chamber allows greater flexibility in terms of sample preparation or pretreatment (e.g. argon ion bombardment, electron bombardment, u.v. irradiation, chemical treatment (e.g. oxidation) etc..). Addition of a quadrupole mass spectrometer facilitates degradation studies and allows close control to be kept of the extraneous atmosphere in the sample region of the spectrometer. Typical operating pressures in the sample region would be $< 10^{-8}$ Torr. The sample area irradiated might typically be 5 mm x 5 mm.

Analyzer. The analyser must typically have a[+] resolution of something approaching one part in 10^4. In the precise energy analysis of the photoelectrons therefore, two types of analyzer have mainly been used (Fig. 7).

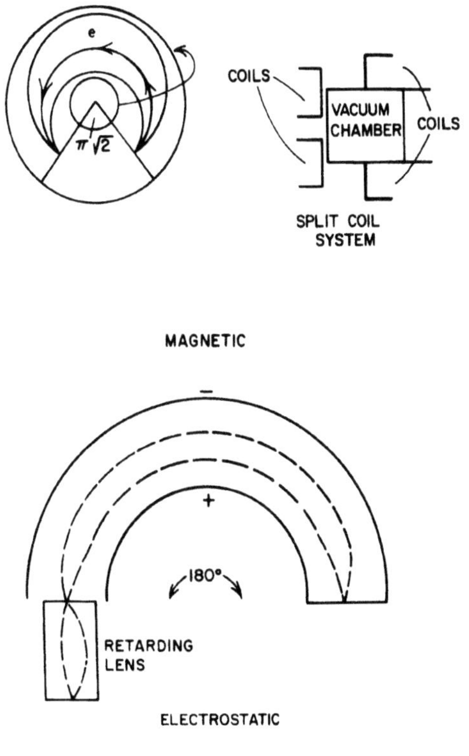

Fig. 7. Electrostatic and magnetic analyzers (schematic).

[+] The main barrier to the development of the technique has been the design of analyzers of sufficiently high resolution and luminosity.

a) Magnetic

These are largely of the double focussing type and are generally made from brass or aluminium typically with a 30 cm. radius. Double focussing is provided by an inhomogeneous magnetic field produced by a set of four cylindrical coils placed about the electron trajectory. The chief advantage of a magnetic analyzer is relative ease of construction, the major disadvantage, the requirement to eliminate stray magnetic fields by employing necessarily bulky Helmholtz coils

b) Electrostatic

Most are based on the hemispherical double focussing design first described by Purcell as long ago as 1938. By employing a retarding field on the electrons before they enter the analyzer, the dimensions of the latter may be considerably reduced and μ metal may also be used for screening. Although more difficult to construct therefore, spectrometers employing electrostatic analyzers can be made more compact. In fact the majority of the commercial instruments available are all of this basic design. Typical dimensions would be 25 cms. mean diameter for the hemispheres which are usually constructed of stainless steel or gold plated aluminium or glass.

c) Detector

The minute electron currents involved means that detection is via counting and most spectrometers employ channel electron multipliers. With most designs of double focussing analyzers their focal plane properties may be exploited by incorporating multi-channel detectors which can give spectacular increases in the rate of data acquisition and this together with the development of X-ray monochromators is the most important development in commercial instruments in the short term. The output from the multiplier is then amplified and fed to counting electronics.

d) Scan and readout

There are basically two ways of generating a spectrum either continuous or step scan. In the continuous mode of operation the field (either electric or magnetic) is increased continuously and the signal from the detector is continuously monitored by a rate meter. If the signal to background and overall count rates are sufficiently high, this is the routine way to obtain a spectrum which may be plotted out directly on an XY recorder. Alternatively, the field may be incremented in small steps and at each setting either a fixed number of counts may be timed or a count can be made for a fixed length of time. Where signal to background is poor then this is the method of choice. It is advisable to have both wide and narrow scan facilities available, the former for carrying out preliminary searches and the latter for detailed study.

3.4 Processes Involved in ESCA

The fundamental processes involved in ESCA are :

a) Photoionization $A + h\nu_1 \rightarrow (A^+)^* + e_1^-$

b) Electronic relaxation by either

(a) X-ray emission $(A^+)^* \rightarrow A^+ + h\nu_2$

or (b) Auger process $(A^+)^* \rightarrow A^{++} + e_2^-$

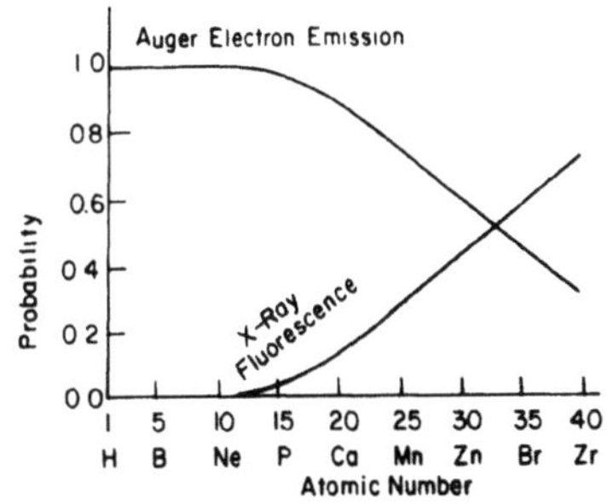

PROBABILITY OF AUGER ELECTRON EMISSION AND X-RAY FLUORESCENCE AS FUNCTION OF ATOMIC NUMBER

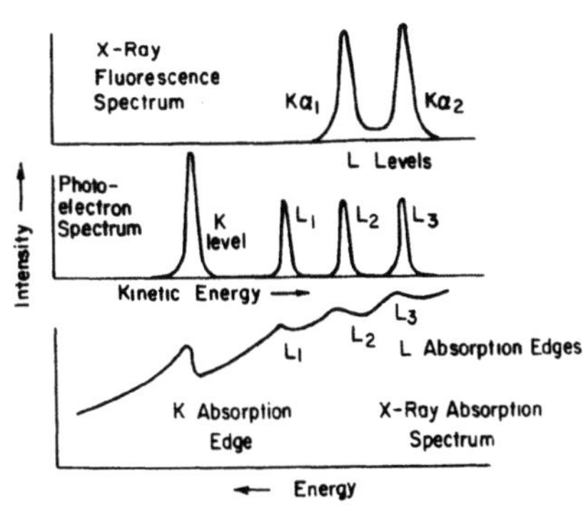

SIMULATED X-RAY FLUORESCENCE, X-RAY ABSORPTION, AND PHOTOELECTRON SPECTRA FOR SAME ELEMENTS

Fig. 8. Auger and fluorescence yields for core hole de-excitation.

The relative probalities for electronic de-excitation by X-ray emission and the Auger process depend on the atomic number of the element concerned and this is outlined in Fig. 8. Since the spectrometer is set up to detect and measure the energies of electrons expelled from the sample both the photoelectron and Auger spectra may be studied.

Typically the lifetimes of the core hole states involved in ESCA are in the range 10^{-13} - 10^{-17} secs emphasizing the extremely short time scales involved. The effective increase in nuclear charge experienced by the valence electrons on removal of a core electron which is almost completely screening, causes an electronic re-organization or relaxation of the hole state which occurs on a time scale short compared with the lifetime of the state (before de-escitation). A simplistic schematic of the time scales involved in ESCA is shown in Fig. 9.

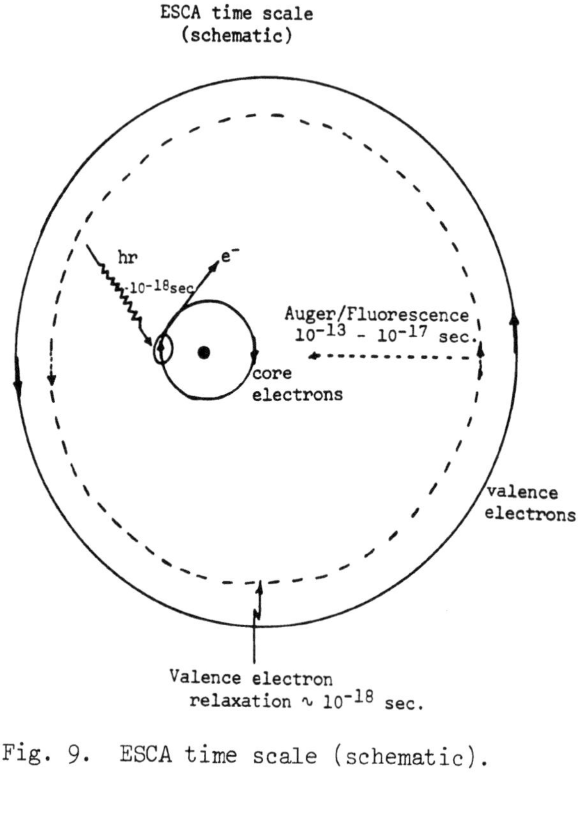

Fig. 9. ESCA time scale (schematic).

3.5 Photoelectric cross sections

The probability for photoemission from a given core level depends on the photoelectric corss section which represents the transition probability per unit time for exciting a photoelectron from a subshell with a photon flux of a given energy. Cross sections are markedly dependent on photon energy and for conventional X-ray sources (e.g. Al$K\alpha_{1,2}$, Mg$K\alpha_{1,2}$) are such that for first row atoms for example, cross sections are much larger for the core levels than for the valence levels. This is illustrated schematically in Fig. 10.

It is clear also that cross sections at a given photon energy depend on the symmetry of the orbitals involved and this is particularly useful in unravelling the complexities of the valence bands of molecular solids and crystals.

For example, it is clear from Fig. 10 that the differential cross section for photoionization of the 2s level of carbon relative to the 2p levels is larger at higher photon energies.[7] This gives a convenient means for distinguishing between σ and π type

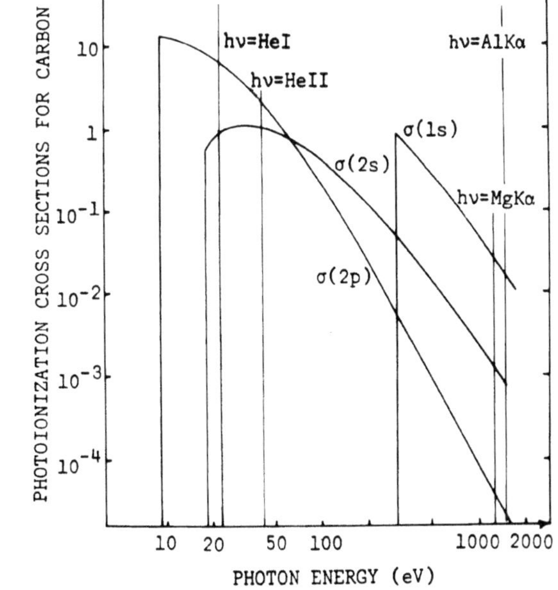

Fig. 10. Subshell photoionization cross sections for carbon.[7]

valence orbitals in organic systems by monitoring the relative changes in peak intensities in going from a low energy (HeI 21.2 eV to high energy $MgK\alpha_{1,2}$ 1253.6 eV) photon source. In addition to being dependent on photon energy, cross sections may also exhibit angular dependence. Two types of angular dependence studies may be envisaged and these are shown in Fig. 11. In the first the analyzer and X-ray source are in a fixed configuration and the

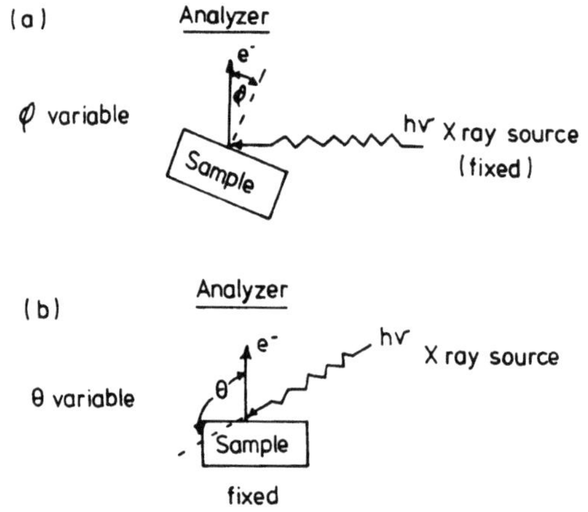

Fig. 11. Angular distribution measurements (schematic).

angular dependence involves the relative orientation of the sample with respect to these. In the second more difficult arrangement the sample is in a fixed configuration with respect to the analyzer and the position of the X-ray source is varied. The studies which have been reported to date on angular dependence of photoelectric cross sections have predominantly referred to the gas phase and have demonstrated the utility of such studies for assignments of valence energy levels.[8] The experimental arrangement has then corresponded to (b) and it has been shown that cross sections conform to an equation of the form[2(a)]

$$\frac{d\sigma}{d\Lambda} = A + B \sin^2 \theta$$

For molecular solids and crystals few attempts have been made to study angular dependencies the reason for this being that other considerations considerably omplicate the analysis. For ESCA studies of solids the most convenient experimental arrangment is (a) in which the angular dependence is introduced by rotating the sample. The observed intensities of the peaks corresponding to photoionization from core levels in addition to being directly dependent on photon energy and angle (via the cross section) also exhibit angular dependence due to photon and electron mean free paths in the solid and also if orientated single crystals are involved there may be specific electron channelling phenomena. In general, therefore, the study of the angular dependence of core levels in molecular solids and crystals may be stated to be still in its infancy. The most usual experimental arrangement is therefore to merely orientate the sample to optimize the count rate on a given core level (e.g. C_{1s}, O_{1s}, F_{1s} etc..) and then operate at this fixed angle ϕ. In our particular experiments ϕ has typically been $\sim 30^{\circ}$ or 45° depending upon whether a conventional or Henke type X-ray source has been employed.

3.6 Energy Considerations

For photo ejection of a core electron in a solid sample of an insulator (in electrical contact with the spectrometer) the energy considerations in the measurements of the electron binding energies are shown in Fig. 12. The reference is the Fermi Level

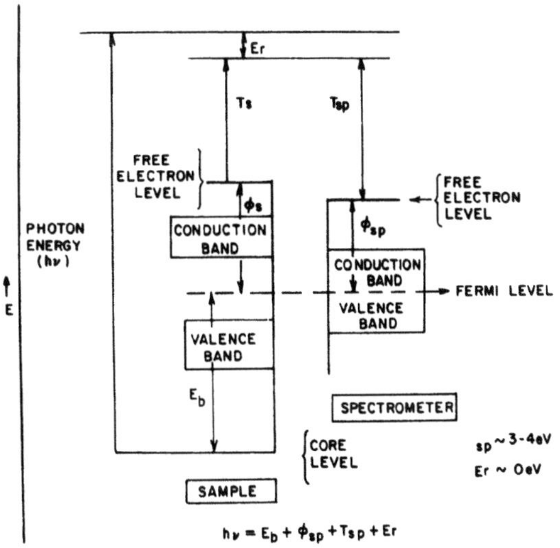

Fig. 12. Energy considerations in measurement of electron binding energies.

and if the work function of the spectrometer is known then the absolute binding energy of a given level may be calculated. This will differ from the absolute binding energy defined with respect to the vacuum level by approximately the work function of the sample.

Conservation of energy requires that

$$h\nu = E_b + \phi_{sp} + T_{sp} + E_r \qquad (1)$$

where $h\nu$ is the photon energy,

E_b the binding energy of the core level (Fermi level as reference),

ϕ_{sp} is the spectrometer work function

T_{sp} is the measured kinetic energy of the photoemitted electron

and E_r is a recoil energy.

Conservation of momentum dictates that for the commonly used X-ray sources E_r may effectively be ignored for atoms other than hydrogen.

More often than not the work function of the sample is not known and therefore energy levels referenced with respect to the Fermi level of the sample are operationally convenient. When we come to consider the theoretical analysis of shifts in core binding energies, shake up transitions etc.., the more convenient energy reference is the vacuum level for an assembly of non-interacting molecules in the gas phase. Relating theory and experiment might therefore be expected to lead to considerable difficulties more particularly since for semi-conductors and insulators it is not easy to locate the Fermi level which lies somewhere between the predominantly filled valence bands and predominantly empty conduction bands. A further problem is that in studying essentially insulating samples the assumption of thermodynamic equilibrium during photemission may not be valid since insufficient charge carriers may be available to maintain overall electroneutrality at the sample surface. So called "sample charging effects" arising from net positive charging of the specimen cause apparent shifts in the binding energy scale (to higher binding energy, due to net retardation of the electrons before entering the spectrometer), commonly occur and depending on spctrometer design can amount to tens of eV but more typically are of the order of 1-2 eV. If absolute binding energies are required therefore sample charging must be detected and allowed for and reliable methods of achieving this end are discussed in a later section.

At this stage it is worthwhile considering briefly the relationship between the binding energies for a given level for an assembly of isolated (e.g. gas phase) molecules with the vacuum level as reference (as appropriate for direct theoretical computation) and those actually measured for the molecular solid. this is best considered in terms of the Born Cycle shown in Fig. 13.

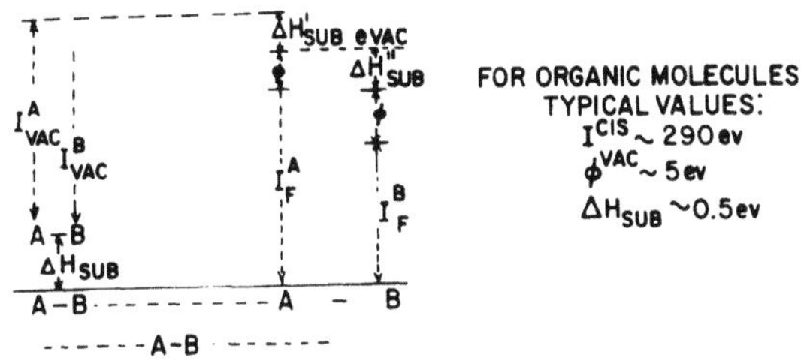

SHIFT IN BE

$$\Delta = (I_A - I_B)_{VAC} = (I_A - I_B)_F + (\Delta H'_{SUB} = \Delta H''_{SUB})$$

$$\Delta H'_{SUB} - \Delta H''_{SUB} = 0$$

SHIFT IN BE FOR DIFFERENT SAMPLES

$$\Delta = (I_A - I_X)_{VAC} = (I_A - I_X)_F + (\phi_A - \phi_X) + (\delta_A - \delta_X) + (\Delta H_X - \Delta H_A) + (\Delta H'_A - \Delta H_X')$$

FOR CLOSELY RELATED MATERIALS IN CONTACT WITH SPECTROMETER

$$(\phi_A - \phi_X) \approx 0$$
$$(\delta_A - \delta_X) = 0$$
$$(\Delta H_X - \Delta H_A) + (\Delta H_A' - \Delta H_X') \approx 0$$
$$(I_A - I_X)_{VAC} \equiv (I_A - I_X)_F$$

∴ CAN UNDERSTAND SHIFTS IN TERMS OF ISOLATED MOLECULES.
(EXCLUDING STRONG INTERMOLECULAR INTERACTIONS, e.g. H BONDING)

Fig. 13. Relationship between energy levels of gaseous and solid samples.

In this case we consider a simple molecular solid AB with core levels of interest on fragments A and B. Considering firstly the RHS of Fig. 13 the measured core binding energies for the solid with Fermi level as reference are I^A_F and I^B_F. For the isolated molecules the measured binding energies (this may be hypothetical if AB represents a high M. Wt. polymer) with respect to the vacuum level as energy reference, are I_{vac}^{A} and I_{vac}^{B}. In most cases the shift in binding energies are of prime interest theoretically rather than the absolute binding energies in which case it is clear from Fig. 13 that the shifts for the two situations are related by differences of lattice sublimation terms involving the system AB with a core hole located on fragment A or B. Since typically for molecular solids, the absolute values of these terms would be expected to be small there should be a good correlation of shifts measured in the gas and solid phases. Studies of appropriate monomer systems in which intermolecular interactions are small shows that experimentally this is realized and computed shifts for the isolated assembly molecules may be directly compared with experimentally determined shifts for the solids. In comparing one sample with another, however, further points arise, namely differences in work function, difference of differences of lattice sublimation terms and any effects due to differences in sample charging (δ).

For closely related materials the samples of which are either sufficiently thin for sufficient charge carriers to be available from say a metallic backing to maintain overall electroneutrality, or for which independent corrections for sample charging effects have been employed, a case can again be made for discussing shifts etc.. in terms of an isolated molecular approach. As will become apparent the complexities of the model systems involved in quantitative discussions of core binding energies for polymers dictates that rigour be sacrificed to computational feasibility and the model employed is such that even if there are small differences in work function for polymers between which shifts are being compared it is still possible to use an isolated molecule approach to discuss the results since such differences may be incorporated into the model parametrization. To put matters in perspective however, whereas the shift in core binding energy for say the C_{1s} levels of PTFE with respect to ethylene is ∿ 7.2 eV the work function difference is only 0.85 eV.[9]

3.7 Linewidths

The measured linewidths for core levels (after taking into account spin orbit splittings if these are not resolved) may be expressed as

$$(\Delta E_m)^2 = (\Delta E_x)^2 + (\Delta E_s)^2 + (\Delta E_{Cl})^2$$

where

ΔE_m is the measured width at half height, the so called full width at half maximum (FWHM),

ΔE_x is the FWHM of the photon source,

ΔE_s is the contribution to the FWHM due to the spectrometer (i.e. analyzer),

ΔE_{Cl} is the natural width of the core level under investigation. (For solids this includes solid state effects not directly associated with the lifetime of the hole state but rather with slightly differing binding energy due to differences in lattice environment.

It has previously been pointed out that $MgK\alpha_{1,2}$ and $AlK\alpha_{1,2}$ are the most useful photon source from the standpoint of keeping the contribution of ΔE_x to the total linewidth small. With well designed magnetic or electrostatic analyzers the contribution ΔE_s can be reduced to negligible proportions so that the major limiting factors in terms of resolution are photon linewidths (which may be reduced by monochromatization) and the inherent width of the level itself. (For solids in which longer range interactions are important e.g. ionic lattices or hydrogen bonded covalent solids, solid state effects can contribute to the overall linewidths)

Some examples of natural linewidths (ΔE_{Cl}) derived from X-ray spectroscopic studies are given in Table 2. The uncertainty principle in the form

$$\Delta E \quad \Delta t. \geqslant \frac{h}{4\pi}$$

Table 2. Natural linewidths for some core levels

	Level	Atom							
		S	A	Ti	Mn	Cu	Mo	Ag	Au
	1s	0.35	0.5	0.8	1.05	1.5	5.0	7.5	54
	$2p_{3/2}$	0.10		0.25	0.35	0.5	1.7	2.2	4.4
Radiative widths	1s	0.04	0.07	0.2	0.33	0.65	3.6	6.0	50
Fluorescence yields	1s	0.1	0.14	0.22	0.31	0.43	0.72	0.8	0.93

L G Parratt, Rev. Mod. Phys., 31, 616 (1959)

shows that for a hole state lifetime of $\sim 6.6 \times 10^{-16}$ sec. the linewidth i.e. uncertainty in the energy of the state is ~ 1 eV.

It is evident from Table 2 that there are large variations in natural linewidths both for different levels of the same element and for the same levels of idfferent elements. These reflect differences in lifetimes of the hole state. From Fig. 8 it is clear that the lifetime is a composite of radiative (fluorescence) and non-radiative (Auger) contributions, the importance of the former increasing with atomic number (approximately as Z^4). This is clearly shown in Table 2. This emphasizes the fact that there is no particular virtue in studying the innermost core level as has been previously pointed out. For gold for example, the 1s level has a halfwidth of ~ 54 eV so that even if a monochromatic X-ray source with the requisite photon energy were available any subtle chemical shift effects we might wish to investigate would be swamped.

As has been previously indicated, lifetimes of core hole states are typically in the range $10^{-13} - 10^{-17}$ sec and the ESCA time scale can therefore be fairly described as sudden with respect to nuclear (but not electronic) motions. Very recent high resolution gas phase ESCA studies of simple molecules using a fine foccussed X-ray monochromation scheme has revealed that vibrational effects are indeed evident for core level photoionizations.[10] For molecular solids and crystals, however, it seems likely that these may well be marked and hence a detailed discussion of this fascinating area of the subject will not be given here.

For typical core levels of interest in the study of polymers e.g. C_{1s}, N_{1s}, O_{1s} etc.. with conventional instrumentation employing unmonochromatized X-ray sources the single largest contribution to the overall linewidths arises from the inherent width of the X-ray source itself ($MgK\alpha_{1,2} \sim 0.7$ eV, $AlK\alpha_{1,2} \sim 0.9$ eV).

3.8 Simple Examples Illustrating Points Discussed in 3.1-7

A typical wide scale ESCA spectrum (i.e. plot of number of electrons of given kinetic energy arriging at the detector in unit time) is shown in Fig. 14 for a PTFE sample.

The "sharp" photoionization peaks due to F_{1s} and C_{1s} levels are readily identified by their characteristic binding energies. The group of three rather broader peaks whose KE are independent of the photon source are readily identified as Auger peaks arising from de-excitation of the F_{1s} hole states. As will become apparent one of the most important attributes of ESCA as a technique is the possibility of providing information relating to surface, subsur-

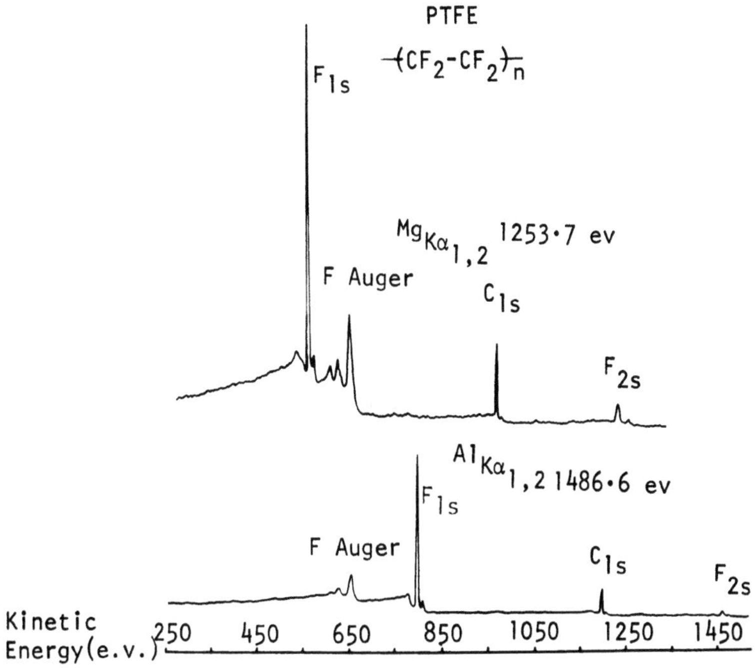

Fig. 14. Wide scan ESCA spectra for PTFE samples studied with $MgK\alpha_{1,2}$ and $AlK\alpha_{1,2}$ radiation.

face or essentially bulk properties, and this depends on differences in escape depths for photoemitted (or Auger) electrons corresponding to different kinetic energies. For the most commonly used soft X-ray sources the penetration of the incident radiation for a solid sample is typically in excess of 1000 Å (dependent on the angle of incidence etc..). However the number of photoemitted electrons contributing to the elastic peak (i.e. corresponding to no energy loss) is determined by the mean free path of the electron in the solid. Thus in general the ESCA spectrum of a given core level consists of well resolved peaks corresponding to electrons escaping without undergoing energy losses, superimposed on a background tailing to lower kinetic energy arising from inelastically scattered electrons. (This is clearly evident in Fig. 14. With an unmonochromatized X-ray source as in this example, there is of course a general contribution to the background arising from the bremsstrahlung).

In the applications to be disccused below we may assume that the X-ray beam is essentially unattenuated over the range of surface thickness from which the photoelectrons emerge. The intensity of electrons of a given energy observed in a homogeneous material be expressed as

$$dI = F\alpha Nke^{-x/\Lambda}dx$$

where F is the X-ray flux, α is the cross section for photoionization in a given shell of a given atom for a given X-ray energy, N is the number of atoms in volume element, k is a spectrometer factor for the fraction of electrons that will be detected and depends on geometric factors and on counting efficienty, Λ is the electron mean free path and depends on the KE of the electron and the nature of the material that the photoelectron must travel through (Fig. 15). The situation which is of common occurence is

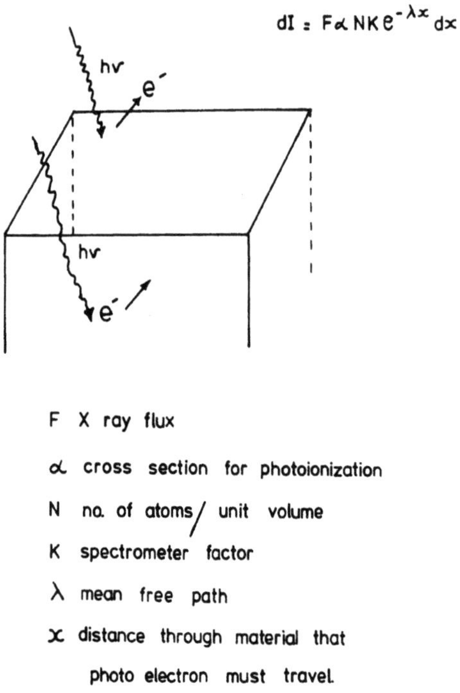

Fig. 15. Intensities of elastic peaks as a function of mean free path etc. for solids.

that of a single homogeneous component or of a surface coating of thickness d on a homogeneous base. This is illustrated in Fig. 16. The intensity for a film A thickness d may be expressed as

$$I^A_{\cdot} = I^A_{\alpha}\,(1-e^{-d/\Lambda_A})$$

whereas for the film B (considered infinitely thick) on the surface of which A is located, the intensity is given as

$$I^B = I_\alpha^B \, e^{-d/\Lambda_B}$$

(with $N_A = N_B$). Fig. 16 shows data pertaining to the escape depth[+]

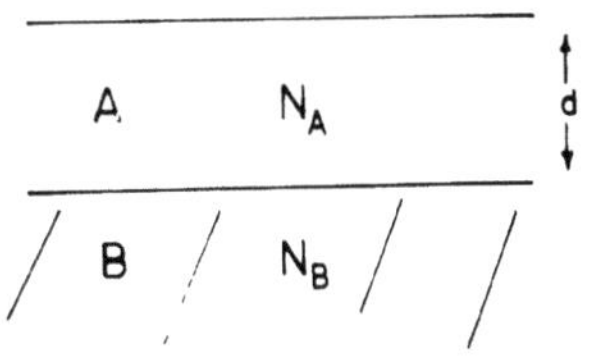

$$I_\alpha = \int_0^\infty F\alpha N K e^{-x/\Lambda} \, dx = F_\alpha N K \Lambda$$

$$I^A = \int_0^d F\alpha N_A K e^{-x/\Lambda_A} \, dx = F\alpha N_A K \Lambda_A (1 - e^{-d/\Lambda_A})$$

$$= I_\alpha{}^A (1 - e^{-d/\Lambda_A})$$

$$I^B = \int_d^\infty F\alpha N_B K e^{-(x-d)/\Lambda_B} \, e^{-d/\Lambda_A} \, dx$$

$$= F\alpha N_B K \Lambda_B \, e^{-d/\Lambda_A}$$

If $N_A \approx N_B \quad \Lambda_A \approx \Lambda_B$

$$I^B = I_\alpha e^{-d/\Lambda_A}$$

Fig. 16. Peak intensities as a function of film thickness.

dependence on kinetic energy for electrons, derived from Auger and ESCA studies on films of known thickness.[11] The striking feature clearly evident in Fig. 17 is that the experimental data for solids, in which the contributions of various scattering processes

---

+ "Mean free path" and "escape depth" have been used interchangeably in this work, they both correspond to the depth at which $^1/e$th of the emitted electrons have not suffered any energy loss.

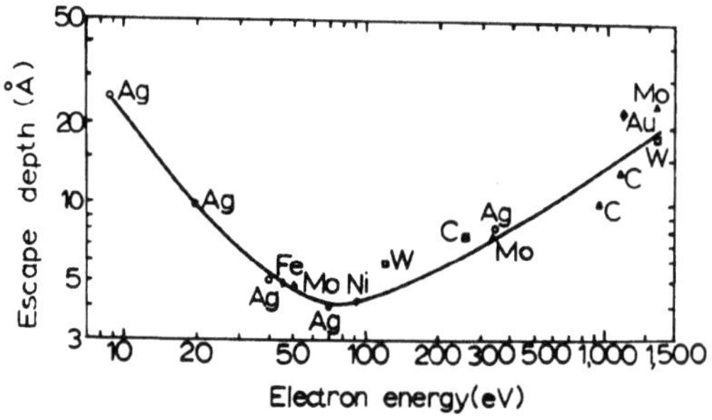

Fig. 17. Escape depth dependence on kinetic energy for electrons.

might be expected to differ, fit closely onto a "universal" curve of escape depth versus kinetic energy. The importance of this will become apparent in the ESCA studies relating to surface fluorination. In general however, it is clear that for photon energy of 1253.7 eV or 1486.6 eV the mean free paths for photoemitted electrons should be < 20 Å.

The dependance of the signal intensity on the escape depth for photoemitted electrons is illustrated by the values listed in Table 3.

The C_{1s} spectrum of benzotrifluoride (Fig. 18) illustrates the large shifts in core levels which can occur.[12a] The electronegative fluorines withdraw electron density from the carbon of the trifluoromethyl group with concomitant large increase in C_{1s} core binding energy. As will become apparent a peak at ∿ 294 eV binding energy may be used to "fingerprint" $-CF_3$ groups.

The C_{1s} and O_{1s} spectra of a film of polyethylene terephthalate (Fig. 19) illustrates that these effects may be as easily detected in polymeric as in monomeric samples. The O_{1s} spectrum clearly showing the two kinds of ester oxygen, and the C_{1s} spectra clearly distinguishing the ester, aliphatic and aromatic carbon levels. Again it can be seen that integrated intensities are

Table 3. Dependence of signal intensity on escape depth

Mean free path λ	50% intensity	90% intensity
7 Å	4.85	16.1
10 Å	6.93	23.0
12 Å	8.3	27.6

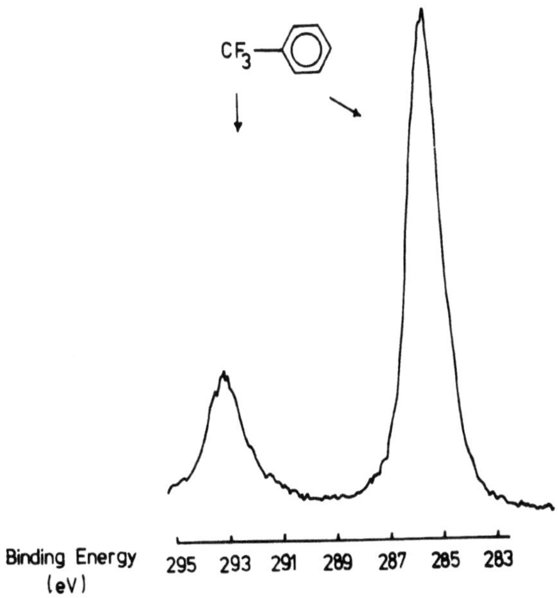

Fig. 18. C_{1s} levels for benzotrifluoride.

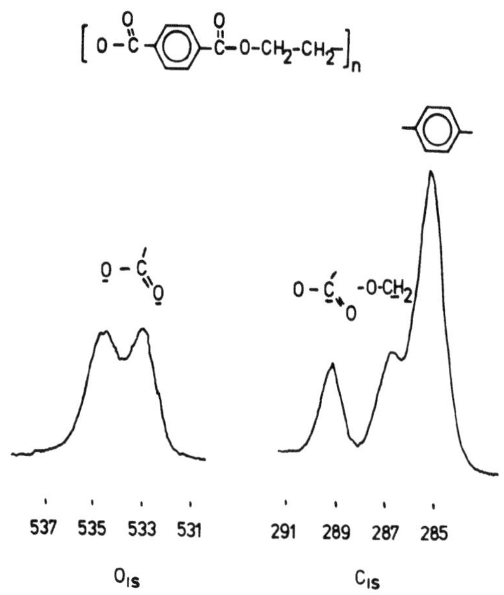

Fig. 19. C_{1s} and O_{1s} levels for polyethylene terephthalate film.

directly proportional to the relative abundances of the various types of environment, namely 1:1 for the oxygen levels and 1:1:3 for the carbon levels.

As a simple example, again illustrating how ESCA can provide direct information on electron distributions and also illustrating the phenomena of shake up processes, Fig. 20 shows the N_{1s} and C_{1s} core levels of diazocyclopentadiene.[12b] The large difference in electron density about the two nitrogens is clearly evidenced by the shift of 2 eV between the two peaks. Clearly evident to the low kinetic energy side of both core level peaks are satellites which are in fact identified as $\pi \rightarrow \pi^+$ shake up transitions for the relevant hole states.

So far we have indicated the importance in ESCA measurements of inter alia - the absolute binding energies and shifts in core levels, relative peak intensities, and their dependence on mean free path, kinetic energy, photoionization cross sections etc. A further possible source of information arises from paramagnetic species due to multiplet splittings. One of the first demonstrations of this due to Siegbahn and coworkers is illustrated in Fig. 21 for the paramagnetic NO and O_2 molecules.[4] The theory in the cases of core ionization from S levels is particularly simple

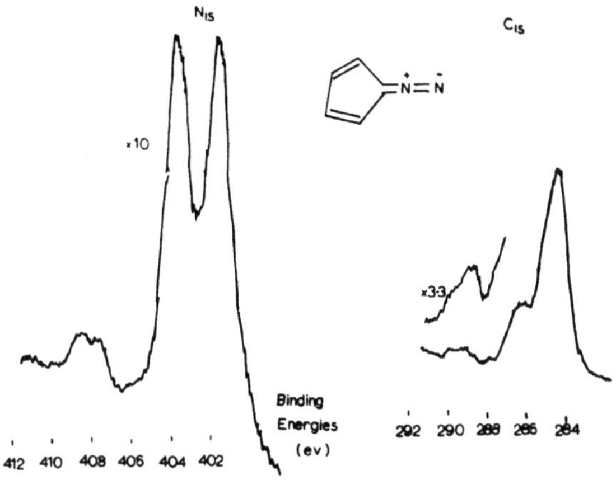

Fig. 20. C_{1s} and N_{1s} levels for diazocyclopentadiene showing shake up structure.

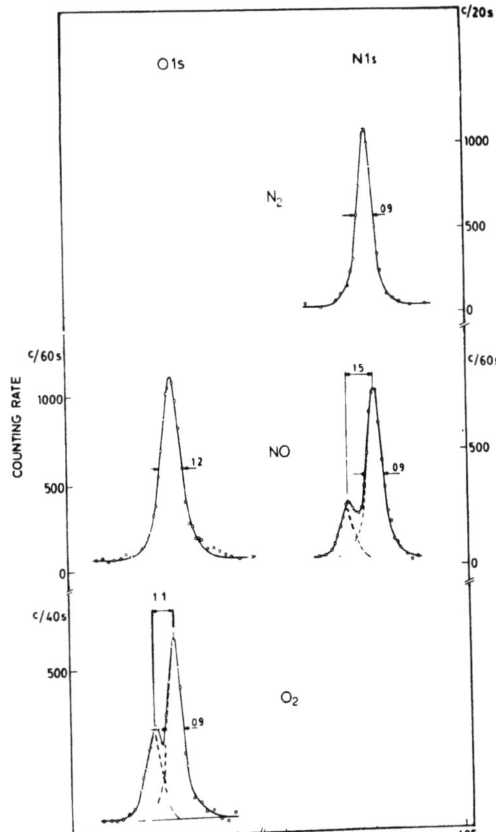

Fig. 21. O_{1s} and N_{1s} levels showing multiplet splittings for the paramagnetic NO and O_2 molecules.[4]

as outlined in Fig. 22. In the case of NO the single unpaired electron is delocalized on nitrogen and oxygen so that the magnitude of the multiplet splitting of the O_{1s} and N_{1s} core levels will depend upon the unpaired spin densities at the two atoms.[12c]

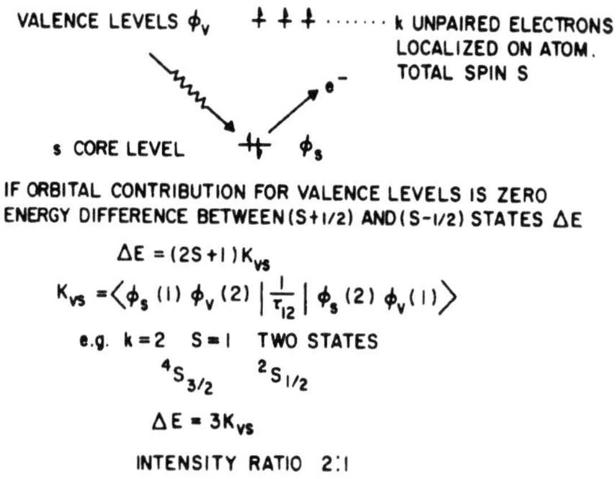

Fig. 22. Multiplet splittings for photoionization form S hole states.

The pronounced satellite for nitrogen indicates that most of the unpaired spin density is on nitrogen and hence the magnitude of multiplet splittings ca provide information comparable to that from ESR studies. The best developed application of multiplet effects is in transition metal chemistry originally pioneered by Fadley and coworkers.[2a] Possible applications in the polymer field might be the study of free radicals localized at the surfaces of polymers.

3.9 Basic Types of Information Available from ESCA Studies

Both the core and valence energy levels of polymers may be studied and the basic types of information available are summarized in Table 4 and with the foregoing discussion are largely self explanatory.

Table 4. Information from ESCA studies.

Core Electrons	Valence Electrons
(1) Binding energy characteristic of a given level of a given element therefore useful for analysis Different KE dependencies for different core levels provides means for analytical depth profiles	(1) Can study valence energy levels of insulators Densities of states for conduction bands of metals (of interest in study of metallised polymers)
(2) Absolute BE may be characteristic of particular structural features (eg CF_3, C=O, $-NH_2$ etc) 'Shifts' can be related to electron distribution	(2) Studies of differential changes in cross section with photon energy provides information on symmetries of orbitals (σ, π etc)

(3) Multiplet Splittings

For paramagnetic species observation of multiplet splittings provides information on spin states of atoms or ions and distribution of unpaired electrons

(4) Shake up and shake off satellites Information on excited states of hole states

IV. APPLICATIONS OF ESCA TO STUDIES OF STRUCTURE, BONDING AND REACTIVITY OF POLYMERS

4.1 Introduction

The great advantage of ESCA as a technique, in being able to study in principle the core and valence levels of any element, (regardless of nuclear properties such as magnetic or electric quadrupole moments), coupled with the low sample requirements, and the ability to study involatile insoluble solids, is nowhere more apposite than in the study of polymers. We have noted in previous sections the particular feature of ESCA in being in principle, capable of providing information concerning the surface and immediate subsurface of solid samples as well as that pertaining to the bulk. This particular feature of the technique is crucial to an understanding of surface treatments of polymers such as oxidation, fluorination, argon ion bombardment etc.., which will be discussed later. As a simple example, however, of how ESCA may routinely be used to monitor surface composition mention might be made of pressed films of PTFE. The spectrum of PTFE recorded earlier (Fig. 14) was obtained from a film pressed from PTFE powder between sheets of aluminium foil at the minimum temperature necessary for coalescence of the powder particles in a hand press ($\sim 200°$). When the procedure is repeated using a higher temperature ($> 300°$) films are produced which visually, bulk chemically and by transmission infrared (TIR) and multiple attenuated total reflectance (MATR) measurements, appear to be the same as those produced at lower temperatures. The ESCA spectra of such films however are very revealing (Fig. 23). It is clear form the appearance of peaks associated with the core levels of both oxygen and aluminium that in the high temperature pressing process a contaminant surface layer of alumina Al_2O_3 is deposited on the PTFE. The thickness of the layer is almost certainly < 10 Å. Although this is a trivial example it can readily be appreciated that if surfaces are mechanically prepared the possibility of contamination is always present.

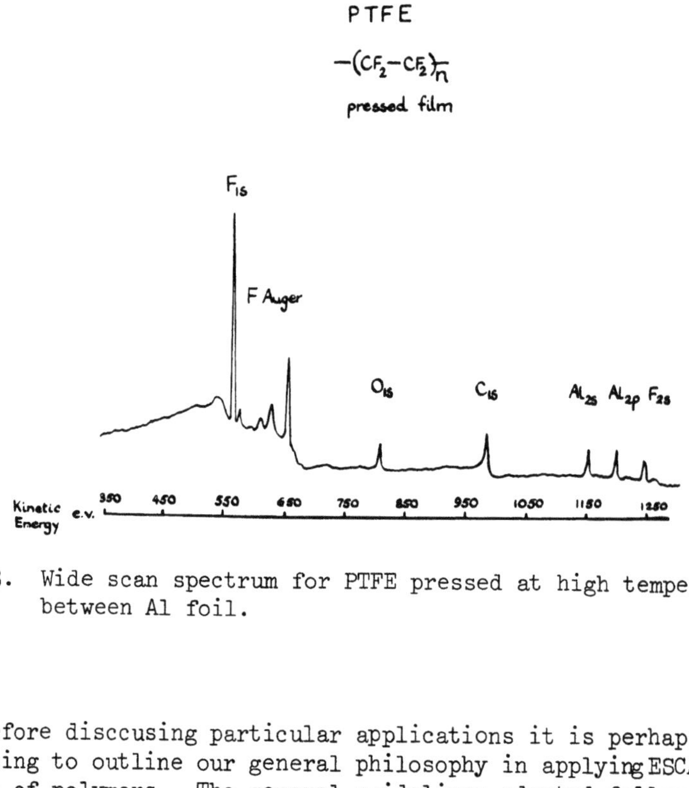

Fig. 23. Wide scan spectrum for PTFE pressed at high temperature between Al foil.

Before disccusing particular applications it is perhaps worth digressing to outline our general philosophy in applyingESCA to studies of polymers. The general guidelines adopted follow directly form our previous development of ESCA as a tool for studying structure and bonding in organic systems in general and halocarbon systems in particular.[13] Namely, to start by studying simple well characterised systems to build up banks of data (on relative peak intensities, absolute binding energies and shifts) from which trends may be discerned and comparison drawn with simple monomers. The simplicity of these systems allows the development of rigorous theoretical models for the quantitative interpretation of the data. These models in turn point to the important electronic factors determining e.g. shifts in core binding energies, from which theoretically sound but less rigorous models may be developed to discuss larger systems for which more detailed computations are not feasible. (This development will be discussed in detail in the lectures). The study both theoretically and experimentally of simple model systems is therefore an important adjunct to the full development of the application of ESCA to studies of complex polymeric systems. It is worthwhile pointing out at this stage that for much of the information of interest from ESCA measurements (e.g. absolute and relative binding energies) the theoretical interpretation is relatively straightforward and in this sense ESCA has a distinct advantage over many other spectroscopic techniques where inevitably the interpretation of data is more often than not in terms of (experimental) analogy with model systems. ESCA data is therefore rather directly related to the electronic structure of a molecular system and this explains to some extent why ESCA and X-ray crystallography have been singled out as the only experimental techniques to be discussed at this summer school.

As a logical first step in the development of ESCA applied to polymers, we have studied simple well characterized homopolymer and copolymer systems.[14-17] These "homogeneous" samples then form a valuable yardstick for gaugingthe utility of ESCA for studies pertaining to information characteristic of the bulk.

In applying the technique to the investigation of simple polymer systems there are several distinct aspects about which one would hope to gain information. Firstly, the gross chemical composition of the polymers. This would include determination of elemental composition and in the case of copolymers the percentage incorporation of comonomers. Secondly, information concerning the gross structure, for example, for copolymers the block and/or alternating or random nature of the linkages. Thirdly, finer detail of structure such as structural isomerism, the nature of end groups, branching sites etc.. Finally deductions made from ESCA measurements, of charge distributions and nature of the valence bands for polymers. It should be emphasized that the data obtainable from ESCA is rather coarser in detail than that often obtainable by more conventional techniques (I.R., N.M.R., inelastic neutron scattering etc..) where information regarding for example, conformational aspects of polymer structure, may often be inferred and which are in principle not amenable to direct study by ESCA. Although, it is worth noting that in favourable cases (see later) detailed interpretation of ESCA spectra can lead to conclusions concerning polymer chain conformations, ESCA should be regarded as a powerful technique providing information complimentary to that from other branches of spectroscopy, but with unique advantages which mean that for many studies of polymer chemistry, such as dynamic studies of thermal or photochemical degradation, and in studies of polymeric films produced at surfaces by chemical reaction (e.g. fluorination, oxidation, etc..) the information derived from ESCA studies is not obtainable by other techniques.

4.2 Areas of Application of ESCA to Polymers

The brief introduction given above should emphasize the fact that ESCA is in principle capable of providing information concerning the surface and immediate subsurface of polymer samples as well as that pertaining to the bulk. It is evident therefore that ESCA has a wide ranging potential in the polymer field.

The application of ESCA to structure and bonding in polymer systems to date largely pertains to fluoropolymer systems. The reason for this are threefold. Firstly our research interests are centred around the halocarbon field (both monomeric and polymeric species) and hence samples and the expertise in preparing them are readily to hand. Even without an underlying research interest in these systems however, there are two further reasons why initial research program into the application of ESCA to polymers, should concentrate initially on fluoropolymer systems. Thus the large shift in (e.g. C_{1s}) core levels induced by fluorine gives the most favourable situation for delineating the likely areas of applicability of ESCA in this field. Finally, fluoropolymers are amongst the technologically most important systems of interest and are often difficult to study by ohter spectroscopic techniques.

The areas of application of ESCA which have so far been demonstrated include the following :

A. Aspects of Structure and Bonding (Static Studies)

a) Gross chemical compositions
 (a) elemental compositions,
 (b) % incorporation of comonomers in copolymers,
 (c) polymeric films produced at surfaces.

b) Gross structural information
 e.g. for copolymers, block, alternating or random nature.
 Domain structure in block copolymers.

c) finer details of structure
 (a) structural isomerisms
 (b) experimental charge distributions in polymers.

d) Valence bands of polymers.

e) Identification of polymers, structural elucidation.

B. Aspects of Structure and Bonding (Dynamic Studies)

a) Surface treatments e.g. CASING

b) Thermal and photochemical degradation

c) Polymeric films produced at surfaces by chemical reaction e.g. fluorination (including the use of ESCA for depth profiling and quantitative measurement of film thickness).

d) Chemical degradation of polymers, e.g. oxidation, nitration etc...

4.3 Sample Preparation

Before discussing the results it is of some relevance to consider the ways in which polymer samples may be prepared for examination by ESCA. When the polymer is available as a powder it is often convenient to study it as such by applying the powder to double sided Scotch tape mounted on the sample probe. The pitfalls to beware of in this approach are that no extraneous signals are observed from the sample backing and also that no chemical reaction occurs between sample and substrate. The incomplete coverage and uneven topography of samples prepared in this way generally lead to lower signal/noise ratios than polymers studied directly as films. The most generally useful methods of preparing polymer films for ESCA studies may be classified as follows :

a) From solution

If the polymer is sufficiently soluble then thin polymer films may be deposited directly on a gold backing (ready for mounting on the sample probe), by conventional dip or bar coating, or spin casting. Since ESCA is such a surface sensitive technique it is important to use clean apparatus and pure solvents containing no involatile residues (e.g. anti-oxidants etc..) which would segregate at the surface on evaporation of the solvent. With readily oxidized systems or with systems with sites capable of hydrogen bonding with extraneous water it is imperative to maintain a suitable inert atmosphere during the slow evaporation of solvent. Solvent entrainment can also be a problem, and indeed the technique lends itself well to studying diffusional problems in polymers.

b) From pressing or extrusion

Because of problems of contamination in solvent casting films it is convenient to study most polymers in the form of pressed or extruded films mounted on a suitable backing (e.g. gold). For elastomers of course it is often possible to "melt" a small amount of the sample and allow it to spread in the form of a thin film on the tip of a sampling probe or to slice a thin film from a larger sample. In preparing samples from powders it is often convenient to press films between sheets of clean aluminium foil at an appropriate temperature and pressure. There are two precautions to be taken in doing this :

(a) the temperature and pressures used should be such that no decomposition or adhesion of surface contamination occurs;

(b) since typically only the top ∿ 50 Å of the sample is studied by ESCA it is important to avoid chemical reaction at the surface during preparation. Thus pressing polyethylene films in air at the minimum temperature and pressure necessary, results in considerable surface oxidation. This may be obviated by pressing in an inert atmosphere (e.g. N_2 or Ar). Surface contamination e.g. hydrocarbon etc., arising for example by inadvertant handling during processing can most readily be removed by careful treatment with an appropriate solvent.

(c) Polymerization in situ

ESCA is particularly suited to dynamic studies and would be useful for monitoring polymerizations carried out in situ at the sample probe by e.g. irradiation ($h\nu$ or e^-) or from impinging beams of precursors generated by e.g. pyrolysis or irradiation.

4.4 Energy Referencing

Of their very nature, most polymers of interest are extremely good insulators and as such in studying samples as thin films there is only a fortuitous possibility that during the ESCA experiment sufficient charge carriers will be available such that the sample is in electrical contact with the spectrometer. Sample charging will therefore occur, resulting in a shift of the energy scale, and some form of referencing back to the Fermi Level is therefore necessary. Referencing of the energy scale is most readily accomplished by depositing a thin coating of a suitable reference (e.g. C_{1s} signal from a hydrocarbon or $4f_{7/2}$ levels of gold with binding energies of 285 eV and 84 eV respectively) and monitoring the core levels of the sample and reference.[6a] To put matters in perspective however, it should be emphasized that with most commercial spectrometers (based on unmonochromatized X-ray sources and retarding lens system, double focussing hemispherical electrostatic analyzers and associated slit systems), sample charging is not too serious a problem and at worst involves only a few eV correction to the energy scale. With slitless designs and a monochromatized X-ray source as conventionally applied in dispersion compensation, sample charging can however reach serious proportions involving shifts in the energy scale of tens of eV for insulating samples. The problem can be alleviated to some extent by use of low energy electron flood guns but this introduces other complications.

4.5 Preliminary Considerations

An interesting observation is that for homopolymers, even in the presence of overall sample charging, with suitable preparation of sample, linewidths for given core levels are comparable

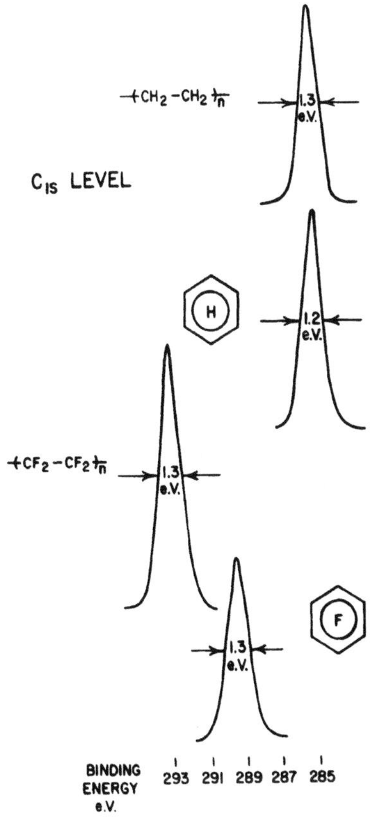

Fig. 24. C_{1s} levels for homopolymers and typical monomers illustrating linewidth similarities.

with those for monomers, Fig. 24, (cf. polyethylene terephthalate, Fig. 19). (Instrumental resolution set such that under comparable conditions ($MgK\alpha_{1,2}$ photon source) the Au $4f_{7/2}$ level has FWHM 1.15 eV).

On this basis, in principle the range of applicability of ESCA in studying homopolymers is essentially the same as in studying simple monomers. With some preparations of thin films, by pressing or more usually with sample prepared by slicing from a larger sample or melted directly onto a probe tip, the less regular nature of the surface is often manifest in slight increase in linewidths arising from non-uniform distribution of charge on the sample. This is often of no consequence since it merely affects the overall resolution, however it may be obviated in many cases by modifying the way in which the film is produced. A simple example illustrating this rather effectively arises from our studies of copolymers of ethylene and tetrafluoroethylene (to be discussed in more detail later on).

The first spectra recorded with these samples (Fig. 25) seemed to indicate that the degree of fluorinated monomer incorporation was not as great as expected. Thus in the carbon-1s spectra the peaks at low binding energy due to $\underline{C}H_2$ groups were always bigger than the higher energy bands due to $\underline{C}F_2$ groups, even for samples where the degree of incorporation of tetrafluoroethylene was reliably known (from conventional carbon and fluorine elemental analysis) to be greater than 50 mole %. This observation is clearly indicative of surface contamination of the samples by hydrocarbon which could conceivably arise in one or more ways which will not concern us here. When the surfaces of the films were cleaned with methylene chloride (by lightly rubbing the surface with a tissue dampened with solvent), the contaminants were removed and the elemental compositions of the cleaned surfaces could be shown (see later) to be the same as those measured by conventional combustion analysis of bulk samples. The effect of this surface cleaning on the carbon-1s spectrum of a copolymer sample is illustrated in Fig. 25, spectrum (a) shows the spectrum of the polymer prior to cleaning and spectrum (b) that of the same sample after it had been cleaned with methylene chloride. Two effects are apparent from inspection of the spectra, firstly the peaks due to hydrocarbon contamination of the surface which occur at low binding energy have been removed by the cleaning process, and secondly both the peaks in the spectrum of the cleaned sample are narrower than those in the original untreated material. This second observation can be explained on the assumption that the surface contamination is unevenly distributed resulting in an uneven distribution of charges at the sample surface and a consequent line broadening, such an uneven distribution of contaminants could arise from handling of the sample.

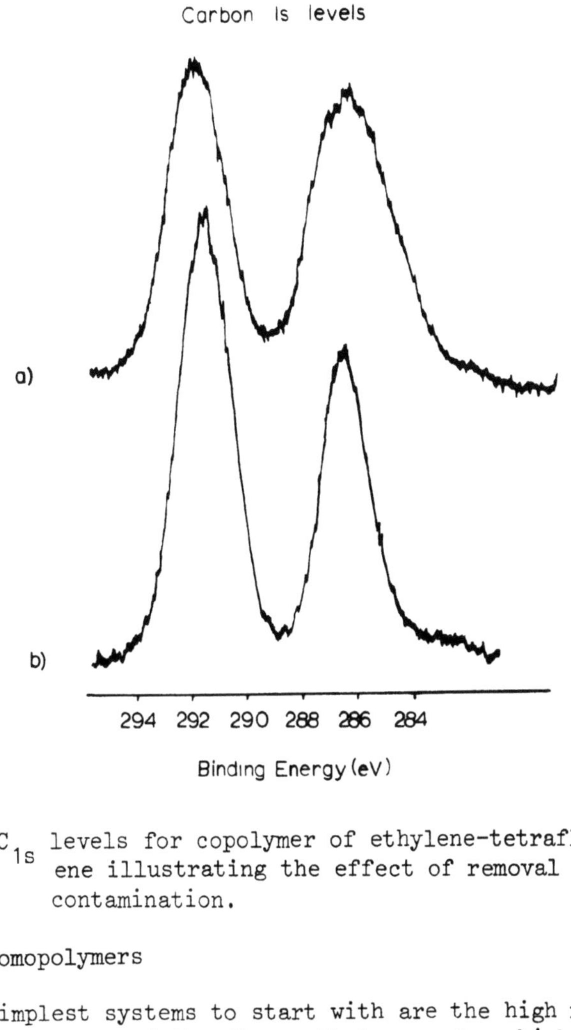

Fig. 25. C_{1s} levels for copolymer of ethylene-tetrafluoroethylene illustrating the effect of removal of surface contamination.

4.6 Homopolymers

The simplest systems to start with are the high molecular weight homopolymers of the fluoroethylenes, for which complications due to branching, end groups, etc., are minimal. The carbon-1s spectra for polyethylene (high density), polyvinyl fluoride, polyvinylidene fluoride, polyvinylene fluoride, polytrifluoroethylene, and polytetrafluoroethylene are presented in Fig. 26, 27 and 28.

Typically measurements of all the core levels of a polymer take $\sim$ 30 mins. whereas under the conditions employed (pressure in sample chamber typically $\sim 5 \times 10^{-7}$ Torr) hydrocarbon build up on the surface only becomes appreciable after several hours. The technique employed for calibration of the energy scale therefore is to measure the core levels of a given polymer immediately on introduction of the sample into the spectrometer when hydrocarbon

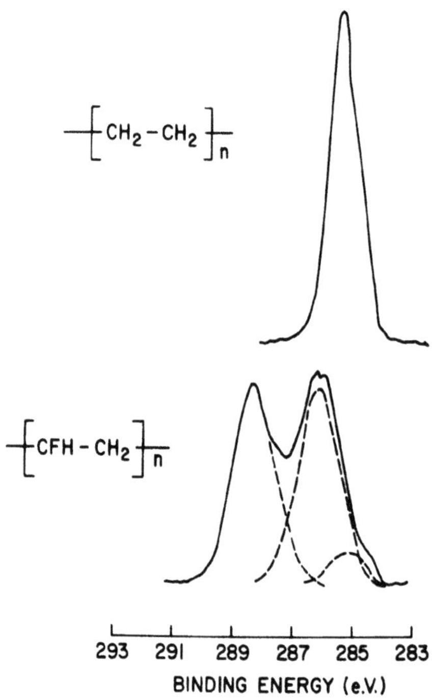

Fig. 26. C_{1s} levels of polyethylene and polyvinylfluoride.

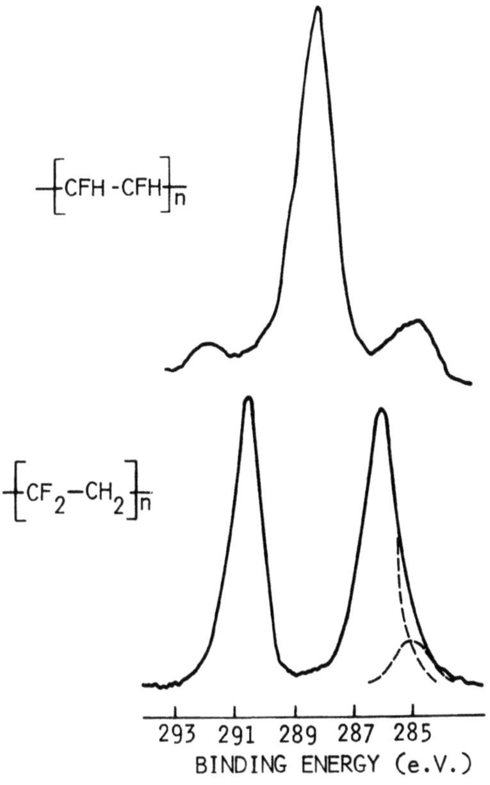

Fig. 27. C_{1s} levels of polyvinylenefluoride and polyvinylidene fluoride.

contamination is unimportant. After several hours in the sample chamber further spectra may then be recorded and the appearance of an extra peak, (clearly evident in the spectra in Figs. 26-28) in the C_{1s} region at 285.0 eV binding energy may then be used to reference the energy scale.

The C_{1s} spectrum of polyvinyl fluoride shows two partially resolved peaks of equal area corresponding to $\underline{C}HF$ and $\underline{C}H_2$ carbons whilst for polyvinylene fluoride in addition to the main peak corresponding to $\underline{C}HF$ carbons and hydrocarbon calibration peak, there is a weak peak at 292.0 eV. This in fact corresponds to $\underline{C}F_2$ type carbon arising from contamination from the fluorocarbon soap $(H(CF_2)_8COO^-NH_4^+)$ used in the emulsion polymerization. For polyvinylidene fluoride well resolved peaks of equal area corresponding to $\underline{C}F_2$ and $\underline{C}H_2$ carbons are evident and for polytrifluoro-

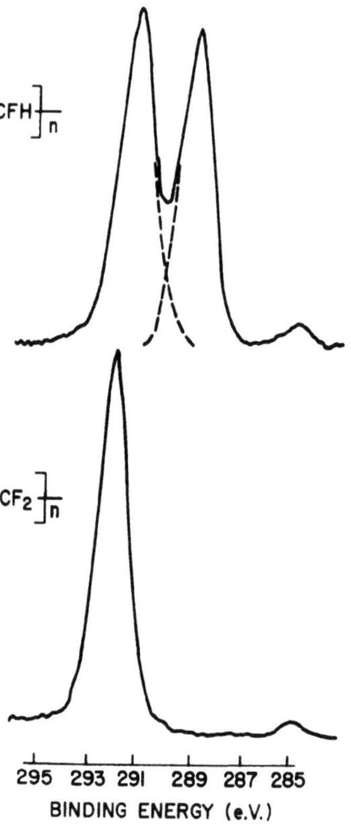

Fig. 28. C_{1s} levels of polytrifluoroethylene and PTFE.

ethylene partially resolved peaks of equal area corresponding to $\underline{C}F_2$ and $\underline{C}FH$ carbons. The data pertaining to these homopolymers are collected in Fig. 29 the assignment of peaks arising from $\underline{C}F_2$, $\underline{C}FH$ and $\underline{C}H_2$ structural units is straightforward. By taking appropriate pairs of polymers it is possible to investigate both primary and secondary effects of replacing hydrogen, Tables 5 and 6. The average primary (2.9 eV) and secondary (0.7 eV) substituent effects are in excellent agreement with those found for monomer systems and their relative constancy emphasizes the characteristic nature of substituent effects. It is interesting to note that these results are also in excellent agreement with detailed non-empirical LCAO MO SCF calculations on neutral model systems and the C_{1s} hole states cf. Table 7. The rapid fall off

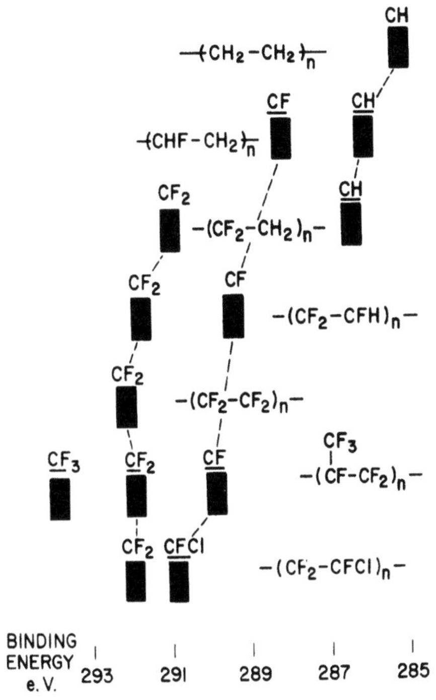

Fig. 29. Substituent effects in some homopolymers.

Table 5. Primary substituent effects for C_{1s} levels

POLYMER PAIRS	SHIFT IN C_{1s} BINDING ENERGY ON REPLACING H BY F eV
$(\underline{C}HFCH_2)_n$, $(\underline{C}H_2CH_2)_n$	3.0
$(\underline{C}F_2CH_2)_n$, $(\underline{C}HFCH_2)_n$	2.8
$(CHF\underline{C}HF)_n$, $(CHF\underline{C}H_2)_n$	2.5
$(CF_2\underline{C}HF)_n$, $(CF_2\underline{C}H_2)_n$	3.0
$(CF_2\underline{C}F_2)_n$, $(CF_2\underline{C}HF)_n$	2.9
$(\underline{C}F_2CHF)_n$, $(\underline{C}HFCHF)_n$	3.2
AVERAGE	2.9

Table 6. Secondary substituent effects for C_{1s} levels

POLYMER PAIRS	SHIFT IN C_{1s} BINDING ENERGY ON REPLACING H BY F (Per Substituent) eV
$(CHF\underline{C}H_2)_n$, $(CH_2\underline{C}H_2)_n$	0.9
$(\underline{C}HFCHF)_n$, $(\underline{C}HFCH_2)_n$	0.4
$(CF_2\underline{C}H_2)_n$, $(CH_2\underline{C}H_2)_n$	0.7
$(\underline{C}F_2CHF)_n$, $(\underline{C}F_2CH_2)_n$	0.8
$(\underline{C}F_2CF_2)_n$, $(\underline{C}F_2CHF)_n$	0.6
$(CF_2\underline{C}HF)_n$, $(CHF\underline{C}HF)_n$	0.9
AVERAGE	0.7

Table 7. Fluorine substituent effects on C_{1s} levels form non-empirical LCAO MO calculations

Effect of fluorine substitution on α and β core binding energies

	Primary (α)	Secondary (β)
CH_3CH_3	(0)	(0)
$CH_2F\text{-}CH_3$	2.96 (2.96)	0.61 (0.61)
$CHF_2\text{-}CH_3$	6.04 (3.08)	1.26 (0.65)
$CF_3\text{-}CH_3$	9.13 (3.09)	1.96 (0.70)
$CH_2{=}CH_2$	(0)	(0)
$CHF{=}CH_2$	2.95 (2.95)	0.58 (0.58)
$CF_2{=}CH_2$	6.07 (3.12)	1.32 (0.74)
$H\text{-}C{\equiv}C\text{-}H$	(0)	(0)
$F\text{-}C{\equiv}C\text{-}H$	3.20	0.86
$H_2C{=}O$	(0)	(0)
$HFC{=}O$	3.41 (3.41)	1.24 (1.24)
$F_2C{=}O$	6.86 (3.45)	2.40 (1.16)
$H\text{-}C{\equiv}N$	(0)	(0)
$F\text{-}C{\equiv}N$	3.3	1.09

in effect of the fluorine substituent provides a crude but immediate manifestation of the σ inductive effect exerted by fluorine. As the degree of fluorine substitution increases in going through the series, the fluorine substituents to some extent compete for the sigma electron drift from the carbon and hydrogen atoms and this is clearly shown by the increase in binding energy for the F_{1s} levels (Table 8). It is gratifying to note that the shifts

Table 8. Binding energies for core levels of some homopolymers

			C_{1s}	$\Delta(C_{1s})$	F_{1s}	$\Delta(F_{1s})$
I	$\{CH_2\text{-}CH_2\}_n$		285.0	(0)	-	-
II	$\{CFH\text{-}CH_2\}_n$	-ÇFH-	288.0	3.0	689.3	(0)
		-ÇH$_2$-	285.9	0.9	-	-
III	$\{CFH\text{-}CFH\}_n$		288.4	3.4	689.3	0.0
IV	$\{CF_2\text{-}CH_2\}_n$	-ÇF$_2$-	290.8	5.8	689.6	0.3
		-ÇH$_2$-	286.3	1.3	-	-
V	$\{CF_2\text{-}CFH\}_n$	-ÇF$_2$-	291.6	6.6	690.1	0.8
		-ÇHF-	289.3	4.3	690.1	0.8
VI	$\{CF_2\text{-}CF_2\}_n$		292.2	7.2	690.2	0.9

in core binding energies are qualitatively in agreement with what might be termed the organic chemists intuitive ideas concerning charge distributions, this trend will be quantified in the next section.

In studying fluoropolymer systems in general in addition to the F_{1s} and C_{1s} core levels and their associated Auger transitions, it may be noted that the F_{2s} levels at the bottom of the valence band are essentially core like so that even with a single photon source the escape depth dependences on kinetic energy span a large range. For $MgK\alpha_{1,2}$ KE's for F_{1s}, C_{1s} and F_{2s} are ∿560 eV, ∿ 960 eV and ∿1220 eV respectively whilst the Auger processes for de-excitation of the F_{1s} and C_{1s} hole states are in the kinetic energy range 610-660 and 240-270 eV respectively. This being the case it is a relative easy matter to use ESCA to "depth profile" samples to investigate their homogeneity. For the homopolymers of the fluorinated ethylenes the C_{1s} spectra consist of two peaks of equal area. Since the KE energies for the photoemitted electrons are virtually identical the sampling depths must be the same and the results therefore suggest a homogeneous sample. This may be confirmed by comparison with data from the F_{1s} level where the sampling depth will be different. Table 9 shows the relative carbon-1s to fluorine-1s peak areas for the four samples (with our particular instrumental arrangement).

Table 9. Relative C_{1s} to F_{1s} peak areas for homopolymers

Polymer	Carbon - 1s peak area: Fluorine-1s peak area	Number of carbon atoms: Number of fluorine atoms	Relative* carbon-1s peak area Fluorine-1s peak area
$(CH_2CHF)_n$	1:1	2:1	1.20
$(CH_2CF_2)_n$	1:2.08	1:1	1.21
$(CHFCF_2)_n$	1:3.15	2:3	1:21
$(CF_2CF_2)_n$	1:3.97	1:2	1.20

*With probable error limits of ± 0.05

The constancy of the ratio demonstrates unambiguously the uniformity of the samples since the escape depth dependencies are significantly different (mean free paths F_{1s} ∿ 7 Å (KE ∿ 564 eV), C_{1s} ∿ 10 Å (KE ∿ 965 eV). The determination of the intensity ratios of core levels for model homogeneous systems is an important precursor in discussing copolymer systems.

4.7 Theoretical Models for a Quantitative Discussion of Results

The theoretical interpretation of ESCA data may be classified into three categories as shown in Fig. 30.[2] Detailed consideration will be given to these in the lectures, however, as far as these notes are concerned we need only outline the major conclusions.

At this stage we consider models for the interpretation of absolute and relative binding energies for core levels. We may note that although the spectra of the homopolymers discussed above can be understood in terms of the qualitative description ad-

1. **Absolute and relative binding energies for core and valence levels.**

2. **Shake up intensities.**

3. **Relative intensities for valence energy level band profiles.**

Fig. 30. Main areas for theoretical interpretations of ESCA data.

vanced, it is essential to establish a reliable quantitative interpretation of observed shifts if headway is to be made with more complicated systems. The details of the theory have been discussed in detail elsewhere and only a brief summary is presented here.[2] The photoionization process itself, at least for systems containing first and second row atoms, can in general be quantitatively described within the Hartree Fock formalism although this is not true for the accompanying shake up and shake off processes for which detailed considerations of electron correlation are necessary.

The energy consideration for discussion of relative or absolute binding energies within the Hartree Fock formalism are shown in Fig. 31. For the C_{1s} level of methane for example, the exact agreement between theory (calculated as an energy difference between neutral molecule and hole state with a basis set approaching the Hartree Fock limit) and experiment indicates that both relativistic and correlation energy considerations are small.[18] By contrast comparison with Koopmans' Theorem show that electronic reorganization accompanying core ionization is substantial (e.g. $\sim$ 14 eV for the C_{1s} level). A further point of interest recently illuminated by detailed theoretical studies is that the magnitude of such relaxation energy is strongly dependent on the electronic environment of a given atom. Fig. 32 illustrates this rather nicely for C_{1s} levels and moreover a relationship between difference in relaxation energy and shift in core binding energy in clearly evident.[19]

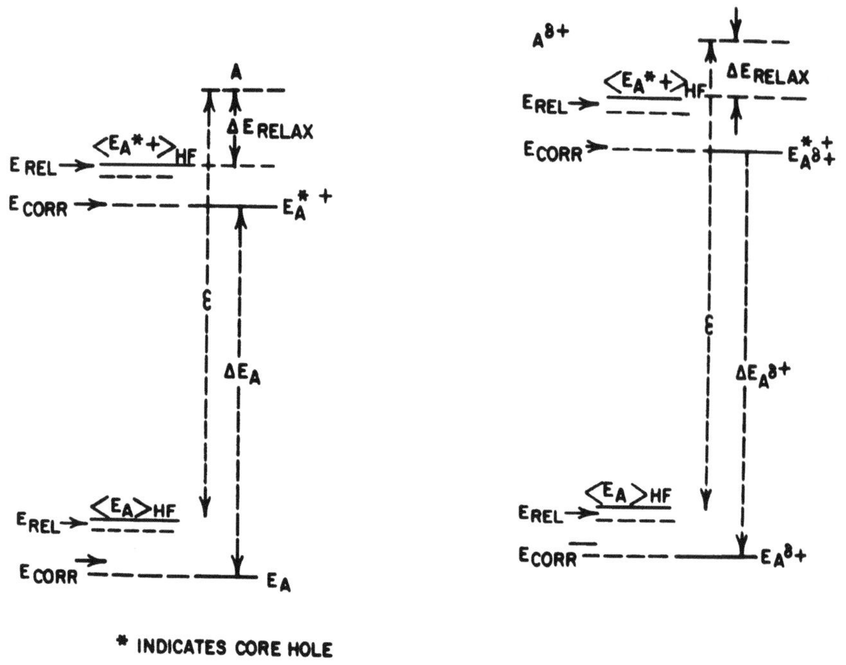

Fig. 31. Energy considerations in the theoretical interpretation of relative and absolute binding energies.

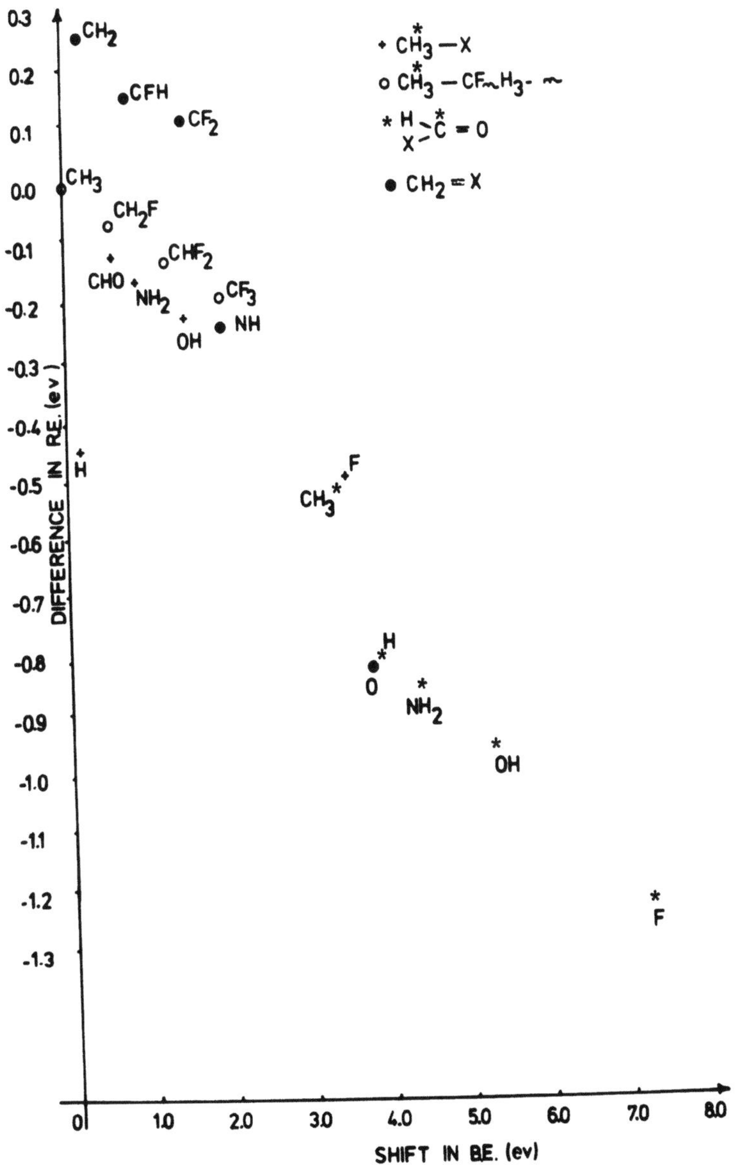

Fig. 32. Shifts in binding energy vs. difference in relaxation energy for C_{1s} levels.

Detailed non-empirical LCAO MO SCF calculations can therefore provide a quantitative discussion of experimental ESCA data,[+] (other than for shake up and shake off phenomena), however, such computations are only feasible on relatively simple systems. In attempting to quantify theoretically the results for polymer systems the great complexity of even model systems renders a rigorous approach impossible. A less sophisticated but still theoretically valid approach based on an expansion for the expression for the Fock eigenvalues (i.e. Koopmans' Theorem) in the zero differential overlap approximation has therefore been developed.[4,6a] By grouping terms which are essentially independent of the local electron density it may be shown that the binding energy of a given core level of an atom in a molecule is related to the overall charge distribution by Equation (2).

$$E_i = E_i^o + kq_i + \sum_{j\neq i} \frac{q_j}{r_{ij}} \qquad (2)$$

The so called "charge potential model" originally developed by Siegbahn et al[4] relates the core binding energy E_i to the charge on atom i and the potential provided by the other charges within the molecule. E_i^o is a reference level and k represents approximately the one centre coulomb repulsion integral between a core and valence electron on atom i. In dealing with complex organic molecules, analysis of experimental data in terms of the charge potential model and all valence electron SCF MO calculations in the CNDO/2 formalism (formally an approximation to a non-empirical treatment) has proved highly rewarding. From studies of series of related molecules values of k and E_i^o may be established for a given core level of a given atom. It is interesting to note that although the charge potential model may be derived from Koopmans' Theorem which does not take into account electronic relaxation, the latter in fact follows a relationship remarkably similar on form to Equation (2) (cf. the data in Fig. 32). The charge potential model with k and E^o determined from experiment overall works extremely well. Fig. 33 shows for example, data for C_{1s} level of a series of heterocyclic molecules.

+ It should be emphasized that in many senses the quantitative discussion of both absolute and relative binding energies is somewhat more straightforward for core than for valence levels. This arises from the fact that the core orbitals being so localized, correlation energy corrections tend to remain constant for a given core level whereas for valence levels of widely differing localization characteristics this is not the case.

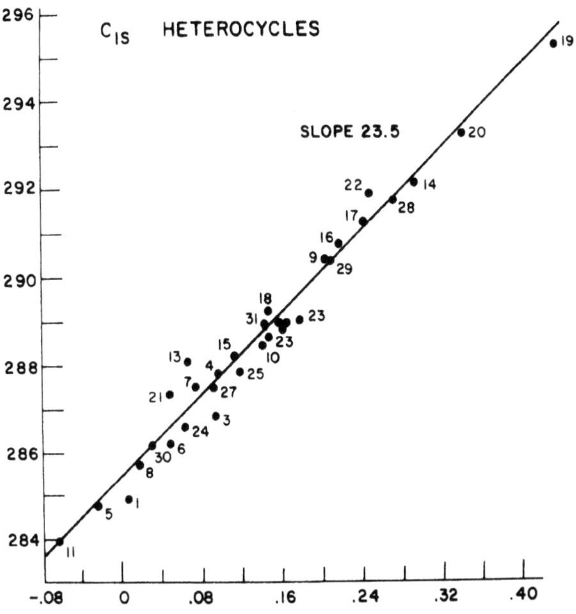

Fig. 33. Charge potential correlation for a series of heterocyclic molecules.

Values for k and E^{o} may readily be established as slope and intercept respectively from the experimental binding energies and calculated charge distributions. The values of k depend on the bonding situation, in Table 10 data pertaining to C_{1s} levels are displayed with the average value of k being $\sim$ 25 eV/unit charge.

The central role of the charge potential model in quantifying experimental data on complex molecules is shown in Fig. 34. Starting on the left hand side of the figure, given geometries and appropriate charge distributions, (e.g. from CNDO/2 SCF MO calculations) experimental shifts on series of model compounds may be used to obtain values of k and E^{o} for a given level of a given element. On the right hand side, if k and E^{o} values are available, then theoretical charge distributions may be used for the assignment of spectra and if peak shapes and widths have been established then theoretically calculated spectra may be simulated. Last but not least if appropriate values of k and E^{o} and geometries are available it is possible to invert the charge potential model to obtain "experimental" charge distributions and

Table 10. k Values for C_{1s} levels of series of molecules obtained from experimental data and CNDO/2 SCF MO charge densities

CHARGE POTENTIAL MODEL

C_{1s} LEVELS

SERIES	k
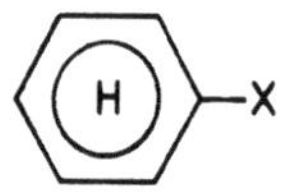	24·6
$\underline{C}Cl_3$–X, $\underline{C}Cl_2$ H – X	28·7
$\underline{C}H_3$ $\underline{C}OX$	25·0
AROMATICS PERHYDRO PERFLUORO	25.0
SIX MEMBERED RING HETEROCYCLES PER H PER F PER Cl	22·4
FLUOROBENZENES	23·5
FIVE MEMBERED RING HETEROCYCLES	25.4

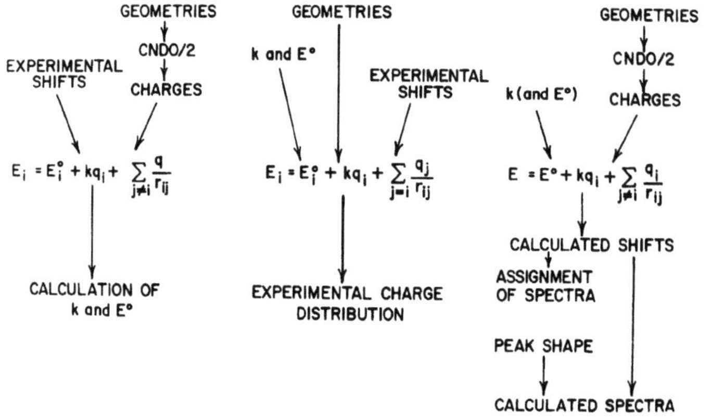

Fig. 34. Central role of the charge potential model in quantifying experimental data.

the application to polymer systems is of particular importance and will be discussed more fully later on.

The short range nature of substituent effects on core binding energies in saturated systems suggests that it may be feasible to quantitatively discuss the results for polymers in terms of calculations on simplified model systems which contain the essential structural features and accommodate all short range interactions. The success of the charge potential model, coupled with CNDO SCF MO calculations of charge distributions in quantitative discussions of data for simple monomers, suggests that this is a feasible approach since this model accounts well for the short range nature of substituent effects.

For polyethylene and polyvinylene fluoride the representative model units were taken as being the monomer linked to other monomer units and then appropriate end groups. In this way substituent effects over three carbon atoms were taken into account. (The calculations showed that longer range effects were negligible so that the model incorporated all of the important short range interactions). All valence electron SCF MO calculations have then been carried out and the electron distributions obtained used, with appropriate values for k's and E°'s, to calculate absolute binding energies for the core levels of the representative structural unit of polymer. The adequacy of the theoretical treatment for polyethylene and polyvinylene fluoride is apparent form Fig. 35. Fig. 35 also includes the results of calcula-

Polymer Models — C_{1s} Binding Energy

$CH_3\ CH_2\ CH_2\ \underline{C}H_2\ \underline{C}H_2\ CH_2\ CH_2\ CH_3$

Calc. 284·9

Observed 285·0

$CH_2F\ CHF\ CHF\ \underline{C}HF\ \underline{C}HF\ CHF\ CHF\ CH_2F$

Calc. 288·6

Observed 288·4

$CF_3\ CF_2\ \underline{C}F_2\ \underline{C}F_2\ CF_2\ CF_3$

Calc. 292·6

Observed 292·0

Fig. 35. Absolute binding energies calculated for model systems of polyethylene, polyvinylene fluoride and PTFE.

tions on a model for PTFE. In this case computer limitations dictated that a smaller model system than optimum be studied, and this accounts for the small discrepancy between calculated and observed binding energies; in fact, recent improvements in computing facilities have allowed calculations on a larger model compound and there is now quantitative agreement between theory and experiment in this case.

For the unsymmetrical monomers polyvinyl fluoride, polyvinylidene fluoride and polytrifluoroethylene the possibility arises of structural isomerism by way of head to tail and/or head to head addition. In fact ^{19}F n.m.r. studies[20] show that structural isomerism does occur in both of the first two and indeed provides information on tacticity which is not available by ESCA studies. Since the C_{1s} spectra of the three polymers are relatively well resolved it is evident that information regarding structural isomerism if available must be encoded in the lineshapes and/or linewidths.

Theoretical calculations on suitable models incorporating the relevant structural features viz. head to tail and head to head linkages gives the results shown in Fig. 36. Considering firstly polyvinyl fluoride , the calculated binding energies for both types of structural arrangements are the same within experimental error and are in fact in excellent agreement with the observed values. It seems clear therefore that for this parti-

$(CHF-CH_2)_n$ C_{1s}

Regular

$CFH_2\ CH_2\ CFH\ CH_2\ \underline{C}FH\ \underline{C}H_2\ CFH\ CH_2CFH\ CH_3$

Calc. 288·1 285·4

Observed 288·0 285·9

Irregular

$CFH_2\ CH_2CH_2CFH\ \underline{C}FH\ \underline{C}H_2CH_2CFH\ CFH\ CH_3$

Calc. 288·0 285·6

Fig. 36. Absolute binding energies calculated for model systems for polyvinyl fluoride.

cular polymer ESCA is unable to provide information on structural isomerism along the chain. (This contrasts with the situation to be described later concerning nitroso rubbers). The same considerations apply to both polyvinylidene fluoride and polytrifluoroethylene, again the theoretical models are in excellent agreement with experiment.

4.8 Copolymers

It is evident from the previous section that substituent effects on core binding energies in polymers can be understood both qualitatively and quantitatively on the same basis as those for simple monomeric systems. With this knowledge it is possible to proceed to more complex systems such as copolymers. The first feature of interest is the determination of compositions, viz. percentage comonomer incorporations. The applicability of ESCA in this field is illustrated by reference to our work on Viton and Kel F type copolymers. Further features of interest arise in studying copolymers of ethylene and tetrafluoroethylene namely the gross structure of the polymers in terms of block, alternating or random features. This leads to a brief discussion of ESCA for studying domain structure in block copolymers. The use of ESCA for studying finer details of structure such as structural isomerism is exemplified by studies of nitroso rubbers.

Before discussing the results, it is interesting to consider how ESCA may be used as a non-desctructive technique, with virtually no sample pre-preparation necessary for routine identification of polymers. As a simple example, we suspected that one of the Kel F samples in our possession was incorrectly labelled. Two rapid ESCA experiments, taking ∿ 20 mins each, in fact identified the polymer as a Viton type polymer. The carbon-1s spectra are shown in Fig. 37. The spectrum of the Viton sample exhibits the characteristic high binding energy peak of the $\underline{C}F_3$ carbon at 194.3 eV and pronounced shoulder on the $\underline{C}F_2$ peak at 292 eV attributable to the $\underline{C}F$ carbon at 290.4 eV. As a double check, the Viton sample also showed no levels attributable to chlorine.

(I) Copolymer Compositions

<u>Viton Polymers</u>

The C_{1s} levels for the parent polyhexafluoropropene and the 30/70 and 40/60 copolymers with vinylidene fluoride are shown in Fig. 38. The binding energies are tabulated in Table 11 and may again be understood in terms of simple substituent effects. In obtaining the binding energies of the $\underline{C}F_2$ and $\underline{C}F$ carbon-1s levels a simple deconvolution is necessary. In any estimation of copo-

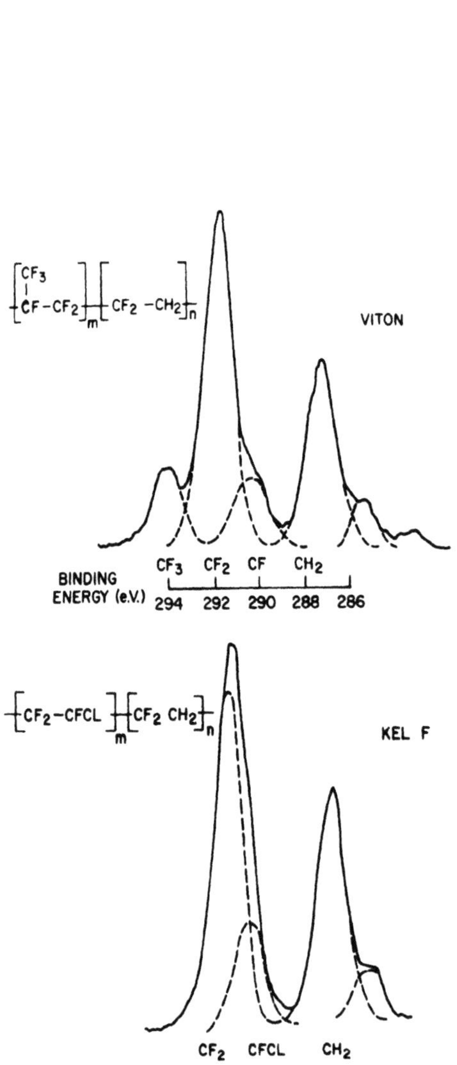

Fig. 37. C_{1s} levels for typical Viton and Kel F type polymers.

lymer compositions, however, it is obviously desirable to avoid even such a minor complication. The procedure adopted, therefore, was to measure the area of the $\underline{C}F_3$ peak, the total area of the ($\underline{C}F_2$ + $\underline{C}F$) peak and the area of the $\underline{C}H_2$ peak. The degree of incorporation of hexafluoropropene (HFP) in the copolymer was then calculated from the percentage of the total area due to C_{1s} levels represented by each peak :

a) The mole percent incorporation of HFP is three times the area of the peak due to $\underline{C}F_3$, on the basis of the stoichiometry of the HFP unit.

b) The area of the peak due to $\underline{C}F_2$ and $\underline{C}F$ (A) is made up of half the total C_{1s} peak area due to vinylene fluoride

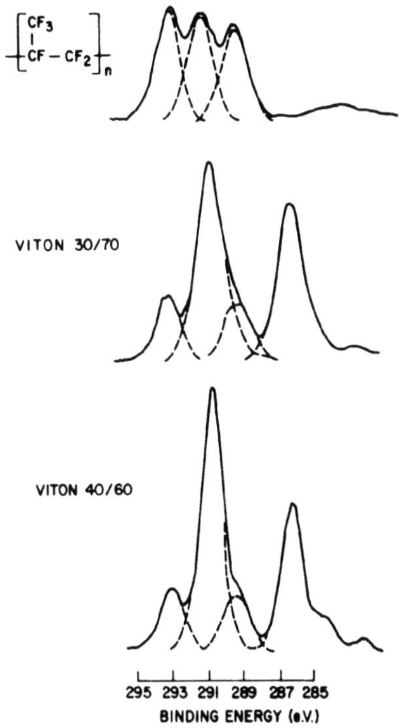

Fig. 38. C_{1s} levels for polyhexafluoropropene and Viton polymers.

Table 11. Binding energies for Viton type polymers

			C_{1s}	$\Delta(C_{1s})$	F_{1s}	$\Delta(F_{1s})$
VIIa	40/60 Viton	CF_3	293.3	8.3	690.2	0.9
		CF_2	291.1	6.1	690.2	0.9
		CF	289.4	4.4	690.2	0.9
		CH_2	286.6	1.6	-	-
VIIb	30/70 Viton	CF_3	293.4	8.4	689.9	0.6
		CF_2	290.9	5.9	689.9	0.6
		CF	289.3	4.3	689.9	0.6
		CH_2	284.6	1.4	-	-
VIII	$\{CF(CF_3)-CF_2\}_n$	CF_3	293.7	8.7	690.2	0.9
		CF_2	291.8	6.8	690.2	0.9
		CF	289.8	4.8	690.2	0.9

($\frac{1}{2}VF_2$) and two thirds the total C_{1s} peak area due to HFP i.e. $A = \frac{1}{2}VF_2 + \frac{2}{3}$ HFP and $100 = VF_2$ + HFP, by definition, hence % HFP = 6(A-50).

c) The area of the peak due to $\underline{C}H_2$ is half the total area due to VF_2 and hence the mole percent incorporation of HFP is given by the expression

$$100-(2 \times \% \text{ Area due to } \underline{C}H_2)$$

Using these three methods to determine the degrees of incorporation of HFP gives an internal check on the reliability of the method, the results are as follows :

% HFP Incorporation

	Method of calculation (a)	(b)	(c)
Sample 40/60	39	42	40
Sample 30/70	33	30	32

The internal consistency is good to within 3% and the values obtained are in good agreement with those quoted for the copolymers investigated.

Kel F Polymers

The C_{1s} spectrum of the parent polychlorotrifluoroethylene consists of a single, broad peak with a flattened top corresponding to overlapping lines from $\underline{C}F_2$ and $\underline{C}FCl$ carbons, Fig. 39. The absolute binding energies and shifts, Table 12, can again be rationalized in terms of simple substituent effects. The C_{1s} spectra for the two copolymers differ considerably, the most noticeable feature being the high proportion of $-\underline{C}H_2$ units in the 30/70 copolymer coupled with the drastically reduced linewidth for the composite line at high binding energy.

In estimating the compositions of these copolymers it is again desirable to avoid reliance on deconvoluted peak areas. Two methods are available : measurement of the total $(-\underline{C}F_2-)$ + $(\underline{C}FCl)$ peak area, and measurement of the $(\underline{C}H_2)$ peak area. Since the total $\underline{C}F_2$ content of each polymer is 50 mole %, the difference between the percentage of the total C_{1s} peak area attributable to $(\underline{C}F_2 + \underline{C}FCl)$ and 50 gives the percentage of the total C_{1s} area due to $\underline{C}FCl$, twice this figure gives the amount of chlorotrifluoroethylene units in the polymer.

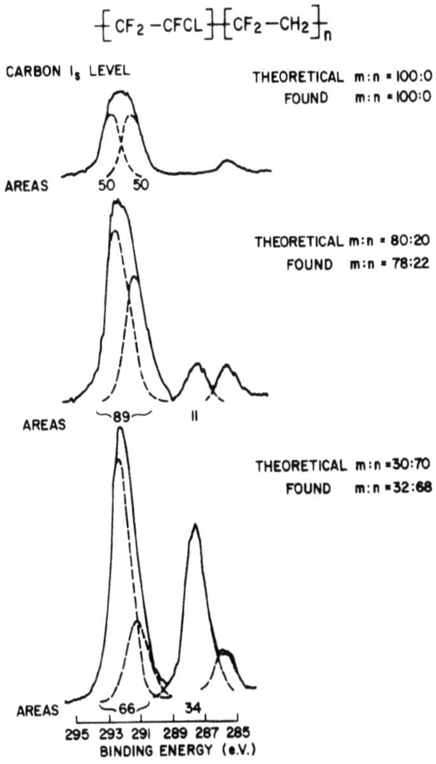

Fig. 39. C_{1s} levels for Kel F polymers.

Table 12. Binding energies for Kel F type polymers

			C_{1s}	$\Delta(C_{1s})$	F_{1s}	$\Delta(F_{1s})$	$Cl_{2p_{3/2}}$	Cl_{2s}
IXa	80/20 Kel-F	$-CF_2-$	291.7	6.7	690.3	1.0	-	-
		CFCl	290.5	5.5	690.3	1.0	201.4	272.8
		CH_2	286.8	1.8	-	-	-	-
IXb	30/70 Kel-F	CF_2	291.5	6.5	690.5	1.2	-	-
		CFCl	290.6	5.6	690.5	1.2	201.6	272.5
		CH_2	286.9	1.9	-	-	-	-
X	$\{CF_2-CFCl\}$	CF_2	291.9	6.9	690.8	1.5	-	-
		CFCl	290.8	5.8	690.8	1.5	201.1	272.2

For the 80/20 and 30/70 copolymers this gives 78% and 32% respectively both being within 2% of the values of 80% and 30% based on elemental analysis. If the areas of the $\underline{C}H_2$ peaks are used, the proportions of chlorotrifluoroethylene so obtained are again 78% and 32%. Thus both methods give exactly the same composition within 2% of the quoted values.

Ethylene/Tetrafluoroethylene Copolymers

The C_{1s} and F_{1s} spectra of a series of ethylene/tetrafluoroethylene copolymers are shown in Fig. 40.[17] From the ESCA data the copolymer compositions may be calculated in two independent ways. Firstly, from the relative ratios of the high to low binding energy peaks in the C_{1s} spectrum attributable to $\underline{C}F_2$ and $\underline{C}H_2$ type environments respectively. This method being analogous to the procedure adopted in the Viton and Kel F examples discussed

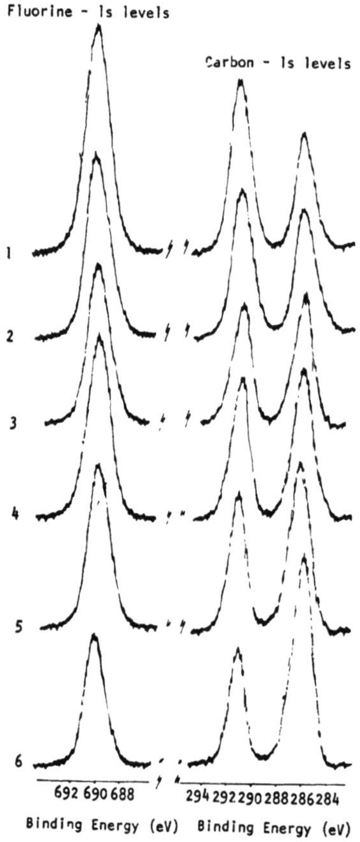

Fig. 40. C_{1s} and F_{1s} spectra of ethylene/tetrafluoroethylene copolymers.

above. Secondly, from the overall C_{1s}/F_{1s} intensity ratios taken in conjunction with the relative intensity ratio established previously for the homopolymers (Table 9). The results of these two analyses from ESCA data and those derived from conventional analysis for carbon and fluorine, and predicted from monomer reactivity ratios are compared in Table 13.

Table 13. Analysis of ethylene/tetrafluoroethylene comonomer incorporations

Sample	Composn monomer mixture	Copolymer composition (mole % C_2F_4)				
	mole % C_2F_4	Predicted monomer reactivity ratio	Calc. from C analysis	Calc. from F analysis	Calc. area ratios of C_{1s} peak: F_{1s} peak	Calc. from C_{1s} (CH_2 peak): C_{1s} CF_2 peak)
1	94	63	61	61	63	62
2	80	53	52	54	52	52
3	65.5	50	49	48	47	46
4	64	50	47	45	44	45
5	35	45	41	40	42	40
6	15	36	-	-	32	31

The agreement between the two sets of ESCA data is striking and demonstrates the uniformity (within the outermost ∿ 50 Å) of the copolymer since the escape depth dependencies for the F_{1s} and C_{1s} levels are significantly different. This is indicative of a largely alternating structure for the copolymers since the PTFE domains of a block copolymer would be expected (on the basis of their lower free energy) to predominate at the surface. In this hypothetical case, estimates of TFE incorporation based on either ESCA method would be unlikely to agree and both procedures would give values higher than those computed from elemental analysis of bulk samples. Comparison of the ESCA data with classical techniques corresponding to bulk analyses reveals overall goog agreement and also demonstrates that for these systems ESCA is competitive with microanalytical techniques in terms of accuracy and reproductability with the added advantage of being non-destructive and much faster.

(II) Structural Details

The examples presented in (I) above demonstrate the utility of ESCA in the determination of copolymer composition, the next question which arises is that of monomer sequence distribution. The case of the ethylene/tetrafluoroethylene copolymers has been examined in detail and provides a good illustration of the potential of the technique in this respect.

Although preliminary observations mentioned above indicate the unlikelihood of essentially block structure, it is desirable to establish whether the monomer units are linked in an alternating, block or random manner and to what extent an analysis of the distributions of the various structural possibilities can be quantified.

The previous studies of the homopolymers of fluorinated ethylene reveals that structural information is most readily derived from absolute binding energies and shifts in C_{1s} core levels. The shifts can be understood both qualitatively (in terms of simple substituent effects) and quantitatively, (in terms of the charge potential model and SCF MO computed charge distribution for appropriate models), and lead to a clear cut distinction between the two extremes of essentially block or alternating structure. Thus the carbon-1s binding energies to be expected for block sequences of ethylene and tetrafluoroethylene are the same as those measured for the repsective homopolymers, viz. 285.0 eV and 292.2 eV. By contrast, for an alternating structure, the C_{1s} binding energies are expected to be 286.3 eV and 291.0 eV, on the basis of (a) the experimentally observed values for polyvinylidene fluoride, (b) predicted values based on substituent effects and (c) theoretical calculations for the model compound shown below.

$$HCF_2CH_2CH_2CF_2\underline{C}F_2\text{-}\underline{C}H_2CH_2CF_2CF_2CH_3$$

291.0 286.3

A clear distinction should therefore be evident if either of these features predominate both in terms of the predicted shifts (7.2 eV for the block and 4.7 eV for the alternating cases respectively) and absolute binding energies for the two major components.

A consideration of the shifts between the two component peaks in the C_{1s} spectra, Fig. 40 illustrates this quite strikingly. The measured shifts in all cases are $\sim$ 5 eV, which establishes that the proedominant structural feature is alternation. However, a close examination of the spectra reveals two features of importance to a complete interpretation of the data. In the first place the total linewidths (FWHM) are commonly greater than 2 eV compared with typical linewidths for individual core levels in homopolymers of ∿ 1.3 ± 0.1 eV. Secondly, the peak shapes are distinctly asymmetric; indeed the observed shape varies with copolymer composition. Both these observations indicate that the experimentally observed spectra are the envelopes of numbers of overlapping peaks arising from different molecular environments. To obtain detailed structural information it is therefore necessary to deconvolute the experimental spectra. Meaningful deconvolutions of spectra require a careful systematic approach which makes use of all the relevant information (chemical and physical) concerning both the sample and the instrumental measurements.[6a] Although in principle there are an infinite number of ways of fitting a spectral envelope with a combination of curves it is often (but not invariably) possible to find a unique solution which fits all the physical and chemical data. This has previously been demonstrated in the deconvolution of complex lineshapes for simple molecules. In the approach to the deconvolution of these spectra the information derived from earlier studies of homopolymers was invaluable. Thus, both linewidth and lineshape for unique carbon-1s levels are closely defined under a given set of experimental conditions. The shape is to a very good approximation Gaussian with a slight tail to the low kinetic energy side due to inelastically scattered electrons, and the full width at half maximum (FWHM) lies in the range 1.2 to 1.4 eV. The spread in linewidths is almost certainly attributable to differences in surface topography giving rise to non-uniform sample charging characteristics.

Calculation of expected carbon-1s binding energies for the various structural features which a priori might be considered as possible components of the overall structure of these copolymers was also indispensable to the complete interpretation of these spectra. The absolute values and shifts for pure block and pure alternating sequences have been discussed, the predicted binding energies for irregular structural features resulting from one or two monomer units in a block of the other comonomer were also computed using model compounds accommodating all the important short range effects. Careful systematic deconvolutions of the six C_{1s} spectra in Fig. 40 were then carried out (the details of the procedure have been described elsewhere), a typical example of a deconvolution being shown in Fi. 41. In practice a set of seven standard lines was required to fit the C_{1s} spectral envelope for these copolymers, the binding energies and intensities of these seven peaks are listed (columns a - g) in Table 14.

The interpretation of binding energies and intensities arising from the deconvolution in general terms is straightforward. The major peaks centred at ∿ 291.1 eV (b) and ∿ 286.3 eV (e) arise from the alternating component to the structure. Whilst those at ∿ 292 eV (a) and ∿ 285.0 eV (g) are accounted for by essentially block runs of tetrafluoroethylene and ethylene respectively.

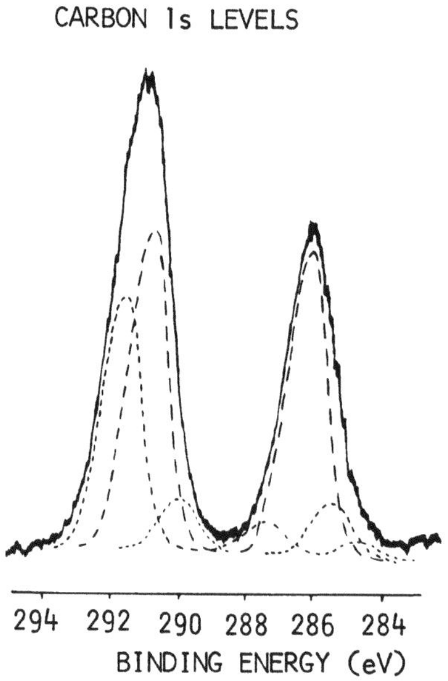

Fig. 41. C_{1s} spectra for a typical ethylen/tetrafluoroethylene copolymer showing the deconvolution into component peaks.

Table 14. Analysis of component peaks for C_{1s} levels for ethylene/tetrafluoroethylene copolymer

Deconvolution of Carbon - 1s Spectra

Binding energies (area ratios)

Sample No.	a	b	c	d	e	f	g
1	292.1(25)	291.2(31)	290.4(5)	287.8(3)	286.3(30)	285.7(5)	285.0(1)
2	292.1(12)	291.0(33)	290.2(7)	287.6(4)	286.3(32)	285.7(6)	285.0(6)
3	292.3(4)	291.0(37)	290.1(5)	287.7(5)	286.4(36)	285.7(9)	285.0(4)
4	292.4(6)	291.1(35)	290.4(4)	287.7(4)	286.3(35)	285.7(9)	285.0(7)
5	292.2(2)	291.0(35)	289.4(3)	287.8(4)	286.4(34)	285.6(17)	285.0(5)
6	292.1(3)	291.1(27)	289.5(1)	287.3(3)	286.4(26)	285.7(30)	285.0(10)

Reference to model compounds suggests that peaks c and f arise from an isolated tetrafluoroethylene unit in an ethylene sequence and two ethylene units near a tetrafluoroethylene unit respectively. The remaining component is that tabulated in column d, from its binding energy this might qualitatively be expected to arise from an isolated ethylene unit in a polytetrafluoroethylene sequence; however, the predicted binding energy from the appropriate model compound (viz. 287.0 eV) is not in particularly good agreement with the value derived by deconvolution viz. an average value of 287.7 eV. This does however correspond very closely to the binding energy expected for carbon-1s levels of carbonyl >C=O groups (cf. CH_3COCH_3 288.0 eV). A detailed experimental analysis of the kinetic energy region appropriate to O_{1s} core levels revealed a peak of low overall intensity with binding energy corresponding to oxygen in a carbonyl environment (binding energy 533.4 eV). The relative intensities for the C_{1s} peaks (d) and corresponding O_{1s} peaks coupled with a knowledge of the escape depth dependence for electrons of differing kinetic energies showed that the small extent of oxidation was localized within the immediate surface (∿ 5Å) of the samples; and almost certainly arose during the production of the polymer films.

The raw data in terms of the separation between the two major components of the C_{1s} spectra revealed the essentially alternating nature of these copolymers without recourse to more refined analysis. Since however kinetic data in terms of monomer reactivity ratios were available for these systems it was of interest to consider a more refined analysis in terms of the detailed deconvolutions, Table 14. Pentad sequences were chosen as the simplest suitable distributions to compute for direct comparison with experimental data, since they are large enough units to accommodate all important short range interactions for the central triads and thus allow a direct comparison between theory and experiment. Sequence probabilities were computed in the standard manner using known reactivity ratios and the ESCA derived copolymer compositions, the probabilities are listed in Table 15 and are shown schematically in Fig. 42.

The core levels corresponding to the central triads of each pentad sequence were assigned to the six distinct peaks obtained (Table 14) in the deconvolution (peaks d arising from surface oxidation of methylene groups were excluded from the analysis for this purpose). With the aid of previous studies on homopolymers and calculations on model compounds this assignment is relatively straightforward and representative examples are given in Table 16 for the more important contributing sequences.

Table 15. Pentad sequence probabilities computed from monomer reactivity ratios

Sample No	1	2	3	4	5	6
Sequence			Probability			
AAAAA	1.8	-	-	-	-	-
BAAAA AAAAB	5.2	.3	-	-	-	-
BAAAB	3.8	.9	.3	.3	-	-
ABAAA AAABA	12.3	2.0	.5	.4	-	-
ABAAB BAABA	17.8	11.3	6.3	5.9	1.6	.3
BBAAA AAABB	.1	-	-	-	-	-
BBAAB BAABB	.2	.4	.5	.5	.4	.3
ABABA	21.0	34.9	33.9	32.9	24.6	9.5
BBABA ABABB	0.4	2.5	5.5	5.8	13.9	16.9
BBABB	-	-	.2	.3	1.9	7.5
AABAA	6.2	1.0	.2	.2	-	-
BABAA AABAB	17.8	11.3	6.3	5.9	1.6	.3
BABAB	12.8	32.2	39.8	40.5	35.4	21.1
ABBAA AABBA	.3	.4	.4	.4	.3	.1
ABBAB BABBA	.4	2.6	5.5	5.8	13.9	16.9
BBBAA AABBB	-	-	-	-	.1	.1
BBBAB BABBB	-	-	.4	.5	3.9	15.0
ABBBA	-	-	.2	.2	1.4	3.4
ABBBB BBBBA	-	-	-	-	.8	6.0
BBBBB	-	-	-	-	.1	2.6

Using these assignments the probabilities for particular sequences were compared directly with the deconvoluted experimental spectra (Table 14), and the results of this comparison are recorded in Table 17.

The remarkable measure of agreement between the two sets of data in Table 17 is considered as most encouraging. Thus the sequence distributions computed from the reactivity ratios show the marked degree of alternation inferred qualitatively from the ESCA data. As a broad generalisation the ESCA data would suggest that when the monomer ratio of tetrafluoroethylene (TFE) to ethylene is greater than unity the "excess" TFE tends to prefer a block arrangement whereas if the ratio is less than one there is a greater tendency for a random distribution of the small percentage of TFE monomers amongst the predominantly ethylene sequences.

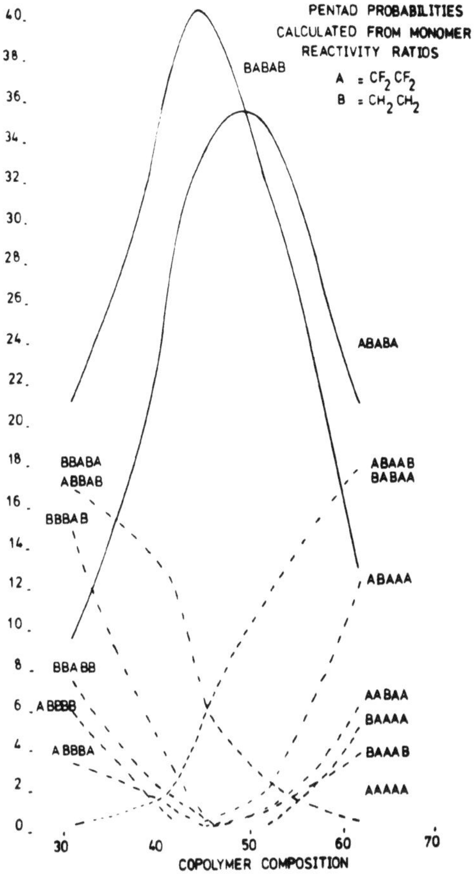

Fig. 42. Main sequence probabilities as a function of composition

Table 16. Assignment of C_{1s} core binding energies as a function of pentad sequences

Sequence		
ABAAA	$-CF_2CF_2(CH_2CH_2CF_2CF_2CF_2CF_2)CF_2CF_2-$	e e b a a a
ABAAB	$-CF_2CF_2(CH_2CH_2CF_2CF_2CF_2CF_2)CH_2CH_2-$	e e b a a b
ABABA	$-CF_2CF_2(CH_2CH_2CF_2CF_2CH_2CH_2)CF_2CF_2-$	e e b b e e
BBABA	$-CH_2CH_2(CH_2CH_2CF_2CF_2CH_2CH_2)CF_2CF_2-$	f e c b e e
BABAA	$-CH_2CH_2(CF_2CF_2CH_2CH_2CF_2CF_2)CF_2CF_2-$	b b e e b a
BABAB	$-CH_2CH_2(CF_2CF_2CH_2CH_2CF_2CF_2)CH_2CH_2-$	b b e e b b
ABBAB	$-CF_2CF_2(CH_2CH_2CH_2CH_2CF_2CF_2)CH_2CH_2-$	e f f e c b
BBBAB	$-CH_2CH_2(CH_2CH_2CH_2CH_2CF_2CF_2)CH_2CH_2-$	g g f e c c

Table 17. Comparison of ESCA and monomer reactivity data for pentad sequences

Sample		Peak Designation a	b	c	e	f	g
1	Expermental	25	31	5	30	5	1
	Theory	26	37	-	37	-	-
2	Experimental	12	33	7	32	6	6
	Theory	8	43	1	45	1	-
3,4	Experimental	5	37	5	37	9	4
	Theory	4	46	2	46	3	1
5	Experimental	2	35	3	34	17	5
	Theory	1	39	6	42	10	3
6	Experimental	3	27	1	26	30	10
	Theory	-	24	10	31	23	12

Experimental from deconvoluted ESCA spectra

Theory computed from pentad probabilities, assignments as in Table

Thus for samples 1 and 5 where the excess of appropriate comonomers is approximately the same the contributions from peak a for sample 1 (representing TFE units linked together) is much larger than for peak g in sample 5 (representing ethylene units linked together). This conclusion is strongly supported by the calculated pentad probabilities shown in Table 15.

These results are qualitatively in good agreement with the physical and more general spectroscopic properties (e.g. TIR) of tetrafluoroethylene-ethylene copolymers. Improvements in resolution and signal/background which will become available with spectrometers employing monochromatized X-ray sources and multiple collector assemblies will considerably extend the scope of ESCA investigations of such copolymer systems. Indeed the results presented here would suggest that it may well be feasible to determine monomer reactivity ratios directly from ESCA data.

(III) Domain Structure in Block Copolymers

In the discussion of the ethylène/tetrafluoroethylene copolymers it was mentioned that the agreement between the various analyses for copolymer composition was evidence against a block structure, since PTFE blocks would be expected to segregate at the surface and hence be over estimated by a surface sensitive technique like ESCA. In a more general sense the differing sampling depths for photoemitted electrons from core levels corresponding to differing kinetic energies suggests that ESCA may be a very useful tool for studying domain structure of block copolymers.[21] A simple illustration of this is provided by a consideration of two block copolymers of poly (dimethylsiloxane) with polystyrene.[22] Two samples were examined as films cast from cyclohexane, their characteristics are listed in Fig. 43.

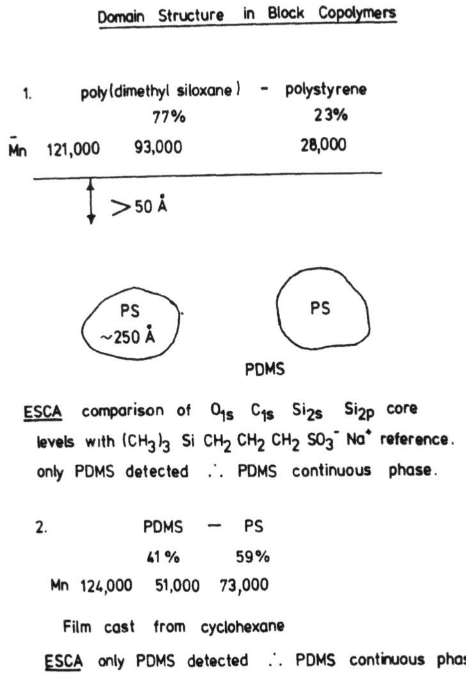

Fig. 43. ESCA studies of block copolymers of polydimethylsiloxane and polystyrene.

Considering firstly sample 1, electron microscopy indicated that PDMS formed the continuous phase, the discrete phase being of PS the domains being of the order of 300 Å in dimension. Contact angle measurements suggested that the surface of the samples were essentially PDMS. ESCA provides a means of investigating this directly.

From studies of simple model systems overall intensity ratios for homogeneous samples were established for the principle core levels pertinent to the copolymer system (viz. I_{α} values for O_{1s}, C_{1s}, Si_{2s}, Si_{2p} core levels). For the block copolymer, comparison of the intensity ratios showed that essentially only the polydimethylsiloxane component was observed. This is an important result for two reasons. Firstly it demonstrates that ESCA is potentially a very valuable technique for identifying the domain structure at the surfaces of block copolymers. Secondly since on the ESCA scale the sample appears to be homogeneously polydimethylsiloxane a knowledge of escape depth dependencies may be used to put a lower limit on the dimension of the domains. A reasonable estimate in this particular case would be > 100 Å. For sample 2 again ESCA reveals that the surface is essentially PDMS which is only explicable in terms of a continuous phase of PDMS and discrete phase of PS cf. Fig. 43.

(IV) Structural Isomerism

At the end of section 4.7 an approach to the investigation of structural isomerism in polymers was outlined based on detection of binding energy shifts between core levels of atoms in the different isomeric components. However, for the systems considered (viz. homopolymers of fluoroethylenes), the shift differences were not large enough to be detected, although structural isomerism was known to occur.[20] Copolymers of trifluoronitrosomethane with fluoroethylenes proved to be amenable to examination by this method and structural isomerism could be clearly demonstrated.[16]

The C_{1s} spectra for copolymers of trifluoronitrosomethane with tetrafluoroethylene, chlorotrifluoroethylene and trifluoroethylene are shown in Fig. 44. The assignment of peaks is straightforward and again the shifts are understandable in terms of simple substituent effects. The 1:2 area ratio for $\underline{C}F_3$ with respect to $\underline{C}F_2$ carbons for the copolymer involving tetrafluoroethylene together with the relevant binding energies demonstrates the 1:1 alternating nature of this copolymer. Similar arguments apply to the other polymers.[16] The possibility of structural isomerism exists for the copolymers with trifluoroethylene and chlorotrifluoroethylene and its presence can be demonstrated by a careful analysis of the ESCA spectra. For example, for the C_{1s} levels of the polymer involving trifluoroethylene although the

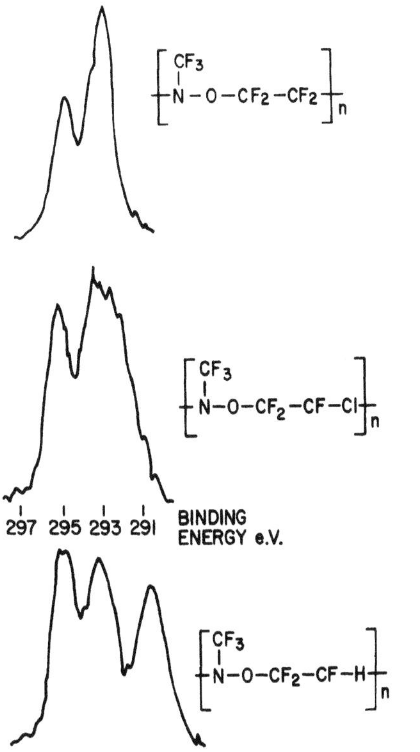

Fig. 44. C_{1s} spectra for some nitroso rubbers.

three peaks are of equal area there are substantial differences in linewidth that are not apparent in the spectrum for the polymer formed from tetrafluoroethylene (Fig. 44). To investigate the possibility that the lineshapes may encode information concerning structural isomersim, calculations were performed on model systems in a manner completely analogous to that described previously; these calculations leading to predictions of binding energy shifts for the possible isomeric structures. The broadened peaks in the C_{1s} spectra were then deconvoluted using the lineshape observed for the trifluoromethyl carbon level as a standard, and in the case of the trifluoroethylene copolymer a satisfactory deconvolution is only possible if the two isomeric structures occur with equal frequency, see Fig. 45. This deconvolution gives experimental shift differences between $\underline{C}F_3$ and $\underline{C}F_2$ levels for the two isomeric structures of 1.6 and 2.2 eV to be compared with the theoretically predicted differences of 1.4 and 2.1 eV (Table 18), the experimental $\underline{C}F_3$ to $\underline{C}FH$ shifts being 4.1 and 4.9 eV (theory 4.0 and 4.3 eV). The deconvolution for the chlorotrifluoroethylene copolymer is less straightforward and leads to the conclusion that-

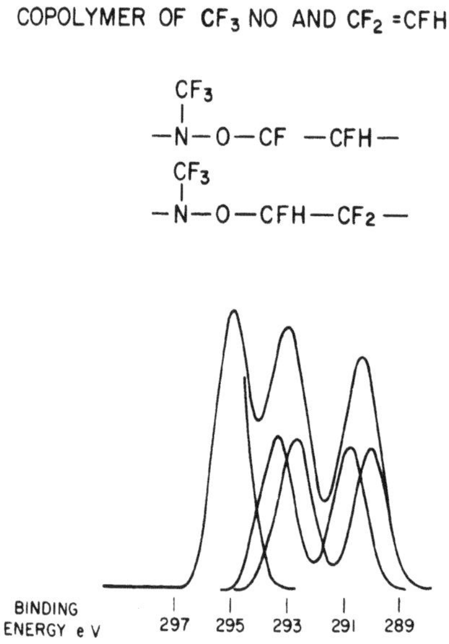

Fig. 45. Deconvolution of C_{1s} spectra for nitroso rubber involving trifluoroethylene.

Table 18. Calculated and experimental shifts for models of the nitroso rubbers

POLYMER MODELS

COPOLYMER			SHIFTS IN RELATIVE B. E. (e.V.) CALCULATED	OBSERVED
CF_3NO and $CF_2=CF_2$	$CF_3 \div CF_2\text{-}N(CF_3)\text{-}O\text{-}CF_2 \div CF_3$	Δ $\underline{C}F_3\text{-}\underline{C}F_2$ (N)	1.7	1.8
		Δ $CF_3\text{-}CF_2$ (O)	1.5	1.8
CF_3NO and $CF_2=CFCl$	$CF_3 \div N(CF_3)\text{-}O\text{-}CFCl \div CF_3$	Δ $CF_3\text{-}\underline{C}FCl$	3.4	3.1
	$CF_3 \div CFCl\text{-}N(CF_3)\text{-}O \div CF_3$	Δ $CF_3\text{-}CFCl$	3.6	3.1
CF_3NO and $CF_2=CFH$	$CF_3 \div CFH\text{-}N(CF_3)\text{-}O\text{-}CF_2 \div CF_2H$	Δ $\underline{C}F_3\text{-}\underline{C}FH$	4.0	4.4
		Δ $\underline{C}F_3\text{-}\underline{C}F_2$	1.4	1.9
	$CF_2H \div CF_2\text{-}N(CF_3)\text{-}O\text{-}CFH \div CF_3$	Δ $\underline{C}F_3\text{-}\underline{C}FH$	4.3	4.4
		Δ $\underline{C}F_3\text{-}\underline{C}F_2$	2.1	1.9

$$\begin{array}{c} CF_3 \\ | \\ \leftarrow N\text{-}O\text{-}CF_2\text{-}CFCl \rightarrow \end{array}$$

- is the major contributing structure and that -

$$\begin{array}{c} CF_3 \\ | \\ \leftarrow N\text{-}O\text{-}CFCl\text{-}CF_2 \rightarrow \end{array}$$

- is the minor structural feature and present to an extent of between 20 and 33%.[16] These conclusions are in reasonable agreement with earlier investigations of the structure of these copolymers which were based on an analysis of the products of pyrolysis of the polymers.[23]

4.9 Miscellaneous Structural Investigations of Polymers

Polyhexafluorobut-2-yne

With the background of theoretical and experimental studies described earlier it became possible to study structure and bonding in polymer systems which had proved intractable by more conventional techniques. An investigation of the product from fluoride ion initiated polymerization of hexafluorobut-2-yne provides a good example of the value of the technique.[24] The rapid polymerization of perfluorobut-2-yne in the presence of fluoride ion has been investigated and described by several groups of workers; the product is an off white, insoluble solid which cannot be pressed into films. The material is not oxidized by $KMnO_4$ in acetone which is known to oxidize C=C bonds in fluorinated systems, neither does the compound display any absorption in the double bond region of the infrared spectrum. At first sight these observations would appear to preclude the linear polyene structure,

$$\left[C(CF_3) = C(CF_3) \right]_n$$

which could reasonably be expected as the reaction product, in favour of an extensively cross linked material or a saturated cyclobutane ladder polymer (cf. Fig. 46).

The carbon-1s spectrum of the polymer is shown in Fig. 47. The hydrocarbon peak at ∿ 285 eV arises from solvent and/or the Scotch tape backing. (It is difficult to remove the last traces

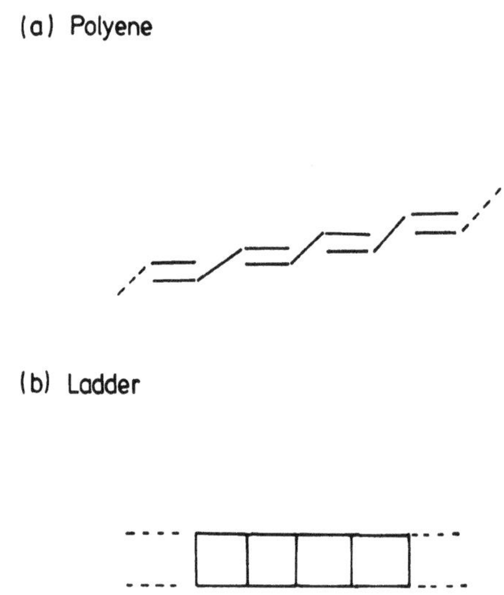

Fig. 46. Possible structures for polymers produced from the fluoride ion initiated polymerization of hexafluorobut-2-yne

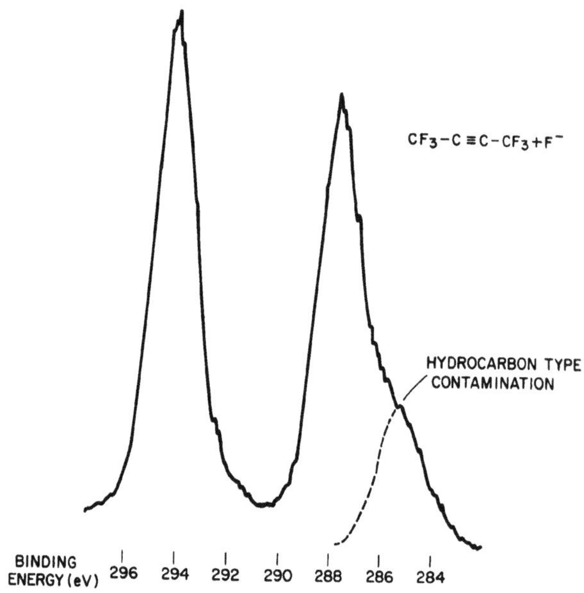

Fig. 47. C_{1s} levels for the fluoride ion initiated homopolymer of hexafluorobut-2-yne.

of solvent used in the reaction). The spectrum for the polymer is therefore extremely simple with just two peaks of approximately equal areas at 287.7 eV and 294.1 eV. From our studies of substituent effects (and since the polymer contains only C and F) the peak at high binding energy may be unambiguously assigned to $-CF_3$ groups. The assignment of the other peak is not as straightforward (but is entirely consistent with a vinylic carbon with attached perfluoroalkyl group) and a complete assignment of the polymer structure can only be made on the basis of a careful examination of the spectrum, illuminated by calculations on model compounds and studies of shake up phenomena in fluorinated systems. Since approximately half the C_{1s} peak area due to the polymer is accounted for by $\underline{C}F_3$ groups and the remaining peak area consists of a single narrow line the structure must be regular and the most probable candidates were the expected linear polyene or the inherently unlikely cyclobutane ladder structure. The question was decided on the basis of two lines of investigation. Firstly calculations on the model compound -

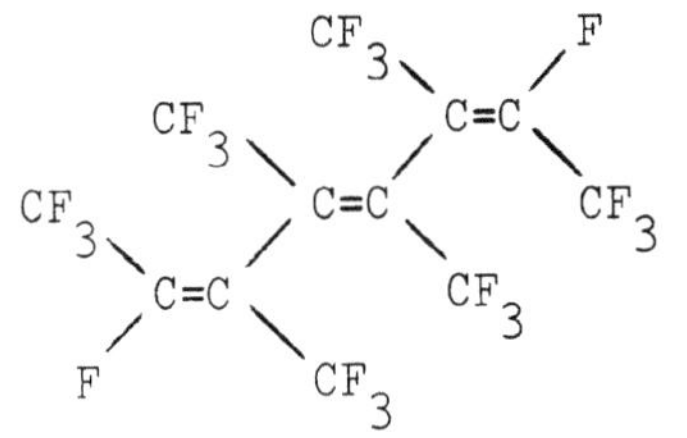

- gave rise to two useful conclusions. When the energy of the model system was computed as a function of the angle of rotation about the single bonds linking the ethylene units the minimum energy conformation was shown to be that with adjacent double bonds perpendicular, this conclusion is eminently reasonable on the basis of consideration of space filling models and invalidates the objection to the polyene structure arising from its colour, the chemical inertness of the material is also accounted for in that in the minimum energy conformation the double bonds are quite effectively shielded by the trifluoromethyl groups (Fig. 48). The predicted binding energies (on the basis of model calculations) for the $\underline{C}F_3$ and vinylic carbon-1s levels are 293.9 and 287.8 eV respectively in excellent agreement with the experimentally observed values of 294.1 and 287.7 eV (Fig. 49). Experimental (ESCA) studies of model compounds also confirms this assignment. The second factor establishing the linear polyene structure arises from the inequality in the areas of the two peaks in the C_{1s} spectrum, the area of the higher binding energy component being slightly greater than that of the peak at lower binding energy. This observation is crucial to the unambiguous differen-

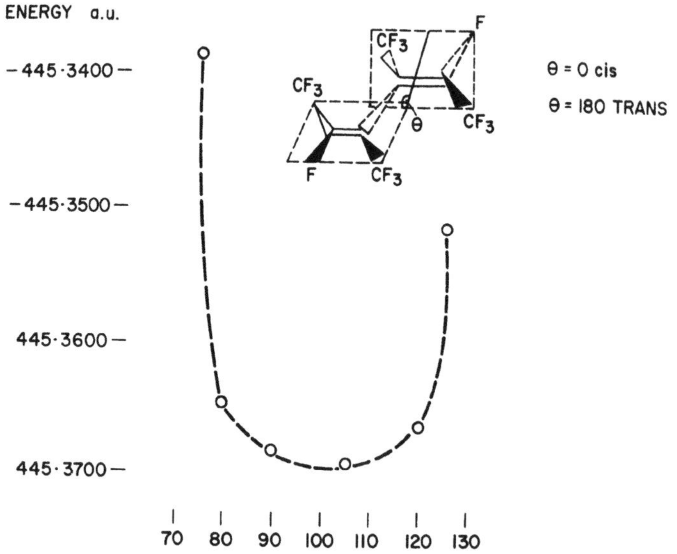

Fig. 48. Computed conformational preference for a model system for the polymer.

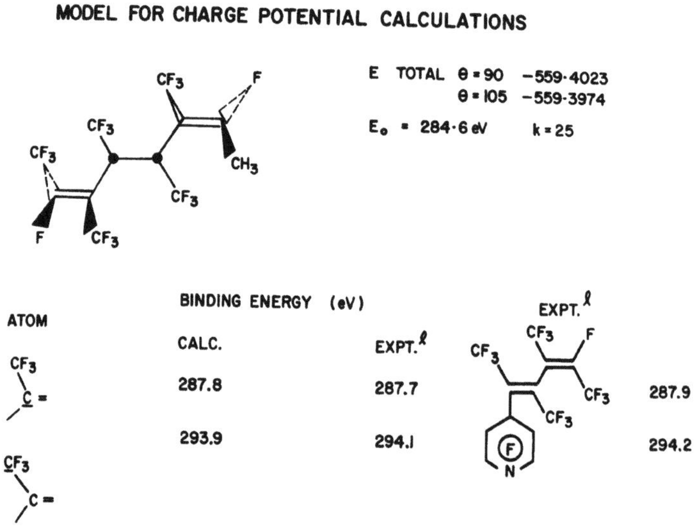

Fig. 49. Computed binding energies and comparison with model compounds for the polymer.

tiation between the linear polyene and the cyclobutane ladder structure. The observed area difference is attributed to a shake-up process associated with the unsaturated units of the linear polyene. Investigation of a large number of unsaturated fluorocarbon systems shows, in all cases, low intensity satellite peaks to the low kinetic energy side of the main photoionization peak for vinylic type carbons.[25] The energy separation in the unconjugated perfluorocyclohexa-1,4-diene as a typical example is ∿ 7 eV (cf. Fig. 50). A satellite at approximately the same

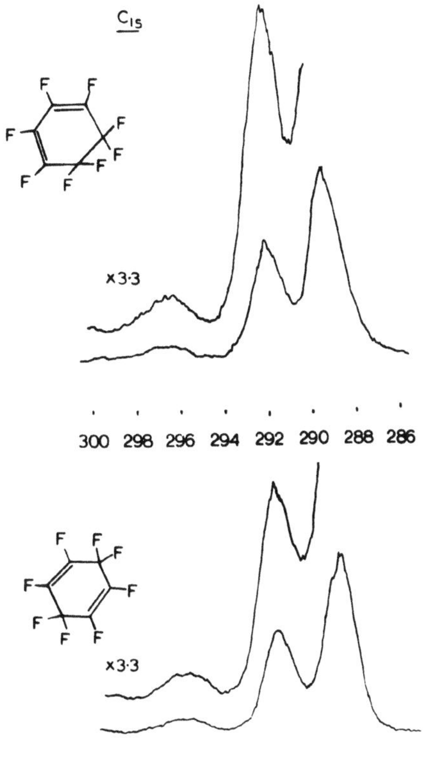

Fig. 50. C_{1s} levels for some unsaturated fluorocarbons showing shake up structure.

energy separation but much reduced in intensity is also apparent in the F_{1s} spectrum (Fig. 51) and this coupled with theoretical investigations of shake up transition probabilities (cf. Fig. 52) with the sudden approximation, identifies the shake up transition as arising from a $\pi - \pi^*$ transition. The energy separation between the principal components of the C_{1s} spectrum of the poly-

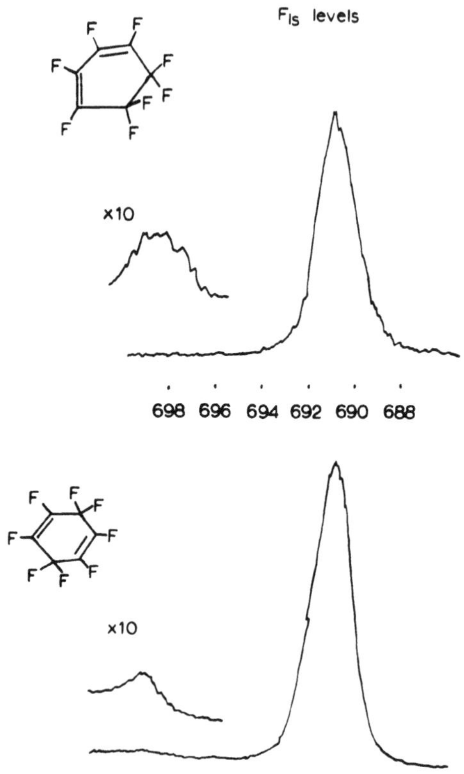

Fig. 51. F_{1s} levels for some unsaturated fluorocarbons showing shake up structure.

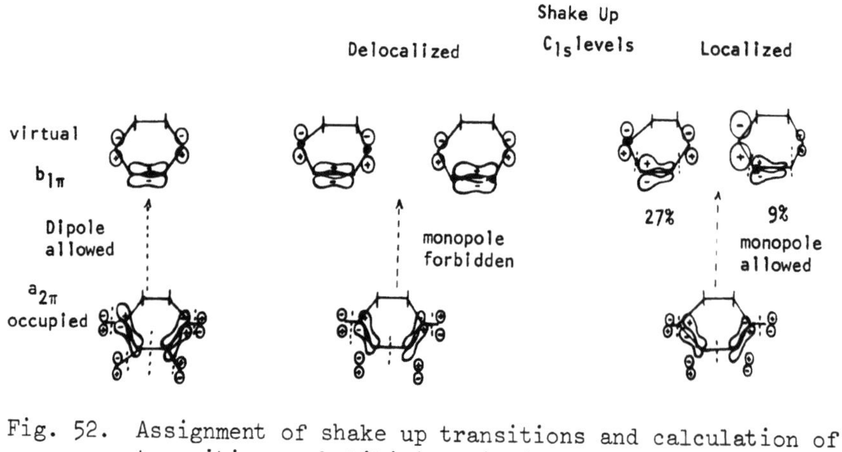

Fig. 52. Assignment of shake up transitions and calculation of transition probabilities within the sudden and ZDO approximations.

mer are 6.4 eV and for a polyene structure the shake up transition arising from $\pi - \pi^*$ excitation accompanying core ionization for the vinylic carbons would fortuitously overlap that arising from the $\underline{C}F_3$ levels. For a ladder structure of course the analogous shake up process would be expected to have a very low intensity and much higher excitation energy. The unequal areas of the two peaks in the C_{1s} region are therefore readily understandable only in terms of a polyene structure for the polymer.

Poly ω-pentachlorocyclopentadienylalk-1-enes

Diels-Alder polymers can be prepared by heating derivatives of pentachlorocylcopentadiene of the form (A). A report in the patent literature[26] describes the preparation of the polymer from

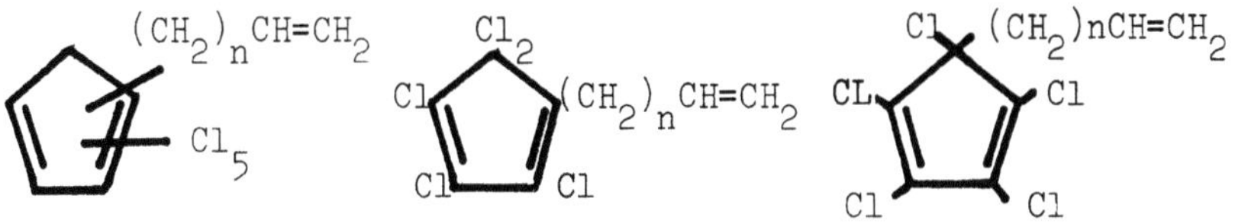

A B C

allyl pentachlorocyclopentadiene and assigns the structure as -

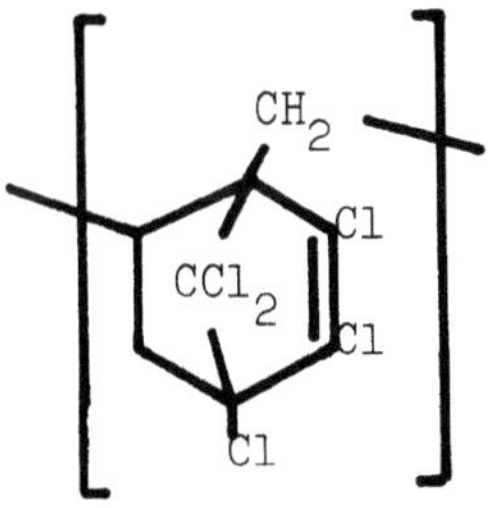

-implying either that the monomer has the structure B (n=1) or that if the monomer has the expected structure C (n=1) then a molecular rearrangement involving a chlorine migration must have occured during or following the polymerisation. Attempts to establish the structure of the monomer by conventional chemical and spectroscopic methods were unsuccessful,[27] however, by means of ESCA the monomer structure was unambiguously assigned as C (n=1).[28] The presence of a $\underline{C}Cl_2$ unit in related systems is indicated by a C_{1s} peak at ca. 289.1 eV, whereas $\underline{C}$-Cl units give C_{1s} peaks at ca.

287.7 eV, shifts of this magnitude are usually fairly easily distinguished, and the absence of a $\underline{C}Cl_2$ unit establishes the structure as C. When polymers from monomers C (N=1 and n=9) were examined by ESCA no peak at 289.1 eV could be detected, and consequently no $\underline{C}Cl_2$ units are present, the reported structure of the polymer must therefore be incorrect and the correct structure must include one or more of the isomeric units represented by :

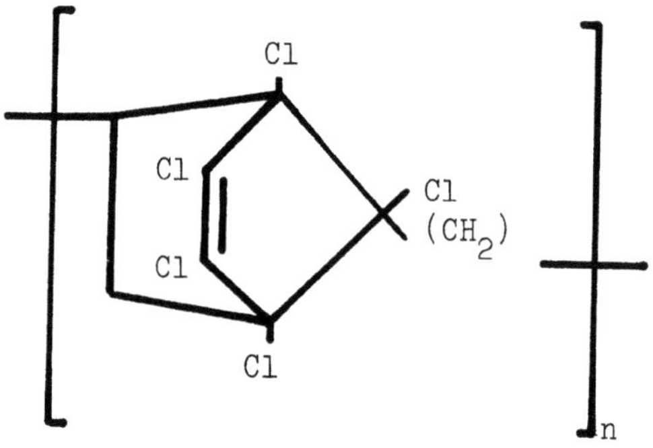

- this reassignment of structure being consistent with other chemical and spectroscopic (i.r. and n.m.r.) data for the polymers and related model compounds.[29]

4.10 Biopolymers

The potential of the ESCA method for studies of systems of biological interest has been discussed,[30] and a variety of studies have been reported.

As an analytical method ESCA can be used to indicate the quantity and quality of protein produced by various strains of protein producing plant without recourse to a laborious extraction procedure.[31] The total protein can be estimated from the N_{1s} intensity; proportions of lysine and arginine can also be obtained since amine and amide N_{1s} shifts are distinguishable; and the intensity of the sulphur signal enables the combined cysteine/cystine/methionine abundance to be obtained. In combination these data give a rapid estimate of both quality and quantity of protein present.

More detailed studies of individual proteins have also been published. For example, the helix-coil transition of poly (γ-benzyl-L-glutamate) has been studied by ESCA. Samples were obtained by freeze drying a 5% solution of the polymer in chloroform (a helix solvent for this polymer) and trifluoroacetic acid/chloroform (50/50) (a coil solvent). The N_{1s} spectra of the two samples were distinctly different, the polymer obtained from chloroform solution gave a sharp peak at binding energy 400.5 eV whereas the sample obtained from the mixed solvent gave a broad peak resolved into two peaks at 399.8 eV (36%) and 400.3 eV (64%), the method of energy referencing was not quoted. The peak at 400.5 eV was assigned to the nitrogen atoms hydrogen bonded with carbonyls in the helix, the peak at 399.8 eV is assigned to the coil domain and that at 400.3 eV to a third domain.[32a] The counting statistics for the spectra presented in this communication were not remarkably good and it will be interesting to see if the present interpretation is borne out when improved spectra are available for these systems.

Studies of the metalloprotein enzyme erythrocuprein lead to the conclusion that the copper atom of this molecule lies at or near its surface whereas the zinc atom is bound at a site deeper in the molecule.[32b]

ESCA has also been used to investigate the binding of magnesium ions by teithoic acids the strongly acidic polymeric constituents of the cell walls of Gram-positive bacteria.[33]

ESCA studies of celluloses are reported to reveal a linear relationship between C_{1s} binding energy and percent crystallinity, and that the shift difference between amorphous and crystalline cellulose amounts to $\sim$ 3 eV.[34]

4.11 Specific Surface Studies

Surface modifications of polymers are of considerable technical importance and the ESCA technique is ideally suited to studies of surfaces; it is not surprising, therefore, to find that some of the earliest applications of ESCA in polymer chemistry were aimed at investigation of specifically surface phenomena. In the first part of this section examples will be presented which show how a relatively straightforward examination of the data provided by ESCA can give an understanding of surfaces not accessible by other methods, and later an example from our own work on surface fluorination of polyethylene will be used to illustrate how a detailed examination of the ESCA data can be used to build up a very detailed picture of the surface.

Surface treatment of wool fibres

ESCA has been used by Millard and coworkers to investigate aspects of some surface treatments of woll fibres.[35-37] Corona discharge and low temperature discharge treatment of woll have been examined as methods of improving surface properties and shrink resistance. Examination of the S_{2p} levels of the materials before and after treatment indicate that before treatment there is predominantly one kind of sulphur environment (binding energy ca. 164 eV) which can be assigned to cystine, together with a minor contributor (binding energy ca. 169 eV). After treatment the higher binding energy peak is considerably increased at the expense of the cystine sulphur peak. The high binding energy peak corresponds with that observed for toluene sulphonic acid and is evidence for oxidation of cystine sulphur to sulphur (VI). Examination of the N_{1s} levels indicated that no oxidation of amine function had occurred in either process. Further evidence for the surface nature of the oxidation was obtained by re-examining the material after it had been laundered about 70 times, when the high binding energy S_{2p} peak had disappeared.[35]

A process for making wool fibres oil and shrink resistant consists of absorbing a fluorocarbon monomer onto the fibre and then subjecting it to a low temperature discharge. The material was examined at various stages of the process; for example directly after treatment and after solvent extraction of excess monomer and unbound homopolymer. The area under the F_{1s} peak in the ESCA spectrum being used as a measure of the fluorocarbon content. Using solubility criteria and oil repellancy ratings, the absolute weight of polymer add-on was estimated to be less than 1%, even so variation in fluorocarbon content consequent on the various treatments could be detected. For example, solvent extraction of the treated fibre resulted in a 50% decrease in the area under the F_{1s} peak, indicating approximately half the surface coating had been removed. By monitoring the silicon-2s and phosphorous-2p levels after treatments with vinyltrimethoxy silane and bis-betachloroethylvinyl phosphate respectively the effectiveness of coating conditions for these reagents could also be investigated.[37]

Alkali metal etching of PTFE

The remarkable physical properties and the difficulties of fabricating PTFE are well known. For many applications of PTFE it is essential to form a good adhesive bond between PTFE and some other material and the low surface energy of the fluorocarbon presents very great difficulties. Several methods of modifying the surface so that good adhesion can be obtained have been developed and one of these consists of a treatment with sodium in liquid ammonia, or sodium naphthalene in tetrahydrofuran (THF). PTFE surfaces modified in this way can be bonded to other materials with an adhesive, or can have another polymer surface grafted on them. Earlier work indicated that the modified surfaces contained carbon-carbon double bonds and a C:F stoichiometry of 1:.1

(see ref. 38). The nature of these surfaces has been examined by ESCA by two groups,[38,39] with qualitatively similar results. The most striking feature immediately evident from the ESCA spectra is the total disappearance of all fluorine signals as a result of the surface treatment, from which it can be concluded that in the outermost 50-100 Å all the fluorines are removed, this layer of course is critical in determining the surface properties. The details of the residual surface are dependant on the exact treatment of the surface after etching with alkali methal and before presentation to the spectrometer. Neither of the groups who have reported investigations were successful in examining the freshly etched surface and it is clear that the surface reacts readily with atmospheric oxygen and oxygen containing solvents. There can be several sorts of oxygen environment (depending on the surface treatment), and carbonyl, carboxyl and oxonium structures have been postulated. The C_{1s} spectra reveal several carbon environments in the range 285 to 288 eV. The freshly etched surface appears to be carbon and must be extensively unsaturated and contain a number of highly reactive sites. U.v. irradiation or heating in the presence of air results in the removal of the surface layer and regeneration of the PTFE surface.

Casing

Technical problems associated with the formation of good adhesive bonds to polymeric materials are not, of course, limited to PTFE. The generality of the problem has initiated a lot of research and it is established that a variety of surface treatments can improve the bonding characteristics of the more common plastics such as, for example, polyethylene. A particular interesting surface treatment is that designated as CASING (Crosslinking by Activated Species of Inert Gases), developed by Hansen and Schonhorn.[40] By exposing polymer samples to activated species of inert gases, produced by electrodeless discharge, it has been shown that dramatic improvements in adhesive bonding may be brought about. The evidence presented suggested that the improvements arose from extensive crosslinking and it was inferred that this occured at the surface.

We have made preliminary studies of the application of ESCA to this process choosing for our study an ethylene/tetrafluoroethylene copolymer. Samples were irradiated with a low energy (2KV) beam of argon ions with a beam current of 5μA for successive periods of 5 seconds and the C_{1s} and F_{1s} core levels monitored. To enable a reasonably complete study in a short time, resolution was sacrificed to sensitivity and the measurement of the core levels of a sample typically took ∿ 2 mins. The results, Fig. 53, are quite striking and it is clear that argon ion bombardment

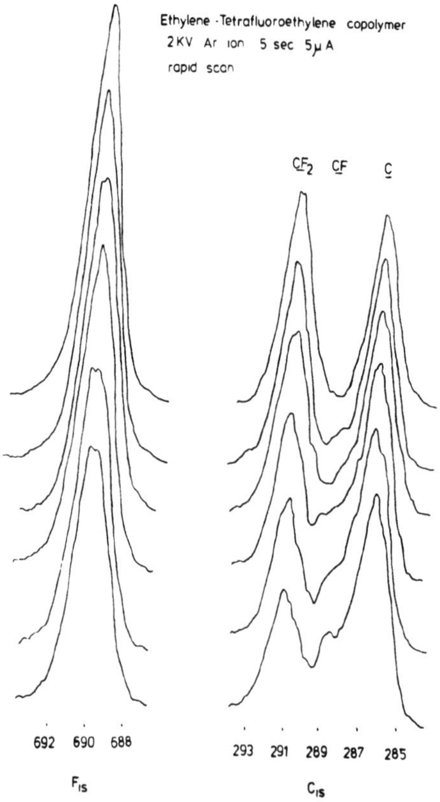

Fig. 53. F_{1s} and C_{1s} levels for argon ion bombarded samples of ethylene/tetrafluoroethylene copolymer.

causes extensive rearrangement in the outermost 50 Å or so of the sample. The main features are as follows. The F_{1s} signal decreases whilst the total C_{1s} signal remains approximately constant in intensity. The original copolymer, with its largely alternating structure of ethylene and tetrafluoroethylene units, exhibits initially the characteristic doublet structure in the C_{1s} region. On successive argon ion treatment the high binding energy peak corresponding to tetrafluoroethylene units progressively decreases in intensity and is replaced by peaks in a region of intermediate binding energy. This is illustrated in Fig. 54. More detailed studies of both these and other core levels will be required to allow a complete elucidation of the structure as a function of depth for this system. However, even with the results from Figs. 53 and 54 a few general conclusions may be drawn. We will consider briefly at this stage however, the possible mechanism for the process.

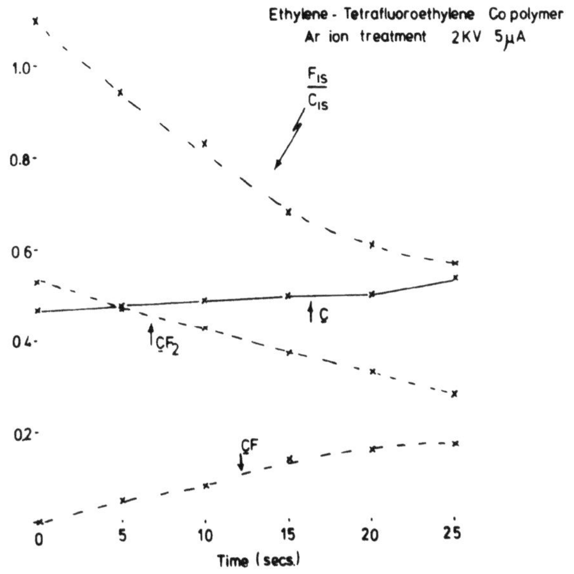

Fig. 54. Peak intensities as a function of irradiation.

From the previously published evidence a plausible mechanism for CASING in the particular case of polyethylene would be as follows. Production of radicals by initial argon ion bombardment (1)

$$M^* + CH_2\text{-}CH_2\text{-} \longrightarrow \text{-}CH_2\text{-}CH\text{-} + H. + M \qquad (1)$$

(M^* is electronically excited species of Ar)

The hydrogen atoms produced in (1) since they are in a fairly rigid matrix, may abstract a hydrogen from a CH_2 group in close proximity. Two situations seem probable. Firstly abstraction from the CH_2 group next to a radical centre, (which would be energetically particularly favourable because of the reduced C-H bond strength), leads to an unsaturated system (2) viz.

$$\text{-}CH_2\text{-}CH\text{-} + H. \longrightarrow \text{-}CH{=}CH\text{-} + H_2 \qquad (2)$$

(It is of cause entirely feasible that collision of M^* with the polymer chain could lead to direct elimination of molecular hydrogen. The crosslinking however, shows that although this could be a contributing pathway to the formation of unsaturated sites, radicals must be involved). Abstraction of a hydrogen from an adjacent chain would lead to radical combination and hence crosslinking (3)

$$-CH_2-\overset{\cdot}{C}H \quad\quad -CH_2-CH-$$
$$\longrightarrow \quad\quad\quad |$$
$$-CH_2-\overset{\cdot}{C}H- \quad\quad -CH_2-CH- \quad\quad (3)$$

For a fully fluorinated system (e.g. PTFE) the corresponding process to (1) might be expected to be energetically less favourable because of the greater bond strength of the C-F as compared to the C-H bond and also because of the low bond dissociation energy of molecular fluorine. Indeed it has been noted that PTFE films require extended treatment to improve the adhesive properties.[40]

In the case of a largely alternating copolymer of ethylene and tetrafluoroethylene the initial process analogous to (1) might therefore be expected to be (4)

$$M^* + -CH_2-CF_2-CH_2-CH_2 \longrightarrow -CF_2CF_2-\overset{\cdot}{C}H-CH_2 + H^\cdot + M \quad (4)$$

The high bond strength for HF suggests that (5) might be an efficient process,

$$-CF_2-CF_2-\overset{\cdot}{C}H-CH_2 + H^\cdot \longrightarrow -CF_2CF=CH-CH_2- + HF \quad (5)$$

although the possibility of direct molecular elimination of HF cannot be discounted. Hydrogen atom abstraction from an adjacent chain corresponding to (6) and (7) could then lead to crosslinked products (8) and (9).

$$-CF_2-CF_2-CH_2-CH_2- + H^\cdot \longrightarrow -CF_2\overset{\cdot}{C}F-CH_2-CH_2- + HF \quad (6)$$
$$\searrow$$
$$-CF_2CF_2-\overset{\cdot}{C}HCH_2- + H_2 \quad (7)$$

$$-CH_2-CF_2-\overset{\cdot}{C}H-CH_2- \quad\quad -CF_2-CF_2-CH-CH_2-$$
$$\longrightarrow \quad\quad\quad | \quad\quad (8)$$
$$-CF_2-CF_2-\overset{\cdot}{C}H-CH_2- \quad\quad -CF_2-CF_2-CH-CH_2-$$

$$\begin{matrix} -CF_2-CF_2-\dot{C}H-CH_2- \\ \\ -CF_2-\dot{C}F-CH_2-CH_2- \end{matrix} \longrightarrow \begin{matrix} -CF_2-CF_2-CH-CH_2- \\ | \\ -CF_2-CF-CH_2-CH_2- \end{matrix} \qquad (9)$$

The ESCA data clearly show the decrease in F_{1s} signal corresponding to loss of fluorine; in the process $\underline{C}F_2$ sites being converted to $\underline{C}F$ type. This would suggest that (5) and (9) are the major product steps. The experiment leads itself to dynamic studies, in which any volatiles produced could be analysed by mass spectrometry in parallel with the ESCA experiments, and such studies will be reported in due course.

The surface of fluorographites

In 1934 Ruff and coworkers[41] showed that graphite could be smoothly fluorinated by the action of fluorine in the temperature range 420-460°C and obtained a grey material with bulk composition $CF_{0.92}$. Since that time numerous investigations have been carried out (particularly by Margrave[42] and Watanabe[43] and their coworkers), into the fluorination of graphite under a wide variety of conditions and fluorinated graphite in the bulk composition range $CF_{0.25}$ - $CF_{1.2}$ have been well authenticated. The material of nominal composition $CF_{1.0}$ often referred to as polycarbon monofluoride is white with highly desirable lubricating properties. Extensive studies have been made of its friction and wear life properties and applications range from such diverse uses as pressure and heat resistant lubricant, to its use in high energy primary cells. For certain applications films were found to have wear lives greater than either graphite or molybdenum disulphide and the coefficient of friction was lower than the former and about the same as the latter. The useful limiting upper temperature is $\sim$ 400°C.[44] Largely as a result of the work of Margrave and coworkers a considerable amount of thermodynamic data has been accumulated on the system and it has also been the subject of several X-ray investigations. Although as recently as ten years ago one standard text[45] considered the structure to be a layer type with fluorine located such that the lattice spacing between the carbon planes increased to 8.17 Å, more recent data suggested a network structure in which each carbon was covalently bonded to three other carbons and a fluorine, the carbons being arranged in puckered inter-connecting six-membered ring with lattice spacing of 5.84 Å between the average planes, Fig. 55.[46]

The reaction of fluorine with graphite is undoubtedly initiated at the surface and is followed by diffusion into the bulk.

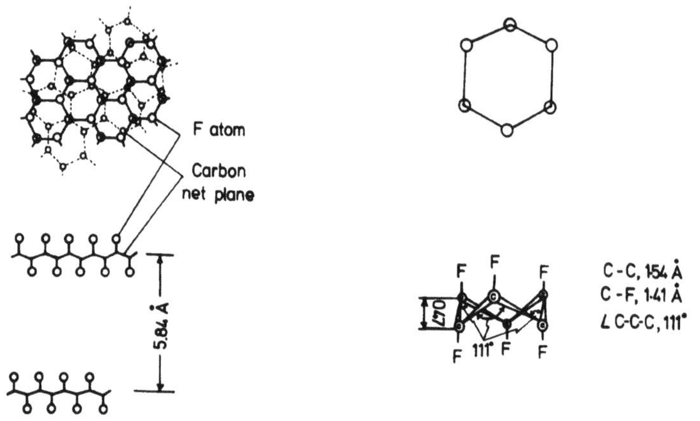

Fig. 55. Crystal structure of graphite monofluoride.

In such circumstances it would be surprising if reaction produced a homogeneous material progressing from surface to subsurface to bulk. Although the recent X-ray data therefore pertaining as it does to the bulk undoubtedly gives a correct overview of the structure it is clear that the surface may well be drastically different and if so have important ramifications as far as the interpretation of friction and wear characteristic are concerned. The surface sensitivity of ESCA is therefore particularly valuable for investigating this in some detail. Fig. 56 shows the C_{1s} and F_{1s} spectra for commercially available samples+ of fluorographite of composition C_1F_1. For comparison, data are also presented for samples of nominal compositions $C_1F_{0.8}$ and $C_1F_{1.2}$. The C_{1s} spectra are immediately revealing. In each case the peak of low intensity at 285.0 eV is identified as arising from hydrocarbon contamination at the surface almost certainly arising from the packaging of the samples. The main peak centred at $\sim$ 290.3 eV by comparison with model compounds previously studied is consistent with tertiary $\geq\underline{C}$-F type environments in a perfluoro system. This would tend to support therefore the major features of the structure proposed on the basis of diffraction studies, Fig. 55. The F_{1s} levels of binding energy 690.3 eV are also entirely consistent with covalently bound $\geq$C-$\underline{F}$ groups and rules out the possibility of a structure based on a loosely bound intercalate. The most significant feature displayed in Fig. 56, however, is the progressive increase in intensity of the shoulder centred around 292.4 eV to the high binding energy of the main C_{1s} peaks.

+ We are indebted to Ozark-Mahoning Co. Special Chemical Division, Tulsa, Oklahoma, for these and other samples which will be discussed in detail elsewhere.

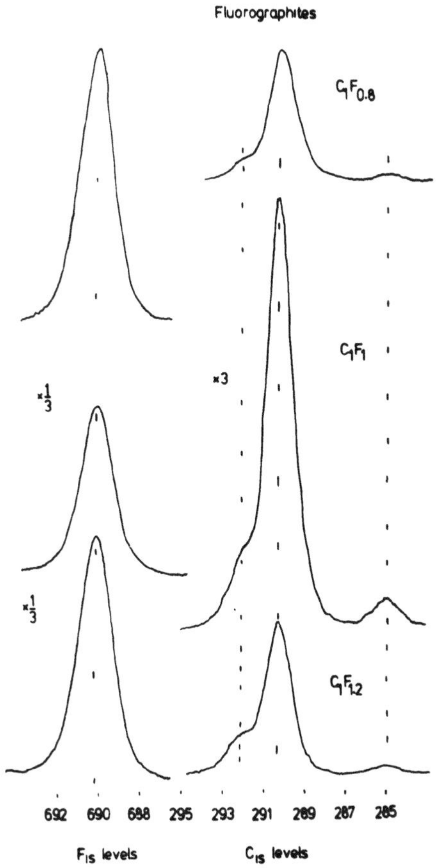

Fig. 56. F_{1s} and C_{1s} levels for some fluorographites.

This feature, together with the relative overall intensities of the C_{1s} with respect to the F_{1s} levels, strongly suggests that the shoulder arises from a differing structure at the surface of the samples. For the direct fluorination of polycrystalline samples of graphite we may envisage fluorination proceeding at basal planes or primatic edges the latter leading to $\underline{C}F_2$ type environments. The various possible $\underline{C}F$ and $\underline{C}F_2$ environments generated may be simulated theoretically as is shown in Fig. 57. The calculated binding energies fall into two groups centred around ∿ 290 eV and 292 eV in excellent agreement with the experimental data. It is interesting to note that the C_{1s} binding energy is almost exactly the same as that for PTFE and we might speculate that the unusual chemical and mechanical properties of fluorographite may well arise from this surface feature. Clearly further discussion must await more definitive studies and these are currently in hand. By studying the direct fluorination of different crystallographic faces of single crystal graphite, ESCA should shed considerable

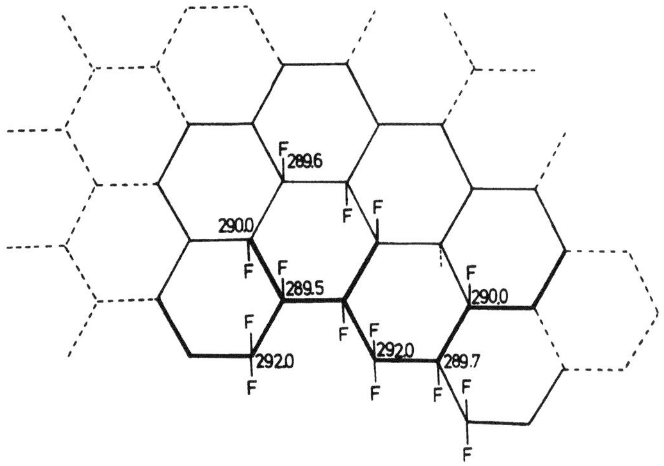

Fig. 57. Calculated binding energies for representative sites in fluorographites from charge potential calculations.

light on the surface structure of fluorographites particularly since for the well ordered system the possibility of carrying out electron channelling experiments exists (cf. ref. 2a). Differences in band structure as a function of composition are also clearly evident and will be referred to in a later section.

Surface fluorination of polyethylene

Direct fluorination of polyethylene has been the subject of several investigations, the main incentive being the possibility of producing a fluorocarbon layer, with its associated desirable chemical and physical properties, on the surfaces of articles fabricated from a relatively cheap and malleable material. Some twenty years ago Rudge[47] published a patent reporting the somewhat surprising result that on reaction with fluorine the hydrogens of polyethylene could be replaced by fluorine in a controlled way to yield a surface coating which appeared, on the basis of elemental analysis, X-ray examination, and its physical and chemical properties to be a form of polytetrafluoroethylene of very low crystallinity. The implications of this disclosure do not appear to have been exploited, although interest in the possibilities of this approach were recently revived by reports of work by Margrave and Lagow.[48] In the period since Rudge's original report several groups have investigated the direct surface fluorination of polymers.[48-51] The feature common to the majority of these previous investigations is that they refer to extensively fluorinated filsm; thus, processes were generally run until no more fluorination would occur and the products were either examined as made or after removal of soluble residual hydrocarbon materials by solvent extraction. Furthermore the investigative techniques employed were such that the results refer essentially to the bulk material. Consequently, little information has been acquired concerning the early stages of the reaction of fluorine with polyethylene films which must, of course, be initiated at the surface. A priori, therefore, one might expect a dependence on depth of stoichiometry (and hence structure) as a function of the fluorinating conditions (e.g. concentration, temperature and duration). Indeed there is evidence for this from studies of extensive fluorination of films of polyvinylfluoride of different thicknesses. Thus, the limiting fluorine contents obtained from the fluorination under similar conditions of films 10 and 100 μ thick were respectively 65 and 55%.[51] Transmission infrared spectroscopic data reported in this investigation clearly show that the fluorination extends to depths of the order of thousands of Angstroms. Although, in general, previous investigations have established overall stoichiometries for the extensively fluorinated films, because of the depth dependence of composition it is not possible to derive unambiguously either the stoichiometry at the surface and immediate subsurface or more particularly the actual molecular structure in this region which is obviously crucial not only to an understanding of the kinetics and mechanism of the fluorination process but also to an understanding of the physical chemical and mechanical properties of the surface. In the particular case of the extensive fluorination of polyethylene single crystals however, there is clear evidence that the surface composition is very close to that of PTFE although any depth dependence of stochiometry might well have been obscured by the solvent treatment used in this work prior to examination of the fluorinated materials.[50] We have carried out a detailed ESCA investigation of the fluorination of pressed films of high density polyethylene, directed primarily at the following points of interest :

(1) Preparation of polymer films;

(2) Establishment of fluorination depth as a function of fluorinating conditions;

(3) Overall stoichiometries as a function of depth for various fluorinating conditions;

(4) Molecular structure of the surface and immediate subsurface;

(5) Kinetics and mechanism of the fluorination process.

The information obtained on all these aspects is too voluminous to be presented in this review ans will be published elsewhere, here we shall summarize the salient features of the argument used in interpretation of the information obtained and the main conclusions.

Before embarking on a detailed study of a surface treatment it is clearly obligatory to establish as much as possible about the nature of the starting surface. To this end polyethylene films from a variety of sources have been examined. Fig. 58 shows the C_{1s} and O_{1s} spectra for commercially produced low density polyethylene films both before (a) and after (b) desiccation. Two features are clearly evident. Firstly, the substantial signals corresponding to the O_{1s} levels exhibit an unresolved structure, one component of which disappears on storing the samples over P_2O_5. Comparison with previously measured binding energies indicates that the higher binding energy component of the O_{1s} levels corresponds to H_2O whilst the lower binding energy peak corresponds to $>C=\underline{O}$, although the peak width suggests that there are probably several differing carbonyl environments. Secondly, the C_{1s} levels show a small shoulder to the high binding energy of the main peak, the shift of $\sim$ 3 eV again being consistent with carbon in $>\underline{C}=O$ environments.

Comparison with MATR measurements employing single crystal Germanium is instructive in this respect since strong absorptions are present in the carbonyl region (but not in the OH region). Careful double beam TIR experiments on the 200μ films also shows the presence of carbonyl groups. The three techniques taken together therefore emphasize that whilst the carbonyl groups are distributed throughout the bulk of the sample, hydrogen bonding involving extraneous water is localized at the surface. The carbonyl groups may be attributable in part to various additives such as anti-oxidants, anti-blocking agents etc... These complications made such materials unsuitable for our study and we next examined films pressed from unstabilized high density polyethylene.+ Direct study of the powdered sample (prior to pressing) by ESCA reveals no trace of O_{1s} signal.

+ We are indebted to Dr. B.W. Cook, I.C.I. Corporate Laboratory for the supply of this material.

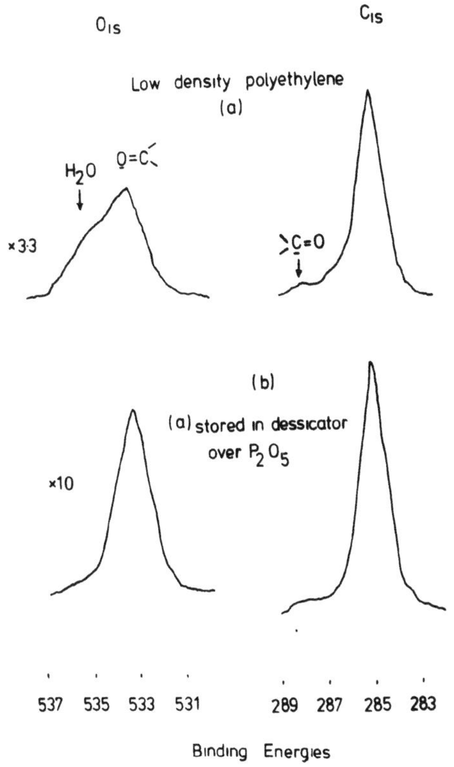

Fig. 58. C_{1s} and O_{1s} levels for low density polyethylene.

In preparing films of high density polyethylene for the fluorination experiments, three methods of preparation have been investigated. In all of these, films were pressed between sheets of clean aluminium foil at the minimum temperature necessary for plastic flow, with a hand press (∼200°C).

(c) The samples were pressed in air;

(d) The samples were pressed in a nitrogen atmosphere;

(e) The polyethylene powder was pumped down (10^{-4} Torr) and let up to a pure nitrogen or argon atmosphere for several cycles before being transferred in an inert atmosphere into the press.

The O_{1s} and C_{1s} spectra corresponding to samples from these three modes of preparation are shown in Fig. 59 and are quite striking. ATR and TIR experiments did not reveal the presence of

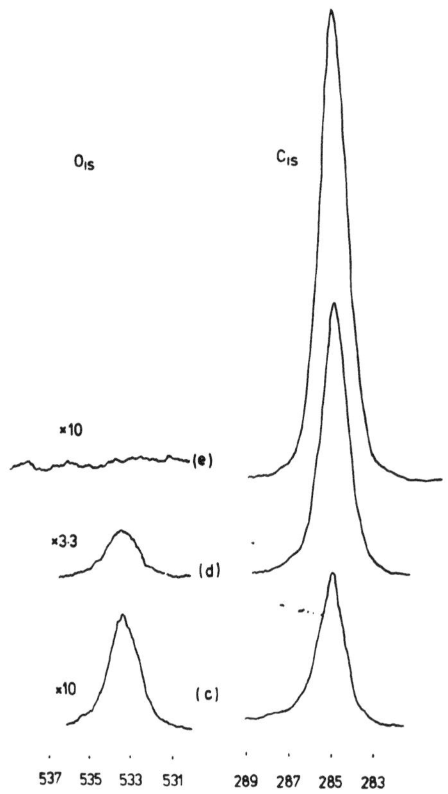

Fig. 59. C_{1s} and O_{1s} levels for pressed films of high density polyethylene.

any oxygen function (-OH, $\geq$C=O, C-O-C etc..) and in fact the spectra were virtually identical. This in itself demonstrates the great power of ESCA in distinguishing minute differences between samples when such differences are localized at or near the surface. Comparison with the data for low density polyethylene and with model monomer systems shows that the O_{1s} signal arises from $\geq$C=O type environments. The three methods of preparation clearly indicate that "unoxidized" surfaces may most readily be prepared by excluding all traces of oxygen during the pressing stage. On leaving samples exposed to the atmosphere for some time hydrogen bonding to the surface C=O groups from extraneous water in the atmonsphere occurs and the O_{1s} peak then acquires the characteristic doublet nature. Again MATR and TIR do not reveal any changes since the hydrogen bonding is localized at the surface.

It is of interest to obtain some information on the relative frequence of $\geq C=O$ features in the oxidized films and this may be achieved as follows. From extensive studies of compounds containing both carbon and oxygen, infinity values (I_α) may be obtained for the C_{1s} and O_{1s} core levels in homogeneous solids.[52] If We make the reasonable assumption that because of temperature gradients, diffusional phenomena, and the relatively low concentration of oxygen (the only available oxygen being that entrained in the powdered polyethylene sample), that in the sample prepared by method (d) oxidation is likely to be limited to the first monolayer (i.e. $\sim$5 Å), we may then with a knowledge of the likely escape depth dependences (Λ_1 and Λ_2 for the O_{1s} and C_{1s} core levels) calculate the surface concentration of $\geq C=O$ structural features. Taling Λ_1 and Λ_2 as 8 Å and 10 Å (corresponding to $MgK\alpha_{1,2}$ exciting radiation) the result obtained is that if all the O_{1s} signals corresponds to $\geq C=\underline{O}$ located in the first monolayer then there is $\sim$ 9% monolayer coverage of this structural feature. Since such films are routinely easy to prepare in a reproducible manner the initial fluorination experiments have been with this material rather than the more difficulty prepared samples produced by method (e). As will become apparent, detailed analysis of structure and bonding of the surface fluorinated films provides strong support for this crude estimate of the extent of surface oxidation.

One question which immediately arises is that of the structure at the surface of polyethylene films produced in this manner. The question of polymer crystallography and morphology has been extensively discussed and documented by Wunderlich.[53] For the most part discussion has centred around samples crystallized from solution or produced from the melt under pressure. There are two considerations which make it difficult to extrapolate from this work. Firstly, our prime consideration is the structure at the surface of samples whereas the published data refer rather to the bulk structure. Secondly, our mode of sample preparation differs considerably from those for which information concerning the bulk structure are available. Of particular relevance to our studies however, are the extensive investigations of Schonhorn[54,55] on nucleation and trans-crystalline growth at polyethylene-metal interfaces. The major points of interest arising from this work are that heterogeneous nucleation and crystallization of polyethylene melts against high energy surfaces such as aluminium, results in marked changes in the surface region morphology; electron microscopy indicating a region of transcrystalline growth extending to a depth of 25-50μ. Schonhorn also commented that the mechanics of peeling the polymer film from the aluminium determines the thickness of polymer remaining on the metal surface and that for thin aluminium, failure should occur closer to the metal polymer interface. In the preparation of the polymer films used in this work, the procedure entailed pressing the high density polyethylene powder between clean sheets of aluminium foil. The aluminium foil was then stripped from the surface. In previous ESCA studies we have investigated in detail the surface of aluminium foil.[56] The main points which emerge from these studies are that the oxide layer in commercially produced annealed (domestic) foil is typically 20 Å thick and that a tenaciously held hydrocarbon type layer is present at the surface which is not readily removed by either degreasing treatment or heating under very high vacuum conditions. ESCA provides a convenient tool therefore for investigating the nature of the peeled surfaces. Fig. 60 shows the O_{1s}, C_{1s} and Al_{2p} levels for the surface of the aluminium foil used for pressing and of the peeled foil appropriate to the polyethylene sample (d) of Fig. 59. (It should be stated at this stage that no trace of Al_{2p} core levels could be detected for the sample d). The most significant feature is that both the aluminium and oxygen core levels are of appreciable intensity in the peeled foil and this can only be interpreted on the basis that failure occurs very close to the surface. From the relative increase in intensity of the peak due to the C_{1s} levels (taken in conjunction with an escape depth of 10 Å for electrons of kinetic energy ∿ 968 eV) a reasonable estimate for the thickness of polymer adhering to the peeled foid would be ∿ 10 Å. In an investigation of the wettability of melt crystallized polyethylene films formed by nucleation at an aluminium (aluminium oxide) surface Schonhorn measured[55] a contact angle of 65 degrees and derived a percentage crystallinity for the surface region of 63.2%. We have investigated therefore the contact angles for both of the peeled surfaces using glycerol as the wetting liquid to correspond to the measurements of Schonhorn. For both surfaces a contact angle of ∿ 66° was obtained which would strongly suggest that our samples are of comparable surface structure to Schonhorn's.

Films of the material d were exposed to an atmosphere of <u>ca</u>. 10% fluorine in nitrogen in a flow system, light was excluded from the reaction vessel. The fluorine was produced by a conventional electrolytic cell and although usual precautions were taken there were almost certainly traces of hydrogen fluoride, oxygen and water in the reaction vessel. In the first instance films were fluorinated for ∿ 1 hr. The C_{1s} spectra revealed that the original peak at 285 eV arising from $-CH_2-CH_2-$ units had been replaced by a complex structure extending to ∿ 292 eV. This together with an intense fluorine 1s peak indicated that extensive fluorination had taken place. This was readily confirmed by MATR experiments with single crystal GE which indicated the presence of C-F groups, and indeed careful double beam TIR sutdies also

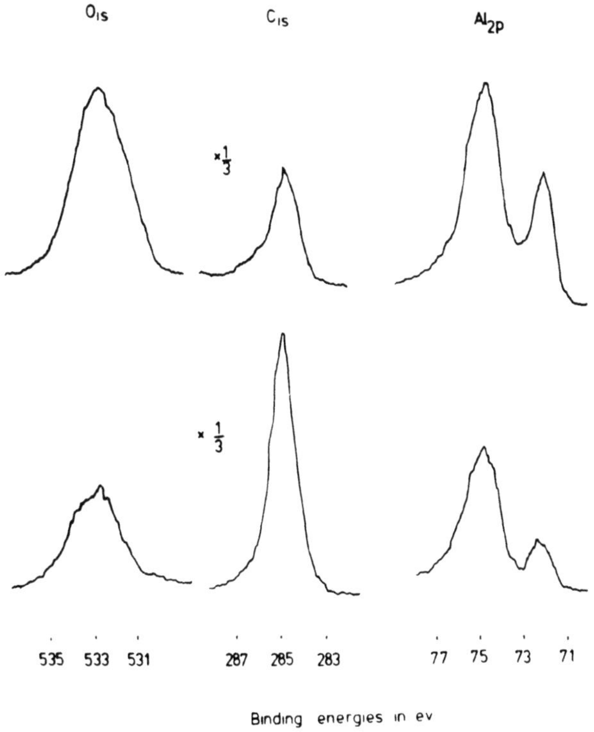

Fig. 60. C_{1s}, O_{1s} and Al_{2p} levels for aluminium foil as used for pressing (upper) and as peeled from polyethylene sample (lower).

indicated this. These results overall indicated therefore that fluorination had proceeded well into the bulk and on the basis of the MATR abd TIR data a reasonable estimate would be > 100 Å. The initial stages of the process were therefore investigated and a convenient range of reaction times established of 0.5 sec-300 secs.

To illustrate the great sensitivity of the technique for studying the initial stages of reaction, Fig. 61 shows the C_{1s}, O_{1s} and F_{1s} core levels for a sample fluorinated for ∿ 0.5 secs. There are several features of interest. Firstly, the appearance of a strong F_{1s} signal; the absolute binding energy being characteristic of a CF type system. Secondly, there is an increase in

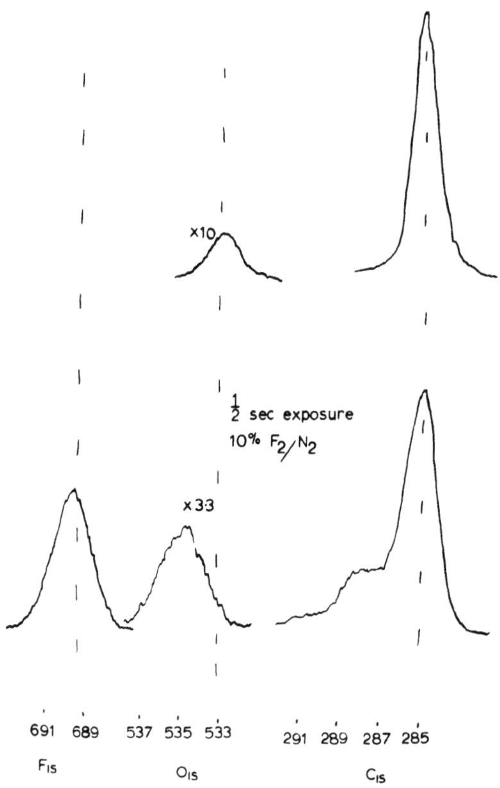

Fig. 61. C_{1s}, O_{1s} and F_{1s} levels for high density polyethylene film before (upper) and after (lower) fluorination for 0.5 sec.

overall intensity of the O_{1s} signal, however it is clear from the characteristic binding energy that this is due to water and most probably arises from hydrogen bonding to the surface carbonyl features from the trace of water impurity in the system. Finally the C_{1s} spectrum shows in addition to the main peak of polyethylene at 285 eV, a broad shoulder centred ∿ 288 eV characteristic of <u>C</u>F type environments and a low intensity tail at higher binding energy centred at ∿ 291 eV attributable to <u>C</u>F_2 type environments. It would clearly be feasible to detect fluorination occurring on a much shorter time scale and a reasonable lower limit with present instrumentation would be ∿ 10^{-3} sec. although of course it would be technically difficult to fluorinate a sample on this time scale. It does indicate, however, that experiments to detect fluorination employing much more dilute fluorinating conditions, would be perfectly feasible. Fig. 62 shows the F_{1s}, O_{1s}

and C_{1s} core levels for films fluorinated for 0.5 sec. and 30 secs. The differences although quite striking go undetected by either MATR or TIR emphasizing the surface nature of the initial reaction. After 30 secs. exposure the original C_{1s} peak has almost completely disappeared, appearing only as a shoulder on a broad peak centred ∿ 286 eV. Further broad peaks are centred ∿ 288.5 eV and 291 eV with a tail extending to ∿ 292 eV.

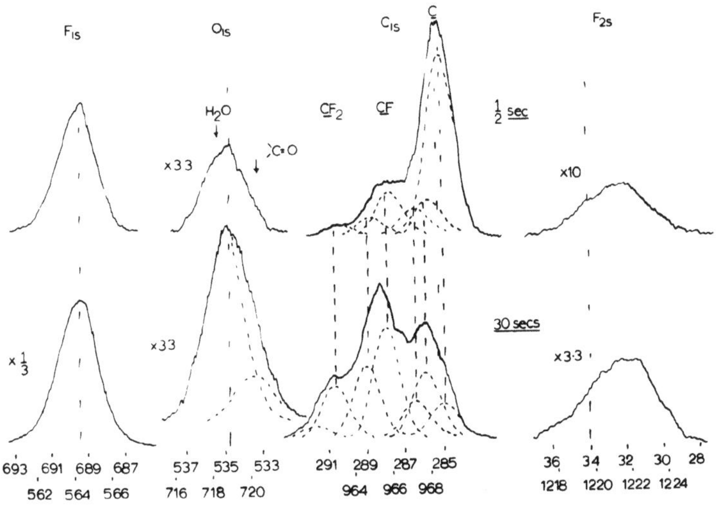

Fig. 62. C_{1s}, O_{1s}, F_{1s} and F_{2s} levels for high density polyethylene films fluorinated for 0.5 sec. and 30 sec. respectively.

Comparison with the binding energies as a function of structural environment allows an assignment of these major features.[16] Firstly the most intense peak at ∿ 288.5 eV corresponds to <u>C</u>F type environments whilst those at ∿ 286 eV and ∿ 291 eV corresponding to carbons subjected to secondary substituent effect from fluorine and from -<u>C</u>F_2- type environments respectively. This data alone therefore demonstrates that extensive fluorination has taken place at the surface and immediate subsurface of the polyethylene films.

These results established that ESCA is capable of detecting surface fluorination at a very early stage, a more refined analysis of the data was then attempted. Firstly the application of ESCA to the investigation of film thickness was examined. In section 3 (h) of this review the escape depth dependence of signal intensity was described. Clearly if an element gives two or more signals from levels with very different escape depths it ought to be possible to use the relative intensities of the signals to calculate the thickness of the surface layer containing that element. For the case of fluorine we can use the F_{1s} and F_{2s}+ signals. It is clear from Fig. 62 that the F_{1s} and F_{2s} levels with kinetic energies of ∿ 564 eV and 1222 eV should correspond to a considerable difference in escape depth dependence. Extrapolation from the generalized curve of a mean free path versus kinetic energy (Fig. 17) for example would suggest ∿8 Å and ∿ 14 Å respectively for the F_{1s} and F_{2s} levels. We can now proceed to use this as a basis for estimating the thickness of the fluorinated polymeric films. Consider (Fig 63) a fluoropolymer film of thickness d formed at the surface of a polyethylene film. (Since in general the films which were fluorinated were in excess of 50μ (i.e. 5 x 10^5 Å) thick it was assumed that the overall thickness of the polymer was infinite compared with typical mean free paths of electrons).

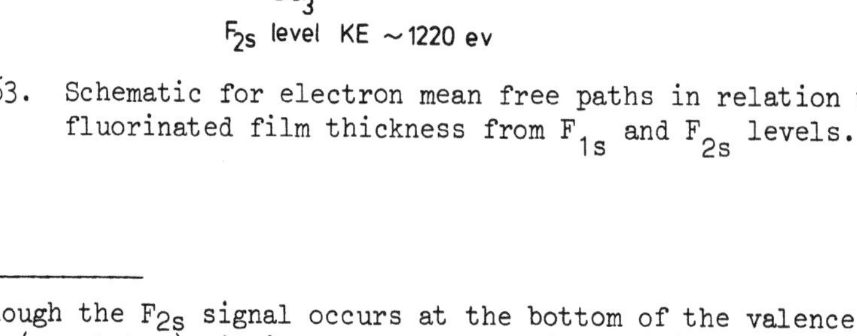

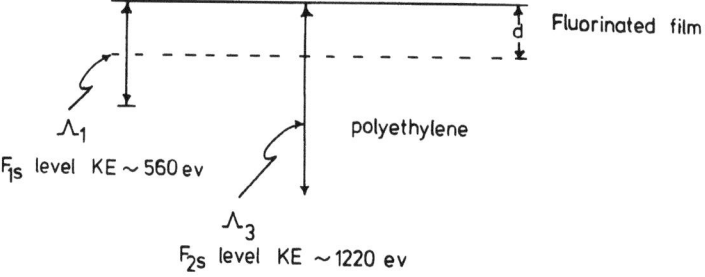

Fig. 63. Schematic for electron mean free paths in relation to fluorinated film thickness from F_{1s} and F_{2s} levels.

+ Although the F_{2s} signal occurs at the bottom of the valence band (see later), it is nevertheless quite distinct and essentially core-like.

The intensity of the F_{1s} and F_{2s} levels are given by Equations (3) and (4),

$$I_{F_{1s}} = I_{\alpha F_{1s}} (1-e^{-d/\Lambda_1}) \quad (3)$$

$$I_{F_{2s}} = I_{\alpha F_{2s}} (1-e^{-d/\Lambda_3}) \quad (4)$$

where the symbols have the usual meanings and Λ_1 and Λ_3 are the electron mean free paths appropriate to the F_{1s} and F_{2s} levels respectively. Now the ratio of infinity values for the F_{1s} and F_{2s} levels es directly proportional to the measured area ratios for the two peaks for effectively infinitely thick films (equation 5). From studies of simple homopolymers a value for the area ratio

$$\frac{I_{\alpha F_{1s}}}{I_{\alpha F_{2s}}} = \frac{LA_{\alpha F_{1s}}}{A_{\alpha F_{2s}}} \quad (5)$$

was readily available and under our experimental conditions it was equal to 10.12. Combination of equations (3), (4) and (5) gives (6), the proportionality constant L cancelling.

$$\frac{I_{F_{1s}}}{I_{F_{2s}}} = \frac{\not{L}A_{F_{1s}}}{A_{F_{2s}}} = \frac{\not{L}A_{\alpha F_{1s}}}{A_{\alpha F_{2s}}} \frac{(1-e^{-d/\Lambda_1})}{(1-e^{-d/\Lambda_3})} \quad (6)$$

This equation relates the measurable properties of the system (viz. the area ratios of the F_{1s} and F_{2s} peaks), and the film thickness and electron mean free paths. Denoting the measured ratios for the fluorinated films by y and the infinity values measured for simple homopolymers by K, Equation (6) may be recast in the form (7).

$$(^y/_K - 1) = {^y/_K}e^{-d/\Lambda_3} - e^{-d/\Lambda_1} \quad (7)$$

At this stage we obtain $^y/_K$ from experiment and the generalized curve of mean free path as a function of kinetic energy gives some idea of the likely range of values for Λ_1 and Λ_3.

Next a computer analysis was used in a self consistent process. Taking the measured intensity ratios, the right hand side of Equation (7) may be evaluated for a range of values of d and of Λ_1 and Λ_3. Tables were constructed corresponding to values of d in the range 2-41 Å at 1 Å intervals and for ratios of Λ_3/Λ_1 from 1-2 in increments of 0.1 with valeus of Λ_1 varying in the range 5-15 Å in 1 Å steps. The data corresponding to 24,200 separate computations for five separate fluorination times is too lengthy to reproduce fully here but sufficient data has been included to illustrate the basic philosophy behind our analysis. This may be done by reference to the two fluorinations so far discussed viz. 0.5 sec. and 30 secs. The corresponding measured values of y were 15.64 and 10.83 respectively yielding values for the LHS of Equation (7) of 0.545 and 0.07. Inspection of the Tables revealed that for the 0.5 sec. run no fit was possible within the range of parameters taken if the ratio of Λ_3 to Λ_1 was < 1.5. The fits within the range of 1.6 - 2.0 for this ratio bracketed the value of Λ_1 in the range 4-12 Å and film thickness d in the range 3-11 Å. Considering the data for the 30 secs. run and limiting the ratio of escape depths to be compatible with those for the 0.5 sec. run, considerably narrows the range of values for Λ_1 for which a fit to the experimental data may be obtained to 5-9 Å with a corresponding range for the film thickness of 22-42 Å. The 30 secs. run, bracketing as it does the escape depth, now further limits the range of film thickness for the 0.5 sec. run to 3-9 Å. It is interesting to note that the ratio of escape depths Λ_3 to Λ_1 suggested from the generalized curve is $\sim$ 1.7, in the middle of the range bracketed by the analysis for just two runs. F_{1s}/F_{2s} area ratios for fluorination experiments corresponding to reaction times of 0.5, 1, 15, 30 and 300 seconds were measured and computer fits to this data further reinforces the analysis presented above. Film thicknesses were therefore computed for values of Λ_1 = 7 Å with Λ_3/Λ_1 = 1.7 Å this gives a value of 11.9 Å for Λ_3, both points fitting very well on the generalized curve previously disccussed. The computed data corresponding to these parameters are shown in Fig. 64. The derived thickness of the fluorinated films are 3.5 Å, 6 Å, 16 Å, 30 Å and 36 Å respectively for reaction times of 0.5, 2, 15, 30 and 300 secs.

For direct comparison of the validity of this data, an independent analysis was made from the C_{1s} spectra. Consider the fluoropolymer film of thickness d formed by reaction of fluorine with polyethylene. Taking standard lineshapes and linewidths and employing the basic philosophy of lineshape analysis previously discussed in detail elsewhere,[6a] the C_{1s} spectra may be unambiguously deconvoluted into component peaks and this is shown for the 0.5 sec. and 30 sec. runs in Fig. 65. The C/F stoichiometry

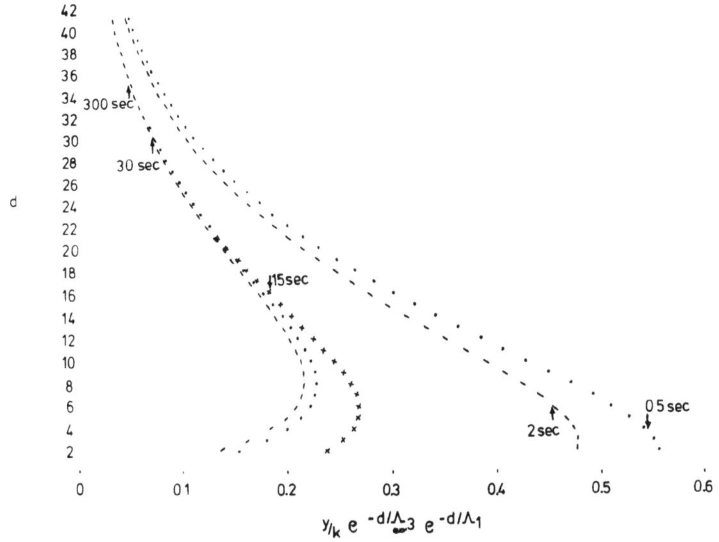

Fig. 64. Estimation of fluorinated film thickness from F_{1s} and F_{2s} levels.

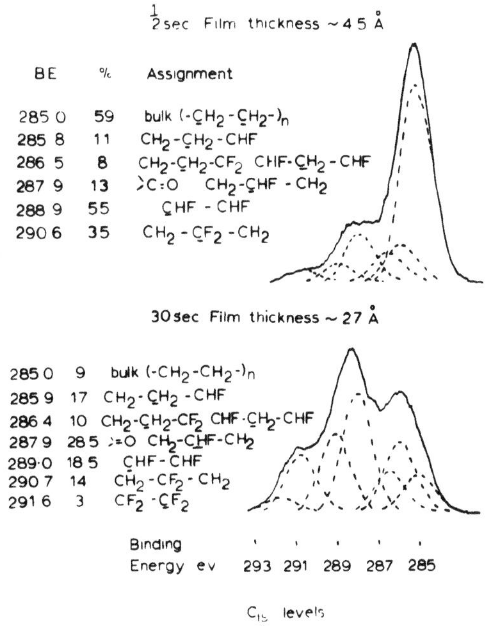

Fig. 65. Deconvolution of C_{1s} spectra for high density polyethylene films fluorinated for 0.5 sec. and 30 sec.

of the fluoropolymer is $\lesssim$ 2 for the films produced in this study (see later), this being the case the peak at binding energy 285.0 eV may be assigned to unreacted polyethylene. The area ratio of this portion of the C_{1s} peak to the remaining higher binding energy part arising from various C_{1s} environments of the fluoropolymer film, forms the basis for estimating the film thickness d.

The analysis proceeds as follows. The intensities of the C_{1s} peaks arising from the fluoropolymer film and polyethylene may be expressed as (8) and (9)

$$\text{Fluoropolymer} \qquad I_{C_{1s}}^{F_{pol}} = I_{\alpha C_{1s}}^{F_{pol}} (1-e^{-d/\Lambda_2}) \qquad (8)$$

$$\text{Polyethylene} \qquad I_{C_{1s}}^{Pet} = I_{\alpha C_{1s}}^{Pet} e^{-d/\Lambda_2} \qquad (9)$$

The measured relative area ratios may therefore be written as (10).

$$\frac{A^{F_{pol}}}{A^{Pet}} = \left(\frac{A_\alpha^{F_{pol}}}{A_\alpha^{Pet}}\right)(e^{d/\Lambda_2} - 1) \qquad (10)$$

Measurements on simple homopolymers show that

$$\left(\frac{A_\alpha^{F_{pol}}}{A_\alpha^{Pet}}\right) \approx 1,$$

(10) therefore reduces to a particularly simple form (11).

$$\frac{A_{C_{1s}}^{F_{pol}}}{A_{C_{1s}}^{Pet}} = (e^{d/\Lambda_2} - 1) \qquad (11)$$

The mean free paths of $\sim$ 7 Å and $\sim$ 12 Å for F_{1s} and F_{2s} levels respectively (corresponding to kinetic energies of $\sim$ 560 eV and $\sim$ 1220 eV), given by the previous analysis and the generalized curve of escape depth dependence on kinetic energies suggests a value of ∿ 10 Å for Λ_2 where the kinetic energy corresponding to photoionization from C_{1s} core levels is ∿ 960 eV. Using the measured area ratios of C_{1s} peaks and taking Λ_2 = 10 Å the derived values of d are 5.5 Å and 24 Å for the 0.5 sec. and 30 sec. runs respectively, in good overall agreement with those derived from the F_{1s} and F_{2s} levels. A comparison for all five runs with the two methods of determination is given in Table 19. The close correspondence between these two sets of data suggests that the logic of the overall analyses is correct. The deviation between

Table 19. Calculated film thickness for surface fluorinated polyethylene films

Experiment	$\frac{A^{F\ pol}_{C_{1s}}}{A_{C_{1s}\ pet}}$	d(Å) From C_{1s} spectra	From F_{1s} and F_{2s} spectra	Av.
0·5 sec.	0·71	5·5	3·5	4·5
2 sec.	1·50	9·0	6·0	7·5
15 sec.	3·76	15·0	16·0	15·5
30 sec.	10·10	24·0	30·0	27·0
300 sec.	15·67	28·0(46)*	36·0	32(41)*

* Revised estimate see text

the two methods of analysis is greatest for large film thicknesses. This was not unexpected since Fig. 64 indicates that the analysis based on the F_{1s} and F_{2s} levels is least sensitive for large d whilst that based on the C_{1s} spectra will be subject to greatest error for a relatively high degree of fluorination since any slight surface contaminant hydrocarbon (B.E. 285.0 eV) arising during sample manipulation after fluorination will tend to lead to d being underestimated.

Having established the depth of fluorination the next question to be examined is that of the stoichiometries of the surface layers. The methods of establishing overall stoichiometries for homogeneous fluoropolymer films using the area ratio for F_{1s}

and C_{1s} levels and component peaks of the C_{1s} region were discussed in section 4(f). The area ratio of the F_{1s} peak to C_{1s} peaks (arising from the fluorinated film) were measured from the spectra, and computed from Equation (12), where d was taken as the average of the two values derived previously, and the infinity ratio $A_{\alpha F_{1s}}/A_{\alpha C_{1s}}$ is that derived from measurements on homopolymers,

$$\frac{A_{F_{1s}}}{A_{C_{1s}}} = \left(\frac{A_{\alpha F_{1s}}}{A_{\alpha C_{1s}}}\right) \frac{(1-e^{-d/\Lambda_1})}{(1-e^{-d/\Lambda_2})} \qquad (12)$$

viz. 2.05.[17] Equation (12) of course applies to equal number of F_{1s} and C_{1s} core levels and therefore dividing the measured area ratios by those calculated for a unit stoichiometry gives the actual stoichiometry. The computed stoichiometries are given in Table 20.

Table 20. C/F stoichiometries for fluorinated polyethylene samples

Expt.	I C/F from C_{1s}/F_{1s} spectra	II C/F from C_{1s} spectra	III C/F from C_{1s} spectra taking surface oxidation into account	Av. I and III
0·5 sec.	2·01	1·60	2·05	2·03
2 sec.	1·43	1·40	1·54	1·48
15 sec.	1·37	1·27	1·40	1·38
30 sec.	1·16	1·12	1·23	1·20
300 sec.	1·17	1·04	1·12 (1·16)*	1·16

Deconvolution of the C_{1s} spectra as previously indicated in Fig. 65 allows an independent means of calculating the stoichiometries if assignments can be made of the component peaks. In this objective earlier work on both homopolymers and copolymers proved invaluable. Fig. 65 also illustrates the detailed assignment of peaks for the 0.5 sec. and 30 sec. fluorination experiments. If we ignore for the time being the contribution from the peak at 287.9 eV arising from surface carbonyl features previously discussed, the stoichiometries may be worked out quite straightforwardly as shown in Table 21 for the 0.5 sec and 30 sec. runs.

Table 21. Evaluation of C/F stoichiometries from data pertaining to C_{1s} levels

Stoichiometries for fluorinated polyethylene films from the C_{1s} spectra

0.5 sec run

Peak	Binding Energy	Structural Feature	Area %	Contribution to overall stoichiometry C	F
1	-	-	-	-	-
2	290.6	$-\underline{C}F_2-$	3.5	0.035	0.070
3	288.9	$-\underline{C}FH-$	5.5	0.055	0.055
4	287.9	$-CFH-$	13.5	0.135	0.135
5	286.5	$-CH_2-$	8	0.08	-
6	285.8	$-CH_2-$	11	0.11	-
				C/F 1.60	

30 sec run

Peak	Binding Energy	Structural Feature	Area %	C	F
1	291.6	$-CF_2-$	3	0.030	0.060
2	290.7	$-CF_2-$	14	0.14	0.28
3	289.0	$-CFH-$	18.5	0.185	0.185
4	287.9	$-CFH-$	28.5	0.285	0.285
5	286.4	$-CH_2-$	10	0.100	-
6	285.9	$-CH_2-$	17	0.170	-
				C/F 1.12	

The preliminary data for the two methods of calculating the stoichiometries are collected in Table 20. Overall there is seen to be a remarkable measure of agreement between the two methods (columns I and II, Table 20) with the relatively minor discrepancies arising at the two extremes. The figures in column III of Table 20 arise from a further refinement of the analysis. The initial films were known to contain carbonyl functions, and from the mode of preparation of the film it seems highly likely that the oxidation will be localized within the first monolayer; which, taking appropriate bond lengths, would correspond to the first ~ 5 Å of the sample, very close in fact to the estimated thickness of the fluorinated film for the 0.5 sec. fluorination experiment. For unit stoichiometry the area ratio of the C_{1s} to O_{1s} peaks for the fluorinated film is given by

$$\frac{A_{C_{1s}}}{A_{O_{1s}}} = \frac{A_{\alpha C_{1s}}}{A_{\alpha O_{1s}}} \frac{(1-e^{-d/\Lambda_2})}{(1-e^{-d/\Lambda_4})} \tag{13}$$

The infinity ratio $A_{\alpha C_{1s}}/A_{\alpha O_{1s}}$ has been established as 0.5 and a value for Λ_4 of 8 Å was taken from the generalized curve of escape depth as a function of kinetic energy (Fig. 17). The sensitivity factor $A_{C_{1s}}/A_{O_{1s}}$ for a film thickness 4.5 Å is then found from Equation (13) to be 0.42 for unit stoichiometry. From the deconvoluted C_{1s} and O_{1s} spectra of the sample fluorinated for 0.5 sec. the area ratio of C_{1s} peaks arising from fluoropolymer to that of the O_{1s} from carbonyl function was measured as 26. Multiplication by the derived sensitivity factor then gives a stoichiometry of C/O of 11.0 for the surface film viz. a surface coverage of ∿ 9%.

If we consider now the original unfluorinated film, (of thickness large compared with d), the area ratios for unit stoichiometry corresponding to the C_{1s} and O_{1s} levels are given by Equation (14) where d corresponds to the film

$$\frac{A_{C_{1s}}}{A_{O_{1s}}} = \frac{A_{\alpha C_{1s}}}{A_{\alpha O_{1s}}(1-e^{-d/\Lambda_4})} \tag{14}$$

thickness at the surface within which the oxidation is confined. This is readily evaluated giving an area ratio for unit stoichiometry of 1.16. The measured area ratio is 52.5 which gives an overall stoichiometry of 61 for C/O. There is no evidence for oxidation accompanying fluorination so that the stoichiometry of the first monolayer should correspond to that established for the 0.5 sec. run viz. 11.0. To give the required stoichiometry therefore for the unfluorinated sample the C_{1s} signal would have to arise from 5-6 monolayers. If the average depth of a monolayer, viz. average separation between adjacent chains (since the surface structure should largely have the chains oriented parallel to the surface), is taken as ∿ 5 Å > 94% of the signal intensity derives from the first 27.7 Å.

We may now turn to the discrepancy between the stoichiometries for the 0.5 sec. fluorination experiment computed by the two methods discussed previously. The overall stoichiometry of C to O for the fluorinated film of 11.0 allows a correction to be computed for the C to F stoichiometry from the C_{1s} sepctra since a percentage of the peak at 287.9 eV assigned to CHF type environments arises from >C=O structural features. With this taken into account the refined C to F stoichiometry is 2.05 in excellent agreement with the figure derived from the C_{1s}/F_{1s} area ratios. A similar analysis for the experimental data leads to the revised figures shown in column III of Table 20. The two methods are seen to be in excellent agreement.

This analysis has been described in some detail up to this point and the reader who has persevered will now appreciate how the details of structure can be built up by careful theoretical analysis of the spectra. Further detailed description of the remainder of this work is presented elsewhere[57] and the remaining conclusions are summarized below.

When the depth of fluorination (using the average values from Table 19) is plotted versus log t, it is evident that the points for the first four experiments in the range 0.5-30 secs. fit on a smooth curve whilst that for the longest run of 300 sec. would appear to indicate too low a fluorinated film thickness than the initial analysis. The most likely cause of this would be a submonolayer contaminant hydrocarbon film arising after the fluorination experiment. Such hydrocarbon contaminant would have the same binding energy (285 eV) as polyethylene and unless allowed for would lead to a low estimate for the fluorinated film thickness. Clearly the longer the reaction and flushing time the greater the chances of contaminating the surface in this way, and further this effect will cause increasing errors with increasing fluorinated film thickness. Blank experiments showed that hydrocarbon surface contaminant does accumulate on fluorocarbon surfaces under our reaction conditions and a detailed analysis of the results leads to the conclusion that for the 300 secs. experiment this contamination amounts to ca. 10% monolayer coverage of the surface with hydrocarbon. Allowing for this in a recalculation of the depth of fluorination gives a value of 46 Å (footnote Table 19) which brings the point for the 300 sec. experiment onto the smooth curve for d versus log t; further the surface stoichiometry for this point can also be recalculated giving the revised value shown in Table 20, column III.

It is to be expected that as the fluorination proceeds into the sample the initially lightly fluorinated surface monolayer will become progressively more fluorinated. Information concerning this aspect of the process can also be derived from a detailed analysis of the data. Figure 66 shows the derived first monolayer stoichiometry as a function of time. This analysis indicates that although for the 300 sec. experiment the overall stoi-

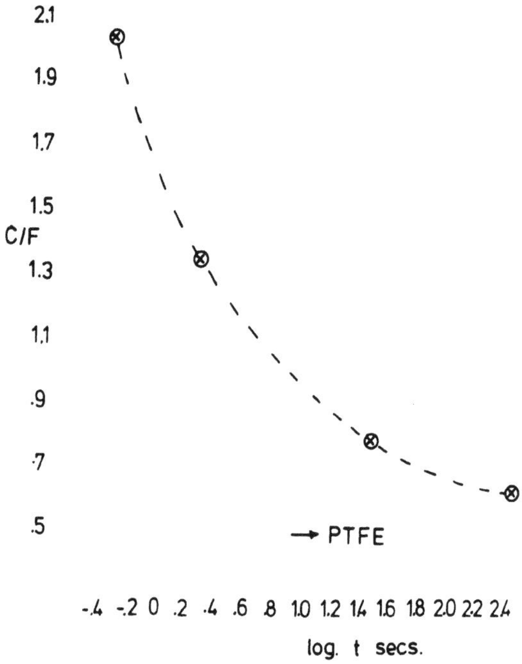

Fig. 66. First monolayer stoichiometry as a function of time for fluorinated polyethylene.

chiometry for the fluorinated layer is 1.16 (Table 20), corresponding to a composition intermediate between $(C_2FH_3)_n$ and $(C_2F_2H_2)_n$, at this stage the surface monolayer has reached a stoichiometry intermediate between that of $(C_2F_3H)_n$ and $(C_2F_4)_n$, which is an entirely reasonable conclusion.

As a final point some comment on the dynamics of the fluorination process can be made. Although the abstraction reaction which initiates the fluorination process has a low activation energy[58] successive replacement of hydrogen by fluorine will be expected to increase the activation energy both from the concomitant strengthening of C-H bonds and adverse polar effects in the transition state. Of some relevance in this connection are the relative rates of fluorination at different sites reported for gas phase studies of 1-fluorobutane.[59] These are shown in Fig. 67. It is clear that introduction of fluorine deactivates both α and β sites. We may use this data as a basis for discussing the fluorination of polyethylene.

Relative rates of fluorination F_2/N_2 at 20°C

$$F - CH_2 - CH_2 - CH_2 - CH_3$$
$$ 0.3 \quad\ \ 0.8 \quad\ \ 1 \quad\ \ 1$$

Fig. 67. Relative rates of abstraction of hydrogen by fluorine for 1-fluorobutane.

For the initial surface fluorination corresponding to the 0.5 sec. experiment, an overall C/F stoichiometry of the surface film of ∿ 2 is observed. Fig. 68 shows the calculated relative rates of fluorination for a four carbon fragment of the polymer chain based on the data for 1-fluorobutane. For a stoichiometry C/F of exactly 2 the possibilities on this simple model would be one CF_2 group or two $\underline{C}HF$ groups in the four carbon chain. It is clear that the preference for the latter should be ∿ 3 x that for the former. The data in Table 22 provides striking confirmation of this.

The two major peaks of approximately equal intensity at 287.9 eV and 285.8 eV corresponds extremely well in binding energy with those previously reported for polyvinyl fluoride (288.0 eV, 289.9 eV). By contrast the peak at 290.6 eV corresponding in binding energy to $\underline{C}F_2$ groups, (cf. 290.8 eV in polyvinylidene fluoride) with adjacent $-CH_2-$ groups represents approximately 0.3 x the intensity of these peaks. It might also be inferred from Fig. 67 that progressive fluorination would tend to proceed relatively uniformly so that for a stoichiometry C/F of 1.0 a $(CHF-CHF)_n$ backbone would predominate. The relatively large increases in intensity of the peaks at ∿ 289.0 eV (Table 22) would tend to support this. Clearly however, a great deal more experimental work is required before a more detailed disccusion is possible.

4.12 Charge Distribution in Polymers

Earlier (section 4.7) a brief outline was given of the application of the charge potential model to the theoretical interpretation of ESCA data for polymers, by computing electron densities for model systems. If values of the parameters K and $E^°$ (Fig. 34) are established for all the relevant core levels of a

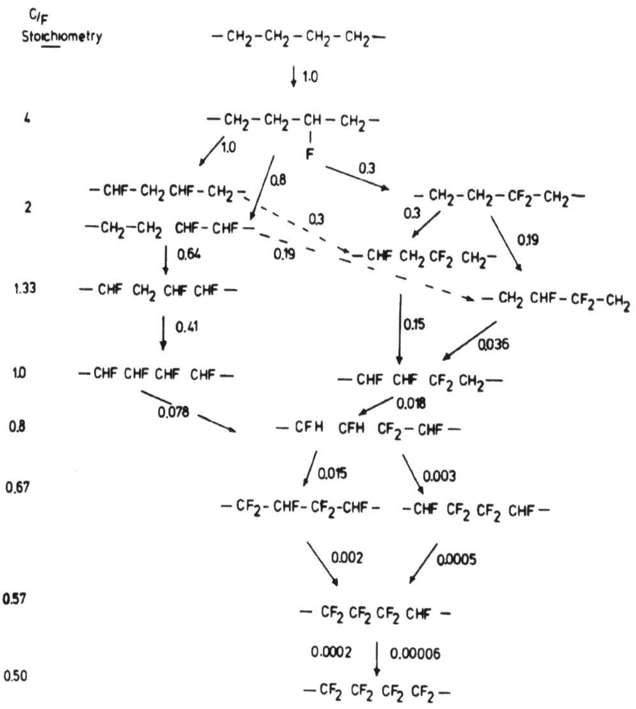

Fig. 68. Schematic for the progress and relative rates of fluorination in a four carbon fragment of a polyethylene chain.

Table 22. Analysis of the component peaks of the C_{1s} spectra for the fluorinated films

C_{1s} levels for fluorinated films of polyethylene (normalized contributions)

Sample	1		2		3		4		5		6	
	BE	%	BE	%	BE	%	BE	%	BE	%	BE	%
0·5 sec.	-	-	290·6	9·6	288·9	14·9	287·9	25·5	286·5	21·3	285·8	28·7
2 sec.	-	-	290·9	12·2	289·0	21·1	287·9	24·4	286·5	14·4	285·9	27·8
30 sec.	291·6	3·3	290·7	17·4	289·0	21·7	287·9	25·0	286·4	12·0	285·9	20·7
300 sec.	291·9	6·5	290·8	17·7	289·1	27·4	288·0	19·4	286·5	18·3	285·9	10·8

system it is possible to invert the charge potential model to obtain experimental charge distributions. This procedure has obvious application in the calculation of charge distributions in molecules that are of such a size that conventional molecular orbital calculations are impracticable. A crude idea of the charge distribution is very often useful as a rule of thumb in discussing the chemistry of complex systems and also for preliminary assignment of other spectroscopic data. The simplicity of the procedure is illustrated in the particular case of CCl_3CF_3 in Figs. 69 and 70. From studies of closely related systems, value of K and E^o may be established for each core level. Knowing the geometry and the measured core binding energies the charge distributions may readily be obtained by solution of the system of linear simultaneous equations, Fig. 70. In fact there is one more equation than unknowns because of the condition of overall electroneutrality. This can in fact be used to advantage to study sample charging.[60]

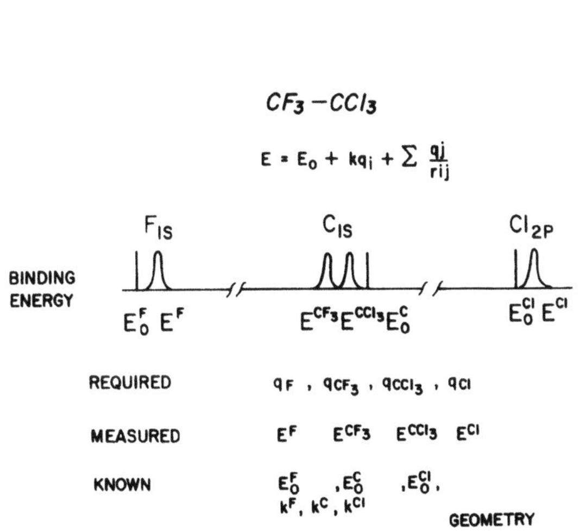

Fig. 69. Schematic illustrating the route to experimental charge distributions from inversion of the charge potential model.

As a practical example of this procedure, Fig. 71 shows the experimental charge distribution in a complex fluorocarbon system which we have recently studied. This typically represents the practicable upper limit for SCF MO calculations since even with a large computer (IBM 360/67) the cpu time required is ∿ 20 mins. By contrast the ESCA experiment and the trivial computer operations necessary to solve the equations can be accomplished in about the same time. For more complex systems therefore the de-

$$\underline{F}_{1s} \quad E^F = E^F + k^F q_F + \frac{2q_F}{r_{F-F}} + \frac{3q_{Cl}}{r_{C-F}} + \frac{q_{CF_3}}{r_{F-CF_3}} + \frac{q_{CCl_3}}{r_{F-CCl_3}}$$

$$C_{1s} \quad E^C_{CF_3} = E_0^C + kq_{CF_3} + \frac{3q_F}{r_{C-F}} + \frac{3q_{Cl}}{r_{C-Cl}} + \frac{q_{CCl_3}}{r_{C-C}}$$

$$\underline{Cl}_{2p} \quad E^{Cl} = E_0^{Cl} + kq_{Cl} + \frac{3q_F}{r_{Cl-F}} + \frac{2q_{Cl}}{r_{Cl-Cl}} + \frac{q_{CF_3}}{r_{Cl-CF_3}} + \frac{q_{CCl_3}}{r_{Cl-CCl_3}}$$

$$\text{ALSO} \quad 3q_F + 3q_{Cl} + q_{CF_3} + q_{CCl_3} = 0$$

Fig. 70. The set of linear simultaneous equations to be solved to obtain experimental charge distributions.

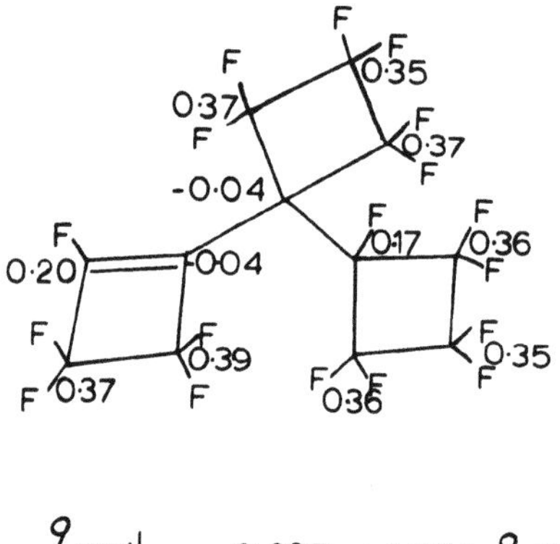

$$q_{expt'l} = 0{\cdot}007 + 1{\cdot}004\, q_{CNDO}$$

Fig. 71. Comparison of experimental and theoretical charge distributions for a complex fluorocarbon.

termination of charge distributions from ESCA data is a quicker and more economical possibility than direct calculation.

For polymer systems this approach is particularly valuable, and to illustrate the great potential in this area, Fig. 72 shows experimental charge distributions determinded for polyvinylene fluoride and polytrifluoroethylene. These show rather effectively the large net migration of electron densities from carbon and hydrogen to fluorine.

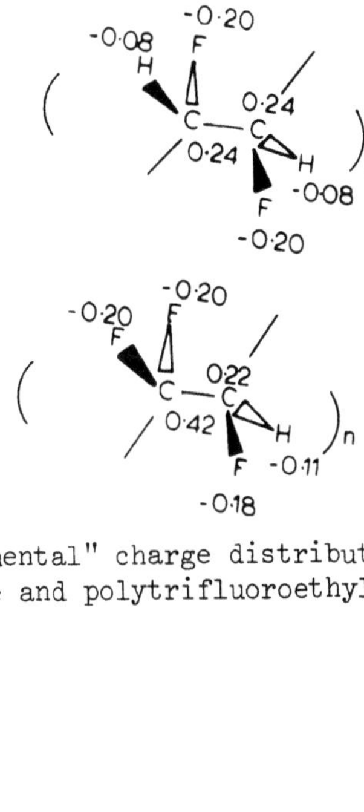

Fig. 72. "Experimental" charge distributions for polyvinylene fluoride and polytrifluoroethylene.

4.13 The Valence Bands of Polymers

In the previous sections emphasis has been placed on the study of the core levels of polymer systems. Information concerning structure and bonding has then been largely inferred from shifts in core binding energies which reflect differences in valence electron distributions. Of obvious interest is the direct investigation of the valence levels of polymers; the derived information being relevant to the detailed interpretation in particular of the electrical properties of samples.

The valence energy levels of simple molecules have been extensively studied in the gas phase by low energy photoelectron spectroscopy. The inherent widths of the exciting radiations which are most commonly used, He (I) and He (II) (photon energies ∿ 21 eV and ∿ 40 eV respectively) are such that in favourable cases vibrational progressions may be resolved which considerably aids assignment. Although the development of ultraviolet photoelectron spectroscopy, (UPS) as the technique has come to be known, has primarily been in the hands of chemists,[61] the application to the study of the valence bands of solids has primarily been the province of the physicist[62] (the technique often being referred to as u.v. photoemission) and has dealt mainly with metals studied as evaporated films under UHV conditions. It is only within the past few years therefore, that interests and differences in instrumentation have converged and the UPS study of the valence bands of polymers is still in its infancy.

In the case of simple molecules the study of the valence energy levels by ESCA has two distinct disadvantages compared with the corresponding UPS measurements. (It should be emphasized however, that comparison between the two is very valuable since differential changes in cross sections with photon energy are useful for assigning the symmetries of occupied orbitals). Firstly, cross sections for photoionization are generally lower than for the longer wavelength photon sources used in UPS, and secondly the resolution is much poorer, (viz. photon linewidths He(I) ∿ 5 meV, $MgK\alpha_{1,2}$ ∿ 800 meV).

In studying involatile materials such as polymers however, these disadvantages are considerably offset. Thus since there are so many vibrational modes possible, resolution becomes less of a problem since even with a He (I) source only broad unresolved bands would be obtained. Inspection of the escape depth versus kinetic energy curve discussed earlier reveals three major problems with using a low energy photon source which do not arise in ESCA examination of valence energy levels. Firstly, with a low energy photon source not all of the valence energy levels may be studied, only the higher occupied levels. Secondly, since the kinetic energy range for electrons will typically be in the range 0-21 eV (HeI), 0-40 eV (HeII) it is clear that this is a region of rapidly varying escape depth (cf. Fig. 17). Surface contamination is therefore very critical, much less so for an X-ray source. Thirdly, in the absence of contamination there are still difficulties of interpreting the data because of marked differences in escape depths which do not arise in ESCA since the escape depth dependence is virtually constant across the valence band. For these reasons it is very convenient to study the valence bands of polymers by ESCA.

A considerable part of this summer school is devoted to the theoretical interpretation of the valence bands of molecular solids and crystals and hence a detailed discussion will not be given in these lecture notes. It is interesting however, to highlight the theoretical (as opposed to experimental) difficulties inherent in interpretations within the Hartree-Fock formalism. The energy considerations in Fig. 31 have previously been commented on with regard to core ionizations. Since valence energy levels can vary quite drastically in their localization characteristics (viz. lone pairs, delocalized π orbitals, localized σ bonding orbitals etc...) differences in both relaxation energies and correlation energies can be appreciable in going from state to state. It is undoubtedly true therefore, that a rather higher level of theoretical sophistication is required to quantitatively describe valence as opposed to core levels. Further consideration of this point will be given in the lectures.

ESCA measurements of valence energy levels could in principle, involve four sets of experiments. Firstly, at a fixed angle with respect to photon source and analyzer at a given photon wavelength. Secondly, angular distribution studies at a given photon energy. Thirdly, as a function of differing photon energies and finally spin polarization measurements may be envisaged using synchrotron radiation for example. To date only the first type of measurement has been made typically with unmonochromatized Al or $MgK\alpha_{1,2}$ or monochromatized $AlK\alpha_{1,2}$ phton sources. In studying solids the analysis of data corresponding to the second, third and fourth type of experiment will be by no means simple. In gross terms, however, the comparison of band profiles for widely differing photon energies (e.g. $MgK\alpha_{1,2}$ and He (I) will be qualitatively very useful in distinguishing essentially pi type from sigma type orbitals. In this connection the theory developed by Siegbahn and Gelius[63] for analyzing differential changes with photon energy based on atomic cross sections weighted by an appropriate population analysis will be particularly valuable and will be considered in the lectures.

To illustrate the utility of ESCA for the study of valence bands of polymers, brief consideration will be given here to some simple homopolymers and copolymers and fluorographites. Fig. 73 shows the measured valence energy levels for PTFE, polyvinylidene fluoride and polyethylene and nicely reveal the ordering of the levels. The large peak at highest binding energy clearly evident in the fluorinated polymers, arises from molecular orbitals essentially F_{2s} in character whilst the prominent peak at lowest binding

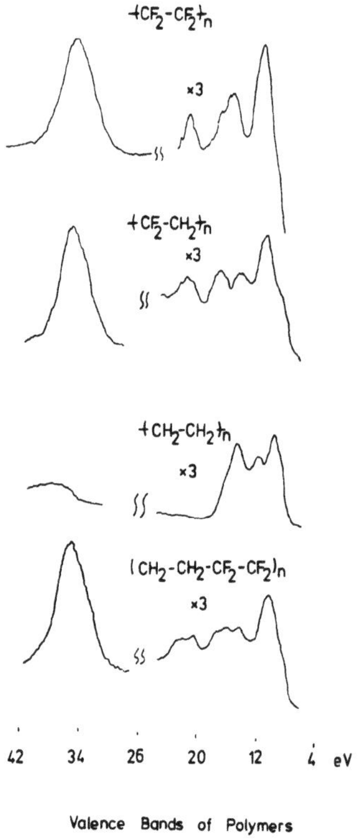

Fig. 73. Valence encrgy levels for some simply polymer systems.

energy for PTFE which is also clearly evident in polyvinylidene fluoride, is assigned to M.O.'s essentially corresponding to fluorine 2p lone pairs. The shoulder at lower binding energy in polyvinylidene fluoride which has its counterpart in polyethylene may then be assigned to carbon hydrogen bonding orbitals (essentially C_{2p} H_{1s}). The assignments for the remaining peaks are then assentially C-F and C-C (2s) bonding orbitals. With an appropriate correction for work function (∿ 5 eV) the binding energies correspond quite nicely to those obtained from UPS studies of simple systems containing the essential structural features. For comparison purposes the valence band for the ethylene-tetrafluoroethylene of composition close to 50-50 is also included. This bears a striking resemblance to that for polyvinylidene fluoride as one might expect on the basis of its largely alternating structure discussed earlier. The particular case of polyethylene has been the subject of recent detailed theoretical and experimental analysis.[64]

Fig. 74 shows data pertaining to some fluorographites and illustrates the wealth of detail concerning the changes in bonding

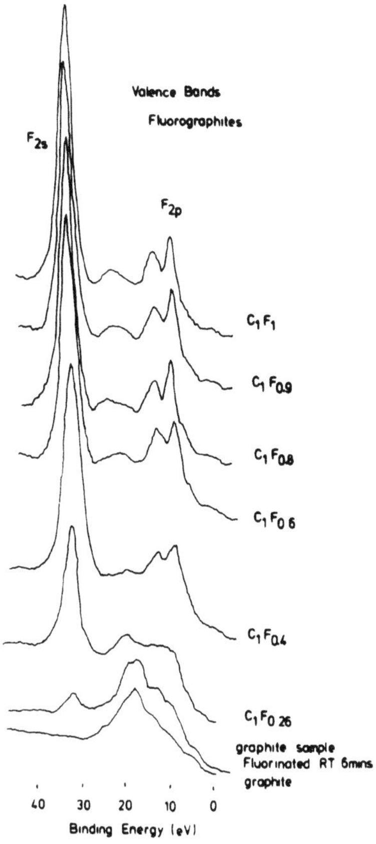

Fig. 74. Valence energy levels for some fluorographites.

patterns in proceeding from graphite to graphite monofluoride. The development of the essentially lone pair F_{2s} and F_{2p} levels is clearly evident.

CONCLUSIONS

It should be evident from this short review that ESCA has a bright future in many areas of polymer chimistry and physics. Since many of the important chemical, physical, mechanical and electrical properties of polymeric systems depend on the electronic structure of the surface and immediate subsurface, the development of a technique capable of providing data relevant to this region is of some considerable importance. The work published to date already points to the important role that ESCA has to play in such widely diverse areas as,polymer degradation (thermal, photochemical, oxidative) and the study of friction, wear and ageing processes. Surface modification of polymers by chemical or physical means is also ideally suited to ESCA examination.

On particular feature of ESCA which should clearly have become evident during the previous discussion is that compared with more conventional techniques (e.g. TIR, MATR, Inelastic Neutron Scattering and NMR) the information provided is at a much coarser level.[65] This in many senses is fortunate since often the information provided by the other techniques is complimentary in nature to that provided by ESCA.[6a] However, in unravelling the complexities of some of the processes previously mentioned (viz., degradation, friction, wear, ageing etc..) the "broad strokes of the impressionist are often much more revealing (at least in the initial stages of development) than the fine detail of the representationalist". In this connection a parallel may be drawn with the early development of gas phase kinetics.[66]

In studying the electrical properties of polymers the evidence to date (both published and unpublished) points to the important role that ESCA can play in studying valence band structure, transport properties and triboelectric phenomena.

We have outlined above the great potential of ESCA for routine studies of structure and bonding with the emphasis being placed on fluoropolymer systems. One reason for the rapid development of the technique in this area as we have previously indicated is that replacement of hydrogen by fluorine in organic systems gives rise to large shifts in C_{1s} core levels, and hence provides a particularly favourable situation for resolving components of complex spectra with limited instrumental resolution. The well characterized line shapes etc.. to some extent alleviate the inherent difficulties of having (by comparison with other spectroscopic techniques) a much less favourable linewidth to shift ratio since line shape analysis becomes a feasible and worthwhile proposition for relatively simple systems. In general the linewidths for C_{1s} levels discussed in the previous sections have largely been determined by the inherent width of the unmonochromatized X-ray source (Full Width at Half Maximum (FWHM) $MgK\alpha_{1,2}$ 0.7 eV, $AlK\alpha_{1,2}$ 0.9 eV). There are indications that the inherent width of C_{1s} core levels may well be $\sim$ 0.1 eV, for gas phase systems and $\sim$ 0.4 eV for the condensed phase. As one example of the improvements in both resolution and signal to background on going from an unmonochromatized $MgK\alpha_{1,2}$ photon source to a slit filtered monochromatized $AlK\alpha_{1,2}$ source, Fig. 75 shows the C_{1s} levels of graphite.+ With an X-ray band pass of 0.4 eV the

+ Obtained using a modified A.E.I. ES 200B instrument.

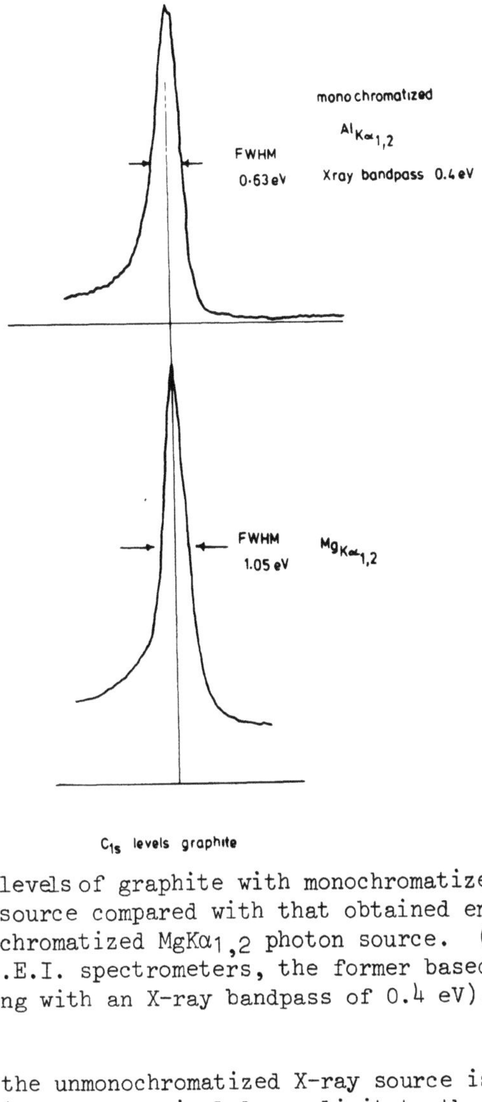

Fig. 75. C_{1s} levels of graphite with monochromatized $AlK\alpha_{1,2}$ photon source compared with that obtained employing an unmonochromatized $MgK\alpha_{1,2}$ photon source. (Both measured on A.E.I. spectrometers, the former based on slit filtering with an X-ray bandpass of 0.4 eV).

linewidth for the unmonochromatized X-ray source is reduced by ∿ 0.4 eV and with a theoretical lower limit to the band pass of 0.2 eV the composite linewidth could undoubtedly be reduced to ∿ 0.45 eV (viz. ∿ 70% less than for the unmonochromatized X-ray sources).

The development of well designed X-ray monochromators[1b] based on conventional X-ray sources (which has already been commercially exploited by at least two instrument manufacturers) allows the possibility of approaching these limits in terms of resolution. The scope of the technique will be enormously increased by such developments. A further advantage accuring from X-ray monochromatization is a dramatic improvement in sensitivity consequent upon removing X-ray satellites ($K\alpha_{3,4}$ etc..) and the bremstrahlung.

The development of multiple collector assemblies as opposed to single channeltron detectors will also allow the ESCA timescale (in terms of the time taken to actually record a spectrum) to be improved by probably two orders of magnitude. ESCA will then be a powerful tool for dynamic monitoring experiments.

It has already been clearly demonstrated in other areas of application of ESCA that considerable additional information can be obtained from angular distribution studies of both core and valence levels.[2a,8] Thus with a close to grazing angle of incidence for the photon beam, surface features may be specifically enhanced[67] whilst in the study of valence levels the angular dependence of photoelectric cross sections can provide valuable information on the symmetries of valence levels.[68]

Exciting possibilities are in prospect with the harnessing of synchrotron radiation for ESCA purposes.[10] With a continuously variable photon energy in the range 0-2.5 KeV and monochromators capable of achieving an adequate photon flux with a band pass in the region of 0.1 eV a whole new horizon is in view for the technique. In the immediate future, definitive experiments on escape depths as a function of kinetic energy will be feasible, as will detailed studies of valence bands of solids for which differential changes in cross section with photon energy will considerably aid in assignments. Analytical depth profiling will also greatly benefit from these developments.

ACKNOWLEDGEMENTS

In putting together these lecture notes based largely on my own research programs, I would like to express my sincere thanks to my colleagues Professor W.K.R. Musgrave and Dr. W.J. Feast, for their help and encouragement. However, none of the work would have been possible without the enthusiasm, skill and perseverance of my research workers, D. Kilcast, D.B. Adams, I. Scanlan, I. Ritchie and J. Peeling. My sincere thanks are also due to Mrs E. McGauley for the arduous task of transcribing my (beautifully) handwritten notes into typescript.

The greatest debt however, must be to Professor Kai Siegbahn and his coworkers at Uppsala for providing the scientific community with such a versatile and exciting technique.

REFERENCES

1. (a) B. Wannberg, U. Gelius and K. Siegbahn, J. Phys. E, 7, 149 (1974).

   (b) U. Gelius, H. Fellner-Feldegg, B. Wannberg, A.G. Nilsson, E. Basilier and K. Siegbahn, Univeristy of Uppsala, Institute of Physics Report UUIP N° 855 (1974).

2. (a) C.S. Fadley, "Theoretical Aspects of X-ray Photoelectron Spectroscopy" in Electron Emission Spectroscopy, Ed. W. Dekeyser, D. Reidel Publishing Co., Dordrecht, Holland, pp. 151-224 (1973).

   (b) U. Gelius, Physica Scripta, 9, 133 (1974).

3. K. Siegbahn et al., Nova Scta R. Soc. Sci., Uppsala Ser. IV, 20 (1967).

4. K. Siegbahn et al., "ESCA Applied to Free Molecules", North Holland Publishing Co., Amsterdam (1969).

5. D.M. Hercules, Analytical Chemistry, 44, 106R (1972).

6. (a) D.T. Clark, "Chemical Aspects of ESCA" in Electron Emission Spectroscopy, Ed. W. Dekeyser, D. Reidel Publishing Co., Dordrecht, Holland, pp. 373-507 (1973) and references therein.

   (b) D.T. Clark, plenary lecture "International Symposium on Polymer Friction and Wear", Ed. L.H. Lee, Plenum Press, to be published Los Angeles (March 1974).

7. Cf. H. Fellner-Feldegg, U. Gelius, K. Siegbahn, C. Nordling and K. Thimm, UUIP Report N° 856, (March 1974).

8. Cf. R.M. White, T.A. Carlson and D.P. Spears, J. Electron Spectroscopy, 3, 59 (1974) and references therein.

9. D.A. Seanor, "Tribo Electrification of Polymers".

10. U. Gelius, S. Svensson, H. Siegbahn, E. Basilier, A. Faxalv and K. Siegbahn, UUIP Report N° 860,(March 1974).

11. J.C. Tracey, "Auger Electron Spectroscopy for Surface Analysis" in Electron Emission Spectroscopy, Ed. W. Dekeyser, D. Reidel Publishing Co., Dordrecht, Holland, pp. 295-372 (1973).

12 (a) D.T. Clark D. Kilcast and W.K.R. Musgrave, J. Chem. Soc. Chem. Comm., 516 (1971).

(b) D.T. Clark, D.B. Adams, I.W. Scanlan and I.S. Woolsey, Chem. Phys. Letters, 25, 263 (1974).

(c) D.T. Clark, Annual Reports, B, The Chemical Society, London, p. 40 (1972), and references therein.

13. Cf. Ref. 4 and D.T. Clark, D.B. Adams and D. Kilcast, J. Chem. Soc. Dis. Faraday Soc. 54, 182 (1972).

14. D.T. Clark and D. Kilcast, Nature Phys. Sci., 233, 77 (1972).

15. D.T. Clark, D. Kilcast, W.J. Feast and W.K.R. Musgrave, J. Pol. Sci., Polymer Chemistry Edn., 11, 389 (1973).

16. D.T. Clark, D. Kilcast, W.J. Feast and W.K.R. Musgrave, J. Pol. Sci., A1, 10, 1637 (1972).

17. D.T. Clark, W.J. Feast, I. Ritchie, W.K.R. Musgrave, M. Modena and M. Ragazzini, J. Pol. Sci., Polymer Chemistry Edn., in press (1974).

18. E. Clementi and H. Popkie, J. Amer. Chem. Soc., 94, 4057 (1972).

19. D.T. Clark, I.W. Scanlan and J. Muller, Theoretica Chim. Acta, (1974) in press.

20. C.W. Wilson, III, and E.R. Santee, Jr., in Analysis and Fractionation of Polymers (J. Polym. Sci. C., 8), J. Mitchel, Jr and F.W. Billmeyer, Jr., Eds., Interscience, New York, p. 97 (1965).

21. H. Kawai, T. Soen, T. Inove, T. Ono and T. Uchide, Memoirs of Faculty of Engineering, Kioto University XXXIII, Pt. 4 (1971).

22. D.T. Clark, J. Peeling and J.M. O'Malley (in preparation).

23. D.A. Barr, R.N. Haszeldine and C.J. Willis, J. Chem. Soc., 1351 (1961) and R.E. Banks, R.N. Haszeldine, H. Sutcliffe and C.J. Willis, J. Chem. Soc. 2506 (1965).

24. R.D. Chambers, D.T. Clark, D. Kilcast and S. Partington, J. Polym. Sci., Polymer Chem. Ed., to be published (1974).

25. D. T. Clark and D.B. Adams (in preparation).

26. E.T. McBee, U.S. Patent, 3, 317497, May 2 (1967).

27. W.L. Dilling, Thesis, Purdue University (1962); Diss. Abstr. B27 (4), 1083 (1966).

28. D.T. Clark, W.J. Feast, M. Foster and D. Kilcast, Nature Physical Science, 236, 107 (1972).

29. W.J. Feast and M. Foster, unpublished; reported in part at the 3rd International Symposium on Polyhalogen Compounds, Barcelona, October (1973).

30. L.D. Hulett and T.A. Carlson, Clinical Chemistry, 16, 677 (1970).

31. M.P. Klein and L.N. Kramer, Impr. Plant. Protein Nucl. Techn., Proc. Symp. 242-52 (1970) IAEA : Vienna, Austria.

32. (a) R. Chujo, K. Sato and A. Nishioka, Polymer Journal, 3, 242 (1972).

(b) G. Jung M. Ottnad, W. Bohnenkamp, W. Bremser and U. Weser, Biochim. Biophys. Acta, 295, 77 (1973); and idem., FEBS Letters, 25, 346 (1972).

33. J. Baddiley, I.C. Hancock and P.M.A. Sherwood, Nature Physical Sciences, 243, 43 (1973).

34. A. Sawatari and T. Takashima, Kami Pa Gikyoshi, 27, 8, (1973); Chem. Abs., 78, 73814p (1973).

35. M.M. Millard, Analytical Chemistry, 44, 828 (1972).

36. M.M. Millard, K.S. Lee and A.E. Pavlath, Textile Research Journal, 42, 307 (1972).

37. M.M. Millard and A.E. Pavlath, ibid., 42, 460 (1972).

38. H. Brecht, F. Mayer and H. Binder, Angew. Makromol. Chem., 33, 89 (1973).

39. W.M. Riggs and R.P. Fedchenko, American Laboratory, 65 (1972).

40. R.H. Hansen and H. Schonhorn, Polymer Letters, 4, 203 (1966).

41. O. Ruff, O. Bretschneider and F. Elert, Z. anorg. allgem. Chem., 217, 1 (1934).

42. Cf. (a) J.L. Wood, R.B. Badachhape, R.J. Lagow and J.L. Margrave, J. Phys. Chem., 73, 3139 (1969).

(b) R.J. Lagow, R.B. Badachhape, P. Ficalora, J.L. Wood and J. L. Margrave, Synthesis in Inorganic and Metal Organic Chemistry, 2 (2), 145 (1972).

43. N. Watanabe, Y. Koyama, A. Shibuya and K. Kuman, Memoirs of Faculty of Engineering, Kyoto University, 33, 15 (1971).

44. R.L. Fusaro and H.E. Sliney, Chem. Abs., 71, 5115s.

45. C.F. Wells, "Structural Inorganic Chemistry", 3rd Edn., Oxford (1964).

46. N. Watanabe and M. Takashima, Abstract p. 19, 7th International Symposium on Fluorine Chemistry, Santa Cruz, California (1973).

47. A.J. Rudge, British Patent, 710, 523 (June 1959).

48. J.L. Margrave and R.J. Lagow, Chem. Eng. News, 48, 40 (1970).

49. K. Tanner, Chimica, 22, 176 (1970).

50. H. Schonhorn, P.K. Gallagher, J.P. Luongo and F.J. Padden Jnr. Macromolecules, 3, 800 (1970).

51. H. Shinohara, M. Twasaki, S. Tsujimwa, K. Watanabe and S. Okazaki, J. Pol. Sci. A1, 10, 2129 (1972).

52. D.M.J. Lilley, Ph. D. Thesis, University of Durham (1973).

53. B. Wunderlich, "Macromolecular Physics", Vol. 1, Academic Press, New York, (1973).

54. H. Schonhorn, Polymer Letters, 467 (1964).

55. idem., Macromolecules, 1, 145 (1968).

56. D.T. Clark and K.C. Tripathi, Nature Physical Sciences, 241, 162 (1973) and idem., ibid., 244, 77 (1973).

57. D.T. Clark, W.J. Feast, W.K.R. Musgrave and I. Ritchie, to be published.

58. R.E. Florin and L.A. Wall, J. Chem. Phys. 57, 4, 1791 (1972).

59. J.M. Tedder, Quart. Rev. Chem. Soc., XIV, 336 (1960).

60. D.T. Clark, D.B. Adams and D. Kilcast, Chem. Phys. Letters, 13, 439 (1972).

61. D.W. Turner, C. Baker, A.D. Baker and C.R. Brundle, "Molecular Photoelectron Spectroscopy", Wiley, New York (1970).

62. W.E. Spicer, "Optical Density of States Ultraviolet Photoelectric Spectroscopy", Proceedings - Electron Density of States Ed. T.H. Bennett, National Bureau of Standards Special Publication, 323, 139-158 (1971). U.S. Government Printing Office, Washington D.C.

63. K. Siegbahn and U. Gelius, J. Chem. Soc. Faraday Disc., 54 (1972).

64. J. Delhalle, J.M. André, S. Delhalle, J.J. Pireaux, R. Caudano and J.J. Verbist, J. Chem. Phys., 60, 595 (1974) and references therein.

65. Cf. ref. 6(a).

66. F.S. Dainton, J. Chem. Soc. (Chem. in Britain), 306 (1973).

67. C.S. Fadley, J. Electron Spectroscopy, 4, (1974) in press.

68. T.A. Carlson, J. Chem. Soc. Faraday Disc., 54 (1972).

ELECTRONIC CORRELATION IN POLYMERS AND MOLECULAR CRYSTALS

Jean-Louis Calais

Quantum Chemistry Group

University of Uppsala

Uppsala, Sweden

WARNING

These notes represent an outline of the lectures to be given during the course at Namur. They are not supposed to be either self-contained or exhaustive, but have to be complemented with the lectures. Their only purpose is to map out the general line of thought to be followed during the lectures and to give some key references from which most other references appropriate to the subject can be reached.

GENERAL ABOUT CORRELATION

Since the very first attempts to apply quantum mechanics to chemical and physical problems it has been recognized that one of the fundamental difficulties is what is usually called the correlation problem. Generally speaking this term is associated with the difficulties involved in going beyond the independent particle model, which is older than wave mechanics itself. The word "correlation" refers to the fact, that in the independent particle model the fact that the motion of the electrons is correlated is only taken care of in an average way, whereas the more detailed correlation between the electrons is treated in calculations which go beyond the independent particle model.

In principle the electronic correlation should be taken care of in all electronic systems : atoms, molecules of various size, crystals of various types, disordered systems. We are naturally very far from such a goal and fortunately there are quite a few number of cases which to a first approximation can be reasonably handled within the independent particle model. There is an enormous literature on many different aspects of the correlation problem. Löwdin[1] wrote a review in 1959 which has become a bit of a classic. In that article he introduced a formal definition of the correlation energy, which has become generally accepted :

$$E_{corr} = E_{exact} - E_{HF} \tag{1}$$

or more specifically :

The correlation energy for a certain state with respect to a specific Hamiltonian is the difference between the exact eigenvalue of the (total) Hamiltonian and its expectation value in the Hartree-Fock approximation.

We notice that what is here called the "exact" eigenvalue refers to the non-relativistic Hamiltonian and should not be confused with what might be obtained from experiment. We also notice that the term "Hartree-Fock approximation" needs further specification.

Ten years after Löwdin's survey a summer school on correlation in atoms and molecules was held at Frascati, near Rome. The proceedings[2] from that meeting, which were published in another volume of Advances in Chemical Physics, show both what had happened during ten years along the lines discussed by Löwdin 1959 and what other methods had given. Despite its name this set[2] of papers contains a lot of material which is valuable also for a solid state physicist.

The correlation problem in solids is the main subject of the review by Hedin and Lundqvist[3]. In that paper they discuss extensively so-called many-body techniques, which have provided some of the more important tools for handling correlation in particular in large systems.

For further references we refer to the various volumes of "Advances of Quantum Chemistry" (Academic Press) and the Symposium Issues of the "International Journal of Quantum Chemistry" (Interscience). Correlation in solids was also one of the topics discussed at an IBM meeting[4] in Schwarzwald in October 1971.

Comparatively little theoretical work on the electronic structure of polymers seems to have been carried out. This is particularly true for treatments including correlation. - The problem of correlation in molecular crystals is intimately connected with the general problem of interatomic and intermolecular forces. - Both for polymers and molecular crystals it is probably fair to say, however, that in a certain sense, no systematic theory of correlation exists. In these lectures we are therefore going to make some "case studies", which will bring out some of the more important aspects of correlation in these systems, which have been noticed so far.

In the first section we will study correlation in linear systems, starting out from the simplest one, the hydrogen molecule. We will also take up a couple of calculations on polyenes and cumulenes. It goes without saying that the very few papers selected and discussed here of course do not cover the whole subject of correlation in linear systems. One-dimensional systems (which is not quite the same thing) formed up till recently the typical playground for the theoretician who was thought to be far from "reality", but this field has recently turned out to be a very hot one, also form the experimental point of view. In this connection the survey and reprint volume of Lieb and Mattis(5) should be mentioned, although we will unfortunately not have time here to discuss any of the very interesting problems taken up in that book.

Molecular crystals and intermolecular forces will be treated in the second section, which will lead up to a discussion of a calculation of the cohesive energy of the neon crystal. Further aspects of electron interaction in polymers and molecular crystals can be found in some papers(6-17) quoted in the list of references.

CORRELATION IN LINEAR SYSTEMS

The hydrogen molecule

In the simplest version of the independent particle model - the MO-LCAO method - both electrons in the hydrogen molecule occupy the same molecular orbital (MO), Ψ_1. The total wave function Ψ can be written as a product of a symmetric space function Φ and an antisymmetric spin function H

$$\Psi(X_1,X_2) = \Phi(\vec{r}_1,\vec{r}_2)\ H\ (\xi_1,\ \xi_2) \tag{2}$$

$$\Phi(\vec{r}_1,\vec{r}_2) = \Psi_1(\vec{r}_1)\ \Psi_1(\vec{r}_2) \tag{3a}$$

$$H(\xi_1,\xi_2) = \frac{1}{\sqrt{2}}\,[\alpha(\xi_1)\beta(\xi_2) - \beta(\xi_1)\alpha(\xi_2)] \tag{3b}$$

The MO is formed from a linear combination of the 1s atomic orbitals of the two hydrogen atoms :

$$\Psi_1 = \frac{1}{\sqrt{2\,(1+S)}}\,(a + b) \tag{4}$$

The normalization constant is determined by the overlap integral S

$$S = \langle a|b\rangle = \int a^*(\vec{r})\,b(\vec{r})\,dv \tag{5}$$

The MO-function (2) is a relatively good approximation for internuclear distances close to the equilibrium. It has however, an important drawback for large internuclear distances, which is clearly seen if we plot the total energy as a function of the internuclear distance R.

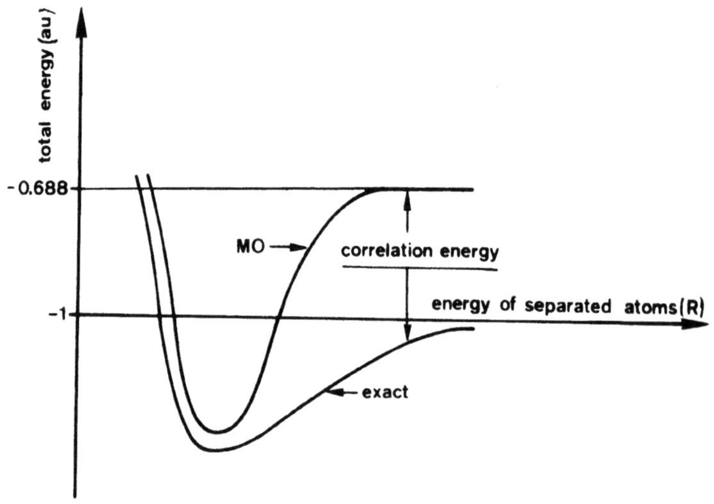

Figure 1. : Sketch of the exact and MO potential energy curves for the hydrogen molecule.

We can see directly on the wave function what is the reason for the strange behaviour of the MO curve for large R. Inserting (4) in (3a) we get

$$\Phi(1,2) = \Psi_1(1)\Psi_1(2) =$$

$$\frac{1}{2(1+S)}\left\{\underbrace{[a(1)b(2) + b(1)a(2)]}_{\text{covalent part}} + \underbrace{[a(1)a(2) + b(1)b(2)]}_{\text{ionic part}}\right\} \quad (6)$$

For large distances the weight of the ionic terms should go to zero whereas in the MO function the covalent and ionic terms have the same weight for all internuclear distances.

A simple way of correcting for this deficience is to introduce the so-called alternant molecular orbitals (AMO), which are semi-localized - not delocalized like the MO's. This is achieved by mixing the occupied MO (4) with the virtual one

$$\Psi_{\bar{1}} = \frac{1}{\sqrt{2\ (1-S)}}\ (a - b) \quad (7)$$

by means of an R-dependent parameter θ :

$$\begin{cases} A = \Psi_1 \cos\theta + \Psi_{\bar{1}} \sin\theta & (8a) \\ \bar{A} = \Psi_1 \cos\theta - \Psi_{\bar{1}} \sin\theta & (8b) \end{cases}$$

These orbitals are not orthogonal

$$<A|\bar{A}> = \int A^{*}(\vec{r})\bar{A}(\vec{r})dv = \cos 2\theta = \lambda \quad (9)$$

We form a new total space function Φ_{AMO} from the AMO's (8), and express this function both in terms of the MO's Ψ_1 and $\Psi_{\bar{1}}$, and of the AO's a and b.

$$\Phi_{AMO}(1,2) = \frac{1}{\sqrt{2\ (1 + \lambda^2)}}\left\{A(1)\bar{A}(2) + \bar{A}(1)A(2)\right\} =$$

$$\sqrt{\frac{2}{1 + \lambda^2}}\left\{\Psi_1(1)\ \Psi_1(2)\ \cos^2\theta - \Psi_{\bar{1}}(1)\ \Psi_{\bar{1}}(2)\ \sin^2\theta\right\} =$$

$$\frac{1}{\sqrt{2\ (1+\lambda^2)}} \cdot \frac{1}{1-S^2}\left\{\ (1-\lambda S)\ \left\{a(1)b(2) + b(1)a(2)\right\} + \right.$$

$$(\lambda - S)\left[a(1)a(2) + b(1)b(2)\right]\Bigg] \qquad (10)$$

The first line of (10) shows that we in a certain sense keep the one-electron picture in the AMO model. By combining the function Φ_{AMO} with the spin function (3b) we see that the AMO model forms a special case of the more general "different orbitals for different spins (DODS) model". The second line of (10) shows that the AMO function can be regarded as a superposition of the two configurations $\Psi_1(1)\ \Psi_1(2)$ and $\Psi_{\bar{1}}(1)\ \Psi_{\bar{1}}(2)$ - one of the standard methods for handling correlation. The last two lines of (10) show that the AMO function can be regarded as a combination of covalent and ionic contributions, with coefficients which depend on R, so that we can get the rigth combination for $R \rightarrow \infty$ by choosing $\lambda = S$. Since for large R we have $S = 0$, we should have $\theta = \pi/4$ for large R.

The fact that the function (10) can be written as a Heitler-London type combination of the two semi-localized orbitals was first recognized by Coulson and Fischer[18]. The AMO method - which can be applied to much more general systems that H_2 - was introduced by Löwdin ; a detailed account can be found in the book by Pauncz[19]. The H_2 case has been analysed in detail by Goscinski and Calais[20].

THE LINEAR CHAIN OF HYDROGEN ATOMS

The AMO method is not restricted to diatomic molecules. Its great merit is that it can be used also for large systems, for which a convential superposition of configurations would be out of question. It seems to be best suited for the treatment of correlation in linear systems. For a discussion about the difficulties associated with the application of the AMO method to three-dimensional crystals we refer to Calais-Sperber[21].

In the lectures we are going to discuss two papers on the linear chain of hydrogen atoms, an AMO-LCAO calculation[22], in which the total energy is minimized with respect to the AMO mixing parameter θ , and a generalization[23] of the method in which θ is determined in a different way.

The mixing parameter θ in Calais' calculation[22] was independent of which MO's were mixed (which in a crystal calculation would correspond to a θ independent of the wave vector $\bar{k}$). The results showed that for large internuclear distances the AMO energy tends to the energy of separated hydrogen atoms as it should - in marked contrast to the MO energy which for large R is -0.34 a.u. instead of -0.5 a.u. This calculation also showed another caracteristic aspect of the simple AMO method. When the internuclear distance gets sufficiently <u>small</u>, one reaches a value such that no

AMO lowering of the energy is possible - the MO function is stable under perturbations of AMO type.

There is also a so-called refined AMO model, in which the mixing parameter θ is allowed to depend on the wave vector $\vec{k}$: $\theta = \theta(\vec{k})$. It is intuitively reasonable that this ought to improve the wave function as compared to the simple AMO method. The MO's close to the Fermi surface should mix more strongly than those far from the Fermi surface (cf. fig. 2).

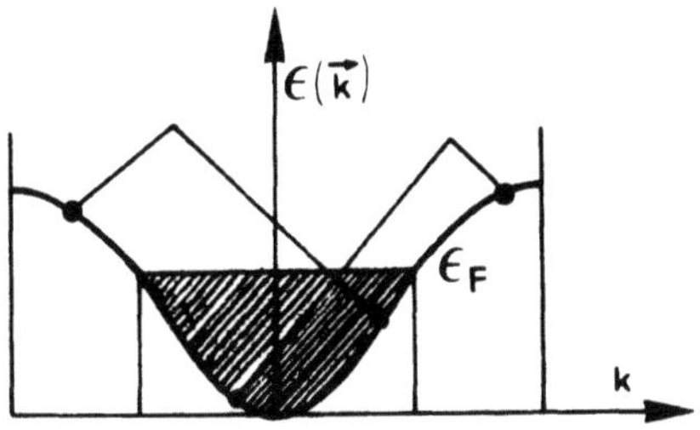

Fig. 2. : Construction of AMO's from pairs of orbitals which are occupied or virtual in the MO approximation.

From the point of view of the one-electron orbital energies the introduction of the AMO's means that we get a chance to split the band. The periodicity of the original lattice gets doubled (cf. fig. 3), and the Brillouin zone consequently becomes half as large. A perturbation of this type has a possibility to open up a gap in the band at the border of the new Brillouin zone. One can derive an integral equation for this gap $\Delta(\vec{k})$ by minimizing the total AMO energy, with respect to the coefficients $\cos\theta(\vec{k})$, or in other ways. Berggren and Martino[23] solved this gap equation in the case of a linear chain of hydrogen atoms with a basis of three Gaussians at each lattice site. The results for two typical internuclear distances are shown in fig. 4.

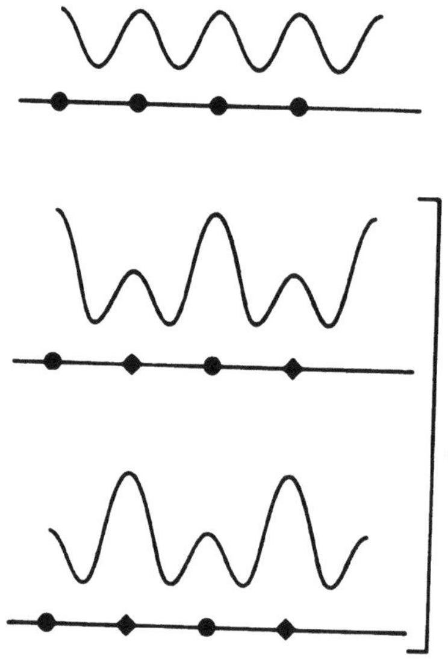

Fig. 3. : Schematic picture showing a delocalized MO and semi-localized AMO's.

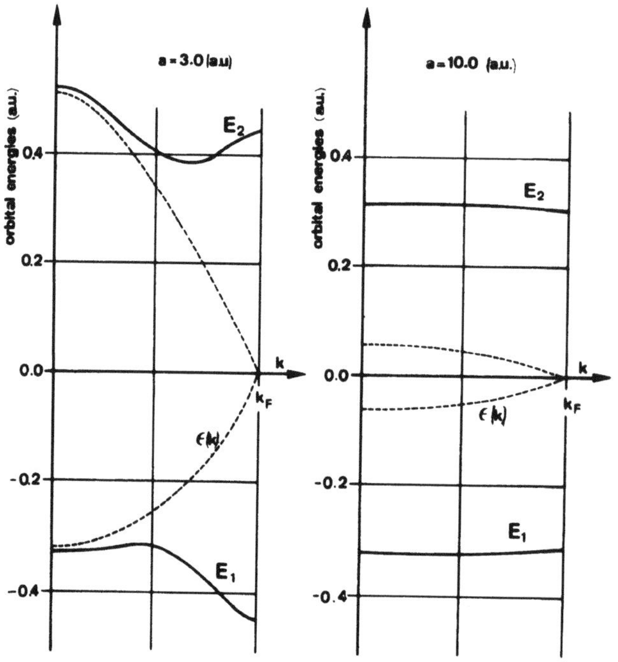

Berggren and Martino also calculated the total energy. Since they had a more flexible wave function they get a more considerable lowering than Calais obtained in the simple AMO case. The results are collected in fig. 5.

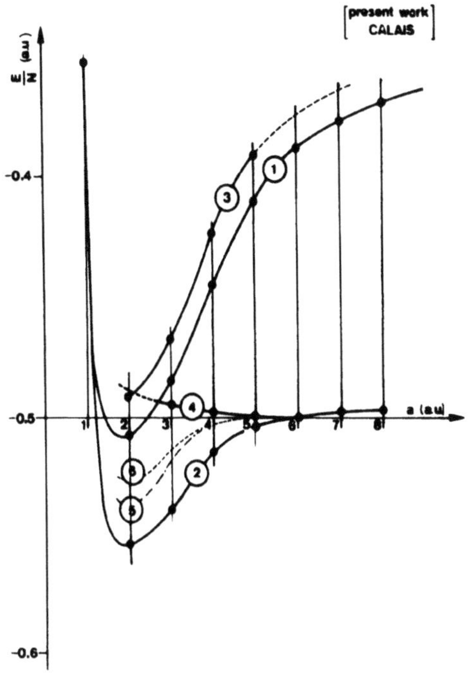

Fig. 5. : Total enrgies/atoms (E/N) for : (I)The paramagnetic state of H_∞ (present calculation), (II) the antiferromagnetic state of H_∞ (present calculation), (III) the paramagnetic state of H_∞ (Calais, Ref.3), (IV) the antiferromagnetic state of H_∞ (Calais, Ref.3), (V) the H_6 ring (Moskowitz, Ref. 12), (VI) the H_{12} ring (Kwo, Ref.13). The heavy dots are calculated points, and dashed lines are extrapolations. The energy of the paramagnetic state for $a = \infty$ is indicated in the upper right corner by two bars — the upper for the result of the present work, the lower for that of Ref.3.

←

Fig. 4. : Effect of pairing on the band structure for two different lattice constants a. The dashed curve is the paramagnetic dispersion $\varepsilon(k)$, which for $|k| > k_F$ has been flipped back onto the first Brillouin zone of the sublattices. The solid line is the dispersion $E_{1,2}(k)$ obtained by the split-band model. The Fermi energy ε_F is chosen as the reference energy.

The bandsplitting which results from the AMO calculation has important implications on the electric and magnetic properties of a system described by an AMO model. A system like the infinite chain of hydrogen atoms, which is a paramagnetic conductor in the MO description, becomes an antiferromagnetic insulator in the AMO model.

Long polyene chains

Kruglyak and Ukrainsky[24] have approached the DODS method in a different way. Like Berggren and Martino[23] they used second quantization. They studied the problem of the polarization of closed shells in an alternant radical due to the field of its unpaired electron. Following the usual procedure in the second quantized version of quantum mechanics they introduced a canonical transformation of the original creation and annihilation operators, which diagonalized the dominating part of the Hamiltonian. The result can be interpreted as a mixing of (originally) occupied and virtual orbitals, i.e. something related to the AMO model. Kruglyak and Ukrainsky applied a self-consistent version of their method to a long polyene radical.

As in the case of the linear chain of hydrogen atoms one obtains a splitting of the original bands into two separate bands. This results means that the restricted Hartree-Fock ("MO") solution is unstable relative to a small perturbation caused by the spin polarization of closed shells in a radical.

Long cumulene chains

Other aspects of the DODS method are discussed by Ukrainsky[25]. They discuss the difference between the unrestricted (UHF) and the extended Hartree-Fock (EHF) method. Different orbitals for different spins are used in both these methods. In UHF one fills a Slater determinant with DODS, and this single determinant is in general not an eigenfunction of the total spin operator S^2, which we should require if the Hamiltonian is independent of spin. The single determinant can however be regarded as a superposition of eigenfunctions of S^2 with different eigenvalues and we can project out the desired component by means of a projection operator as described by Löwdin[19]. If the spin orbitals of this projected determinant are optimized we have the EHF or better the projected Hartree-Fock method (PHF=EHF).

For a system with a large number of electrons the distinction between UHF and PHF tends to get washed out[19][26]. Ukrainsky[25] shows that the ground state energy, the energies of the lowest excitations and the SCF equations for large systems are the same in the UHF and PHF methods.

The cumulenes C_NH_4 have a forbidden gap of about 1eV in their optical spectra. Such a gap appears in the R(estricted) HF model if one assumes alternating bond distances. There is however, a drawback associated with this model. A torsion barrier of the end groups CH_2 persists even for large N, which seems rather unnatural. If the bond lengths are assumed to be equal the RHF method does not give any gap.

The DODS model provides a way out of these difficulties. The connection between the gap and the torsion barrier disappears. A gap appears, but the torsion barrier tends to zero for large N.

CORRELATION IN MOLECULAR CRYSTALS

General about intermolecular forces

The cohesive energies of molecular crystals are very small compared to other types of crystals. The cohesion in such crystals is due to a balance between attractive van der Waal's forces and repulsive overlap forces[17].

We have both an intramolecular and an intermolecular correlation problem. The latter is intimately connected with the specific coordination of electronic motions known as dispersion or van der Waal's forces.

Eisenschitz and London[27-29] were the first ones to treat these forces by means of quantum mechanics. Via second order perturbation theory they obtained the famous series in inverse powers of the internuclear distance R, starting with R^{-6}. The unperturbed function for a pair of hydrogen atoms at large separation R is the product of their 1s orbitals and the perturbation is in atomic units

$$H' = -\frac{1}{r_{b1}} - \frac{1}{r_{a2}} + \frac{1}{r_{12}} + \frac{1}{R} \qquad (11)$$

The same result is also obtained with a variational method as shown by Slater-Kirkwood[30]. - For details we refer to Margenau-Kestner[6], which also contains a very interesting history of intermolecular forces.

Mavroyannis and Stephen[31] showed an important relation between the van der Waal's constant C, multiplying R^{-6}, and the polarizabilities α at imaginary frequencies, of two systems interacting at larges distances

$$C_{I\,II} = \frac{3}{\pi} \int_0^\infty \alpha_I(i\omega)\alpha_{II}(i\omega)d\omega \tag{12}$$

Since it is possible to construct upper and lower bounds to polarizabilities, this formula gives a possibility of finding upper and lower bounds of van der Waal's constants. For a survey of the recent development we refer to Goscinski[32].

The cohesive energy of the neon crystal

Linderberg[33] has proposed a method for handling correlation in molecular crystals, which is centered around the concept of generalized susceptibility $\alpha(\vec{k},\vec{k}'; i\omega)$. An integral equation is derived, which relates that quantity for the whole crystal to the corresponding quantity for the molecular constituents.

The generalized susceptibility or polarizability can be expressed in terms of the quantities

$$\rho_{\vec{k}} = \sum_j e^{-i\vec{k}.\vec{r}_j} \quad ; \quad S_{\vec{k}} = -\sum_g Z_G\, e^{-i\vec{k}.\vec{r}_g} \tag{13}$$

which appear in the Fourier expansion of the potential of the basic Hamiltonian

$$-\sum_{i,g} \frac{Z_g}{r_{ig}} + \frac{1}{2}\sum_{i\neq j}\frac{1}{r_{ij}} + \frac{1}{2}\sum_{g\neq h}\frac{Z_g Z_h}{R_{gh}} =$$

$$\frac{1}{2}\sum_{\vec{k}}\left(V_{\vec{k}}\,(\rho_{\vec{k}} + S_{\vec{k}})\,(\rho_{-\vec{k}} + S_{-\vec{k}}) - N - \sum_g Z_g^2\right) \tag{14}$$

Here N is the total number of electrons, Z_g the charge of nucleus g and

$$V_{\vec{k}} = \frac{1}{\Omega}\int V(\vec{r})\, e^{-i\vec{k}.\vec{r}}\, dv \tag{15}$$

Lindner and Goscinski[34] have pointed out that the generalized polarizability can be expressed in the form

$$\alpha(\vec{k},\vec{k}';i\omega) = \sum_s \left(\frac{<0|\rho_{\vec{k}}|s><s|\rho_{-\vec{k}}|0>}{E_s - E_o - i\omega} + \frac{<0|\rho_{-\vec{k}}|s><s|\rho_{\vec{k}}|0>}{E_s - E_o + i\omega} \right) \quad (16)$$

where $|0>$ and $|s>$ are the eigenstates of the total Hamiltonian with the energies E_o and E_s respectively. Lindner and Goscinski[34] have also given an explicit formula for the correlation part of the potential energy

$$U_{corr} = \frac{1}{2} \sum_{\vec{k}} V_{\vec{k}} \left(\sum_{s\neq 0} <0|\rho_{\vec{k}}|s><s|\rho_{-\vec{k}}|0> - N \right) =$$

$$\frac{1}{2} \sum_{\vec{k}} V_{\vec{k}} \left\{ \frac{2}{\pi} \int_0^{\infty} <0|\rho_{\vec{k}}(H-E_o)\left[(H-E_o)^2 + \omega^2\right]^{-1} P_{\rho-\vec{k}}|0> d\omega - N \right\} \quad (17)$$

We notice that (16) and (17) are not restricted to molecular crystals, but refer to many-electron systems in general.

Linderberg's treatment[33] of the cohesive energy of molecular crystals gives an expression for this quantity which is a sum of a Hartree-Fock and a correlation term. The evaluation of the correlation term leads to dispersion forces corrected for many-atom interactions and intra-atomic (-molecular) correlations.

The method has been applied to the neon crystal[35]. This crystal has been studied from many points of view, but Linderberg's calculation seems to be one of the few based on an accurate quantum mechanical description.

REFERENCES

(Notice : this list of references is not meant to be in any way exhaustive. Its only aim is to provide further background for the lectures and to make it possible, via some survey articles, to go further in the study of the correlation problem.)

1. P.-O. Löwdin, "Correlation in Quantum Mechanics"; Adv. Chem. Phys. II, 207 (1959).

2. R. Lefebvre, C. Moser (ed.), "Correlation Effects in Atoms and Molecules"; Adv. Chem. Phys. XIV, (1969).

3. L. Hedin, S. Lundqvist, "Effects of Electron-Electron and Electron-Phonon Interactions on the One-Electron States of Solids", Solid State Phys. 23, 1 (1969).

4. F. Herman, N.W. Dalton, T.R. Koehler (ed.), "Computational Solid State Physics"; Proceedings of an International Symposium held October 6-8, 1971 in Wildbad, Germany; Plenum Press New York and London 1972.

5. E.H. Lieb, D.C. Mattis, "Mathematical Physics in One Dimension. Exactly Soluble Models of Interacting Particles"; Academic Press, New York and London 1966.

6. H. Margenau, N.R. Kestner, "Theory of Intermolecular Forces"; International Series of Monographs in Natural Philosophy, Volume 18; Pergramon Press, Oxford 1969.

7. L. de Brouckère, M. Mandel, "Dielectric Properties of Dilute Polymer Solutions"; Adv. Chem. Phys. I, 77 (1958).

8. D.S. McClure, "Electronic Spectra of Molecules and Ions in Crystals. Part I. Molecular Crystals"; Solid State Phys. 8, 1 (1959).

9. H.C. Wolf, "The Electronic Spectra of Aromatic Molecular Crystals"; Solid State Phys. 9, 1 (1959).

10. H.S. Jarret, "Electron Spin Resonance Spectroscopy in Molecular Solids"; Solid State Phys. 14, 215 (1963).

11. N. Saito, K. Okano, S. Iwayanagi, T. Hideshima, "Molecular Motion in Solid State Polymers"; Solid State Phys. 14, 344 (1963).

12. L. Jansen, "Quantum Chemistry and Crystal Physics. Stability of Crystals of Rare Gas Atoms and Alkali Halides in Terms of Three-Atom and Three-Ion Exchange Interactions"; Adv. Quantum Chem. 2, 119 (1965).

13. A. Keller, "Polymer Crystals"; Repts. Progr. Phys. 31, 623 (1968).

14. D.P. Craig, S.H. Walmsley, "Excitons in Molecular Crystals. Theory and Applications"; Frontiers in Chemistry, W.A. Benjamin, Inc., New York and Amsterdam 1968.

15. L.Meyer, "Phase Transitions in van der Waal's Lattices"; Adv. Chem. Phys. XVI, 343 (1969).

16. T. Kihara, "Multipolar Interactions in Molecular Crystals";

Adv. Chem. Phys. XX, 1 (1971).

17. A.J. Kitaigorodsky, "Molecular Crystals and Molecules"; Vol. 29 of "Physical Chemistry" (E.M. Loebl, ed.). Academic Press, New York and London 1973.

18. C.A. Coulson, I. Fischer, "Notes on the Molecular Orbital Treatment of the Hydrogen Molecule", Phil. Mag. 40, 386 (1949).

19. R. Pauncz, "Alternant Molecular Orbital Method"; Studies in Physics and Chemistry 4; W.B. Saunders Co; Philadelphia and London 1967.

20. O. Goscinski, J.-L Calais, "Some Comments on the AMO Method for the Ground State of the Hydrogen Molecule"; Arkiv Fysik 29, 135 (1965).

21. J.-L. Calais, G. Sperber, "A Study of the AMO Method as Applied to the Lithium Metal. I. Review, Results and Discussion"; Int. J. Quantum Chem. 7, 501 (1973).

22. J.-L. Calais, "Different Bands for Different Spins. II. Application to a Linear Chain of Hydrogen Atoms", Arkiv Fysik 28, 511 (1965).

23. K.-F. Berggren, F. Martino, "Different Orbitals for Different Spins in an Infinite Chain of Hydrogen Atoms", Phys. Rev. 184, 484 (1969).

24. Y.A. Kruglyak, I.I. Ukrainsky, "Study of the Electronic Structure of Alternant Radicals by the DODS Method", Int. J. Quantum Chem. 4, 57 (1970).

25. I.I. Ukrainsky, "Electronic Structure of Long Cumulene Chains"; Int. J. Quantum Chem. 6, 473 (1972).

26. W.H. Adams, "Correlation Effects in the Alternant Molecular Orbital Approximation"; J. Chem. Phys. 39, 23 (1963).

27. R. Eisenschitz, F. London, "Über das Verhältnis der van der Waalschen Kräfte zu den homöopolaren Bindungskräften"; Z. Physik 60, 491 (1930).

28. F. London, "Zur Theorie und Systematik der Molekularkräfte"; Z. Physik 63, 245 (1930).

29. F. London, "Über einige Eigenschaften und Anwendungen der Molekularkräfte"; Z. Physik. Chem. B11, 222 (1930).

30. J.C. Slater, J.G. Kirkwood, "The van der Waals Forces in Gases";

Phys. Rev. 37, 682 (1931).

31. C. Mavroyannis, M.J. Stephen, "Dispersion Forces"; Mol. Phys. 5, 629 (1962).

32. O. Goscinski, "Upper and Lower Bounds to Polarizabilities and van der Waals Forces. I. General Theory"; Int. J. Quantum Chem. 2, 761 (1968).

33. J. Linderberg, "Dispersion Energy and Electronic Correlation in Molecular Crystals"; Arkiv Fysik 26, 323 (1964).

34. P. Lindner, O. Goscinski, "Generalized Polarizabilities and Energy Expressions"; Int. J. Quantum Chem. 4S, 251 (1971).

35. J. Linderberg, F.W. Bystrand, "Cohesive Energy of Solid Neon"; Arkiv Fysik 26, 383 (1964).

AB-INITIO SCF-LCAO HARTREE-FOCK CALCULATIONS AND THE DETERMINATION OF CORRELATION CORRECTIONS IN THREE-DIMENSIONAL CRYSTALS

T. C. Collins

Aerospace Research Laboratories

Wright-Patterson Air Force Base, Ohio 45433

A formalism for calculating a Hartree-Fock crystal is given in detail. This includes a description of the basis set used and integral approximations needed in the calculation. The results of the calculation of the ground and excited states are discussed. For the ground state the first order density matrix is compared with experiment by investigating the x-ray structure factors and the Compton profiles. The cohesive energy, lattice constant, and bulk modulus of the Hartree-Fock results are also compared with experiment to judge the correctness of this model. The Hartree-Fock excited stated plus correlation corrections are included. The corrections include relaxation of the electron and hole states by using two models. One is the screened exchange plus coulomb hole approximation where several approximations for the dielectric function are studied. The second model is the electronic polaron model where both the relaxation effects and a new scattering state are discussed. The correction to the electronic excited (or vitual) states so that they see the correct potential (namely a V^{N-1}) is given in detail.

I. INTRODUCTION

Over the last three or four years there have been two very important steps made in the calculation of the electronic systems of crystals. The first step was the correct Hartree-Fock model solutions of the crystalline systems.[1-4] That is no free electron gas approximation [5-7] was used and the equations with the non-local exchange operator were solved. The other important step was the inclusion of the correlation effects ; [8-15] that is the relaxation of the hole and electron plus the effect of the attraction of the electron and hole being put into the calculation. These two steps make it possible to eliminate the parameters one had to insert into his model and adjust to fit experimental results. One can now calculate to his desired degree of accuracy from first principles the electronic structure of a large number of crystalline systems.

We will devide the subject matter covered into four parts. The first part contains a brief description of the Hartree-Fock model and the usual approximations to the Hartree-Fock crystal exchange operator one made in the past. The first part will also contain the description of a particular Hartree-Fock calculation ; [4] namely a SCF-LCAO method with Gaussian lobe functions used as the basis set is given. How the crystal symmetry was used and the integral approximations utilized is given in this section also. The second section will contain the ground state properties which can be compared with experiment. We will compare the calculated cohesive energies, virial coefficients, lattice constant, bulk modulus, x-ray structure factors and Compton profiles. A brief look at Hartree-Fock excitations will conclude this section.

The next step, that is obtaining correlated excitation energies, are given in section IV. Here the derivation of the screened exchange plus coulomb hole approximation [10] and the electronic polaron model [11] will be outlined. Along with this, the results of adding these corrections to the Hartree-Fock excitation energies are presented for a series of crystals. All of the above work presented in section IV are corrections to the Dyson Equation. The eigenvalues of this equation are solutions to the N+1 or N-1 systems, but what is most often measured are excitations to the N system. When localized excitations are present in the system, one finds that the N+1 or N-1 solutions are far from experiment. Thus in section V solutions to the N system will be given under the name of $\hat{O}\hat{A}\hat{O}$ formalism.[15] A short conclusion of the problems in crystalline calculations, and the outlook for the next few years will be the subject of section VI

II. HARTREE-FOCK CRYSTAL MODEL

In this section we will start with a brief description of the Hartree-Fock model.[16] The Hamiltonian for the electronic system is:

$$H = \sum_{i=1}^{N} p_i^2/_{2m} - e^2 \sum_{i,b} \frac{Zb}{|\vec{r}_i - \vec{R}_b|} + e^2 \sum_{i<j} \frac{1}{|\vec{r}_i - \vec{r}_j|} \qquad (2.1)$$

In this model one assumes the trial wave function is an anti-symmetric product of the N one-electron functions. This is the so-called Slater Determinant.

$$\Psi(x_1,\ldots,x_N) = \frac{1}{\sqrt{N!}} \det\{\psi_i(x_j)\} = \sqrt{N!}\hat{A}_N \prod_{i=1}^{N} \psi_i(x_i) \qquad (2.2)$$

The ϕ_i's are one-electron functions and x stands for space and spin coordinates.

$$\hat{A}_N = (N!)^{-1} \sum_P (-1)^P \hat{P} \qquad (2.3)$$

where the $\hat{P}$'s are the permutation operators. The main idea of the Hartree-Fock model is to find the best single determinant which will minimize the total energy of the electronic system. One uses the variational principle with the constraint that the orbitals, ϕ_i's, remain orthogonal. The orthogonality is insured by using Lagrange undetermined multipliers, λ_{ij}, and one varies the following equations :

$$\delta_{\psi^*}\{<\Psi|H|\Psi> - \sum_{ij} <\psi_i|\psi_j>\lambda_{ij}\} = 0 \qquad (2.4)$$

where

$$<\Psi|H|\Psi> = N!< \prod_{i=1}^{N} \psi_i|\hat{A}_N H \hat{A}_N| \prod_{j=1}^{N} \psi_j > \qquad (2.5)$$

Using $[\hat{A}_N, H] = 0$ and $\hat{A}_N^2 = \hat{A}_N$, one obtains

$$<\Psi|H|\Psi> = \sum_P (-1)^P < \prod_{i=1}^{N} \psi_i |H\hat{P}| \prod_{j=1}^{N} \psi_j >$$

$$= \frac{-\hbar^2}{2m} \sum_{i=1}^{N} \int \psi_i^*(x)\, \nabla_r^2\, \psi_i(x)\, dx - e^2 \sum_{i=1}^{N} \int \psi_i^*(x) \sum_{b=1}^{n} \frac{Z_b}{|\vec{r}-\vec{R}_b|} \psi_i(x) dx$$

$$+ 1/2\, e^2 \sum_{i,j}{}' \int \psi_i^*(x)\, \psi_j^*(x') \frac{1}{|\vec{r}-\vec{r}\,'|} \psi_j(x')\, \psi_i(x)\, dx\, dx'$$

$$- 1/2\, e^2 \sum_{i,j}{}' \int \psi_i^*(x)\, \psi_j^*(x') \frac{1}{|\vec{r}-\vec{r}\,'|} \psi_i(x')\, \psi_j(x)\, dx\, dx' \qquad (2.6)$$

The restriction of $i \neq j$ may be removed because the $i = j$ terms cancels. This will lead to having N + 1 or an excess electron solution for the unoccupied orbitals as will be show later.

Performing the variation one finds :

$$[-\frac{-\hbar^2}{2m} \nabla_r^2 - e^2 \sum_b \frac{Z_b}{|\vec{r}_i-\vec{R}_b|}]\psi_i(x_1) + e^2 \sum_j dx2\, \psi_j^*(x_2) \frac{1}{|\vec{r}_1-\vec{r}_2|}$$

$$\otimes \; [1-\hat{P}_{12}] \; \psi_j(x_2) \, \psi_i(x_1) = \sum_j \lambda_{ij} \, \psi_j(x_1) \tag{2.7}$$

In the case of double occupancy of the space part of the orbitals, one can make a unitary transformation on the ϕ_i' s which diagonalizes the λ_{ij} matrix.

$$\left[\frac{-\hbar^2}{2m} \nabla^2_{r_1} - e^2 \sum_{b=1}^{n} \right] \frac{Z_b}{|\vec{r}_1 - \vec{R}_b|} \Psi_i(x_1) + e^2 \sum_j \, dx_2 \, \Psi_j(x_2) \frac{1}{|\vec{r}_1 - \vec{r}_2|}$$

$$\otimes \; [1-\hat{P}_{12}] \; \Psi_j(x_2) \, \Psi_i(x_1) = \varepsilon_i \, \Psi_i(x_1) \tag{2.8}$$

The first-order density matrix is defined as :

$$\rho(x,x') = \Psi(x,x_2,\ldots x_N) \, \Psi^+ (x',x_2,\ldots x_N) \, dx_2 \ldots dx_N \tag{2.9}$$

In the Hartree-Fock model, the first order density matrix reduces to :

$$\rho(x,x') = \sum_{i=1}^{N} \Psi_i(x) \, \Psi_i(x') \tag{2.10}$$

the Fock-Dirac density matrix.[17] Substituting (2.10) into (2.8) one obtains

$$F(x_1) \; \Psi_i(x_1) = \left[\frac{-\hbar^2}{2m} \nabla^2_{r_1} - e^2 \sum_{b=1}^{n} \frac{Z_b}{|\vec{r}_1 - \vec{R}_b|} \right] \Psi_i(x_1) + e^2 \, dx_2$$

$$\frac{1}{|\vec{r}_1 - \vec{r}_2|} \otimes [1-\hat{P}_{12}]_{x_2 = x_2'} \, \rho(x_2,x_2') \, \Psi_i(x_1) = \varepsilon_i \, \Psi_i(x_1) \tag{2.11}$$

The subscript $x_2 = x_2'$ means perform the operation then remove the prime. Eq. (2.11) gives the basic form of obtaining self-consistent solutions. That is make a guess of the ψ_i's, form ρ of Eq. (2.10) ; then form F of Eq. (2.11) to obtain new ψ_i's.

The physical meaning of the eigenvalues, ε_i, are defined as the energy to remove an electron from the system if relaxation is not important. One has

$$E_N^{HF} - E_{N-1,i}^{HF} = \varepsilon_i \tag{2.12}$$

if the orbitals are not allowed to adjust to the absence of the removed electron. This is called Koopmans' theorem.[18]

What about the states of the system that are not ionization energies. That is other excited states of the system. To study this, expand the ψ_i's in terms of a basis set. This leads to the alphabet mixes which are very common in the literature. For example, in calculation crystalline properties one uses orthogonalized plane waves, (OPW),[19] linear combination of atomic orbitals (LCAO),[20] and sometimes mixed basis sets of plane waves and Gaussians (PWG).[21] The correct basis set is LCMBF which stands for linear combination of my basis functions. One finds in order to obtain convergence, that it requires more basis functions than particules in the system. One forms the matrix of the Fock operator with this basis and diagonalizes this matrix. One obtain the N lowest eigenvalues which are the Hartree-Fock values. The other values above these "occupied" levels are called virtual levels. Lets look at the kth solution where $k > N$.

$$\varepsilon_k = \langle k| \{\frac{-\hbar^2}{2m} \nabla_1^2 - e^2 \sum_{b=1}^{n} \frac{Z_b}{|\vec{r}_1 - \vec{R}_b|}\}|k\rangle$$

$$+ e^2 \int dx_1 dx_2\, \Psi_k^*(x_1) \{ \sum_{j=1}^{N} \frac{\Psi_j^*(x2)\,\Psi_j(x2)}{|\vec{r}_1 - \vec{r}_2|}\} \Psi_k(x_1)$$

$$- e^2 \int dx_1 dx_2\, \Psi_k^*(x_1) \{ \sum_{j=1}^{n} \frac{\Psi_j^*(x2)\,\Psi_k(x2)}{|\vec{r}_1 - \vec{r}_2|}\} \Psi_j(x_1) \qquad (2.13)$$

The last two terms do not cancel the self-energy term because $j \neq k$. Thus the virtuals see a V^N potential whereas the occupied orbitals see a V^{N-1} potential.

Until the last few years one did not solve the Hartree-Fock model for crystals because the exchange operator made the problem too difficult. Most of the calculations of electron energy bands of crystals make the "free-electron" approximation. Multiplying the exchange term of Eq. (2.11) by

$$[\Psi_i^*(x_1') \Psi_i(x_1')] / [\Psi_i^*(x_1') \Psi_i(x_1')] \quad x_1 = x_1'$$

one obtains :

$$\{ -e^2 \sum_j \int dx_2 \frac{\Psi_i^*(x_2)\, \Psi_i(x_1)\, \Psi_i(x_2)\, \Psi_j(x_1)}{\Psi_i^*(x_1)\, \Psi_i(x_1)\, |\vec{r}_1 - \vec{r}_2|}\} \Psi_i(x_1) = V_{ex}^i(x_1) \Psi_i(x_1) \qquad (2.14)$$

The next step taken by Slater [5] was to perform a weighted average of this term. The weight was $\psi_i(x_1)\ \psi_i(x_1)\ /\ \rho$, and the result is:

$$V^S_{ex}\ (\vec{r}) = -6\ \{\frac{3}{8\pi}\rho\ (\vec{r})\}^{1/3} \tag{2.15}$$

One has the possibility of making this approximation in the Fock operator as Slater did, or one can make the approximation in the total energy as Gaspar,[6] and Kohn and Sham[7] did. The results for the exchange become :

$$E_{ex} = -1/2\ C \int dx\ \rho^{4/3} \tag{2.16}$$

where C are the constant terms in Eq. (2.15). Performing the variation gives :

$$-1/2\ C\ (4/3\,\rho^{1/3})\ \Psi_i(x_1) = -2/3(C\,\rho^{1/3})\ \Psi_i\ (x_1) \tag{2.17}$$

Thus one obtains 2/3's the operator obtained by Slater. This has lead to a large number of ways to get the coefficient to $\rho^{1/3}$. This is called the X_α method

$$V_{x\alpha} = -\alpha\, C\ \rho^{1/3} \tag{2.18}$$

One can also play the same game with the same results for defining the meaning of the eigenvalues. Also one can not take the weighted average, and the results are called the Liberman Exchange Approximation.[22]

$$V^{L,i}_{ex}\ (x_1) = -8\ F(\eta_i)\{\frac{3}{8\pi}\rho\ (\vec{r})\}^{1/3} \tag{2.19}$$

where

$$F\ (\eta_i) = 1/2 + \frac{1-\eta i}{4\eta_i}\ln\frac{1+\eta i}{1-\eta_i} \quad ; \quad \eta_i = K_i/K_F$$

The choice for K_i and K_F are generally

$$K_i = [\,\varepsilon_i - V(\vec{r})]^{1/2} \quad \text{and} \quad K_F = \{3\pi\ \rho\ (\vec{r})\}^{1/3}$$

The thing to remember is that all of these approximations are made only for the exchange term and only the exchange term. It turns out in most cases these are poor approximations. So lets look and see what doing the correct exchange does.

Before we give the results of the Hartree-Fock Crystal Calculations in the next section, we will describe here the method plus the approximations made in the calculations.[4] The basis functions are spatially localized functions centered at various locations in the crystal. Very long-range basis functions do not need to be employed because of the periodicity of the lattice. For example a crystalline diamond computation needs only a few s and p like functions. For s-symmetry basis functions, contracted sets of primitive Gaussian functions[23] are centered on the various atomic locations :

$$\psi(\vec{r}) = \sum_{\alpha} A_{\alpha}\, e^{-\alpha|\vec{r}|^2} \tag{2.20}$$

The lobe functions of p symmetry are constructed as the differences of two contracted Gaussian lobes, centered $\pm R_{\alpha}$ from the origin of the p functions. This displacement is designed to reproduce the dipolar angular dependence of the atomic p functions :

$$P_x(\vec{r}) = \sum_{\alpha} A_{\alpha}\{e^{-\alpha|\vec{r}-\vec{R}|^2} - e^{-\alpha|\vec{r}+\vec{R}|^2}\} \tag{2.21}$$

where

$$|\vec{R}_{\alpha}| = {}^{C}/_{\sqrt{\alpha}} \quad ; \quad 0.005 \leqslant C \leqslant 0.01$$

Atomic studies have established the lobe displacement range given above. A series of SCF calculations on the Ar atom were performed, allowing C to vary over the above range with the results of total energy change of only 0.001 a.u. The value in diamond used is 0.09 in our calculations. On top of the contracted set of Gaussian fully uncontracted functions were used to allow for maximal distortion from atomic character.

A very important step is to fully utilize crystal symmetry. Again using diamond as the example, the atom at (0,0,0) will be labeled type 1, and the atom at (1/4, 1/4, 1/4) will be labeled type 2. The coordinates respect to the basis vectors are :

$$\hat{t}_1 = 1/2\ a\ (\hat{j}+\hat{k})\ ;\ \hat{t}_2 = 1/2\ a\ (\hat{i}+\hat{k})\ ;\ \hat{t}_3 = 1/2\ a\ (\hat{i}+\hat{j}) \tag{2.22}$$

There are twenty four rotation-reflection operators which transform each atom into an equivalent atom. The inversion about the point $\vec{\tau} = (1/8,\ 1/8,\ 1/8)$ interchanges atoms of type 1 with atoms of type 2.

As was shown in Eq. (2.11), only the first-order density matrix,

$$\rho(x,x') = \sum_{i=1}^{N} \Psi_i(x)\, \Psi_i^{*}(x')$$

is needed in the Hartree-Fock formalism. This density matrix can be expanded using our local basis set (assuming double occupancy of the Hartree-Fock orbitals)

$$\rho(\vec{r},\vec{r}\,') = \sum_{\alpha,\beta,a,b} P^{ab}_{\alpha\beta}\, \psi_\alpha(\vec{r}-\vec{R}_a)\, \psi^*_\beta(\vec{r}\,'-\vec{R}_b) \tag{2.23}$$

For closed-shell ground states (semiconductors and insulators), the first-order density matrix has full crystal symmetry. Thus

$$\rho(\vec{r},\vec{r}\,') = \rho(\vec{r}\,',\vec{r}) = \rho(\mathfrak{X}\vec{r}, \mathfrak{X}\vec{r}\,') = \rho(T+\vec{r},T+\vec{r}\,') \tag{2.24}$$

Now operating on any product of two functions

$$\psi_\alpha(\vec{r}-\vec{R}_a)\ \psi_\beta(\vec{r}-\vec{R}_b)$$

with all of the symmetry operators of the crystals will produce a set of product functions. By taking all possible values of α and β on each site and by taking different values for $|\vec{R}_a-\vec{R}_b|$, we can generate distinct product sets containing all possible product pairs for $|\vec{R}_a-\vec{R}_b|$ less than a given value. We only need to acquire information about one member of each product set, since information about all other members may then be generated by using crystal symmetry operators. Also the symmetry requirement of Eq (2.24) says that the $P^{ab}_{\alpha\beta}$ in Eq (2.23) will be the same (except for sign) for all the pairs of the functions of s and p character in one product set. This gives

$$\rho(\vec{r},\vec{r}\,') = \sum_I P_I \sum_{\substack{(\alpha\beta ab)\\ \in I}} \varepsilon^{ab}_{\alpha\beta}\ \psi_\alpha(\vec{r}-\vec{R}_a)\ \psi_\beta(\vec{r}-\vec{R}_b) \tag{2.25}$$

where I runs over the different products sets which are labeled symmetry sets, $\varepsilon = \pm 1$, and where $(\alpha\beta ab)$ labels the different produce pairs of a symmetry set. It should be noted that P_I's do not have complete variational freedom since ρ must be idempotent and must satisfy the normalization condition

$$\int \rho(\vec{r},\vec{r})\, d\vec{r} = 2\ N\ Z \tag{2.26}$$

N is the number of unit cells, each with a total electronic charge of 2Z.

Table I lists the members of the first six symmetry sets. These sets include all of the one-center and nearest-neighbor two-center contributions to the first-order density matrix. The symmetry analysis reduces 68 coefficients to 6. Table II gives a representative member of each symmetry set for the first six shells of atoms. The total number of symmetry-independent coefficients can be found by multiplying the number of symmetry-independent s-s local basis function products by $1/2\ N_s\ (N_s + 1)$, the s-p products by $N_s\ N_p$, etc., where there are N_s (N_p) separate basis functions of s (p) character.

Table I. First six symmetry sets for diamond symmetry. The s stands for an LBF of s symmetry, while x, y, and z are p-symmetry LBF's. The second atom's coordinates are (0, 0, 0), (1/4, 1/4, 1/4), (1/4,-1/4,-1/4), (-1/4, 1/4,-1/4), and (-1/4, -1/4, 1/4) relative to the basis vectors given in Eq (4). The first atom is at (0, 0, 0).

Set	Atom	[μν]
1	1	xx + yy + zz
2	1	ss
3	2	xx + yy + zz
	3	xx + yy + zz
	4	xx + yy + zz
	5	xx + yy + zz
4	2	xy + yx + xz + zx + yz + zy
	3	-xy - yx - xz - zx + yz + zy
	4	-xy - yx + xz + zx - yz - zy
	5	xy + yx - zx - xz - yz - zy
5	2	xs - sx + ys - sy + zs - sz
	3	xs - sx - ys + sy - zs + sz
	4	-xs + sx + ys - sy - zs + sz
	5	-xs + sx - ys + sy + zs - sz
6	2	ss
	3	ss
	4	ss
	5	ss

For example, good calculations can be done using six shells of atoms in diamond with four s-like basis functions and three-with p-like character. One then has 270 symmetry-independent coefficients characterizing the first-order density matrix. It will be shown later that only 270 one-electron integrals of each type need be done, and that all of the two-electron integrals can be collected in a symmetric array $(270)^2$.

In order to determine the symmetry-independent coefficients of the first-order density matrix, the method of Roothan is used.

For each local basis function $\psi_\alpha(\vec{r}-\vec{R}_a)$, where ψ_α is a contracted set of Gaussian lobe functions, construct a Bloch function

$$\Phi_\alpha^{\vec{k}}(\vec{r}) = \frac{1}{\sqrt{N}} \sum_a e^{i\vec{k}.\vec{R}a} \psi_\alpha(\vec{r}-\vec{R}_a) \qquad (2.27)$$

where $\vec{k}$ labels the Brillouin zone point and N is the number of unit cells in the crystal. Separate Bloch functions are needed for each of the two atoms in the unit cell. The one-electron wave functions Ψ_n^k are then expanded in terms of the Bloch functions

$$\Psi_n^k(\vec{r}) = \sum_\alpha C_{n\alpha}^{\vec{k}} \Phi_\alpha^{\vec{k}}(\vec{r}) \qquad (2.28)$$

where n is the band index. The coefficients $C_{n\alpha}^k$ are adjusted to minimize the total energy. The following Hamiltonian and overlap matrices which are block diagonal in $\vec{k}$ are obtained :

$$\sum_\beta H_{\alpha\beta}^{\vec{k}} C_{n\beta}^{\vec{k}} = \varepsilon_n^{\vec{k}} \sum_\beta \Delta_{\alpha\beta}^{\vec{k}} C_{n\beta}^{\vec{k}} \qquad (2.29)$$

$$\Delta_{\alpha\beta}^{\vec{k}} = \int \Phi_\alpha^{\vec{k}+}(\vec{r})\, \Phi_\beta^{\vec{k}}(\vec{r})\, d\vec{r}$$

$$= \sum_b U_{\alpha\beta}(\vec{R}_a,\vec{R}_b)\, e^{i\vec{k}.(\vec{R}_a-\vec{R}_b)} \qquad (2.30)$$

$$U_{\alpha\beta}(\vec{R}_a,\vec{R}_b) = \int \psi_\alpha(\vec{r}-\vec{R}_a)\, \psi_\beta(\vec{r}-\vec{R}_b)\, d\vec{r} \qquad (2.31)$$

$$H_{\alpha\beta}^{\vec{k}} = \sum_b H_{\alpha\beta}(\vec{R}_a,\vec{R}_b)\, e^{i\vec{k}.(\vec{R}_b-\vec{R}_a)} \qquad (2.32)$$

$$H_{\alpha\beta}(\vec{R}_a,\vec{R}_b) = \int d\vec{r}_1\, \psi_\alpha^*(\vec{r}_1-\vec{R}_a) \{ -\nabla_1^2 - 2 \sum_c \frac{Z}{|r_1-R_c|} + \int dr_2 [2\frac{\rho(\vec{r}_2,\vec{r}_2)}{r_{12}} - \frac{\rho(\vec{r}_2,\vec{r}_2)}{r_{12}} \hat{P}_{12}] \} \psi_\beta(\vec{r}_1-\vec{R}_b) \qquad (2.33)$$

Equations (2.30) and (2.32) need only be summed over atoms b owing to translational symmetry. Note that the sum over b is over all atoms of type 1 or type 2, depending on whether the Bloch function Φ_β^k is for atom type 1 or type 2. In Eq. (2.33), c sums over all atoms and P_{12} is the permutation operator which interchanges $\bar{r}_1$ with $\bar{r}_2$. Matrix elements where the atom at $\bar{R}_a$ is of type 2 can be obtained from those where the atom at R_a is of type 1 by application of the inversion operator about (1/8, 1/8, 1/8).

Table II. Representative members of the diamond symmetry sets for the first six shells of atoms. The s stands for an s-symmetry LBF, while x, y, and z are p-symmetry LBF's. The coordinates are relative to the basis vectors given in Eq. (2.22)

Shell	Atom	
1	(0, 0, 0)	ss, xx
2	(1/4, 1/4, 1/4)	ss, sx, xx, sy
3	(1/2, 1/2, 0)	ss, sx, sz xx, zz, xy, xz
4	(3/4, 1/4, -1/4)	ss, sx, sy xx, yy, xy, yz
5	(1, 0, 0)	ss, sx, sy xx, yy, xy, yz
6	(3/4, 3/4, 1/4)	ss, sx, sz xx, zz, xy, xz

The matrix eigenvalue problem (Eq. (2.29)) is solved by performing a Choleski decomposition on each positive definite overlap matrix $\Delta^{\bar{k}}$

$$\Delta = L\,L^{+} \;;\quad L_{ij} = 0 \text{ for } i<j \tag{2.34}$$

After this decomposition is performed, a single matrix diagonalization yields the desired eigenvalues and eigenvectors :

$$[L^{-1}\;H\;L^{-1^{+}}]\;[L^{+}\,C] = \lambda\,[L^{+}C] \tag{2.35}$$

Since in Eqs. (2.31) and (2.33) one has integrals which involve local basis function product pairs as in $\rho(\vec{r},\vec{r}\,')$, one can use the same decomposition into symmetry sets to reduce the number of integrals which need to be done. All of the operators in the Hamiltonian, as well as the overlap matrix (the unit operator) have full crystalline symmetry. Therefore, none of the one-electron operators can change the symmetry of the function it operates on. Thus all of the one-electron integrals may be classified into the same symmetry sets as $\rho(\vec{r},\vec{r}\,')$. Only a single one-electron integral of each type for each different symmetry set need be done. The two-electron integrals in Eq. (2.33) involve localized basis pairs multiplying

$\rho(\vec{r},\vec{r}')$. Thus, if k symmetry-independent coefficients characterize $\rho(\vec{r},\vec{r}')$ in Eq. (2.25), then all of the two-electron integrals can be stored in a K by K array. The two-electron part of Eq. (2.33) is summed over $\vec{R}_b$, α and β for $\vec{R}_a$ fixed to give the array,

$$A_{IJ} = \frac{1}{N} \sum_{(\alpha\beta ab)} \sum_{(\gamma\delta ed)} \int d\vec{r}_1 \, d\vec{r}_2 \, \psi_\alpha (\vec{r}_1-\vec{R}_a) \, \psi^*_\gamma(\vec{r}_2-\vec{R}_c)$$

$$\otimes \frac{2-\hat{P}_{12}}{r_{12}} \psi_\delta (r_2-R_d) \, \psi_\beta(r_1-R_b). \qquad (2.36)$$

Element A_{IJ} is thus proportional to the total Coulomb-plus-exchange integral of all members of symmetry set I with all members of symmetry set J. The constant of proportionality is the reciprocal of the number of cells in the crystal. The symmetry of Eq. (2.36) guarantees the symmetry of the array, so that $A_{IJ} = A_{JI}$. In practice, the centers for the atoms a, b, c, and d are chosen such that they all are within M shells of each other and atom a is taken at zero. The two-electron integral contribution in Eq. (2.33) is now given by

$$\frac{1}{W_I} \sum_J P_J \, A_{IJ} \qquad (2.37)$$

where W_I is the number of members of symmetry set I for fixed $\vec{R}_a$. Note that in evaluating Eq. (2.36), a single member of symmetry set I, say μ, can be chosen and integral for that member can be multiplied by the number of members of I with $\vec{R}_a$ at zero. In addition, one can use the set of symmetry operations {X}which leave μ invariant, for a further reduction. The members of symmetry set J can be decomposed into subsets which transform into each other under the operations {X} . Only integrals between μ and one member of each subset need be calculated. Finally, one has the permutation symmetry that the Coulomb integral for $(ab\alpha\beta:cd\gamma\delta)$ is equal to the exchange integral for $(ac\alpha\gamma:bd\beta\delta)$ and $(ad\alpha S:bc\beta\gamma)$. All of these considerations greatly reduce the number of integrals which must be calculated.

In order to construct a new first-order density matrix from the self-consistent results, one in principle needs an integral over the occupied Hartree-Fock eigenfunctions in the first Brillouin zone :

$$\rho(\vec{r},\vec{r}') = \int_{BZ} d\vec{k} \sum_{n(\text{filled})} \Psi_n^{\vec{k}} (\vec{r}) \, \Psi_n^{\vec{k}^+} (\vec{r}) \qquad (2.38)$$

In practice, one replaces the Brillouin-Zone integral by a weighted sum over a numerical mesh in an irreductible sector of the Brillouin zone :

$$\rho(\vec{r},\vec{r}') = \sum_{\vec{k}} W_{\vec{k}} \sum_{n(\text{filled})} \Psi_n^{\vec{k}}(\vec{r})\Psi_n^{\vec{k}+}(\vec{r}') \qquad (2.39)$$

A natural weight to use for a given mesh point is one proportional to that volume of the first zone which is closer to that mesh point than to any other mesh point.

A very important computational simplification that is employed in these calculations is the three approximations[24] of some of the less important two-electron integrals. It is found that all of the integrals involving the basis function product $\psi_a(1)\ \psi_b(1)$ can be zeroed or approximated if the charge distribution

$$n_{ab}(1) = \psi_a(1)\ \psi_b(1)$$

is small enough and if the zeroing or approximating is charge conserving. A charge conserving integral approximation has the property that the total nuclear charge for which electron-nuclear integrals are evaluated for any basis function product should be equal and opposite to the total electronic charge for which electron-electron coulomb integrals are evaluated for the same basis function product. This charge conserving property becomes increasingly important as the electronic system becomes larger because of the long range of the coulomb interaction.

One needs some rough measure of the strength of a charge distribution. When ψ_a and ψ_b are both basis functions of s symmetry, the overlap is an adequate measure. However, when p or higher basis functions are involved, the total charge can be misleading. For example, when ψ_a and ψ_b are on the same site and x and y symmetry respectively, the product charge distribution is large even though the total charge is zero. A good measure would be obtained by integrating the absolute value of $\psi_a(1)\ \psi_b(1)$. For computational simplicity ψ_a and ψ_b are reoriented on their respective sites to obtain the maximum overlap. This is called "pseudo-overlap", and it is used to measure the strength of the charge distribution. Also the pseudo-overlap is taken to be unity for all single-site basis function pairs.

The first integral approximation consists of zeroing all one and two-electron integrals which involve "small" basis function products $\psi_a(1)\ \psi_b(1)$. A "small" basis function product is one whose pseudo-overlap, S_{ab}, is less than some tolerance, τ_S. This approximation is charge conserving. The charge density can be expanded as

$$\rho(\vec{r}) = \sum_{cd} A_{cd}\ \psi_c^*(\vec{r})\psi_d(\vec{r})$$

so that the total electronic charge is

$$N = \sum_{cd} A_{cd}\, Q_{cd} \quad ; \quad Q_{cd} = \int \psi_c^*(\vec{r})\, \psi_d(\vec{r}) d\vec{r}.$$

In this approximation, the true overlap, Q_{cd}, is zeroed for "small" products. Consequently, by wave function normalization during the SCF calculation, the "large" basis function products alone carry the correct total electronic charge.

$$N = \sum_{\{c,d|S_{cd} > \hat{1}_s\}} A_{cd}\, Q_{cd}$$

Charge conservation for matrix element of "small" products is trivially satisfied because both electron-nuclear and electron-electron integrals are zeroed.

The next approximation for a two-electron integral over four basis functions replaces the many integrals over primitive gaussians by a single, or several, gaussian integrals. Only if the result exceeds some tolerance are the many integrals over primitive gaussians evaluated.

An electron repulsion integral has the form

$$I = \int \psi_a^*(\vec{r}_1)\, \psi_b(\vec{r}_1)\, \frac{1}{r_{12}}\, \psi_c^*(\vec{r}_2)\, \psi_d(\vec{r}_2)\, d\vec{r}_1 d\vec{r}_2$$

This gives the coulomb interaction energy of two charge contributions :

$$I = \int N_{ab}(\vec{r}_1)\, \frac{1}{r_{12}}\, N_{cd}(\vec{r}_2)\, d\vec{r}_1 d\vec{r}_2$$

where (for two s-symmetry basis functions)

$$N_{ab}(\vec{r}) = \psi_i(\vec{r}-\vec{R}_b)\, \psi_j(\vec{r}-\vec{R}_b) = \sum_{k=1}^{m_i m_j} C_k\, e^{-ak|\vec{r}-\vec{R}_k|}$$

Here we have used the well known fact that the product of two gaussians is again a gaussian. Hence an integral over four basis functins, each containing m_i primitive gaussians, consists of a sum of $(m_i m_j m_k m_l)$ integrals. The approximation consists of replacing each multi-term s-s charge distribution by a single gaussian.

$$\tilde{N}_{ab}(\vec{r}) = \vec{q}\left(\frac{\gamma}{\pi}\right)^{3/2} e^{-\gamma(\vec{r}-\vec{R})}$$

such that the total charge of $N_{ab}(r)$ is correct :

$$q = \sum_i c_i\, (\pi/a_i)^{3/2}$$

and the center of charge of N_{ab} is correct :

$$qR = \sum_i C_i \, (\pi/a_i)^{3/2} \, \vec{R}_i$$

The exponent parameter γ is adjusted to give the best least square fit to $N_{ab}(r)$:

$$\frac{\delta}{\delta r} \int [q \, (\frac{\gamma}{\pi})^{3/2} \, e^{-\gamma|\vec{r}-\vec{R}|^2} - N_{ab} \, (\vec{r})]^2 \, d\vec{r} = 0$$

When calculating p-like functions each lobe has to be done.

The third approximation like the first, zeros selected electron repulsion integrals. An integral involving a relatively small charge distribution N_{ab} with a large distribution N_{cd} must be kept, however one should be able to zero an integral involving the small N_{ab} with a small N_{cd} if careful attention is given to charge balance. A rough estimate of the importance of the integral is given by the pseudo-overlap product $S_{ab}S_{cd}$. If this product is smaller than some tolerance, we zero the integral. If nothing further were done, we would loose the balance between electronic and nuclear charge. The restoration of this charge balance must be done. Now :

$$\rho(\vec{r},\vec{r}\,') = \sum C_{ab} \, \psi_a(\vec{r}) \, \psi_b^*(\vec{r}\,')$$

For each basis function product in term, we add up the total electronic charge, T_{cd}, for which two-electron integrals were calculated.

$$T_{cd} = \sum C_{ab} \, Q_{ab} \, F_{cd}^{ab}$$

where

$$Q_{ab} = \int \psi_a(\vec{r}) \, \psi_b^*(\vec{r}) \, d\vec{r}$$

and

$$F_{cd}^{ab} = \begin{cases} 1 & \text{if } S_{ab} \cdot S_{cd} > \tau \\ 0 & \text{if } S_{ab} \cdot S_{cd} < \tau \end{cases}$$

The nuclear integrals involving this basis function product,

$$\langle \psi_a \, | \, \frac{Z}{|\vec{r}-\vec{R}|} \, | \, \psi_b \rangle$$

are all scalled by the factor $T_{ab/N}$, there being N electron in the system.

III. CALCULATED RESULTS OF THE CRYSTAL HARTREE-FOCK MODEL

In this section we will spend most of the time looking at the ground state properties of the model. This will include the total energy (also cohesive energy), the virial coefficient, the equilibrium lattice constants, bulk modulus, x-ray form factors and compton profiles. We will also show the Hartree-Fock eigenvalues. This correction to these eigenvalues will be the subject of the next two sections.

Table III. The total energy per cell and cohesive energy of several example compounds in the Hartree-Fock approximation. The values are given Ryds.

Compound	Total Energy	Cohesive Energy	Exp.
Diamond[4]	-75.752/atom	0.38/atom	0.56/atom
BN[25] (cubic)	-158.499/atom pair	0.65/atom pair	
LiF[26]	-214.18/atom pair	0.53/atom pair	0.69/atom pair
Neon[26]	-257.10/atom pair	0.02/atom pair	

Table III has examples of the total energy calculation and cohesive energy in the Hartree-Fock approximation. It is interesting to note that both in diamond and LiF well over 50% of the cohesive energy is accounted for by this model. In the case of neon it is found to be slightly bound; although this amount is in the numerical noise of the calculation. A measure of the accuracy of the Hartree-Fock model calculations is the ability of the calculations to achieve the correct value of the virial coefficient, namely $-2T/V = 1$. The values from different compounds are presented in Table IV, and as one can see there are very good values.

Some very interesting results are found in Table V. First of all the Hartree-Fock model gives lattice constants which agrees well with experiment. Next the crystal values reflect the same type of results calculated on molecular systems. For example most molecules calculated in the Hartree-Fock approximations have their equilibrium lattice constant slightly less than experiment.

Table IV. The virial coefficient, $-2T/_V$, value for the Hartree-Fock model.

Compound	Diamond[4]	BN(cubic)[25]	LiF[26]	Neon[26]
$-2T/_V$	1.0001	1.0012	0.99915	0.99967

Table V. The equilibrium lattice constant of the Hartree-Fock model Calculations the units are in Å.

Compound	Cal.	Exp.
Diamond[28]	3.547	3.567
LiF[26]	3.972	4.02
Li[27]	3.72	3.479

Whereas, in the case of Li_2, the molecular Hartree-Fock calculation shows the value to be larger than experiment. These results are also shown in Fig. 1. From the total energy calculations of Fig. 1, one can also obtain the bulk modulus.[28] The bulk modulus, B, is defined by the relations

$$U(a) = 1/2\, B\, \delta^2 \qquad (3.1)$$

where U(a) is the energy per unit cell, δ is the dilation, and a is the lattice constant. Near equilibrium,

$$U(a) = 1/2 \left.\frac{d^2U(a)}{d\,a^2}\right|_{a=a_o} (a-a_o)^2 \qquad (3.2)$$

Using the relation for cubic crystals

$$\delta = 3(a-a_o)\,/_{a_o} \qquad (3.3)$$

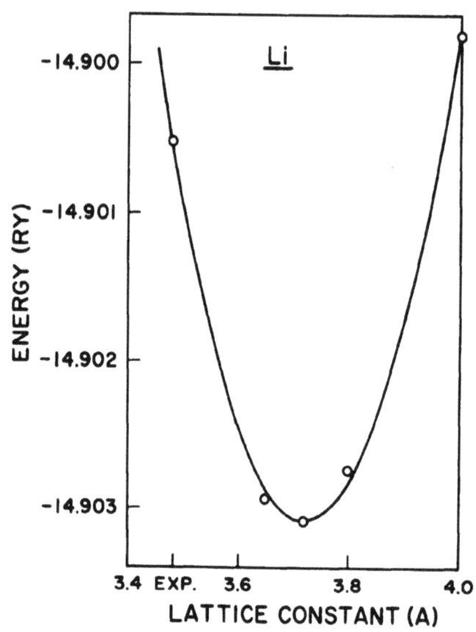

Fig. 1. The least-squares-fitted parabolas of total energy as a function of lattice constant. The calculated points are denoted by the dots.

one obtains the usual relation for B :

$$B = \frac{a_o^2}{q} \frac{d^2 U(a)}{d a^2} \Big|_{a=a_o} \qquad (3.4)$$

Thus one direct way of obtaining the bulk modulus is to determine the energy per unit cell for a number of lattice constants, fit this data to some curve, and then determine the second derivative at equilibrium. Several examples are given in Table VI. Here again force constants calculated for molecules are almost always slightly higher and the bond lengths are usually slightly less than experiment.[29]

Table VI. The bulk modulus of the Hartree-Fock model and experiment.

Compound	Cal.	Exp.
Diamond[28]	4.38×10^{12} dyn/cm^2	4.42×10^{12} dyn/cm^2
LiF[26]	7.54×10^{11} dyn/cm^2	6.71×10^{11} dyn/cm^2
Li[27]	1.11×10^{11} dyn/cm^2	1.21×10^{11} dyn/cm^2

Next we turn our attention to the first-order density matrix. The Fourier transform of the charge density are obtained from

$$\rho_o(\vec{k}) = \frac{1}{\Omega} \int e^{i\vec{k}.\vec{r}} \rho(\vec{r},\vec{r})\, d\vec{r} \qquad (3.5)$$

These values can be obtained experimental from x-ray structure factors. The values for several examples are found in Table VII. Here again one finds very good agreement with experiment without any correlation. Similarly, the electron momentum distribution is obtained from

$$\rho_1(\vec{k},\vec{k}) = \frac{1}{(2\pi)^3\, \Omega} \int e^{i\vec{k}.\vec{r}} \rho_1(\vec{r},\vec{r}\,') e^{-i\vec{g}.\vec{r}\,'} d\vec{r} d\vec{r}\,' \Big|_{\vec{g}=\vec{k}} \qquad (3.6)$$

From the above, the radial momentum distribution is then

$$I_o(k) = \int k^2 \rho(\vec{k},\vec{k})\, d\sigma \tag{3.7}$$

where $d\sigma$ is an element of solid angle in mementum space. Within the framework of the impulse approximation [31,32] the intensity of the Compton profile J(q) is given by

$$J(q) = 1/2 \int_{|q|}^{\infty} \frac{I_o(k)}{k}\, dk \tag{3.8}$$

Table VII. The experimental and calculated structure factors in electrons per unit cell. The Debye-Waller corrections were removed. The reciprocal lattice vectors are labelled by the integers h, k and ℓ.

hkl	Diamond[4]		Si[30]		LiF[26]	
	Cal.	Exp.	Cal.	Exp.	Cal.	Exp.
111	3.25	3.32	10.751	10.739	4.98	4.84
220	1.94	1.98	8.613	8.651	5.72	5.71
311	1.69	1.66	8.024	8.024	2.36	2.37
222	0.08	0.14	0.108	0.169	4.60	4.61

An example of the results obtained[30] are shown in Fig. 2. The SCF Hartree-Fock results are remarkably close to the experimental results. Correlation effects on J(q) are apparently small for crystalline solids. In all cases calculated, the general trend is for the calculated profiles are several percent above experiment for small momentum transfer. This is accompanied with under shooting experiment at large momentum transfer. This same type of results are found in atomic calculations.[33]

In Table VIII, there are given several eigenvalues differences of diamond and LiF. One finds, contrary to the ground state properties, relatively poor agreement with experiment. It will now be the subject of the next two sections to find out what is missing and how to correct these values.

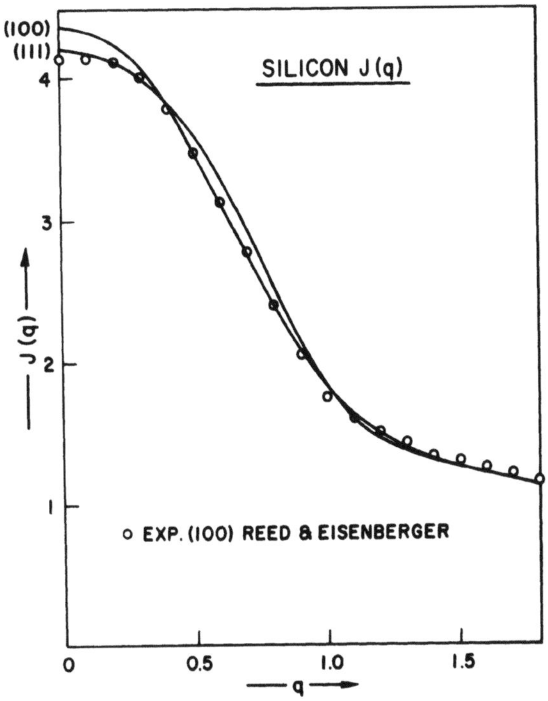

Fig. 2. Experimental (dots) and Hartree-Fock model (line) Compton profiles

Table VIII. Figenvalue differences for diamond and LiF in the Hartree-Fock calculations along with experiment. The values are given in eV.

	Diamond[4]		LiF[24]	
	HF	Exp.	HF	Exp.
Direct Gap at Γ	13.7	7.3	22.9	13.6
Indirect min Gap	12.1	5.6		
Gap at X	22.2	12.5		
Gap at L	22.8			

IV. CORRELATION CORRECTIONS : LONG RANGE

Various methods have been developed for adding correlation to Hartree-Fock excitations. The eigenvalues of the Fock operator were shown to represent ionization energies if the system did not relax (Koopmans' Theorem).[18] So the first step will be to include relaxation effects. We will outline the method proposed by Hedin[9] and Lundqvist[10] first, and we will also look at some results of the model of the electronic polaron.[11]

The method of Hedin and Lundqvist is called screened exchange plus coulomb hole (SECH). To start, look at the single particle theory. The one-electron states u_k and corresponding energies ε_k satisfy the equation

$$[h(\vec{r}_1) + V(\vec{r}_1)]\; u_k(\vec{r}_1) + \int \Sigma\; (\vec{r}_1,\vec{r}_2)\; U_k(\vec{r}_2)d\vec{r}_2 = {}_k U_k(\vec{r}_1) \quad (4.1)$$

where h is the kinetic energy plus the interaction of the electron with the nuclei, V is the Coulomb repulsion due to the other electrons, and Σ is the self energy which contains all the exchange and correlation effects. Eq. (4.1) applies whenever the self energy is independent of frequency, which is generally not the case. In the SECH method, we expand the self energy in powers of a dynamically screened interaction $w(\vec{r}_1,\vec{r}_2,t_1-t_2)$, rather than a bare instantaneous Coulomb interaction, and keep the first term. Since the screened interaction is a sharply peaked function of t_1-t_2, we replace it by its integrated value times a delta function, which corresponds to an averaged instantaneous interaction. When the Fourier transform with respect to t_1-t_2 is then taken, we obtain the frequency-independent SECH approximation for the self energy

$$\Sigma_{SECH}(\vec{r}_1,\vec{r}_2) = -\rho(\vec{r}_1,\vec{r}_2)\, W(\vec{r}_1,\vec{r}_2,\omega=0) + 1/2\, \delta(\vec{r}_1,\vec{r}_2)\, [\, W(\vec{r}_1,\vec{r}_2,\omega=0)-V(\vec{r}_1,\vec{r}_2)]$$

$$\rho(\vec{r}_1,\vec{r}_2) = \sum_{k=1}^{N} U_k^*(\vec{r}_2)U_k(\vec{r}_1) \tag{4.2}$$

$$V(\vec{r}_1,\vec{r}_2) = \frac{e^2}{|\vec{r}_1-\vec{r}_2|}$$

where ω is the frequency, e is the charge of an electron, and the screened interaction W is defined in terms of the dielectric response function ε

$$W(\vec{r}_1,\vec{r}_2,\omega) = \int V(\vec{r}_1,\vec{r}_2)\varepsilon^{-1}(\vec{r}_3,\vec{r}_2,\omega)d\vec{r}_3 \tag{4.3}$$

The first term in Eq. (4.2) leads to a statically screened exchange (SE) while the second produces a Coulomb hole (CH).

If the self energy were given by

$$\Sigma_{HF}(\vec{r}_1,\vec{r}_2) = -\rho(\vec{r}_1,\vec{r}_2)\, V\,(\vec{r}_1,\vec{r}_2)$$

then Eq. (4.1) would become the Hartree-Fock (HF) equation which has only an unscreened exchange term. Let us denote the HF one-electron wave functions and energies by $u_{\vec{k}n}$ and $\varepsilon_{\vec{k}n}$ respectively where k is a reciprocal space vector restricted to the first zone and n is a band index. Since we are already in possession of a set of HF wave functions and energies, we can use first order perturbation theory to obtain corrections $\Delta\varepsilon_{\vec{k}n}$ to the HF energies due to using Σ_{SECH} in Eq. (4.1). Considering the HF Hamiltonian to be unperturbed and letting the difference between Σ_{SECH} and Σ_{HF} be the perturbation, the correction to the energy is then given by

$$\Delta\varepsilon_{\vec{k}n} = -\int u_{\vec{k}n}(\vec{r}_1)\rho(\vec{r}_1,\vec{r}_2)\,[W\,(\vec{r}_1,\vec{r}_2,\omega=0)-V(\vec{r}_1,\vec{r}_2)]\,u_{\vec{k}n}(\vec{r}_2)d\vec{r}_1d\vec{r}_2$$

$$+\ 1/2\int u^*_{\vec{k}n}(\vec{r}_1)\delta(\vec{r}_1,\vec{r}_2)\,[W(\vec{r}_1,\vec{r}_2,\omega=0)-V(\vec{r}_1,\vec{r}_2)]\,u_{\vec{k}n}(\vec{r}_2)d\vec{r}_1d\vec{r}_2 \tag{4.4}$$

These energy shifts, which are correletion corrections, have been obtained by using Green function theory to go beyond the HF approximation.

Expressing the bare screened interactions in terms of their Fourier transforms, we have

$$v(\vec{r}_1,\vec{r}_2) = \frac{1}{(2\pi)^3}\int v(q)\, e^{i\vec{q}.(\vec{r}_1,\vec{r}_2)}d\vec{q} \qquad (4.5)$$

$$W(\vec{r}_1,\vec{r}_2,\omega) = \frac{1}{(2\pi)^6}\int W(\vec{q},\vec{q}',\omega)e^{i\vec{q}.\vec{r}_1-i\vec{q}'.\vec{r}_2}\, dqdq' \qquad (4.6)$$

$$W(\vec{q},\vec{q}',\omega) = v(\vec{q})\varepsilon^{-1}(\vec{q},\vec{q}',\omega) \qquad (4.7)$$

$$v(\vec{q}) = \frac{4\pi e^2}{q^2} \qquad (4.8)$$

The HF states, which are calculated using the LCAO method, have the form

$$U_{\vec{k}n}(r) = \frac{1}{\sqrt{N}}\sum_{i\nu} b_{ni}(\vec{k})\, e^{i\vec{k}.\vec{R}_\nu}\psi_i(\vec{r}-\vec{R}_\nu) \qquad (4.9)$$

where the sums are over atomic orbitals ϕ_i and direct lattice vectors $R_{\vec{\nu}}$ the b_{ni}'s are the coefficients of the Bloch functions associated with the atomic orbitals and N is the number of unit cells in the crystal. Using Eq. (4.9), the Fourier transform of the dielectric function can be written as

$$\varepsilon(\vec{q},\vec{q}',\omega) = (2\pi)^3\delta(\vec{q}'-\vec{q}) - \sum_{\vec{K}_2} P(\vec{q},\vec{q}+\vec{K}_2,\omega)v(\vec{q}+\vec{K}_2)\delta(\vec{q}'-\vec{q}-\vec{K}_2) \qquad (4.10)$$

where P is the polarization and $\vec{K}_2$ is a reciprocal lattice vector.

We now make the approximation of keeping only the $\vec{K}_2=0$ term in Eq. (4.10), in which case the inverse dielectric function is readily obtained

$$\varepsilon^{-1}(\vec{q},\vec{q},\omega) = \frac{(2\pi)^3}{\varepsilon(\vec{q},\omega)}\,\delta(\vec{q},\vec{q}') \qquad (4.11)$$

$$\varepsilon(\vec{q},\omega) = 1-\frac{1}{(2\pi)^3}\, v(\vec{q})\, P(\vec{q},\vec{q},\omega) \qquad (4.12)$$

$$P(\vec{q},\vec{q},\omega) = \frac{2(2\pi)^3}{V_c}\sum_{k,n,n'}\frac{\eta_{\vec{k}+\vec{q}+\vec{K}_1,n} - \eta_{\vec{k},n'}}{\varepsilon_{\vec{k}+\vec{q}+\vec{K}_2 n}-\varepsilon_{\vec{k}n'}-\omega}|\langle\vec{k}n'|e^{-i\vec{q}.\vec{r}}|\vec{k}+\vec{q}+\vec{K}_1,n\rangle|^2 \qquad (4.13)$$

The factor of 2 in front of the sum is due to spin, V_c is the volume of the crystal, $\eta_{\vec{k}n}$ is the occupation number (either 0 or 1) of the state $u_{\vec{k}n}$, $\vec{K}_1$ is the reciprocal lattice vector such that $\vec{k}+\vec{q}+\vec{K}_1$ is in the first zone, and $|\vec{k}n\rangle = u_{\vec{k}n}$

Using Eqs. (4.5-9) and (4.11), the energy shifts can be written as

$$\Delta\varepsilon_{\vec{k}n} = -\frac{1}{Vc} \sum_{\substack{\vec{k},\vec{k}',n' \\ (\text{occupied})}} |\langle \vec{k}'n'| e^{-i(\vec{k}-\vec{k}'+\vec{K}).\vec{r}} | \vec{k}n\rangle|^2 \, v(\vec{k}-\vec{k}'+\vec{K})$$

$$\times \left[\frac{1}{\varepsilon(\vec{k}-\vec{k}'+\vec{K},\omega=0} -1\right] + \frac{1}{2(2\pi)^3} \int \left[\frac{1}{\varepsilon(\vec{q},\omega=0)} -1\right] v(\vec{q})d\vec{q} \qquad (4.14)$$

where the sum on $\vec{K}$ is over reciprocal lattice vectors. These energy shifts are state-dependent due to the $|\vec{k}n\rangle$ dependence in the first term, which is the difference between a screened and unscreened exchange. The second term, the Coulomb hole, is constant in our diagonal approximation for the dielectric function, however, if the full dielectric matrix rather than just the diagonal part is used, then the Coulomb hole term is also state-dependent. Making the approximation that the dielectric function $\varepsilon(q,\omega=0)$ falls off to 1 outside of the first zone, we keep only the reciprocal lattice vector $\vec{K}$ in the above sum such that $\vec{k}-\vec{k}'+\vec{K}$ is the first zone.

The dielectric function used in this calculation[34,35] is the diagonal part of the one appearing in Eq. (4.10), which is called the random phase approximation (RAP) dielectric function. Since we are using first order perturbation theory to obtain the energy shifts, the dielectric function must be computed with HF wave functions and energies. The HF eigenvectors and energies were initially determined at 20 points in 1/48 of the Brillouin zone, and then the eigenvectors (the coefficients b_{ni} in Eq. (4.9) were promoted to obtain the HF wave fuctions at 341 points throughout the zone. These 341 zone points were used to compute the diagonal part of the RPA at 20 points in 1/48 of the first zone, with the 6 occupied bands and lowest 12 conduction bands being included in the computation. As shown by Table IX the RPA appears to be well converged after 12 conduction bands.

A simpler expression for the dielectric function is provided by the Penn model[36] which is a semi-empirical model that takes on the experimental value for the optical dielectric constant when $\vec{q} = 0$. Fig. 3 compares the RPA computed with HF wave functions and energies (HF RPA), the RPA computed with HF wave functions and SECH correlated energies (SECH RPA), and the Penn model dielectric function. We see that the HF RPA is smaller than the experimental value as represented by the Penn model, with the difference being due partly to the uncorrelated energy differences that appear in the denominator of Eq. (4.13). The SECH RPA curve shows that correcting these HF energy differences makes up less than half of the difference between the HF RPA and the Penn model. Three factors which contribue to the remaining difference between the HF RPA and experiment are the

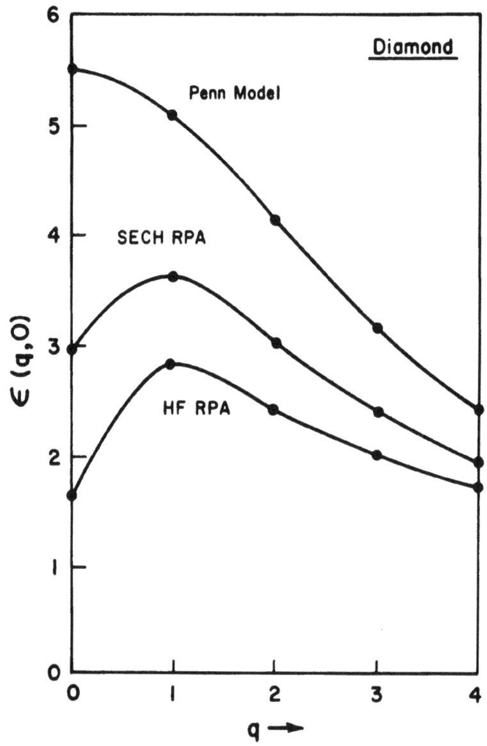

Fig. 3 Diamond dielectric function along the Δ axis.

uncorrelated wave functions appearing in the matrix elements in Eq. (4.13), the neglect of higher order polarization terms beyond the radom phase approximation, and the neglect of the off diagonal terms of the dielectric matrix (the $K_2 \neq 0$ terms in Eq. (4.10) which correspond to local field effects. Since we have computed only the diagonal part of the RPA thus far, the Penn model will be used to calculate the SECH energy shifts, however, a comparison will be made with the energy shifts obtained using the HF RPA and the SECH RPA.

Table IX. RPA for diamond at $\bar{q} = \frac{\pi}{2a}(1,00)$ versus total number of conduction bands used in the calculation.

Bands	$\varepsilon(\vec{q},0)$
1	1.885
2	2.348
3	2.668
4	2.760
5	2.776
6	2.791
7	2.803
8	2.812
9	2.817
10	2.820
11	2.821
12	2.822

Fig. 4 shows the uncorrelated HF energy bands which yield energy differences that are too large in comparison with experiment The corresponding HF eigenvectors were determined at 341 points throughout the Brillouin zone and then used to compute the SECH energy shifts for the 6 occupied bands and first 10 conduction

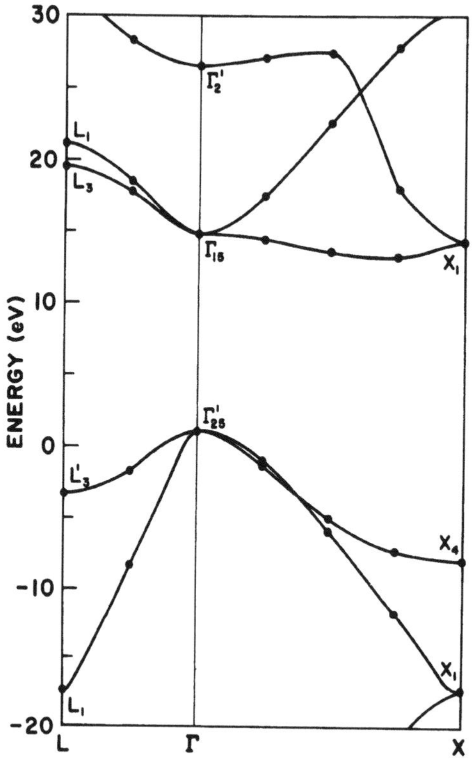

Fig. 4. Uncorrelated Hartree-Fock energy bands for diamond.

bands at 20 points in 1/48 of the first zone. For core and valence bands, the first term in Eq. 14 is larger than the second, producing positive energy shifts, while for conduction bands, the Coulomb hole term dominates and the energy shifts are negative. Thus the correlation corrections raise the occupied bands and lower the conduction bands with a resulting decrease in energy differences. Energy shifts, obtained with the Penn model dielectric function, are shown in Table X for several points in the first zone. These shifts are quite state-dependent, varying by as much as 2 eV over the zone and by much as 3 eV (in magnitude) over bands. In general the correlation corrections tend to flatten the bands, e.g., the shifts for the Δ_5 and Δ'_2 valence bands increase while the shift for Δ_1 valence band decreases as one moves out from the center of the zone along the Δ axis.

Table X. SECH energy shifts for diamond at several points in the first Brillouin zone. Shifts are in eV.

$\vec{q}$	Valence			Conduction		
	Γ_1	Γ'_{25}	Γ'_{25}	Γ_{15}	Γ_{15}	Γ'_2
$\frac{\pi}{2a}$(000)	5.53	2.45	2.45	-3.61	-3.61	-4.95
	Δ_1	Δ'_2	Δ_5	Δ_1	Δ_5	Δ'_2
(100)	5.52	2.66	2.57	-3.69	-3.81	-4.96
(200)	5.45	3.32	3.10	-3.84	-4.44	-4.98
(300)	5.23	4.09	3.51	-4.01	-4.98	-4.55
	X_1	X_1	X_4	X_1	X_3	X_1
(400)	4.77	4.77	3.68	-4.24	-5.23	-4.24
	Λ_1	Λ_1	Λ_3	Λ_3	Λ_1	Λ_1
(111)	5.48	3.37	2.78	-4.00	-4.14	-5.03
	L'_2	L_1	L'_3	L_3	L_1	L'_2
(222)	5.09	4.72	3.14	-4.29	-4.51	-5.33

Table XI gives a comparison of SECH band gap changes ΔE_g obtained with the HF RPA , and the Penn model dielectric funtions. We see that the diagonal parts of the HF RPA and SECH RPA produce respectively about 2/3 and 5/6 of the correlation obtained with the Penn model. Thus the correlated energies used in the SECH RPA make up about half of the difference between the HF RPA and Penn model results. The SECH RPA gap changes are relatively close to the Penn model results but still differ from them by an eV or more.

Table XI. Comparison of diamond band gap changes ΔE_g obtained with HF RPA, SECH RPA, and Penn model dielectric functions. Gap changes are in eV.

	(HF RPA) ΔE_g	(SECH RPA) ΔE_g	(Penn) ΔE_g
$\Gamma'_{25} \rightarrow \Delta_1$	4.3	5.5	6.5
$\Gamma'_{25} \rightarrow \Gamma_{15}$	4.0	5.1	6.1
$X_4 \rightarrow X_1$	5.4	6.7	7.9
$X_4 \rightarrow X_3$	6.1	7.6	8.9
$L'_3 \rightarrow L_3$	5.0	6.3	7.4

Fig. 5 shows the SECH correlated energy bands while Table XII compares the corresponding SECH correlated energy differences, obtained with the Penn modeln to the uncorrelated HF energy gaps at a few symmetry points. We see that the SECH correlation corrections significantly reduce the HF gaps, producing changes of almost 9 eV in some cases. The table also makes comparisons with experimental values for some of the energy differences and we see that the SECH results are in close agreement with the experimental indirect and direct band gaps. The agreement is best for the indirect gap which is the only piece of "hard" experimental information available for diamond. Thus the SECH method provides most of the correlation needed in diamond for the top of the valence band.

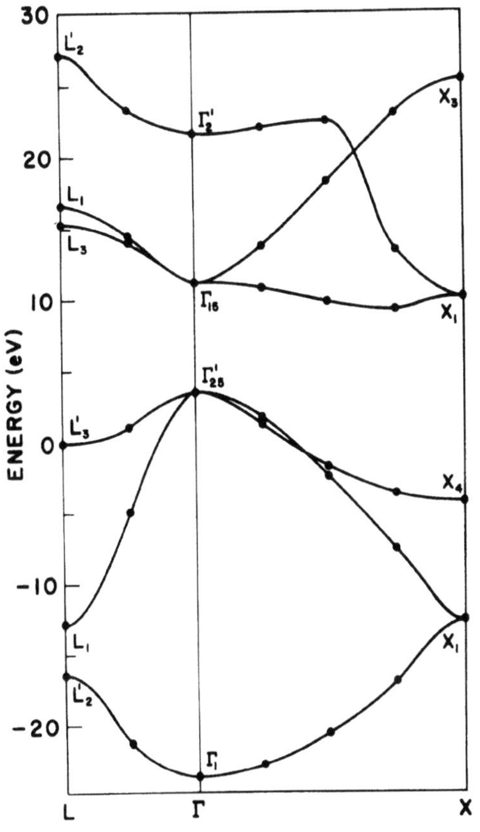

Fig. 5. SECH correlated energy bands for diamond.

Table XII. Diamond HF and SECH energy differences in eV. Some experimental values are also given.

	HF	SECH	Experiment
$\Gamma'_{25} \rightarrow \Delta_1$	12.1	5.6	5.5 - 5.6
$\Gamma'_{25} \rightarrow \Gamma_{15}$	13.7	7.6	7.3 - 7.4
$\Gamma_1 \rightarrow \Gamma'_{25}$	30.3	27.2	24.2[c]
$\Gamma'_{25} \rightarrow \Gamma'_2$	25.5	18.1	
$X_4 \rightarrow X_1$	22.2	14.3	12.5 - 12.6
$X_4 \rightarrow X_3$	38.5	29.6	
$L'_3 \rightarrow L_3$	22.8	15.3	
$L'_3 \rightarrow L_1$	24.3	16.7	
$L'_3 \rightarrow L'_2$	35.6	27.2	

The Penn model dielectric function is derived as an approximation to the diagonal part of the RPA, however, it is then forced to assume the experimental value for the optical dielectric constant when $\vec{q} = 0$. The fact that the Penn model gives such good correlated energy differences indicates that if one uses a dielectric function having proper (experimental) $\vec{q} = 0$ behavior, then the SECH method works well in the case of diamond. However it does not give all of the corrections needed in the case of LiF.[37] First of all, as show, in Fig. 6, the comparison of the different approximations of the dielectric function are farther apart. Looking at Table XIII, we will return to LiF excitation energies in the next section as we add short range correlation effects.

Table XIII. LiF gap energies in the HF approximation, SECH approximation, and the experimental fundamental gap.

	HF	SECH	Experiment
$\Gamma_{15} \rightarrow \Gamma_1$	22.9	17.9	13.6
$X'_5 \rightarrow X'_4$	31.2	26.0	
$L'_3 \rightarrow L'_2$	25.0	19.9	

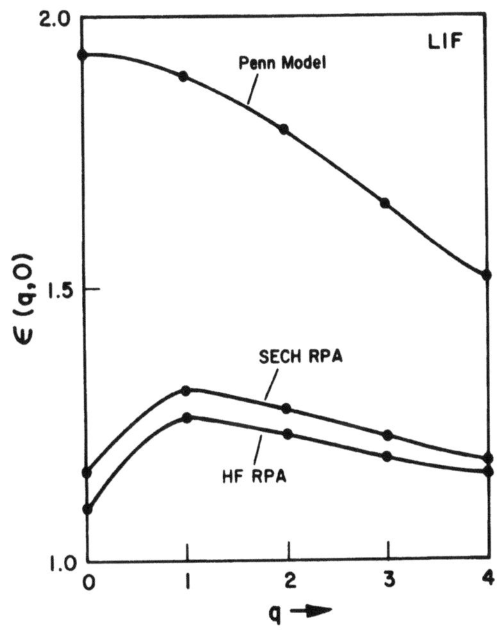

Fig. 6. LiF dielectric function along the Δ axis.

One can also obtain the same type of shifts described above by using the electron polaron model.[11] In fact one calculated approximately the same values found above. Our attention in this case will be to a new resonance state which comes out of the many body effects.[38] The system is described by

$$H = \varepsilon_{\vec{k}} \hat{C}^{+}_{\vec{k}} \hat{C}_{\vec{k}} + \hbar \omega_{ex} \sum_{\vec{q}} \hat{b}^{+}_{\vec{q}} \hat{b}_{\vec{q}} + e \left[\frac{2\pi\hbar\omega_{ex}}{V_c} (1 - {}^{1}/_{\varepsilon_\infty}) \right]^{1/2} \sum_{\vec{q}} \frac{i}{|\vec{q}|} \{ \hat{b}^{+}_{-\vec{q}} - \hat{b}_{\vec{q}} \} \hat{C}^{+}_{\vec{k}+\vec{q}} \hat{C}_{\vec{k}} \tag{4.15}$$

Here ε_k is the "bare" energy ; $\hat{c}^{+}_{k}, \hat{c}_k$ are the fermion creation and annihilation operators ; $h\omega_{ex}$ is the energy of the longitudinal exciton (assumed to be $\vec{q}$ independent) ; $\hat{b}^{+}_{q}, \hat{b}_q$ are the boson creation and annihilation operators ; Vc is the crystal volume , and ε_∞ is the optical dielectric constant. To second-order perturbation theory, the ground-state energy of the electron is

$$\varepsilon'_{\vec{k}} = \varepsilon_{\vec{k}} + \sum_{\vec{k}} \frac{|V_{\vec{k}}|^2}{\varepsilon_{\vec{k}} - \hbar\omega_{ex} - \varepsilon_{\vec{k}-\vec{k}}} \tag{4.16}$$

with

$$V_k = e \left[\frac{2\pi\hbar\omega_{ex}}{V_c} (1 - {}^{1}/\varepsilon_\infty) \right]^{1/2} \frac{i}{|\vec{k}|} \int \psi^2(\vec{r})\, e^{i\vec{k}.\vec{r}}\, d\vec{r} \tag{4.17}$$

$\phi(r)$ is the Wannier function for the band in question.

We now find a coupling constant for the electron-exciton interaction. We set

$$V_{\vec{k}} = -i \frac{\hbar\omega_{ex}}{|\vec{k}|} \left(\frac{\hbar}{2m\omega_{ex}} \right)^{1/4} \left(\frac{4\pi\alpha}{V_c} \right)^{1/2} \int \psi^2(\vec{r})\, e^{i\vec{k}.\vec{r}}\, d\vec{r} \tag{4.18}$$

Then one has

$$\alpha = \frac{e^2}{2\sqrt{\hbar/_{2m\omega_{ex}}}} \left(\frac{1}{\hbar\omega_{ex}} \right) (1 - {}^{1}/_{\varepsilon_\infty}) \tag{4.19}$$

The Hamiltonian in Eq. (4.15) is formally equivalent to the Fröhlich polaron Hamiltonian, within the effective mass approximation, and we just use the results of Ref. 39. The optical absorption due to the interaction of electron and exciton field is found to be

$$\Gamma(\Omega) = \left(\frac{1}{\varepsilon_o\, c n}\right) \frac{(\Omega-\omega_{ex})\, \mathrm{Im}\,[X(\Omega-\omega_{ex})]}{(\Omega-\omega_{ex})^4 - 2(\Omega-\omega_{ex})^2\, \mathrm{Re}\,[X(\Omega-\omega_{ex})] + |X(\Omega-\omega_{ex})|^2} \tag{4.20}$$

where

$$X(\Omega-\omega_{ex}) = (\Omega-\omega_{ex})^2 + i(\Omega-\omega_{ex})\, Z\,(\Omega-\omega_{ex}) \tag{4.21}$$

$Z(\Omega-\omega_{ex})$ is the impedance, Ω the frequency of the incident light, ε_o the permittivity of free space, n the index of refraction for medium, and c the velocity of light. We also obtain the energy-loss equation

$$-\,\mathrm{Im}(1/\varepsilon) = \frac{\mathrm{Im}\,[X(\Omega-\omega_{ex})]}{[(\omega_{ex}-\Omega)^2+1]^2 - 2\,[(\omega_{ex}-\Omega)^2+1]\,\mathrm{Re}[X(\Omega-\omega_{ex})] + |X(\Omega-\omega_{ex})|^2} \tag{4.22}$$

For $\alpha \leqq 1$ the optical absorption has a threshold at Ω = band gap $+\omega_{ex}$. The maximum occurs for Ω = band gap + 1.20 ω_{ex} and then decreases slowly (see Fig. 7b). This is due to the interaction of the electron and exciton field. Physically we have light creating and electron-hole pair which scatters, emitting an exciton (long.) plus an electron-hole pair. An interesting change occurs if one computes the energy loss. Here the peak occurs at Ω = band gap + 1.25 ω_{ex}. Thus there is a relative shift in peak position of absorption and energy loss. This is seen experimentally in Fig. 7a.

In this section we have been concerned with the long range correlation corrections. However as pointed out by the LiF example there is more to be added before one gets excellent agreement with experiment. Thus in the next section we will deal with this problem.

V. CORRELATION CORRECTIONS : SHORT RANGE

In this section we want to consider two effects. The first is to obtain corrections to the fact that in many crystals, some of the excitations are local in nature and polarize the near electrons.[12,13] The other effects is to include the hole-electron interactions which arise from these local excitations.[15] This makes it important that one calculates excitations of the N-system instead of the N $\pm$1 systems.

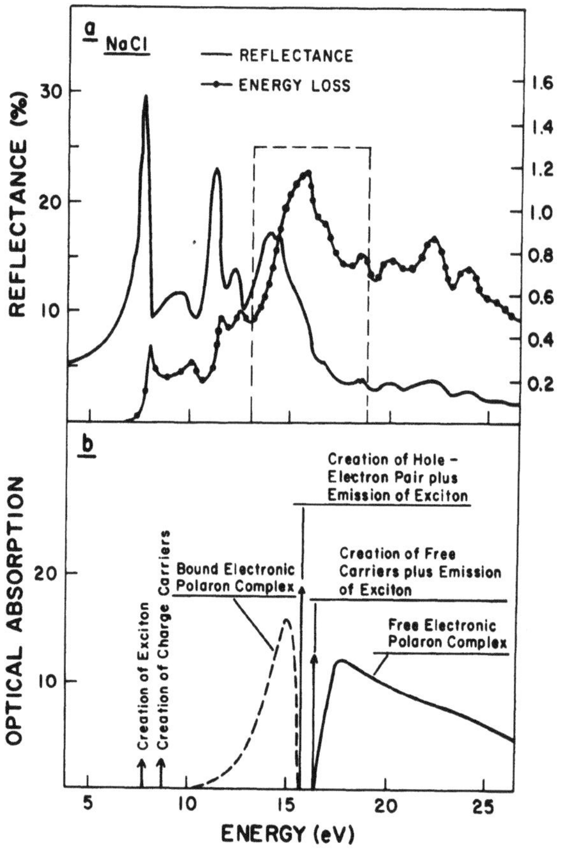

Fig. 7. (a) Experimental reflectance and energy loss for NaCl.
(b) The optical absorption for the bound electronic polaron excited state, and for the free electronic polaron excited state. (arbitrary units).

Ref. 12 has absorption of the soft x-ray by insulators with a forbidden exciton transition. Let's look at, e.g. Li K shell transitions to the lowest conduction band (which is s-like at Γ). Having calculated the Hartree-Fock model, it is found that the absorption peak is off by about 20 eV. Using the electronic-polaron model discussed in the previous section, one finds a net correction of 5.5 eV. In this case only ~ 25 % of the difference between theory and experiment. Clearly additional effects which go beyond band theory are needed. The most likely source of such an effect is that the core hole is localized rather than being in a band state. What is needed then is an estimate of the size of the energy correction due to the presence of a localized hole in the 1s orbital of Li in LiF. The principal effects will be these :

(1) The occupied 1s orbital on the Li ion containing the 1s hole is polarized by the hole.

(2) The valence-electron cloud on the fluorine ions surronding the Li ion with the 1s hole will be polarized from the band orbitals they are in.

(3) The local hole in the Li 1s shell will have a coulomb attraction for the electron excited to the conduction band. This gives the change from a V^N potential seen by a conduction electron in the Hartree-Fock theory to that of a V^{N-1} potential. This is of course the physical case. Such effects will include the formation of excitons, both free and bound.

Effects (1) is simple to estimate. Compute the difference in the energy needed to remove a Li^+ 1s electron from a Li^+ ion is free space using Koopmans' theorem and using the relaxed Hartree-Fock final state. This correction is found to be 1.5 eV. Effect (2) is calculated using the Mott-Littleton model to compute the effect of a static hole on a Li^+ ion on the surronding F^- ions.[8] The size of this effect is about 3.0 eV. Effect (3) is most difficult to evaluate. We do this using an atomic model. We compute the energy of an extra electron on a Li^+ ion ; this corresponds to the usual band model in the Hartree-Fock limit and for the same excited state of the ion in a model in which a hole is present in the core. The difference in the energy of the outer electron corresponds to the effect of the hole on that electron. The energy correction is

$$[E(1s;Li^{++}) - E(1s;\ 2p;Li^{+})] - [E(1s^2,Li^{+})-E(1s^2 2p,Li)] \approx 10\,eV \tag{5.1}$$

This value seems reasonable in that careful analysis of the conduction wave functions which contribute most to the main ε_2 peak were found to be dominantly Li^+ atomic virtual states in character. Thus the correction due to the hole localization is estimated to be about 14.5 eV and the total correction from all effects is estimate to be 20.0 eV.

In Ref. 13 a similar procedure was used for the valence and conduction band corrections. We calculate using a series of atomic Hartree-Fock calculations the difference in energies needed to remove an electron from the system or to add on an electron to the system (Koopmans' Theorem) and from the differences in energies for the self-consistent calculations for the two states. That is, this correction Σ^{local} is found, for example, for the creation of a valence hole on the 2p F^- ion to be (going from F^- to F)

$$\Sigma^{local}_{P.hole} = -\epsilon^{F^-}_{2p} + E^{F-}_{HF} - E^{F}_{HF} \approx 3.6 \text{ eV} \qquad (5.2)$$

Similarly, calculation for adding an extra electron into the Li^+ - ion site, where it is presumed the conduction electron to be largely localized gives a correction to be about - 1.0 eV. These corrections plus the shifts of the last section give a band of about 14 eV. If on top of this we add in the electron-hole interaction, the results of Fig. 8 are obtained.

Lets see now how to formulate a simple excitation operator which will generate eigenvalues which will be excitations to the N system.[15] In obtaining an excitation Hamiltonian consider to make sure that the excited states or virtual states see the correct type field. A localizing operator for the occupied space orbitals has been suggested by Adams[40] and Gilbert[41] which will give the same charge density as the Hartree-Fock charge density. Namely on has

$$A = \hat{\rho}\hat{A}\hat{\rho} \qquad (5.3)$$

where $\hat{\rho}$ is the charge density operator of the Hartree-Fock operator which has the property

$$\hat{\rho}^2 = \hat{\rho} = \hat{P} \qquad (5.4)$$

$\hat{A}$ is an arbitraty operator which is chosen to help reduce the computational problem. If ϕ_i is a virtual orbital, one has

$$A\psi_i = \hat{\rho}\hat{A}\hat{\rho}\psi_i = 0 \qquad (5.5)$$

On the other hand if ϕ_i is an occupied orbital, one has

$$A\psi_i \neq 0 \qquad (5.6)$$

Clearly the use of $\hat{\rho}$ as a projection onto the occupied space can be extended to form a projection operator onto the virtual space. One has simply

$$1 = \hat{\rho} + (1-\hat{\rho}) = \hat{P} + \hat{O} \qquad (5.7)$$

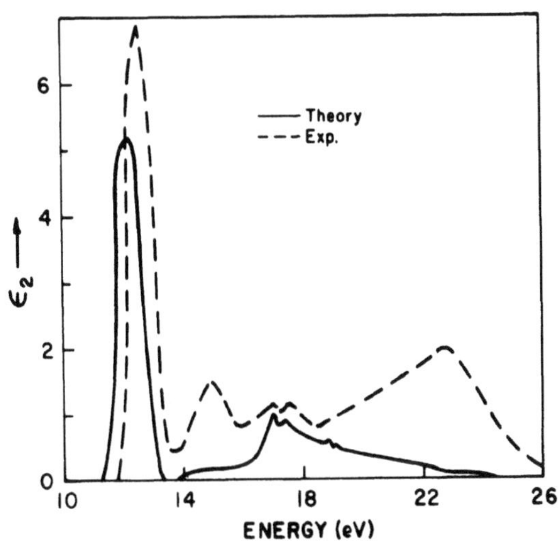

Fig. 8. The imaginary part of the LiF dielectric function from experiment and the calculated results for this quantity using corrected Hartree-Fock bands which have been corrected for both long and short range correlation.

and one can form an operator B of the form

$$B = \hat{O}\hat{A}\hat{O} \tag{5.8}$$

Here again $\hat{A}$ is an operator which is at the present an arbitrary operator. In obtaining correlation energy in Ref. 42, a substantial improvement of the rate of convergence of the configuration interaction calculation over the use of normal solutions to the Fock equation. In this calculation $\hat{A}$ was a three-dimensional square well. The best results were obtained with a radius of the square well at three atomic units and a depth of one Ryd. A similar basis set was generated in Ref. 43, and used in a linked-cluster many-body perturbation theory of Brueckner[44] and Goldstone[45] calculation. In this calculation the operator $\hat{A}$ was chosen to have the form $-\beta/r$ with $\beta = 1$.

Let us now turn to the main objective ; that is to formulate an $\hat{A}$ such that when $\hat{O}\hat{A}\hat{O}$ added to the Fock operator forms an operator whose eigenvalues are the excitations of the system being investigated. The removal of an electron of a core state of an atom will leave behind a coulomb potential of the form

$$V(\vec{r}_1) = \int \psi_c^*(\vec{r}_2) \frac{e^2}{|\vec{r}_1 - \vec{r}_2|} \psi_c(\vec{r}_2) d\vec{r}_2 \tag{5.9}$$

There will be some change in the charge density. However since we are looking at core states, the nuclear attraction is the major force and the removed orbital or hole will retain nearly the same shape ; however the outer shells may relax substantially. One has also an exchange potential, but since we are looking at excitations from the s-like core orbitals going to excited states in this first study, this is ignored. The major change in the remaining electron charge density will come from the outer electron orbits. This gives relaxation and causes the energy change between the ionized atom and the ground state to be less than the Hartree-Fock eigenvalue. This can be taken into account by obtaining self-consistent solutions of both the atom and ion and taking the total energy difference. Or one could calculate to a high degree of accuracy the energy change caused by the mixing of the top orbitals through second order perturbation theory where the perturbation is the potential given in Eq. (5.9).

This is one more major effect which will be discussed more fully later. That is the excited electron will not see the exact potential of Eq. (5.9), but this potential will be screened by the outer electrons and thus somewhat reduced.

It is useful at this point to develop specific mathematical formulation of the ideas presented previously. We begin by defining a system Hamiltonian for an n-particle system in terms of general one body operators f_i and two body operators g_{ij}. We find

$$H = \sum_{i=1}^{N} f_i + 1/2 \sum_{\substack{i,j=1 \\ i \neq j}}^{N} g_{ij} \tag{5.10}$$

In terms of a Slater determental type wave function of the form

$$\Psi(\vec{r}_1 \ldots, \vec{r}_N) = (N!)^{-1/2} \det (\psi_i(\vec{r}_i)) = (N!)^{1/2} \hat{A}_N \prod_{i=1}^{N} \psi_i(\vec{r}_i) \tag{5.11}$$

Here $\hat{A}_N$ is the antisymmetrizing operator. The energy, E, is given as

$$E = \langle\Psi|H|\Psi\rangle = \sum_{i=1}^{N} \langle i|f_1|i\rangle + \frac{1}{2} \sum_{i,j=1}^{N} \{ \langle ij|\, g_{12}|\, ij\rangle - \langle ij|g_{12}|ji\rangle\} \tag{5.12}$$

If the orbitals are not permitted to relax, the energy needed to remove an electron in state ϕ_n, is given as

$$\Delta E_N = \langle N|f_1|N\rangle + \sum_{i=1}^{N} \{ \langle iN|g_{12}|iN\rangle - \langle iN|g_{12}|Ni\rangle\} \tag{5.13}$$

Let us assume Ψ is the system ground state and the orbitals ϕ are chosen so that the system energy is stationary one finds ϕ's satisfy the Hartree-Fock equation

$$\{f_1 + \sum_{i=1}^{N} \int |\psi_i(2)|^2 g_{12} d\tau_2\} \psi_j(1) - \sum_{i=1}^{N} \psi_i(1) \int \psi_i^*(2)\, \psi_j(2) g_{12} d\tau_2 = \varepsilon_j\, \psi_j(1) \tag{5.14}$$

The solutions to (5.14) fall into two classes, for $i \leq N$ for which one finds that ΔE_i as given by (5.13) is exactly equal to ε_i given by (5.14). This is a statement of Koopmans' theorem. However, there are solutions to (5.14) for which $a > N$. For such solutions the eigenvalues are not ionization energies of the N-particle system. These orbitals here are termed virtual and are labeled by a, b, etc. It is noted that the (1) includes space and spin degrees of freedom and that integration implies summation on spin variables.

Consider a ϕ_a solution to (5.13) and also the energy needed to remove the electron in ϕ_a from a N-body system where ϕ_a replaces ϕ_N say in (5.11) and the ϕ_i where we adopt the convention i,j are always less than or equal to n, are orbitals occupied in the ground state and solutions to (5.13). We find

$$\Delta E_a = \langle a|f_1|a\rangle + \sum_{i=1}^{N-1} \{ \langle ai|g_{12}|ai\rangle - \langle ai|g_{12}|ia\rangle \} \tag{5.15}$$

However the ε_a for (5.14) is given by

$$\varepsilon_a = \langle a|f_1|a\rangle + \sum_{i=1}^{N-1} \{ \langle ai|g_{12}|ai\rangle - \langle ai|g_{12}|ia\rangle \} \tag{5.16}$$

demonstrating the prior discussed failure of Koopmans' theorem for the virtual orbitals.

Consider now an operator of form $\hat{O}\hat{A}\hat{O}$, where we pick $\hat{A}$ to be given by, here

$$\hat{O}\hat{A}\hat{O} = \hat{O}\,\{(-1)\,\langle N|g_{12} - g_{12}\hat{P}_{12}|N\rangle\}\,\hat{O} \tag{5.17}$$

Here $\hat{P}_{12}$ is defined such that

$$\langle a|\hat{O}\,\langle N|g_{12}\hat{P}_{12}|N\rangle\hat{O}|a\rangle = \langle aN|g_{12}|Na\rangle \tag{5.18}$$

One may add $\hat{O}\hat{A}\hat{O}$ to (5.14) without disturbing the ground state solution so the stationary condition is satisfied. If this is done one gets an equation for a new set of orbitals ϕ_i or ϕ_a of the form

$$[\hat{F} + \hat{O}\hat{A}\hat{O}]\,\psi_i = \varepsilon_i\psi_i \tag{5.19}$$

For this equation we find a ψ_a of $\hat{A}$ is as in (5.17)

$$\varepsilon_a = \langle a|f_1|a\rangle + \sum_{i=1}^{N-1} \{ \langle ai|g_{12}|ai\rangle - \langle ai|g_{12}|ia\rangle \} \tag{5.20}$$

Thus for (5.19) and (5.17) one has a Koopmans' theorem for the virtual orbitals as well as occupied orbitals.

It is further simple to show that in the unrelaxed orbital limit (Koopmans' limit) the difference in the eigenvalues of (5.14) (5.17) are excitations of the N-particle system for single electrons. Assume electron in state N is excited to state a. Call this energy for excitation as ΔE_N^a. Using (5.12) one has if E_g is ground state energy and E_N^a is excited state energy

$$E_N^a - E_g = \Delta E_N^a \; \langle a|f_1|a\rangle + \sum_{i=1}^{N-1} \{ \langle ai|g_{12}|ai\rangle - \langle ai|g_{12}|ia\rangle \}$$

$$-\langle N|f_1|N\rangle - \sum_{i=1}^{N-1} \{\langle Ni|g_{12}|Ni\rangle - \langle Ni|g_{12}|iN\rangle\} = \varepsilon_a - \varepsilon_N \tag{5.21}$$

Therefore the eigenvalue differences correspond to the excitation energies of the N-particle system for excitation from state $|N\rangle$. Please note the choice of excitation orbital $|N\rangle$ is not special since any of the occupied states can be chosen to be $|N\rangle$.

Finally we consider the proper variational determination of the virtual ϕ_a. We show these are properly the solutions of (5.19) (5.17). Consider a given set of orbitals ξ_i, $i< N-1$ for the self consistent ground state Fock equation. Choose the ξ_a so that the energy of the state

$$\Psi_N^a = (N!)^{1/2} \hat{A}_N [\xi_a \prod_{i=1}^{N-1} \xi i] \tag{5.22}$$

is stationary with respect to ξ_a, ξ_a^*. If we require ξ_a be orthogonal to the occupied orbitals, one has for example

$$\delta_{\xi_a}^* \{ E_N^a - \sum_{\substack{i=1\\ i=a}}^{N-1} [\int \xi_a \xi_i^* d\tau - \delta_{ia}] \} = 0 \tag{5.23}$$

with

$$E_N^a = \sum_{i=1}^{N-1} \langle i|f_1|i\rangle + \langle a|f_1|a\rangle + 1/2 \sum_{i,j=1}^{N-1} \{ \langle ij|g_{12}|ij\rangle -$$

$$\langle ij|g_{12}|ji\rangle \} + \sum_{i=1}^{N-1} \{ \langle ai|g_{12}|ai\rangle - \langle ai|g_{12}|ia\rangle \} \tag{5.24}$$

The variation produces for the ξ_a the equation

$$[f_1 + \sum_{i=1}^{N-1} \int |\xi_i(2)|^2 g_{12} d\tau_2] \xi_a(1) - \sum_{i=1}^{N-1} \xi_i(1) \int \xi_i^*(2)\xi_a(2)$$

$$g_{12}\, d\tau_2 = \lambda_a \xi_a(1) \tag{5.25}$$

We see the form for the expectation value this operator with ξ_a is the same as for (5.19) (5.17) for ϕ_a. Thus then if we show ϕ_a to be the same as the ξ_a of (5.19) (5.17) are those which minimize the system energy for unrelaxed ϕ_i, $i \leq N-1$.

The proof is to consider the matrix of the operator $\hat{B}$ defined by Eq. (5.25) with respect to the solution ϕ_a of (5.19) and (5.17). We need to evaluate $\langle\phi_a|\hat{B}|\phi_b\rangle$. Now we know the operator in (5.19) is given as

$$\hat{F} = \hat{B} + \int |\psi_N(2)|^2 g_{12} d\tau_2 - \psi_N(1) \int \psi^*_N(2) g_{12} \hat{P}_{12} \qquad (5.26)$$

Therefore if $\hat{P}_{12}$ interchanges coordinates 1 and 2,

$$\begin{aligned}\langle a|\hat{B}|b\rangle &= \varepsilon_a \delta_{ab} - \langle aN|g_{12}|bN\rangle + \langle aN|g_{12}|Nb\rangle - \langle a|\hat{O}\hat{A}\hat{O}|b\rangle \\ &= \varepsilon_a \delta_{ab} - \langle aN|g_{12}|bN\rangle + \langle aN|g_{12}|Nb\rangle + \langle aN|g_{12}|bN\rangle - \\ &\quad \langle aN|g_{12}|Nb\rangle = \varepsilon_a \delta_{ab} \qquad (5.27)\end{aligned}$$

Thus we find that $\hat{B}$ is diagonal in the solution to (5.17) (5.19), hence, the energy is minimized as desired.

The results of atomis He are given in Table XIV. One finds very close agreement between experiment and the calculated values obtained using $\hat{O}\hat{A}\hat{O}$ formalism described above. Here $\hat{A}$ is defined by Eq. (5.9) as in all results presented in this section. The He atomic system should be the best for this choice of $\hat{A}$ because there is only one electron orbital which can screen and relax left in the system. Also since in the ground state the two electrons have opposite spin and the excited state measured has ^1S total spin, the exchange term not included in Eq. (5.9) is zero. It should also be noted with the basis used in this calculation only one bound Hartree-Fock state was obtained, and it was approximately 5 eV higher in energy where as the mean deviation from experiment was only 0.6 eV.

Moving up the atomic table, the next example system investigated was atomic Li excited states and the results are given in Table XV. The first four rows give excitation of the outer electron from the 2S to 3S-6S states. In this system the exchange neglected term is not zero but as can be seen by the comparisons has little effect. In fact the mean deviation of the first four rows is only ~ 0.2 eV. This means that the screening and relaxation effects are very small or tend to cancel each other. In the 5th through 9th row the electron is removed from the inter orbit. One finds that relaxation about the hole in the lower orbit appears to be the major factor. The hole electron effect causes a change of about 5 volts whereas the relaxation effects appears to be larger than 8 eV.

Table XIV. The excitation energies of He atom given in eV The first column of numbers contains the experimental values taken from Ref. 46. The second column of numbers is the calculated values using $\hat{O}\hat{A}\hat{O}$ where $\hat{A}$ is equal to V in Eq. (5.9). The third column of numbers is the bound Hartree-Fock eigenvalue differences.

Trans. States	Exp (eV)	$\hat{O}\hat{A}\hat{O}$ (eV) Δ eigenvalues	Hartree-Fock (eV) Δ eigenvalues
$1S^2 \Rightarrow 1S2S$	20.61	20.03	24.95
$1S^2 \Rightarrow 1S3S$	22.91	23.36	
$1S^2 \Rightarrow 1S4S$	23.67	24.34	
$1S^2 \Rightarrow 1S5S$	24.00	24.71	

Table XV. The excitation energies of Li atom given in eV. The first column of numbers contains the experimental values taken from Ref. 46. The second column of numbers is the calculated values $\hat{O}\hat{A}\hat{O}$ where $\hat{A}$ is equal to V in Eq. (5.9). The third column of numbers is the bound Hartree-Fock eigenvalue differences.

Trans. States	Exp. (eV)	$\hat{O}\hat{A}\hat{O}$ (eV) Δ eigenvalues	Hartree-Fock (eV) Δ eigenvalues
$1S^2sS \Rightarrow 1S^2 3S$	3.37	2.85	5.11
$1S^2 2S \Rightarrow 1S^2 4S$	4.34	4.15	5.18
$1S^2 2S \Rightarrow 1S^2 5S$	4.75	4.72	
$1S^2 2S \Rightarrow 1S^2 6S$	4.96	5.01	
$1S^2 2S \Rightarrow 1S2S^2$	54.3[a]	62.14	67.25
$1S^2 2S \Rightarrow 1S2S3S$		62.79	
$1S^2 2S \Rightarrow 1S2S4S$		66.14	
$1S^2 2S \Rightarrow 1S2S5S$		66.85	
$1S^2 2S \Rightarrow 1S2S6S$		67.14	

[a]Value taken from: J.A. Breaden and A.F. Burr, Rev. Mod. Phys. 39, 125 (1967)

In the next example system, namely excitations of atomic Be, one sees a slightly different type of trend in the results of the results of the first four rows of Table XVI. That is the calculated OAO values underestimate the excitations from experiment with a mean deviation of 0.7 eV. Thus screening of the hole-electron interaction which is not included in the calculation is done mainly by the remaining 2S electron. One has to also add correlation effects to obtaining the unrelaxed eigenvalue which accounts for some of this errors. However when the excitation is from the lower orbit relaxation of the hole is again the major factor as can be seen in row five of Table XVI.

Table XVI. The excitation energies of Be atom given in eV. The first column of numbers contains the experimental values taken from Ref. 46. The second column of numbers is the calculated values using $\hat{O}\hat{A}\hat{O}$ where A is equal to V in Eq.(5.9). The third column of numbers is the bound Hartree-Fock eigenvalue differences.

Trans. States		Exp. (eV)	OAO (eV) Δ eigenvalues	Hartree-Fock (eV) Δ eigenvalues
$1S^2 2S^2$ =>	$1S^2 2S3S$	6.78	5.53	8.20
$1S^2 2S^2$ =>	$1S^2 2S4S$	8.09	7.11	8.32
$1S^2 2S^2$ =>	$1S^2 2S5S$	8.59	7.78	
$1S^2 2S^2$ =>	$1S^2 2S6S$	8.84	8.13	
$1S^2 2S^2$ =>	$1S2S^2 3S$	108.5[a]	124.15	128.48
$1S^2 2S^2$ =>	$1S2S^2 4S$		127.21	
$1S^2 2S^2$ =>			128.05	
$1S^2 2S^2$ =>			128.41	

[a]Value taken from : J. A. Bearden and A. F. Burr, Rev. Mod. Phys. 39, 125 (1962)

VI. CONCLUSION

One is able to conclude that the ground state properties of the Hartree-Fock crystal model are very good. The excitations can be calculated by including simple extensions of the Hartree-Fock model achieveng outstandingly good comparison with experiment. Unlike the local density approximations where it is clear what is being calculated, the formalism outlined here is well defined. If further extensions are needed to obtain better accuracy it can be done. There are no adjustable parameters.

References

1. A. B. Kunz, Phys. Stat. Sol. 36, 301 (1969).
2. F. E. Farris and H. J. Monkhorst, Phys. Rev. Letters 23, 1026 (1969).
3. L. Dagens and F. Perrot, Phys. Rev. B5, 641 (1972).
4. R. N. Euwena, D. L. Wihlhite and G. T. Surrat, Phys. Rev. 7B, 818 (1973).
5. J. C. Slater, Phys. Rev. 81, 385 (1951).
6. R. Gaspar, Acta Phys. Sci. Hung. 3, 263 (1954).
7. W. Kohn and L. J. Sham, Phys. Rev. 140, A1133 (1965).
8. W.B. Fowler, Phys. Rev. 151, 657 (1966).
9. L. Hedin, Phys. Rev. 139, A796 (1965).
10. L. Hedin and S. Lundqvist, Solid State Physics, Vol. 23 (Academie, New York, ed. by F. Seitz, D. Turnbull, and H. Ehrenreich) (1969).
11. A. B. Kunz, Phys. Rev. B6, 606 (1972).
12. A. B. Kunz, D. J. Michish and T. C. Collins, Phys. Rev. Letters 31, 756 (1973).
13. D. J. Michish, A. B. Kunz and T. C. Collins, Phys. Rev. B9, 4461 (1974).
14. J. T. Devreese, A. B. Kunz and T. C. Collins, Solid State Commun. 11, 673 (1972).
15. T. C. Collins, A. B. Kunz and P. W. Deutsh, Phys. Rev. Sept. (1974).
16. V. Fock, Z. Physik, Vol 61, 1261 (1930).
17. P. A. M. Dirac, Proc. Camb. Phil. Soc. 26, 376 (1930).
18. T. Koopmans, Physica 1, 104 (1933).
19. C. Herring, Phys. Rev. 57, 1169 (1940).
20. J. C. Slater, Phys. Rev. 36, 57 (1930).
21. R. N. Euwena, Phys. Rev. B4, 4332 (1971).
22. D. A. Liberman, Phys. Rev. 171, 1 (1966).
23. H. Preuss, Z. Naturforach 11, 823 (1956).
24. D. L. Wilhite and R. N. Euwema, J; Chem. Phys. 15 Jun 74.
25. R. N. Euwema, G. T. Surratt, D. L. Wilhite and G. G. Wepfer
26. G. T. Surratt, R. N. Euwema, D. L. Wilhite and G. G. Wepfer, Phys. Rev. (tobe published).

27. G. T. Surratt, R. N. Euwema, D. L. Wilhite and G. G. Wepfer, Phys. Rev. (to be published).
28. G. T. Surratt, R. N. Euwema, D. L. Wilhite, Phys. Rev. B8, 4019 (1973).
29. M. Krauss, "Compendium of ab initio Calculations of Molecular Energies and Properties", NBS Technical Note No. 438 (U.S. GPO, Washington, D. C. (1971).
30. R. N. Euwema, G. T. Surratt, G. G. Wepfer and D. L. Wilhite, Phys. Rev. (to be published).
31. G. F. Chew and G. C. Wick, Phys. Rev. 85, 636 (1952).
32. G. G. Wepfer, R. N. Euwema, G. T. Surratt and D. L. Wilhite, Int. J. Q. M. SYM 7, 613 (1973).
33. R. N. Euwema and G. T. Surratt, Solid State Communication (to be published).
34. N. E. Brener and T. C. Collins, "Proceedings of the 12th International Conference on the Physics of Semiconductors", Stuttgart, Germany, July 74.
35. N. E. Brener, Phys. Rev. (to be published).
36. D. R. Penn, Phys. Rev. 128, 2093 (1963).
37. N. E. Brener, Phys. Rev. (to be published).
38. T. C. Collins, A. B. Kunz and J. T. Devreese, Int. J. Quan. Chem. Symp. No. 7, 551 (1973).
39. J. T. Devreese, J. De Sitter, and M. Govaerts, Phys. Rev. B5, 2367 (1972).
40. W. H. Adams, J. Chem. Phys. 34, 89 (1961); 36, 2009 (1962).
41. T. L. Gilbert in "Molecular Orbitals in Chemistry, Physics, and Biology", ed. P. O. Löwdin and P. Pullman (Academie Press, New York, 1964).
42. A. B. Kunz, Phys. Stat. Sol. B46, 697 (1971).
43. D. F. Scofield, N. C. Dutta, and C. M. Dutta, Int. J. Quan. Chem. 6, 9 (1972).
44. K. A. Brueckner, Phys. Rev. 97, 1353 (1955).
45. J. Goldstone, Proc. Roy. Soc. (London) A239, 267 (1957).
46. C. E. Moore, "Atomic Energy Levels", Circular 467, National Bureau of Standards.

ELECTRONIC STRUCTURE CALCULATIONS ON CRYSTALS AND POLYMERS

Frank E. Harris

Department of Physics

University of Utah

Salt Lake City, Utah

1. CLASSICAL MADELUNG SUMS

A. Introduction

Consider a macroscopic periodic lattice whose interior cells are neutral. Let $\vec{r}_i$ be the position of the origin of Cell i; let $\vec{s}_m$, m = 1, ..., d be the positions within each cell of charges q_m. Let q_o be the charge, if any, at eqch cell origin. Then

$$V(\vec{r}_i) = q_o \sum_{j\neq i} |\vec{r}_j-\vec{r}_i|^{-1} + \sum_j \sum_{m=1}^{d} q_m |\vec{r}_j-\vec{r}_i + \vec{s}_m|^{-1} \quad (1)$$

The individual summations are divergent if extended infinitely, but may be combined to give q convergent result. Care must be taken in defining this result. We will consider relevant factors, plus the relationships between various methods of calculation.

B. Choice of sample

The potential at an interior point of a three-dimensional sample will be independent of the sample size and shape if built from repeating units of zero net charge, dipole moment, and quadrupole moment. The repeating unit can always be chosen to meet these conditions, if it is permitted to have deviations at the surface of the crystal.

Examples :

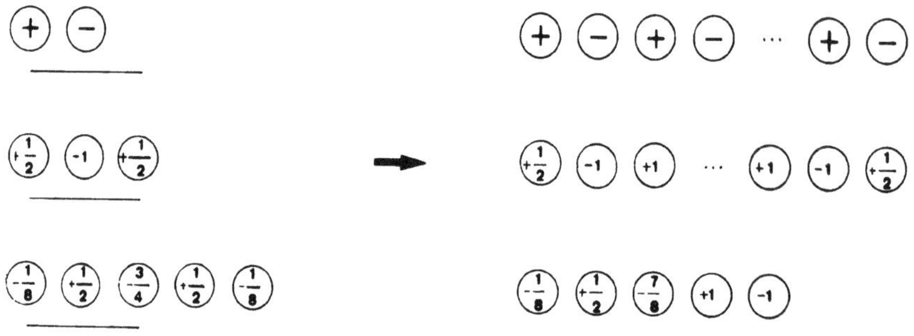

Note that in these one-dimensional examples deviations occur only at the ends.

Even though zero charge, dipole and quadrupole moment suffice for size and shape independence of $V(\vec{r}_i)$, the value of $V(\vec{r}_i)$ is still incompletely fixed. Different distributions with these moments vanishing can still differ by surface double-layers of charge.

Example--body-centered cubic structure :

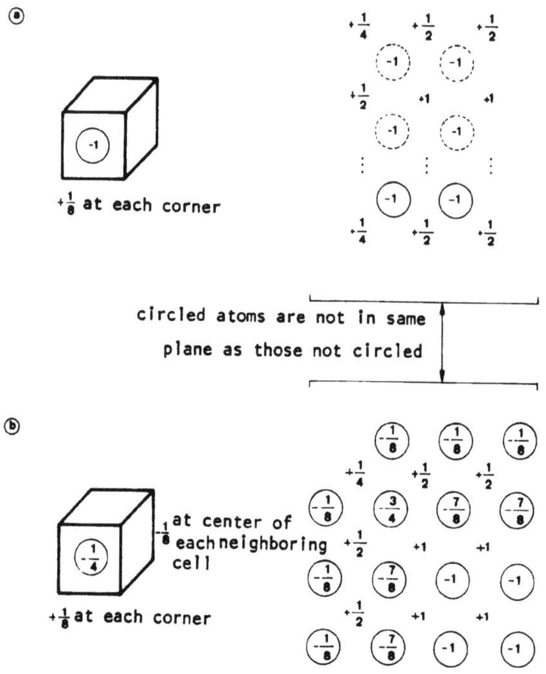

Difference between (a) and (b).

	$-\frac{1}{8}$	$-\frac{1}{8}$	$-\frac{1}{8}$
$-\frac{1}{8}$	$+\frac{1}{4}$	$+\frac{1}{8}$	$+\frac{1}{8}$
$-\frac{1}{8}$	$+\frac{1}{8}$	0	0

This is a dipolar double layer.

Calculation of the effect of this double layer : Let repeating unit have trace per unit volume of its second moment tensor

$$T = \frac{1}{2V_r} \sum_{\text{unit}} q_i r_i^2$$

Since the dipole moment vanishes this trace is independent of the origin with respect to which it is computed. For a total sample of a macroscopic sphere of radius R the total second moment trace will have the value $\frac{4}{3}\pi R^3 T$, and the difference of two samples will have a difference of this trace given by

$$\frac{4\pi R^3}{3} \Delta T = \frac{4\pi R^3}{3} (T_2 - T_1)$$

But also

$$\frac{4\pi R^3}{3} \Delta T = \frac{1}{2} \sum_{\text{layer}} q_i (R + u_i)^2 ,$$

where $R + u_i$ is the distance from the center of the sample to the i-th difference charge, and $u_i/R \ll 1$. The potential due to this charge difference at the center of the sphere is

$$V_2 - V_1 = \sum_{\text{layer}} \frac{q_i}{R + u_i}$$

Expanding the formulas for $\frac{4}{3}\pi R^3 \Delta T$ and $V_2 - V_1$,

$$\frac{4}{3}\pi R^3 \Delta T = \frac{1}{2} \sum_i q_i R^2 + R \sum_i q_i u_i + \ldots$$

$$V_2 - V_1 = \frac{1}{R} \sum_i q_i - \frac{1}{R^2} \sum_i q_i u_i + \ldots$$

$$\frac{4\pi R^3}{3} \Delta T \approx - R^3 (V_2 - V_1)$$

so,

$$V_2 - V_1 = - \frac{4\pi}{3} (T_2 - T_1) \qquad (2)$$

Thus the potential depends upon the trace of the second moment tensor per unit volume for the repeating unit from which the crystal is considered constructed.

C. Ewald method

The Ewald method is based on the introduction of the following integral transform for the potential :

$$\frac{1}{r} = \frac{2}{\sqrt{\pi}} \int_0^\infty e^{-r^2 z^2} dz$$

The result is that $V(\vec{r}_i)$ can be written

$$V(\vec{r}_i) = \frac{2q_o}{\sqrt{\pi}} \sum_{j \neq i} \int_0^\infty e^{-|\vec{r}_j - \vec{r}_i|^2 z^2} dz + \sum_j \sum_{m=1}^{d} \frac{2q_m}{\sqrt{\pi}} \times \int_0^\infty e^{-|\vec{r}_j - \vec{r}_i + \vec{s}_m|^2 z^2} dz \qquad (3)$$

Convergence of the lattice sums is improved by use of the relationship

$$\sum_j e^{-|\vec{r}_j+\vec{s}|^2 z^2} = \frac{\pi^{3/2}}{V_o} \sum_\mu z^{-3} e^{-k_\mu^2/4z^2 + i\vec{k}_\mu \vec{s}} \tag{4}$$

to convert terms of the general form exp $(-z^2)$ to the form exp $(-z^{-2})$. Equation (4) is used in Equation (3) for small z. However, we note that the j summation does not converge uniformly in the neighborhood of $z = 0$, and therefore cannot be interchanged with integration in equation (3) to enable us of equation (4).

However, we may break the z integral into the three ranges $(0,\varepsilon)$, (ε,Z), and (Z,∞), and then make the interchange of summation and integration for the range (ε,Z) only. We then take the limit $\varepsilon \to 0$ and see when the integration range $(0,\varepsilon)$ and other terms involving ε make no contribution. After some manipulation, we have

$$V(\vec{r}_i) = \sum_{m=o}^{d} g_m \bar{V}_m + \Delta V, \tag{5}$$

with

$$\bar{V}_m = \frac{4\pi}{V_o} \sum_{\mu\neq o} k_\mu^{-2} e^{-k_\mu^2/4z^2 + i\vec{k}_\mu \vec{s}_m}$$

$$+ \sum_j |\vec{r}_j+\vec{s}_m|^{-1} \operatorname{erfc}(z|\vec{r}_j+\vec{s}_m|) - \frac{\pi}{V_o z^2} \tag{6}$$

$$\Delta V = \sum_{m=o}^{d} q_m \left(\frac{\pi}{V_o \varepsilon^2} - \frac{4\pi}{V_o} \sum_{\mu\neq o} k_\mu^{-2} e^{-k_\mu^2/4\varepsilon + i k_\mu s_m} \right.$$

$$\left. + \sum_j \frac{2}{\sqrt{\pi}} \int_o^\varepsilon e^{-|\vec{r}_j+\vec{s}_m|^2 z^2} dz \right) \tag{7}$$

Applying the condition of charge neutrality, we obtain

$$\Delta V = \lim_{\varepsilon\to o} \sum_{m=o}^{d} q_m \sum_j \frac{2}{\sqrt{\pi}} \int_o^\varepsilon e^{-|\vec{r}_j+\vec{s}_m|^2 z^2} dz$$

However, we note that ε cannot go to zero more rapidly than r_{max}^{-1}, where r_{max} is the distance from $\vec{r}_i$ to a point on the sample surface.

The j summation in the expression for ΔV contains of the order of r_{max}^3 terms, each potentially of magnitude r_{max}^{-1}, and it is therefore not even obvious that the expression will converge to a finite result. However, if we expand in powers of the components of $\vec{s}_m$, we may find that ΔV goes to zero if all components of all moments through the second vanish.

Note also that $\overline{V}_m$ vanishes if averaged over the unit cell. This implies that $\overline{V}_m = 0$ for a uniform distribution of charge, which in turn means that $\overline{V}_m$ for a point-charge lattice is actually the potential for that lattice plus a compensating uniform background. But because the sample must be chosen to have a vanishing second moment, the overall sample choice may not be that most intuitively perceived.

D. FOURIER METHOD

The Fourier representation method for evaluating Madelung sums is based on the integral representation

$$r^{-1} = \frac{1}{2\pi^2} \int \frac{d\vec{k}}{K^2} e^{-i\vec{k}.\vec{r}} , \qquad (8)$$

and leads to the expression

$$V(\vec{r}_i) = \frac{q_o}{2\pi^2} \sum_{j\neq i} \int \frac{d\vec{k}}{k^2} i^{i\vec{k}.(\vec{r}_j-\vec{r}_i)}$$

$$+ \sum_j \sum_{m=1}^{d} \frac{q_m}{2\pi^2} \int \frac{d\vec{k}}{k^2} e^{-i\vec{k}.(\vec{r}_j-\vec{r}_i+\vec{s}_m)} \qquad (9)$$

We now introduce tha lattice orthogonality relation

$$\sum_j e^{-i\vec{k}.(\vec{r}_j-\vec{r}_i)} = \frac{8\pi^3}{V_o} \sum_\mu \delta\,(\vec{k}-\vec{k}_\mu) \qquad (10)$$

and, after interchanging the summation and integration, we obtain

$$V(\vec{r}_i)=\frac{q_o}{2\pi^2}\frac{d\vec{k}}{k^2}\left(\sum_j e^{-i\vec{k}.(\vec{r}_j-\vec{r}_i)}-1\right)+$$

$$\sum_{m=1}^{d}\frac{q_m}{2\pi^2}\int\frac{d\vec{k}}{k^2}\sum_j e^{-i\vec{k}.(\vec{r}_j-\vec{r}_i+\vec{s}_m)} \qquad (11)$$

$$=\frac{q_o}{2\pi^2}\int\frac{d\vec{k}}{k^2}\left(\frac{8\pi^3}{V_o}\sum_\mu \delta(\vec{k}-\vec{k}_\mu)-1\right)+$$

$$\frac{8\pi^3}{V_o}\sum_{m=1}^{d}\frac{q_m}{2\pi^2}\int\frac{d\vec{k}}{k^2}e^{-\vec{k}.\vec{s}_m}\sum_\mu \delta(\vec{k}-\vec{k}_\mu)$$

$$=\frac{q_o}{2\pi^2}\left(\frac{8\pi^3}{V_o}\sum_\mu\frac{1}{k_\mu^2}-\int\frac{d\vec{k}}{k^2}\right)+$$

$$\sum_{m=1}^{d}\frac{q_m}{2\pi^2}\cdot\frac{8\pi^3}{V_o}\sum_\mu\frac{e^{-i\vec{k}_\mu.\vec{s}_m}}{k_\mu^2} \qquad (12)$$

A careful consideration of the region about $\vec{k}=0$ indicates that due to a lack of uniform convergence the summation and integration cannot be interchanged for small k, with the result that the integrations in Eq. (11), and thus the sums in Eq. (12), do not contain the point $k_\mu = 0$. It can also be shown that the region about $\vec{k} = 0$ gives no contribution if <u>all</u> moments through the second vanish.

The quantity $[(8\pi^3/V_o)\sum_{\mu\neq o} k_\mu^{-2} - \int k^{-2}\, d\vec{k}]$ approaches a definite limit whose value is proportional to $v_o^{-1/3}$, with the value dependent upon the lattice symmetry. Giving the first term of Eq. (12) the symbol q_oC, we have

$$V(\vec{r}_i) = q_o\, C + \sum_{m=1}^{d}\frac{4\pi q_m}{V_o}\sum_{\substack{\mu \\ (k_{\mu\neq o})}}\frac{e^{-i\vec{k}_\mu.\vec{s}_m}}{k_\mu^2} \qquad (13)$$

The last term of Eq.(13) can be interpreted as due to the lack of uniformity with which the charges q_1, ..., q_m (of net charge $-q_o$ per unit cell) are distributed in each unit cell.

E. TWO-DIMENSIONAL PERIODICITY

We now consider how the Fourier representation methods described in the foregoing section can be applied to systems possessing periodicity in only two dimensions. Let us take z as the aperiodic direction. Lattice sums therefore now refer to points of a planar lattice, which we shall take as situated in the x-y plane. However, note that we continue to consider three-dimensional charge distributions which are not restricted to lie in a single plane. We proceed as in the foregoing section, but now use the lattice orthogonality relation appropriate to two-dimensional periodicity, namely

$$\sum_j e^{-i\vec{k}.(\vec{r}_j-\vec{r}_i)} = \frac{4\pi^2}{a_o} \sum_\mu \delta(\vec{k}_2 - \vec{k}_{2\mu}) \tag{14}$$

Here $\vec{k}_2$ is a vector with x and y components which is the projection of $\vec{k}_2$ on the x-y plane.

The result corresponding to Eq. (12) is

$$V(\vec{r}_i) = \frac{q_o}{2\pi^2} \int dk_z \left(\frac{4\pi^2}{a_o} \sum_\mu{}'' \frac{1}{k_z^2 + k_{2\mu}^2} - \int \frac{d\vec{k}_2}{k_z^2 + k_2^2} \right) + \sum_{m=1}^{d} \frac{8\,m}{2\pi^2} \frac{4\pi^2}{a_o} \int dk_z \sum_\mu{}'' \frac{e^{-i\vec{k}_{z\mu}\cdot\vec{s}_{2m} - ik_z s_{zm}}}{k_z^2 + k_{2\mu}^2} \tag{15}$$

Here $\vec{k}_{2\mu} = 0$ is to be omitted if $k_z \approx 0$. If we rearrange to treat $\vec{k}_{2\mu} = 0$ separately, we can write

$$V(\vec{r}_i) = \frac{q_o}{2\pi^2} \int dk_z \left(\frac{4\pi^2}{a_o} \sum_\mu{}' \frac{1}{k_z^2 + k_{2\mu}^2} - \int \frac{d\vec{k}_2}{k_z^2 + k_2^2} \right) + \sum_{m=o}^{d} \frac{q_m}{2\pi^2} \left(\frac{4\pi^2}{a_o}\right) \int{}' \frac{dk_z\, e^{-ik_z s_{zm}}}{k_z^2} + \sum_{m=1}^{d} \frac{q_m}{2\pi^2} \left(\frac{4\pi^2}{a_o}\right) \int dk_z \sum_\mu{}' \frac{e^{-i\vec{k}_{z\mu}\cdot\vec{s}_{2m} - ik_z s_{zm}}}{k_z^2 + k_{2\mu}^2}$$

We note that because of charge neutrality the second integral in Eq. (16) will not diverge.

Applying contour integration techniques, we can reduce Eq. (16) to the form

$$V(\vec{r}_i) = \frac{q_o}{2\pi^2}\Big[\frac{4\pi^2}{a_o}\sum_{\mu}{}' \frac{1}{k_{2\mu}} - \pi\int\frac{d\vec{k}_2}{k_2}\Big]$$

$$- \frac{2\pi}{a_o}\sum_{m=1}^{d} q_m |s_{zm}|$$

$$+ \frac{2\pi}{a_o}\sum_{m=1}^{d} q_m \sum_{\mu}{}' \frac{e^{-i\vec{k}_{2\mu}\cdot\vec{s}_{2m} - k_{2\mu}|s_{zm}|}}{k_{2\mu}} \tag{17}$$

This formula has the following interpretation : The first term describes the potential of a lattice of charges q_o balanced by a uniform plane of compensating charge in the lattice plane. The second term describes the changes in $V(\vec{r}_i)$ if for each m a plane of charge q_m/a_o per unit area is moved a distance s_{zm} from the lattice plane. The final term gives the effect of changing the charge distribution for each m form a uniform plane to actual charges q_m at points s_{2m} within the plane.

The lattice term $[(4\pi^2/a_o)\sum_{\mu}{}' k_{2\mu}^{-1} - \pi\int k_2^{-1}\, d\vec{k}_2]$ can be evaluated from formulas already presented for three-dimensional periodicity.

The region near $\vec{k} = 0$ gives no contribution if the repeating unit has zero net charge and dipole moment.

F. ONE-DIMENSIONAL PERIODICITY

We finally consider systems with but one dimension of periodicity. This time we take z as the periodic direction, and use the notations introduced in the preceding section. The lattice orthogonality relation now takes the form

$$\sum_j e^{-i\vec{k}.(\vec{r}_j - \vec{r}_i)} = \frac{2\pi}{a}\sum_{\mu}\delta(k_z - k_{z\mu}) \tag{18}$$

and leads to the following equation for $V(\vec{r}_i)$:

$$V(\vec{r}_i) = \frac{q_o}{2\pi^2} \int d\vec{k}_2 \left[\frac{2\pi}{a} \sum_\mu{}'' \frac{1}{k_2^2 + k_{z\mu}^2} - \int \frac{dk_z}{k_2^2 + k_z^2} \right]$$

$$+ \sum_{m=1}^{d} \frac{q_m}{2\pi^2} \left(\frac{2\pi}{a}\right) \int d\vec{k}_2 \sum_\mu{}'' \frac{e^{-i\vec{k}_2 \cdot \vec{s}_{2m} - ik_{z\mu} s_{zm}}}{k_2^2 + k_{z\mu}^2} \qquad (19)$$

Because a uniform line of charge yields a divergent potential at points on the line we cannot obtain $V(\vec{r}_i)$ as a uniform-background limit plus nonuniformity terms. A better limiting case is a set of charges $+q_o$ at lattice points with compensating charges $-q_o$ halfway between the lattice points. As a consequence of the above the first term of Eq.(19) is, as presently written, divergent. We therefore add to it, and subtract from the last term, the quantity

$$\frac{-q_o}{\pi a} \int d\vec{k}_2 \sum_\mu{}'' \frac{(-1)^\mu}{k_2^2 + k_{z\mu}^2}$$

(In the last term we write $-q_o$ as $\sum_{m=1}^{d} q_m$). Then we obtain

$$V(\vec{r}_i) = \frac{q_o}{2\pi^2} \int d\vec{k}_2 \left[\frac{2\pi}{a} \sum_\mu \frac{1 - (-1)^\mu}{k_2^2 + k_{z\mu}^2} - \int \frac{dk_z}{k_2^2 + k_z^2} \right]$$

$$+ \frac{1}{\pi a} \sum_{m=1}^{d} q_m \int d\vec{k}_2 \sum_\mu \frac{e^{-i\vec{k}_2 \cdot \vec{s}_{2m} - ik_{z\mu} s_{zm}} - (-1)^\mu}{k_2^2 + k_{z\mu}^2}$$

It is necessary to keep the μ summations inside the $\vec{k}_2$ integration to avoid a misinterpretation of the Fourier representation integral. It is possible to evaluate the μ sum and (in circular coordinates) the $\vec{k}_2$ integral. (The $\vec{k}_2$ integral leads to a series expansion, but not the same expansion which whould have resulted if we had illegitimately interchanged the $\vec{k}_2$ integral with the μ summation). The result for $V(\vec{r}_i)$ is as follows :

$$V(\vec{r}_i) = \frac{-4q_o}{a} \left(\ln 2 - \frac{1}{2}\gamma\right) + \frac{4}{a} \sum_{m=1}^{d} q_m \quad x$$

$$\left(\sum_{\mu=1}^{\infty} (-1)^\mu \cos\left[\mu\pi\left(1 - \frac{2s_{zm}}{a}\right)\right] K_o\left(\frac{2\mu\pi\, s_{zm}}{a}\right) - \frac{1}{2}\ln\left(\frac{2s_{zm}}{a}\right)\right) \qquad (20)$$

Here K_0 is a Bessel function of imaginary argument, and γ is Euler's constant (0.57721...).

The quantity $-(4q_0/a)\ \ln 2$ is the potential of the alternating line of charge referred to as the limiting case for one-dimensional periodicity, as may be seen by expanding $\ln 2 = 1 - 1/2 + 1/3 - \ldots$ The remainder of Eq. (20) describes deviations from the alternating line, and is a better formulation than a straight lattice sum because the factor K_0 gives exponential convergence.

REFERENCES

P.P. Ewald 1921. Ann. Physik 64, 253.

O. Emersleben 1923. Physik. Z. 24, 97.

O. Emersleben 1950. Z. Angew. Math. Mech. 30, 252.

H.M. Evjen 1932. Phys. Rev. 39, 675.

B.R.A. Nijboer and F.W. DeWette 1957. Physica 23, 309.

M.P. Tosi 1964. Solid State Physics, F. Seitz and D. Turnbull, eds. (Academic Press, N.Y.), pp 1-120.

R.N. Euwema and G.T. Surratt 1974. Private Communication.

F.E. Harris and H.J. Monkhorst 1970. Phys. Rev. B2, 4400.

J. Sherman 1932. Chem. Rev. 11, 93.

E.F. Bertaut 1952. J. Phys. Radium 13, 499.

F.E. Harris 1972. J. Chem. Phys. 56, 4422.

2. QUANTUM-MECHANICAL ELECTRONIC ENERGY CLACULATIONS

A. INTROCUTION

We start with a consideration of systems with three-dimensional periodicity. We restrict consideration to the zero-order problem familiar in quantum chemistry, namely that in which the nuclei are regarded as fixed in position, in which relativistic and magnetic effects are ignored, and which has a Hamiltonian which we write as follows :

$$H = \sum_{i=1}^{ZNd} \left(-\frac{1}{2}\nabla_i^2\right) + \sum_{1 \leq i<j \leq ZNd} h\,(\vec{r}_i,\vec{r}_j) \tag{21}$$

We work in atomic units, and assume the system to have a total of ZNd electrons. The first term of Eq. (21) describes the electronic kinetic energy, while the remaining term describes the potential energy of the system. For simplicity we shall take a system whose nuclei all have unit charge, so Z = 1, and the potential energy then takes the form

$$h(\vec{r}_1,\vec{r}_2) = r_{12}^{-1} - \frac{1}{Nd} \sum_{\mu} \sum_{u=1}^{d} |\vec{r}_1-\vec{R}_\mu-\vec{s}_u|^{-1} - \frac{1}{Nd} \sum_{\mu} \sum_{u=1}^{d} |\vec{r}_2-\vec{R}_\mu-\vec{s}_u|^{-1} + \frac{1}{N^2d^2} \sum_{\mu\mu'} \sum_{\mu\mu'}{}'' |\vec{R}_\mu-\vec{R}_{\mu'}+\vec{s}_u-\vec{s}_{u'}|^{-1}$$

Next we must select an appropriate electronic wavefunction. We discuss at this time only calculations in the Hartree-Fock approximation, and shall specify a many-electron wavefunction as a product of doubly-occupied crystal orbitals. These orbitals will be built from a crystal-orbital basis. There are at least four different bases which xhould be considered. We list them here :

(i) $|\vec{k}_i> = \sum_\mu \sum_u e^{i\vec{k}.(\vec{R}_\mu+\vec{s}_u)} \phi_i(\vec{r}-\vec{R}_\mu-\vec{s}_u)$ LCAO

(ii) $|\vec{k}_i> = \sum_\mu \sum_u e^{i\vec{k}.\vec{r}} \phi_i(\vec{r}-\vec{R}_\mu-\vec{s}_u)$ MPW

(iii) $|\vec{k}_i> = \sum_\mu \sum_u e^{i(\vec{k}+\vec{k}_i).\vec{r}} \phi\,(\vec{r}-\vec{R}_\mu-\vec{s}_u)$ Mixed

(iV) $|\vec{k}_i> = e^{i(\vec{k}+\vec{k}_i).\vec{r}}$ PW

The first, marked "LCAO", is a linear combination of atomic orbitals, combined with phase factors to form fuction of the Bloch Type. The second, marked "MPW", may be described as formed of modulated plane waves. This function also has the symmetry required by Bloch's theorem, but we note that it is not periodic in $\vec{k}$. The third basis form has a mixture of atomic-orbital and plane-wave-expansion character, while the fourther is simply a plane wave expansion. All the atomic orbitals in the above forms are as yet unspecified, but the method we shall describe permits them to be chosen freely, and they may be either STO's or Gaussian orbitals. Observe also that the notation we are using causes the $\vec{k}$ value of the basis functions to be indicated within the kets denoting the basis functions.

The different basis forms have different features. The LCAO form can produce only Nd linearly independent functions and will automatically provide features characteristic of filled bands. The MPW and mixed bases are easier to calculate with than the LCAO basis but retain the atomic-orbital features of the LCAO basis. The mixed basis proves to be an easier way of enlarging a basis set than is the MPW basis. The PW basis does not converge rapidly in comparison to the others. If we compare the use of the LCAO and MPW bases with the same atomic orbitals, we note that they will produce different effective mixing of functions of higher angular momenta at each atom.

After selecting a basis, final crystal orbitals are formed as linear combinations,

$$|\vec{k}>_\alpha = \sum_i C_{i\alpha}(\vec{k})|\vec{k}_i>$$

For a single band system, the energy is given in the restricted Hartree-Fock approximation by the formula

$$<E> = \frac{2NV_o}{(2\pi)^3}\int d\vec{k}\ \frac{<\vec{k}|-\frac{1}{2}\nabla^2|\vec{k}>}{<\vec{k}|\vec{k}>}$$

$$+\frac{2N^2V_0^2}{(2\pi)^6}\int d\vec{k}\,d\vec{k}'\ \frac{<\vec{k}\vec{k}'|h|\vec{k}\vec{k}'> - \frac{1}{2}<\vec{k}\vec{k}'|h|\vec{k}'\vec{k}>}{<\vec{k}|\vec{k}>\ <\vec{k}'|\vec{k}'>} \qquad (22)$$

where the crystal orbitals have been determined to minimize the energy. In terms of the basis orbitals and the coefficients $C_{i\alpha}$, we may write (suppressing the subscript α , which is not needed when there is only one band)

$$<E> = \frac{2V_0}{(2\pi)^3 d}\int d\vec{k}\sum_{ij} C_i^*(\vec{k})<\vec{k}_i|-\frac{1}{2}\nabla^2|\vec{k}_j>C_j(\vec{k})$$

$$+\frac{2V_0^2}{(2\pi)^6 d}\int d\vec{k}d\vec{k}'\sum_{ijmn} C_i^*(\vec{k})C_j(\vec{k})C_m^*(\vec{k}')C_n(\vec{k}')$$

$$\times [\langle \vec{k}_i \vec{k}'_m | h | \vec{k}_j \vec{k}'_n \rangle - \frac{1}{2} \langle \vec{k}_i \vec{k}'_m | h | \vec{k}'_n \vec{k}_j \rangle] \tag{23}$$

where we have chosen the scaling of $|\vec{k}\rangle$ so that

$$\langle \vec{k} | \vec{k} \rangle = \sum_{ij} C_i^* (\vec{k}) \langle \vec{k}_i | \vec{k}_j \rangle C_j (\vec{k}) \tag{24}$$

The above expressions lead to the Hartree-Fock equations :

$$\sum_j F_{ij}(\vec{k}) C_j(\vec{k}) = \varepsilon(\vec{k}) \sum_j S_{ij}(\vec{k}) C_j(\vec{k})$$

$$S_{ij}(\vec{k}) = \frac{1}{Nd} \langle \vec{k}_i \vec{k}_j \rangle$$

$$F_{ij}(\vec{k}) = \frac{1}{Nd} \langle \vec{k}_i | -\frac{1}{2} \nabla^2 | \vec{k}_j \rangle + \frac{2}{d} \sum_{mm} \int d\vec{k}' \times$$

$$C_m^* (\vec{k}') C_n(\vec{k}') [\langle \vec{k}_i \vec{k}'_m | h | \vec{k}_j \vec{k}'_n \rangle - \frac{1}{2} \langle \vec{k}_i \vec{k}'_m | h | \vec{k}'_n \vec{k}_j \rangle]$$

The occupied orbitals are to be chosen from the solutions to the Hartree-Fock equations which minimize the energy. These will be the orbitals whose orbital energies $\varepsilon(\vec{k})$ are within the Fermi surface, of energy $\varepsilon = \varepsilon_F$ = constant.

B. MATRIX ELEMENT REDUCTION

The process of matrix element reduction will be illustrated for the simple prototype function

$$|\vec{k}\rangle = e^{i\vec{k}.\vec{r}} \sum_{\mu} \phi(\vec{r}-\vec{R}_{\mu}) \tag{25}$$

It will prove convenient to express many of the results in terms of the Fourier transform of a lattice sum of orbital products :

$$\Phi(\vec{q}) = \sum_{\mu} [\phi^*(\vec{r}) \phi(\vec{r}-\vec{R}_{\mu})]^T (\vec{q})$$
$$= \sum_{\mu} \langle \phi(\vec{r}) | e^{i\vec{q}.\vec{r}} | \phi(\vec{r}-\vec{R}_{\mu}) \rangle \tag{26}$$

The overlap integral is

$$\langle \vec{k} | \vec{k} \rangle = \langle e^{i\vec{k}.\vec{r}} \sum_{\mu} \phi(\vec{r}-\vec{R}_{\mu}) | e^{i\vec{k}.\vec{r}} \sum_{\nu} \phi(\vec{r}-\vec{R}_{\nu}) \rangle$$
$$= \sum_{\mu\nu} \langle \phi(\vec{r}-\vec{R}_{\mu}) | \phi(\vec{r}-\vec{R}_{\nu}) \rangle$$

$$= \sum_{\mu\nu} < \phi(\vec{r})| \ \phi \ (\vec{r}+\vec{R}_\mu-\vec{R}_\nu)>$$

$$= N \sum_{\mu} <\phi(\vec{r})| \ \phi \ (\vec{r}-\vec{R}_\mu)>$$

$$= N \ \Phi \ (o) \tag{27}$$

The kinetic energy integral is

$$<\vec{k}|-\frac{1}{2}\nabla^2|\vec{k}> = <e^{i\vec{k}.\vec{r}}\sum_{\mu}\phi(\vec{r}-\vec{R}_\mu)| -\frac{1}{2}\nabla^2|e^{i\vec{k}.\vec{r}}\sum_{\nu}\phi(\vec{r}-\vec{R}_\nu)>$$

$$= \frac{1}{2}k^2<\sum_{\mu}\phi(\vec{r}-\vec{R}_\mu)|\sum_{\nu}\phi(\vec{r}-\vec{R}_\nu)>$$

$$+ <\sum_{\mu}\phi(\vec{r}-\vec{R}_\mu)|-\frac{1}{2}\nabla^2|\sum_{\nu}\phi(\vec{r}-\vec{R}_\nu)> - <\sum_{\mu}\phi(\vec{r}-\vec{R}_\mu)|i\vec{k}.\nabla|\sum_{\nu}\phi(\vec{r}-\vec{R}_\nu)>$$

The last term vanishes, and thus we have the final form

$$<\vec{k}|-\frac{1}{2}\nabla^2|\vec{k}> = \frac{1}{2}k^2 \ N \ \Phi \ (o) + N\sum_{\mu} < \phi(\vec{r})| -\frac{1}{2}\nabla^2| \ \phi \ (\vec{r}-\vec{R}_\mu)> \tag{28}$$

The Coulomb terms are now reduced, using a Fourier representation formula of the general form

$$\int f(\vec{r}_1)g(\vec{r}_2) \ r_{12}^{-1} \ d\vec{r}_1 \ d\vec{r}_2 = \frac{1}{2\pi^2}\int f^T(\vec{q}) \ g^T(-\vec{q}) \ \frac{d\vec{q}}{q^2} \tag{29}$$

The introduction of Fourier representation techniques will produce convergence problems analogous to those already discussed for the evaluation of classical Madelung sums. We ignore temporarily these convergence problems, because it can be shownthat an infinitesimal region of space about $\vec{q} = 0$ must be omitted from the Fourier integrals and that the omitted region will produce no contribution to the energy if the crystal repeating unit is suitably chosen, namely to have vanishing moments through order two.

Starting with the r_{12}^{-1} term, we have

$$<\vec{k}\vec{k}'|r_{12}^{-1}|\vec{k}\vec{k}'> = \sum_{\mu\nu\lambda\sigma} < e^{i\vec{k}.\vec{r}_1 + i\vec{k}'.\vec{r}_2} \ \phi(\vec{r}_1-\vec{R}_\mu) \ \phi \ (\vec{r}_2-\vec{R}_\nu)| \ x$$

$$r_{12}^{-1} | e^{i\vec{k}.\vec{r}_1 + i\vec{k}'.\vec{r}_2} \phi (\vec{r}_1 - \vec{R}_\lambda) \phi (\vec{r}_2 - \vec{R}_\sigma) >$$

$$= \sum_{\mu\nu\lambda\sigma} < \phi (\vec{r}_1 - \vec{R}_\mu) \phi (\vec{r}_2 - \vec{R}_\nu) \mid r_{12}^{-1} | \phi (\vec{r}_1 - \vec{R}_\lambda) \phi (\vec{r}_2 - \vec{R}_\sigma)>$$

$$= \frac{1}{2\pi^2} \sum_{\mu\nu\lambda\sigma} \int \frac{d\vec{q}}{q^2} [\phi^*(\vec{r} - \vec{R}_\mu) \phi (\vec{r} - \vec{R}_\lambda)]^T (\vec{q}) [\phi^*(\vec{r} - \vec{R}_\nu) \phi(\vec{r} - \vec{R}_\sigma)]^T (-\vec{q})$$

$$= \frac{1}{2\pi^2} \sum_{\mu\nu\lambda\sigma} \int \frac{d\vec{q}}{q^2} e^{i\vec{q}.(\vec{R}_\mu - \vec{R}_\nu)} [\phi^*(\vec{r}) \phi(\vec{r} + \vec{R}_\mu - \vec{R}_\lambda)]^T (\vec{q}) \times$$

$$[\phi^*(\vec{r}) \phi (\vec{r} + \vec{R}_\nu - \vec{R}_\sigma)]^T (-\vec{q})$$

Moving the λ and σ summations to the transform brackets and recognizing that a sum over λ of $\vec{R}_\lambda - \vec{R}_\mu$ is the same as a sum over $\vec{R}_\lambda$ alone, and making similar observations with respect to $\vec{R}_\sigma - \vec{R}_\nu$, we have

$$<\vec{k}\vec{k}' | r_{12}^{-1} | \vec{k}\vec{k}'> = \frac{1}{2\pi^2} \sum_{\mu\nu} \int \frac{d\vec{q}}{q^2} e^{i\vec{q}.(\vec{R}_\mu - \vec{R}_\nu)} \Phi(\vec{q}) \Phi (-\vec{q})$$

Now, using the lattice orthogonality relation given in Eq. (10), we obtain delta functions when summing over μ, and the ν sum simply yields a factor N. Thus,

$$<\vec{k}\vec{k}' | r_{12}^{-1} | \vec{k}\vec{k}'> = \frac{N}{2\pi^2} \left(\frac{8\pi^3}{V_0}\right) \int \frac{d\vec{q}}{q^2} \sum_\mu \delta(\vec{q} - \vec{q}_\mu) \Phi (\vec{q}) \Phi (-\vec{q})$$

$$= \frac{N}{2\pi^2} \left(\frac{8\pi^3}{V_o}\right) \sum_\mu{}' \frac{\Phi(\vec{q}_\mu) \Phi (-\vec{q}_\mu)}{q_\mu^2} \quad (30)$$

The prime is to remind us that the point $\vec{q}_\mu = 0$ is to be omitted from the summation.

Continuing to the electron-nuclear interaction integrals, the equivalent of Eq. (29) is

$$\int f(\vec{r}) \frac{1}{|\vec{r} - \vec{R}|} d\vec{r} = \frac{1}{2\pi^2} \int e^{-i\vec{q}.\vec{R}} f^T(\vec{q}) \frac{d\vec{q}}{q^2} \quad (31)$$

and a process similar to that used in obtaining Eq. (30) leads to

$$\frac{1}{N} \langle \vec{k}\vec{k}' | \sum_{\mu} |\vec{r}_1 - \vec{R}_\mu|^{-1} | \vec{k}\vec{k}' \rangle$$

$$= \frac{1}{N} \langle \vec{k}' | \vec{k}' \rangle \cdot \frac{N}{2\pi^2} \left(\frac{8\pi^3}{V_o}\right) \sum_{\mu}{}' \frac{\Phi(\vec{q}_\mu)}{q_\mu^2}$$

or

$$\frac{1}{N} \langle \vec{k}\vec{k}' | \sum_{\mu} |\vec{r}_1 - \vec{R}_\mu|^{-1} | \vec{k}\vec{k}' \rangle = \frac{N}{2\pi^2} \left(\frac{8\pi^3}{V_o}\right) \sum_{\mu}{}' \frac{\Phi(\vec{q}_\mu)\Phi(o)}{q_\mu^2} \quad (32)$$

$$\frac{1}{N} \langle \vec{k}\vec{k}' | \sum_{\mu} |\vec{r}_2 - \vec{R}_\mu|^{-1} | \vec{k}\vec{k}' \rangle = \frac{N}{2\pi^2} \left(\frac{8\pi^3}{V_o}\right) \sum_{\mu}{}' \frac{\Phi(o)\Phi(-\vec{q}_\mu)}{q_\mu^2} \quad (33)$$

Finally, the nuclear-nuclear term reduces just as in the classical problem, giving

$$\frac{1}{N^2} \langle \vec{k}\vec{k}' | \sum_{\mu\nu}{}' |\vec{R}_\mu - \vec{R}_\nu|^{-1} | \vec{k}\vec{k}' \rangle$$

$$= \frac{N}{2\pi^2} \left[\frac{8\pi^3}{V_o} \sum_{\mu}{}' \frac{1}{q_\mu^2} - \int \frac{d\vec{q}}{q^2} \right] \Phi(o)\ \Phi(o)$$

$$= N\ C\ \Phi(o)\ \Phi(o) \quad (34)$$

Looking next at the exchange contribution,

$$\langle \vec{k}\vec{k}' | r_{12}^{-1} | \vec{k}'\vec{k} \rangle = \sum_{\mu\nu\lambda\sigma} \int \phi^*(\vec{r}_1 - \vec{R}_\mu)\ \phi(\vec{r}_1 - \vec{R}_\lambda)\ e^{i(\vec{k}' - \vec{k}).\vec{r}_1} \times$$

$$r_{12}^{-1}\ \phi^*(\vec{r}_2 - \vec{R}_\nu)\ \phi(\vec{r}_2 - \vec{R}_\sigma)\ e^{-i(\vec{k}' - \vec{k}).\vec{r}_2}\ d\vec{r}_1\ d\vec{r}_2$$

$$= \frac{1}{2\pi^2} \sum_{\mu\nu} \int \frac{d\vec{q}}{q^2}\ e^{i(\vec{k}' - \vec{k} + \vec{q}).(\vec{R}_\mu - \vec{R}_\nu)} \times$$

$$\Phi(\vec{q} + \vec{k}' - \vec{k})\ \Phi(-\vec{q} - \vec{k}' + \vec{k})$$

$$= \frac{N}{2\pi^2} \left(\frac{8\pi^3}{V_o}\right) \int \frac{d\vec{q}}{q^2} \sum_{\mu} \delta(\vec{q} + \vec{k}' - \vec{k} - \vec{q}_\mu)\ \Phi(\vec{q} + \vec{k}' - \vec{k})\ \Phi(-\vec{q} - \vec{k}' + \vec{k})$$

$$= \frac{N}{2\pi^2} \left(\frac{8\pi^3}{V_o}\right) \sum_\mu \frac{1}{|\vec{q}_\mu - \vec{k}' + \vec{k}|^2} \Phi(\vec{q}_\mu)\, \Phi(-\vec{q}_\mu) \tag{35}$$

There is no reason to omit the point $\vec{q}_\mu = 0$ from the sum. In fact, $\vec{q}_\mu = 0$ gives the exchange energy associated with a uniform distribution of charge.

Finally, collecting the energy contributions, we have

$$\langle \vec{k}\vec{k}' | h | \vec{k}\vec{k}' \rangle = \frac{N}{2\pi^2} \left(\frac{8\pi^3}{V_o}\right) \sum_\mu{}' \frac{1}{q_\mu^2} [\Phi(\vec{q}_\mu)\, \Phi(-\vec{q}_\mu)$$

$$- \Phi(\vec{q}_\mu)\, \Phi(o) - \Phi(o)\, \Phi(-\vec{q}_\mu)] + N\, C\, \Phi(o)\, \Phi(o)\ ;$$

$$\langle \vec{k}\vec{k}' | h | \vec{k}'\vec{k} \rangle = \frac{N}{2\pi^2} \left(\frac{8\pi^3}{V_o}\right) \sum_\mu \frac{1}{|\vec{q}_\mu + \vec{k}' - \vec{k}|^2} \Phi(\vec{q}_\mu)\, \Phi(-\vec{q}_\mu)$$

If we now examine the various energy contributions as to their relative sizes, we note that some will have no reason to be small, while others should be smaller because they refer to the description of a deviation from uniformity in the electron distribution, while others will be smaller still because they describe the interaction between the nonuniform parts of the distributions of two electrons. Schematically, we have

E = kinetic energy + lattice structure potential

+ free-electron exchange + (e-n) nonuniformity

+ (e-e Coulomb) nonuniformity + (e-e exchange) nonuniformity

The first three of these are largest; the next is intermediate in size, and the last two are small. Incidentally, we also note that the formulas for the electron-interaction nonuniformity terms shows that the contributions for Coulomb and exchange will be comparable in size.

C. EVALUATION OF Φ

Evaluation of $\langle E \rangle$ depends crucially upon the calculation of $\Phi(\vec{q})$ for $\vec{q} = \vec{q}_\mu$, where the $\vec{q}_\mu$ are the points of the reciprocal lattice. (If we had used the LCAO wavefunction we would have found that we would have needed $\Phi(\vec{q})$ for general values of $\vec{q}$.)

The first observation to be made is that $\Phi(\vec{q})$ can itself be reduced to a lattice sum. Using the convolution theorem, we have

$$[\phi(\vec{r})\ \phi(\vec{r}-\vec{R}_\mu)]^T(\vec{q}) = \frac{1}{(2\pi)^3}\int d\vec{p}\ \phi^{T^*}(\vec{p})[\phi(\vec{r}-\vec{R}_\mu)]^T(\vec{q}-\vec{p})$$

Setting $[\phi(\vec{r}-\vec{R}_\mu)]^T(\vec{q}-\vec{p}') = e^{i(\vec{q}-\vec{p}).\vec{R}_\mu}\phi^T(\vec{q}-\vec{p})$, and summing over $\vec{R}_\mu$ with the aid of the lattice orthogonality relation given in Eq.(10), we have

$$\Phi(\vec{q}) = \frac{1}{8\pi^3}\left(\frac{8\pi^3}{V_o}\right)\int d\vec{p}\ \phi^{T^*}(\vec{p})\ \phi^T(\vec{q}-\vec{p})\sum_\mu \delta(\vec{q}-\vec{p}-\vec{p}_\mu)$$

$$= \frac{1}{V_o}\sum_\mu \phi^{T^*}(\vec{q}-\vec{p}_\mu)\ \phi^T(\vec{p}_\mu) \qquad (36)$$

We are thus able to evaluate $\Phi(\vec{q})$ for any basis atomic orbitals whose transforms are available. This class of basis orbitals includes both Slater-type orbitals and Gaussians :

$$(e^{-\zeta r})^T = \frac{8\pi\zeta}{[q^2+\zeta^2]^2}$$

$$(e^{-ar^2})^T = \left(\frac{\pi}{a}\right)^{3/2} e^{-q^2/4a}$$

Transforming to a scale in which the reciprocal lattice is of unit dimensions, and considering a cubic lattice and 1s STO's, the sum in Eq. (36) is of the form

$$\Phi(\vec{\nu}) = \sum_{\vec{\mu}} \frac{1}{[\mu_1^2+\mu_2^2+\mu_3^2+\delta^2]^2[(\mu_1-\nu_2)^2+(\mu_2-\nu_2)^2+(\mu_3-\nu_3)^2+\delta'^2]^2} \qquad (37)$$

where $\delta = a\zeta/2\pi$, $\delta' = a\zeta'/2\pi$, and $\vec{\nu} = (\nu_1, \nu_2, \nu_3) = a\,\vec{q}/2\pi$, with $\underline{a}$ the (direct-space) lattice parameter. The sum is now over zero and positive and negative integer values of μ_1, μ_2, μ_3, including

$\mu_1 = \mu_2 = \mu_3 = 0.$

Direct summation of Eq. (37) is an acceptable way to evaluate $\Phi(q)$, but for some values of ζ and ζ' will not be extremely efficient. Since $\Phi(\vec{q})$ is so central to electronic structure studies, it pays to search for more rapid evaluation methods.

One possibility is to perform analytically the sum in one of the three dimensions, after which the remaining summations may be undertaken numerically. Two principal tools for handling these sums are the Feynman identity, which enables different factors in a denominator to be combined into the square of a single factor, and the contour integration methods based on the introduction of a function like ctn (πz), which has poles at integer values of z. The Feynman identity is

$$\frac{1}{ab} = \int_0^1 du \frac{1}{[au + b(1-u)]^2} \tag{38}$$

The contour integration method depends on finding a contour surrounding a series of poles of ctn (πz), so that the function $f(z)$ ctn (πz) will produce a series of residues $f(n)/\pi$ whose sum can be evaluated from the integral along the contour, plus the effects of any other poles within it.

Another possibility, not yet completely implemented, is the use of Ewald techniques to accelerate convergence. An integral representation for $\exp(-\zeta r)$ suitable in this connection is

$$e^{-\zeta r} = \frac{\zeta}{2\pi^{1/2}} \int_0^\infty ds\, s^{-3/2}\, e^{-r^2 s - \zeta^2/4s} \tag{39}$$

Use of this expression for $\phi(\vec{r})$ and $\phi(\vec{r}-\vec{R}_\mu)$, followed by breaking of the integration into two parts $(0,Z)$ and (Z,∞), leads to

$$\Phi(\vec{q}) = \left[\int_0^Z ds \int_0^Z ds' + \int_0^Z ds \int_Z^\infty ds' + \int_Z^\infty ds \int_0^Z ds' + \int_Z^\infty ds \int_Z^\infty ds'\right]$$

$$\times \sum_\mu \frac{\zeta\zeta'}{4\pi} (ss')^{-3/2}\, e^{-\zeta^2/4s - \zeta'^2/4s'} \langle e^{-sr^2} | e^{i\vec{q}\cdot\vec{r}} | e^{-s'|\vec{r}-\vec{R}_\mu|^2} \rangle$$

If in the first three integration regions we introduce the transformation given in Eq. (4), the integrals can be evaluated to obtain sums like Eq. (37), but with exponential damping factors of the exp $(-\mu^2/Z)$. The last integration region gives a contribution which converges exponentially, as exp $(-\mu^2 Z)$, but unfortunately has not yet been reduced to an attractively simple form not involving a numerical integration.

A last possibility worthy of mention is that of reductions involving the use of Fourier transformations in one dimension at a time. For example, a summation such as

$$\sum_{\mu_3} \frac{1}{\mu_1^2 + \mu_2^2 + \mu_3^2 + \delta^2}$$

is equivalent to a sum of the form

$$\sum_{\mu_3} e^{-|\mu_3|\sqrt{\mu_1^2 + \mu_2^2 + \delta^2}}$$

where the convergence is not that of $e^{-\mu_3\delta}$ but that of $e^{-|\mu_3|\sqrt{\mu_1^2 + \mu_2^2 + \delta^2}}$. For small δ this can make a real difference.

D. ONE-DIMENSIONAL SYSTEMS

The techniques applicable to three-dimensional crystals can also be used when a system has one-dimensional periodicity. We illustrate by considering a linear array of nuclei of unit nuclear charge, and take

$$|\vec{k}> = \sum_{\mu} e^{ikz} \phi(\vec{r} - \vec{R}_\mu)$$

where now the $\vec{R}_\mu$ are displacements on a line which we take in the z direction, and the Bloch wave vector is now a scalar, being restricted to the z direction. Just as before, we find it convenient to express many of the energy contributions in terms of a quantity

$$\Phi(\vec{q}) = \sum_{\mu} < \phi(\vec{r})| \; e^{i\vec{q}\cdot\vec{r}} \; | \phi(\vec{r} - \vec{R}_\mu) >$$

and the overlap and kinetic energy formulas are the same as derived for three-dimensionally periodic systems :

$$<\vec{k}|\vec{k}> \;=\; N\;\Phi\;(o)$$

$$<\vec{k}\,|-\tfrac{1}{2}\,\nabla^2|\vec{k}> \;=\; \tfrac{1}{2}\,k^2\;N\;\Phi(o) \;+\; N\sum_{\mu} <\phi(\vec{r})\,|\,-\tfrac{1}{2}\,\nabla^2|\;\phi\;(\vec{r}-\vec{R}_\mu)>$$

The potential energy can now be reduced as before, and ignoring possible convergence questions we find

$$\begin{aligned} <\vec{k}\vec{k}'|h|\vec{k}\vec{k}'> \;=\; & \frac{N}{2\pi^2}\sum_{\mu}\int\frac{d\vec{q}}{q^2}\;\Phi(\vec{q})\Phi(-\vec{q})\;e^{-i\vec{q}.\vec{R}_\mu} \\ & -\frac{N}{2\pi^2}\sum_{\mu}\int\frac{d\vec{q}}{q^2}\;\Phi(\vec{q})\Phi(o)\;e^{-i\vec{q}.\vec{R}_\mu} \\ & -\frac{N}{2\pi^2}\sum_{\mu}\int\frac{d\vec{q}}{q^2}\;\Phi(o)\Phi(-\vec{q})\;e^{-i\vec{q}.\vec{R}_\mu} \\ & +\frac{N}{2\pi^2}\sum_{\mu\neq o}\int\frac{d\vec{q}}{q^2}\;\Phi(o)\Phi(o)\;e^{-i\vec{q}.\vec{R}_\mu} \end{aligned} \qquad (40)$$

$$\begin{aligned} <\vec{k}\vec{k}'|h|\vec{k}'\vec{k}> \;=\; & \frac{N}{2\pi^2}\sum_{\mu}\int\frac{d\vec{q}}{q^2}\;\Phi(\vec{q}+\vec{k}'-\vec{k})\;\times \\ & \Phi(-\vec{q}-\vec{k}'+\vec{k})\;e^{-i(\vec{q}+\vec{k}'-\vec{k}).\vec{R}_\mu} \end{aligned} \qquad (41)$$

We now interchange the μ summation and the $\vec{q}$ integration, we use Eq. (18) to convert the summation into a sum of delta functions, and we perform the q_z integration. Letting $\vec{q}_2$ refer to a two-dimensional vector of Cartesian components $(q_x\,,\;q_y)$ we have

$$<\vec{k}\vec{k}'|h|\vec{k}\vec{k}'> = \frac{N}{2\pi^2}\left(\frac{2\pi}{a}\right)\int d\vec{q}_2\sum_{\mu}\frac{1}{q_2^2+q_{\mu z}^2}\;\times$$

$$[\,\Phi(\vec{q}_\mu)\Phi(-\vec{q}_\mu)\;-\;\Phi(\vec{q}_\mu)\;\Phi(o)\;-\;\Phi(o)\;\Phi(-\vec{q}_\mu)]$$

$$+ \frac{N}{2\pi^2} \int d\vec{q}_2 \left[\frac{2\pi}{a} \sum_\mu \frac{1}{q_2^2 + q_{\mu z}^2} - \int \frac{dq_z}{q_2^2 + q_z^2} \right] \Phi(o)\Phi(o) \tag{42}$$

$$< \vec{k}\vec{k}'|h|\vec{k}'\vec{k} >= \frac{N}{2\pi^2} \left(\frac{2\pi}{a}\right) \int d\vec{q}_2 \frac{\Phi(\vec{q}_\mu)\Phi(-\vec{q}_\mu)}{q_2^2 + (q_{\mu z} + k - k')^2} \tag{43}$$

where $\vec{q}_\mu$ means $(q_x , q_y , q_{\mu z})$. The exchange contribution given in Eq. (43) can only be simplified if we make use of relationships occurring because the nuclei are on a line, and hence we shall not undertake further discussion. However, the Coulomb contribution in Eq. (42) must, as in the classical case, be rearranged into a termwise convergent form.

It is necessary to add to the first integral on the left side of Eq. (42), and then to subtract from the second integral, a quantity which will eliminate the divergences at small q_2 when $q_{\mu z} = 0$. The term to be added and subtract must have at small q_2 a form similar to $q_2^{-2}\ \Phi(o)\Phi(o)$; at large q_2 it must not destroy the existing convergence of the various terms. (The first integral converges at large q_2 because of the properties of $\Phi(\vec{q})$; the second converges because of the behavior of the difference between the μ sum and the q_z integral). We could, as in the classical case, add and subtract an expression descriptive of an array of point charges midway between the lattice points, but we presently have a better natural alternative, namely to add subtract a quantity more similar to $q_2^{-2}\Phi(q_x , q_y , o)\Phi(-q_x, - q_y, o)$ or to $q_2^{-2}\Phi(q_x , q_y , o)\Phi(o)$.

If we contemplate the use of STO's, a form suggested by the above considerations is $[\alpha^2/q_2^2 (q_2^2 + \alpha^2)]\ \Phi(o)\Phi(o)$, where α may be adjusted to mimic $\Phi(q_x , q_y , o)\Phi(o)$. We then reduce the last integral of Eq. (42) with the help of the formula

$$\int d\vec{q}_2 \left[\frac{2\pi}{a} \sum_\mu \frac{1}{q_2^2 + q_{\mu z}^2} - \int \frac{dq_z}{q_2^2 + q_z^2} - \frac{2\pi}{a} \frac{\alpha^2}{q_2^2(q_2^2 + \alpha^2)} \right]$$

$$= - \frac{4\pi^2}{a} \ln (\alpha a)$$

The result is

$$< \vec{k}\vec{k}'|h|\vec{k}\vec{k}' > = \frac{N}{2\pi^2} \left(\frac{2\pi}{a}\right) \int d\vec{q}_2 \sum_\mu \frac{1}{q_2^2 + q_{\mu z}^2} \times$$

$$[\Phi(\vec{q}_\mu)\Phi(-\vec{q}_\mu) - \Phi(\vec{q}_\mu)\Phi(o) - \Phi(o)\Phi(-\vec{q}_\mu) +$$

$$\frac{\delta_{\mu o}\,\alpha^2}{q_2^2+\alpha^2}\,\Phi(o)\Phi(o)] \quad - \frac{2N}{\alpha}\ln(\alpha a)\Phi(o)\Phi(o) \qquad (44)$$

If we expect to use GTO's, it may be more advantageous to add and subtract a quantity of the form q_2^{-2} ex $(-q_2^2/\alpha^2)$. In that case, we use the formula

$$\int d\vec{q}_2 \;[\frac{2\pi}{a}\sum_\mu \frac{1}{q_2^2+q_{\mu z}^2} - \int \frac{dq_z}{q_2^2+q_{\mu z}^2} - \frac{2\pi}{a}\,\frac{e^{-q_2^2/\alpha^2}}{q_2^2}]$$

$$= \frac{2\pi^2}{a}[Y - 2\ln(\alpha a)]$$

This leads to

$$<\vec{k}\vec{k}'|h|\vec{k}\vec{k}'> = \frac{N}{2\pi^2}(\frac{2\pi}{a})\int d\vec{q}_2 \sum_\mu \frac{1}{q_2^2+q_{\mu z}^2}[\Phi(\vec{q}_\mu)\Phi(-\vec{q}_\mu)$$

$$- \Phi(\vec{q}_\mu)\Phi(o) - \Phi(o)\Phi(-\vec{q}_\mu) + \frac{\delta_{\mu o}\, e^{-q_2^2/\alpha^2}}{q_2^2}] \qquad (45)$$

$$+ \frac{N}{a}(Y - 2\ln(\alpha a)).$$

REFERENCES

R.A. Bonham, J.L. Peacher, qnd H.L. Cox Jr. 1964. J. Chem. Phys. 40, 3083.

F.E. Harris and H. Monkhorst 1969. Phys. Rev. Lett. 23, 1026.

F.E. Harris and H. Monkhorst 1971. Solid State Commun. 9, 1449.

L. Kumar and H. Monkhorst 1974. J. Phys. F (in press).

F.E. Harris, L. Kumar and H. Monkhorst 1971. Int. J. Quantum Chem. 5S, 527.

F.E. Harris, L. Kumar and H. Monkhorst 1972. J. Phys. (Paris) 33, 99.

F.E. Harris, L. Kumar and H. Monkhorst 1973. Phys. Rev. B 7, 2850.

J.L. Calais and G. Sperber 1972. J. Phys. (Paris) 33, 205.

J.L. Calais and G. Sperber 1973. Int. J. Quantum Chem. 7, 501, 521, 537.

F.E. Harris 1972. J. Chem. Phys. 56, 4422.

L. Kumar, H. Monkhorst and F.E. Harris 1974. Phys. Rev. B 9, 4084.

NON-EMPIRICAL MOLECULAR ORBITAL THEORY FOR MOLECULAR CRYSTALS

D. P. SANTRY

Department of Chemistry

Mc Master University

Hamilton, Ontario, Canada L8S 4M1

INTRODUCTION

The theory discussed in these lectures is based on single determinant SCF (Self-Consistent-Field) molecular orbital theory. It differs from the conventional tight binding approach only in the method used to solve the crystal SCF matrix equation. The usual approach (1,2) is to factorize the crystal Fock operator by the introduction of translational symmetry adapted basis functions. This procedure is relatively straightforward and works well for polymers and one-dimensional crystals but is rather complicated when applied to three-dimensional crystals. The approach under discussion here attempts to side-step these problems by directly calculating the crystal density matrix[3,8]. It uses the molecular character of molecular crystals to advantage by means of matrix partitioning methods and SCF perturbation theory.[9]

The theory was originally developed in terms of semi-empirical molecular orbital theory for crystals in which all the lattice molecules are symmetry related. Later, it was generalized to include crystals with crystallographically independent sets of molecules. This version of the semi-empirical theory has been applied to a wide range of crystals.

The primary quantities calculated are the crystal density matrix and all quantities calculable therefrom. In the original versions of the theory no attempt was made to calculate the crystal orbitals or band structures. It was found that these quantities could be calculated, if required, on completion of the density matrix calculation by means of wave-vector dependent unitary transformations of various interaction matrices generated as by-products of the main calculation.[5] These data may then be used to calculate density of states plots[6] or exciton energies[7] for the crystal.

An important advantage of working with molecular crystals is that the theoretical results are open to direct unambiguous experimental investigation. One area gaining increasing importance in this regard is the experimental determination of charge distributions for molecules in crystals. Once the accuracy of both the experimental and theoretical methods have been increased, it will be possible to make a direct comparison between theoretical, Hartree-Fock, and experimental charge distributions. This comparison will provide a detailed check of the Hartree-Fock approximation.

The calculations on molecular crystals can provide valuable data for other areas of theoretical chemical research. Studies on the simpler hydrogen bonded crystals will provide a theory of hydrogen bonding within environments close to those prevailing in real chemical systems. The results from these calculations should prove useful to theories of hydrogen bonded liquids. Similarly, the results from exciton calculations should provide useful data for theorists working on time-dependent crystal excited state phenomena.

All of the applications of the theory so far have been at the semi-empirical level. Although this type of study has a useful role, the stage of development has been reached where non-empirical calculations on some of the simpler crystals are required. These would serve to check the results and conclusions from the simpler semi-empirical calculations, and also provide the standard theoretical data against which the experimental results will eventually be compared.

The non-empirical method, developed by O'Shea and Santry,[8] for calculating the crystal density matrix is presented here. For ease of presentation, the perturbation expansions are only taken to the second order. In practice this should be just sufficient for many purposes. More accurate work may require the third order contributions, but these can be easily derived. As an example of an algebraic application of the SCF perturbation approach, the intermolecular charge transfer within a crystal lattice is analysed.

THEORY

The objective here is to solve Roothaan's SCF matrix equation[10]

$$FC = SCE \tag{1}$$

for a molecular crystal. Where, in equation (1), F and S are, respectively, the Fock and Overlap matrices for the entire crystal. They are referred to a basis set of atomic orbitals centered on all the atoms of all the molecules in the crystal lattice. C is the coefficient matrix for the expansion of the crystal orbitals in terms of these atomic orbitals and E is the orbital energy matrix. The energies from the diagonal of this matrix constitute the one-electron band structure of the crystal.
The crystal orbitals are normalized through the matrix equation

$$\tilde{C}SC = 1 \qquad (2)$$

The crystal density matrix, P, is given by

$$P_{\mu\nu} = 2\sum_{i}^{occ} C_{\mu i}C_{\nu i} \qquad (3)$$

where μ and ν label atomic orbitals and the summation is over all occupied crystal orbitals.

Equation (1) is solved for the crystal density matrix, P, in preference to the crystal orbitals, C. The density matrix for an effectively infinite crystal is itself effectively infinite in dimension. Accordingly, it cannot be handled in a single entity but must be broken down into smaller units, or submatrices. This can easily be achieved in the case of a molecular crystal by ordering the basis set so that all atomic orbitals associated with a given lattice molecule are collected in a group. Under this arrangement the density matrix can be naturally partitioned into intramolecular and intermolecular submatrices. These submatrices occupy the diagonal and off-diagonal positions, respectively.

In general, if there is more than one chemical species in the lattice, the off-diagonal submatrices will not be square. However, for much of the present discussion we shall assume the simplest crystal structure, in which there is only one molecule per unit cell. This assumption greatly simplifies the notation and clarifies the underlying principles. Generalization of the theory to the many-molecules unit cell is relatively straightforward.[4]

A single label, T, is sufficient to identify a given lattice molecule if there is only one molecule per unit cell. This label, which is a vector, points to the Tth unit cell in the crystal. A submatrix of a given crystal matrix, U, may be written as

$$^{TW}U = \left\{ {}^{TW}U_{\mu\nu} \right\} \qquad (4)$$

μ and ν label atomic orbitals associated with molecules T and W, respectively.

The full crystal density matrix, P, is built up by the calculation of the component submatrices, ${}^{TW}P$. Fortunately, not all of these submatrices are required for the complete specification of P. All that is required for a crystal with perfect translational symmetry is one intramolecular submatrix for an arbitrarily chosen molecule together with a complete set of intermolecular submatrices involving this molecule and the remainder of the lattice. In practice, relatively few intermolecular matrices would be required since, for molecular crystals at least, ${}^{TW}P$ would be expected to tent stongly to zero with increasing molecular separation. This number could be further reduced by making use of any elements of rotational symmetry present in the crystal space group.

Equation (1) is solved for the crystal density matrix by the SCF perturbation method outlined in the Appendix. Expanding this equation in a perturbation series, to the second order, we find

$$(F^{(0)}+F^{(1)}+F^{(2)})(C^{(0)}+C^{(1)}+C^{(2)}) = (S^{(0)}+S^{(1)}+S^{(2)})(C^{(0)}+C^{(1)}+C^{(2)})(E^{(0)}+E^{(1)}+E^{(2)}) \quad (5)$$

In order to make use of the molecular character of molecular crystals, $F^{(0)}$, $S^{(0)}$, and $E^{(0)}$ are chosen to be block diagonal with the appropriate matrices for properly oriented but hypothetically non-interacting lattice molecules along their diagonals. $C^{(0)}$ can be assumed, under this definition, also to be block diagonal. The submatrices are simply the molecular orbital coefficients for properly oriented but independent molecules. These matrices satisfy the zero order equation

$$F^{(0)}C^{(0)} = S^{(0)}C^{(0)}E^{(0)} \quad (6)$$

or

$${}^{TT}F^{(0)}\ {}^{TT}C^{(0)} = {}^{TT}S^{(0)}\ {}^{TT}E^{(0)}, \qquad T = 1,\ldots N \quad (7)$$

where N is the number of unit cells in the crystal. Leaving aside, for the moment, the definition of the remaining orders of F and S, we proceed with the evaluation of the crystal density submatrices. These are given by, to the second order

$${}^{TW}P = {}_{TW}\,{}^{TT}P^{(0)} + {}^{TW}P^{(1)} + {}^{TW}P^{(2)} \quad (8)$$

where δ_{TW} is the Kronecker delta. Note, the semi-empirical version of the theory takes P to the third order to check on the convergence.

The first order correction, $P^{(1)}$, to the density matrix is given, in the notation of the Appendix, by

$$P^{(1)}_{\mu\nu} = 2 \sum_i^{occ} \sum_k^{all} A_{ki}(C^{(0)}_{\mu i} C^{(0)}_{\nu k} + C^{(0)}_{\mu k} C^{(0)}_{\nu i}) \tag{9}$$

The summations over i and k refer to the zero order crystal orbitals. If i and k both label occupied orbitals then, from the Appendix,

$$A_{ki} = -\frac{1}{2} S^{(1)}_{ki} \tag{10}$$

where

$$S^{(1)}_{ki} = \sum_{\mu\nu}\sum C^{(0)}_{\mu k} S^{(1)}_{\mu\nu} C^{(0)}_{\nu i} \tag{11}$$

The summations over μ and ν include all the functions in the crystal basis set.

If i and k label occupied and vacant orbitals, respectively, then

$$A_{ki} = F^{(1)}_{ki}/(\varepsilon^{(0)}_i - \varepsilon^{(0)}_k) \tag{12}$$

Here

$$F^{(1)}_{ki} = F^{(1)}_{ki} - S^{(1)}_{ki}\varepsilon^{(0)}_i \tag{13}$$

where $\varepsilon^{(0)}_i$ is the ith zero orbital energy.

Equation (10) can be readily partitioned into submatrix equations by means of the block diagonal form of $C^{(0)}$. Thus,

$$^{TW}P^{(1)}_{\mu\nu} = 2 \sum_i^{occ_T} \sum_k^{all_W} {}^{WT}A_{ki}\, {}^{T}C^{(0)}_{\mu i}\, {}^{W}C^{(0)}_{\nu k} + 2 \sum_i^{occ_W} \sum_k^{all_T} {}^{TW}A_{ki}\, {}^{W}C^{(0)}_{\nu i} \times {}^{T}C^{(0)}_{\mu k} \tag{14}$$

The summation $\sum_i^{occ_T}$ is over all the occupied molecular orbitals localized on the Tth molecule to the zero order. ^{WT}A is the WT^{th} submatrix of A.

Equation (14) constitutes the first order solution to the crystal SCF matrix equation. All that is required for the calculation of the various ^{TW}P's are the corresponding first order submatrices, $^{TW}F^{(1)}$ and $^{TW}S^{(1)}$.

The expression derived for $P^{(2)}$ in the Appendix can easily be partitioned into submatrix equations for P in much the same way. The final result is,

$$^{TW}P^{(2)}_{\mu\nu} = 2 \sum_{i}^{occ_W} \sum_{k}^{all_T} {}^{TW}\mathcal{B}_{ki}\, {}^{T}C^{(0)}_{\mu k}\, {}^{W}C^{(0)}_{\nu i} + 2 \sum_{i}^{occ_T} \sum_{k}^{all_W} {}^{WT}\mathcal{B}_{ki}\, {}^{W}C^{(0)}_{\nu k}\, {}^{T}C^{(0)}_{\mu i} + 2 \sum_{X}^{crystal} \sum_{i}^{occ_X} \sum_{k}^{vac_T} \sum_{\ell}^{vac_W} {}^{TX}A_{ki}\, {}^{WX}A_{\ell i}\, {}^{T}C^{(0)}_{\mu k}\, {}^{W}C^{(0)}_{\nu \ell} \tag{15}$$

where, i and k both label occupied molecular orbitals

$$^{TW}\mathcal{B}_{ki} = -\frac{1}{2} \sum_{X}^{crystal} \left[\sum_{\ell}^{vac} {}^{WX}A_{i\ell}\, {}^{TX}A_{k\ell} - \sum_{\ell}^{all} {}^{WX}S^{(1)}_{i\ell}\, {}^{XT}S^{(1)}_{\ell k} \right] \tag{16}$$

and if i and k label occupied and vacant orbitals, respectively

$$^{TW}\mathcal{B}_{ki} = {}^{TW}\mathcal{F}^{(2)}_{ki}/(\varepsilon_i^{(0)} - \varepsilon_k^{(0)}) + \sum_{X}^{crystal} \{ \sum_{\ell}^{vac} {}^{TX}\mathcal{F}^{(1)}_{k\ell}\, {}^{XW}A_{\ell i} - \sum_{\ell}^{occ} {}^{TX}A_{k\ell}\, {}^{XW}\mathcal{F}^{(1)}_{\ell i} - \sum_{\ell}^{all} {}^{TX}S^{(1)}_{k\ell}\, {}^{XW}\mathcal{F}^{(1)}_{\ell i} \}/(\varepsilon_i^{(0)} - \varepsilon_k^{(0)}) - \sum_{\ell}^{occ} {}^{TX}A_{k\ell}\, {}^{XW}S^{(1)}_{\ell i} \tag{17}$$

Equation (15) constitutes the second order solution to equation (1) Both equations (14) and (15), anticipating the next section, must be solved iteratively since the evaluation of $\mathcal{F}^{(1)}$ and $\mathcal{F}^{(2)}$ requires $P^{(1)}$ and $P^{(2)}$, respectively. The second order density matrix has a constant, non-iterative, component. This component arises from various products of, previously calculated, first-order quantities, equations (15), (16) and (17). The evaluation of this contribution requires the lattice summation, $\sum_{X}^{crystal}$, and leads to the explicit inclusion of three-molecule interactions.

All of the second-order density submatrices can be calculated once the corresponding Fock submatrices, $^{TW}F^{(2)}$, are known.

DEFINITION OF THE OVERLAP SUBMATRICES

The TW^{th} submatrix of the overlap matrix S is given by

$$^{TW}S = ({}^{TW}S_{\mu\nu}) = (\int \chi^{T}_{\mu}(1) \chi^{W}_{\nu}(1) d\tau_1) = ((\mu_T | \nu_W)) \tag{18}$$

where χ^{T}_{μ} is the μth atomic orbital associated with the Tth molecule. From our choice of the zero order solution for the crystal

$$^{TW}S^{(0)} = \delta_{TW}\, {}^{TW}S \tag{19}$$

All elements of S not included at the zeroth order are defined to be part of the first order perturbation. Thus

$$^{TW}S^{(1)} = (1 - \delta_{TW})\, {}^{TW}S \tag{20}$$

and

$$ {}^{TW}S^{(n)} = 0 \qquad \text{if } n > 1 \tag{21}$$

DEFINITION OF THE FOCK SUBMATRICES FOR THE CRYSTAL

The Fock matrix, F, for the crystal is given by

$$F = H + G \tag{22}$$

where the Hamiltonian matrix is given by

$$H_{\mu\nu} = (\mu|-\tfrac{1}{2}\nabla^2|\nu) - \sum_{\beta}^{atoms} Z_{\beta}(\mu|R_{\beta}^{-1}|\nu) \tag{23}$$

Here

$$(\mu|-\tfrac{1}{2}\nabla^2|\nu) = \int\chi_{\mu}(1)-\tfrac{1}{2}\nabla^2\chi_{\nu}(1)d\tau_1 \tag{24}$$

and

$$(\mu|R^{-1}|\nu) = \int\chi_{\mu}(1)R_{\beta 1}^{-1}\chi_{\nu}(1)d\tau_1 \tag{25}$$

Z_{β} is the nuclear charge on the βth atom and the summation in equation (21) includes all atoms in the crystal lattice

The electron repulsion matrix, G, is given by

$$G_{\mu\nu} = \sum_{\sigma\lambda}\sum P_{\sigma\lambda}\{(\mu\nu|\sigma\lambda) - \tfrac{1}{2}(\mu\lambda|\sigma\nu)\} \tag{26}$$

The summations include all atomic orbitals in the crystal basis set, and

$$(\mu\nu|\sigma\lambda) = \iint\chi_{\mu}(1)\chi_{\nu}(1)r_{12}^{-1}\chi_{\sigma}(2)\chi_{\lambda}(2)d\tau_1 d\tau_2 \tag{27}$$

The next step is to partition H and G into submatrices. Starting with H, we find

$$ {}^{TW}H_{\mu\nu} = (\mu_T|-\tfrac{1}{2}\nabla^2|\nu_W) - \sum_{X}^{crystal}\sum_{\beta}^{X} {}^{X}Z_{\beta}(\mu_T|{}^{X}R_{\beta}^{-1}|\nu_W) \tag{28}$$

The summation over X includes all molecules in the crystal and the summation over β all atoms in the Xth molecule. ${}^{X}Z_{\beta}$ is the nuclear charge of the βth atom of the Xth molecule.

Similarly, the TW^{th} submatrix of G is given by

$$ {}^{TW}G_{\mu\nu} = \sum_{XY\sigma}\sum\sum^{X}\sum_{\lambda}^{Y} {}^{XY}P_{\sigma\lambda}\{(\mu_T\nu_W|\sigma_X\lambda_Y) - \tfrac{1}{2}(\mu_T\lambda_Y|\sigma_X\nu_W)\} \tag{29}$$

The summations over X and Y include all molecules in the crystal and the summations over σ and λ are limited to the basis functions associated with molecules X and Y, respectively. For convenience, the integrals in the curly bracket can be represented by a single quantity, M, where

$$^{TW}M^{(XY,TX)}_{\mu\nu:\sigma\lambda} = (\mu_T\nu_W|\sigma_X\lambda_Y) - \frac{1}{2}(\mu_T\lambda_Y|\sigma_X\nu_W) \tag{30}$$

The electron repulsion submatrix may be written as

$$^{TW}G_{\mu\nu} = \sum_{XY\sigma}\sum^{X}\sum_{\lambda}^{Y} {}^{XY}P_{\sigma\lambda}\, {}^{TW}M^{(XY,TX)}_{\mu\nu:\sigma\lambda} \tag{31}$$

We now consider the definition of the perturbation operators. The form of the full crystal Fock and Hamiltonian operators allows considerable latitude in the definition of these quantities. In order to optimize the convergence of the perturbation expansion, we include two contributions in the first order Fock operators. The first is the static field arising from the zero order charge distribution of the lattice molecules. The second is the first order polarization field arising from the effect of this static field.

By definition the zero order Fock submatrices, which are the free molecule Fock matrices, are given by

$$\begin{aligned} ^{TW}F^{(0)}_{\mu\nu} &= \delta_{TW}[-\tfrac{1}{2}(\mu_T|-\tfrac{1}{2}\nabla^2|\nu_T) - \sum_{\beta}^{T} {}^{T}Z_\beta(\mu_T|{}^{T}R_\beta^{-1}|\nu_T) \\ &\quad + \sum_{\sigma}^{T}\sum_{\lambda}^{T} {}^{TT}P^{(0)}_{\sigma\lambda}\, {}^{TT}M^{(TT,TT)}_{\mu\nu:\sigma\lambda}], \end{aligned} \tag{32}$$

$^{TT}P^{(0)}$ is the zero order intramolecular density matrix for the Tth molecule. Similarly, by definition

$$\begin{aligned} ^{TW}F^{(1)}_{\mu\nu} &= (1-\delta_{TW})(\mu_T|-\tfrac{1}{2}\nabla^2|\nu_W) - \sum_{X}{}'\sum_{\beta} {}^{X}Z_\beta(\mu_T|{}^{X}R_\beta^{-1}|\nu_W) + \\ &\quad \sum_{X}{}'\sum_{Y\sigma}\sum^{X}\sum_{\lambda}^{Y} {}^{XY}P^{(0)}_{\sigma\lambda}\, {}^{TW}M^{(XY,TX)}_{\mu\nu:\sigma\lambda} + (1-\delta_{TW})\sum_{\sigma\lambda}\sum {}^{TT}P^{(1)}_{\sigma\lambda}\, {}^{TT}M^{(TT,TT)}_{\mu\nu:\sigma\lambda} \\ &\quad + \sum_{X}{}'\sum_{Y\sigma}\sum^{X}\sum_{\lambda}^{Y} {}^{XY}P^{(1)}_{\sigma\lambda}\, {}^{TW}M^{(XY,TX)}_{\mu\nu:\sigma\lambda} \end{aligned} \tag{33}$$

The primed lattice sum, $\sum_X{}'$, in the nuclear electron attraction contribution is to exclude the term with X = T when W = T. In the case of the electron repulsion integral sum, over M, it is to exclude the term X = Y = T when T = W. Under this definition of $F^{(0)}$ and $F^{(1)}$ all Hamiltonian matrix elements are accounted for, so that $^{TW}H^{(n)} = 0$ for all $n > 1$. All remaining higher order Fock submatrices, therefore, contain only polarization terms.

$$^{TW}F_{\mu\nu}^{(n)} = (1-\delta_{TW})\sum_{\sigma}^{T}\sum_{\lambda}^{T} {}^{TT}P_{\sigma\lambda}^{(n)}\ {}^{TT}M_{\mu\nu:\sigma\lambda}(TT,TT)$$

$$+ \sum_{X}{}' \sum_{Y}\sum_{\sigma}^{X}\sum_{\lambda}^{Y} {}^{XY}P_{\sigma\lambda}^{(n)}\ {}^{TW}M_{\mu\nu:\sigma\lambda}(XY:TX) \quad \text{if } n > 1 \tag{34}$$

With the definition of the Fock submatrices completed, the density submatrices may be calculated by means of equations (14) and (15). Although the present theory is intended for the theoretical study of molecular crystals, no use has as yet been made of translation symmetry. In fact, as it stands, the theory is applicable to molecular aggregates, both finite and infinite, in any arbitrary configuration. However, the calculation of the density submatrices for such systems would be difficult as it would entail the solution of a large number, as many as there are different pairs of molecules in the aggregate, of coupled submatrix equations. This coupling arises through the presence of all density submatrices for the system in each and every Fock submatrix. Thus, it is necessary to estimate all the density submatrices for the aggregate in order to calculate any given density submatrix.

We now show how the translational symmetry of a lattice simplifies the calculation for a molecular crystal by partially decoupling these matrix equations.

THE INTRODUCTION OF CRYSTAL SYMMETRY

If a crystal has perfect translational symmetry, any two submatrices of a given crystal matrix, U, will be equal if the difference between their unit cell identifiers are equal :

$$^{XY}U = {}^{TW}U \tag{35}$$

if

$$Y - X = W - T \tag{36}$$

Such matrices may be said to be translationally equivalent. In order to take advantage of this property, it is appropriate to identify submatrices by means of relative vector labels, such as those shown in Fig. 1, in preference to the absolute TW labels. The submatrices of equation (35) would thus be written as $^{\underline{\tau}}U$ and $^{\rho}U$, respectively. The matrix ^{WT}U would be written $^{\bar{\rho}}U$.

The presence of lattice symmetry simplifies the calculation of the crystal density matrix in two ways. First, it reduces the number of submatrices that need to be calculated to only those that are translationally inequivalent. Second, the translational symmetry can be used to simplify and partially decouple the equations for the density submatrices.

We start by simplifying the Hamiltonian submatrices through the introduction of lattice sums. Decoupling problems need not be considered here since the matrix equations are coupled only through the electron repulsion terms. It can be seen that, because of the translational symmetry, it is unnecessary to label the nuclear charge in eq. (33) with respect to its unit cell, X. Thus, order of summation over X and β can be reversed, Z_β taken out the X summation and the whole term replaced by a lattice sum, V:

$$ {}^{\rho}V_{\mu\nu} = \sum_{\beta} Z_\beta \sum_{\zeta}{}' \, (\mu_T | {}^{\zeta}R_\beta^{-1} | \nu_{T+\rho}) \tag{37} $$

where ζ and ρ are the vector labels shown in Fig. 1 and T now labels a molecular arbitrarily chosen as the origin. The prime, as before, omits the origin term when ρ = 0.

Turning now to the electron repulsion matrices, the M terms can be most conveniently expressed as ;

$$ {}^{\rho}M^{(\tau,\zeta)}_{\mu\nu:\sigma\lambda} = {}^{TW}M^{(XY,TX)}_{\mu\nu:\sigma\lambda} \tag{38} $$

The matrix equations are coupled through the presence of all density submatrices in each electron repulsion submatrix to all orders except the zeroth. A partial decoupling can be achieved by replacing these density submatrices by a complete set of their translational equivalents. This step decreases the number of coupled equations to be solved from $\frac{1}{2}N(N+1)$ to N, where N is the number of molecules in the crystal. It also leads to the appearance of lattice sums ofelectron repulsion integrals in the theory.

Taking the last term in equation (34) for ${}^{TW}F^{(n)}$, as the simplest example, express all matrices in the relative vector label notation of Fig. 1.

$$ \sum_{X}{}' \sum_{Y}\sum_{\sigma}{}^{X} \sum_{\lambda}{}^{Y} \, {}^{XY}P^{(n)}_{\sigma\lambda} \, {}^{TW}M^{(XY,TX)}_{\mu\nu:\sigma\lambda} = \sum_{\zeta}{}' \sum_{\tau}\sum_{\sigma}\sum_{\lambda} {}^{\tau}P^{(n)}_{\sigma\lambda} \, M^{(\tau,\zeta)}_{\mu\nu:\sigma\lambda} \tag{39} $$

Since ${}^{\tau}P_{\sigma\lambda}$ is independent of ζ, this may be written as

$$ \sum_{\tau}\sum_{\sigma}\sum_{\lambda} {}^{\tau}P^{(n)}_{\sigma\lambda} \sum_{\zeta}{}' \, {}^{\rho}M^{(\tau,\zeta)}_{\mu\nu:\sigma\lambda} = \sum_{\tau}\sum_{\sigma}\sum_{\lambda} {}^{\tau}P^{(n)}_{\sigma\lambda} \, {}^{\rho}\Gamma^{(\tau)}_{\mu\nu:\sigma\lambda} \tag{40} $$

where ${}^{\rho}\Gamma^{(\tau)}_{\mu\nu:\sigma\lambda}$ is the following lattice sum :

$$ {}^{\rho}\Gamma^{(\tau)}_{\mu\nu:\sigma\lambda} = \sum_{\zeta}{}' \, {}^{\rho}M^{(\tau,\zeta)}_{\mu\nu:\sigma\lambda} \tag{41} $$

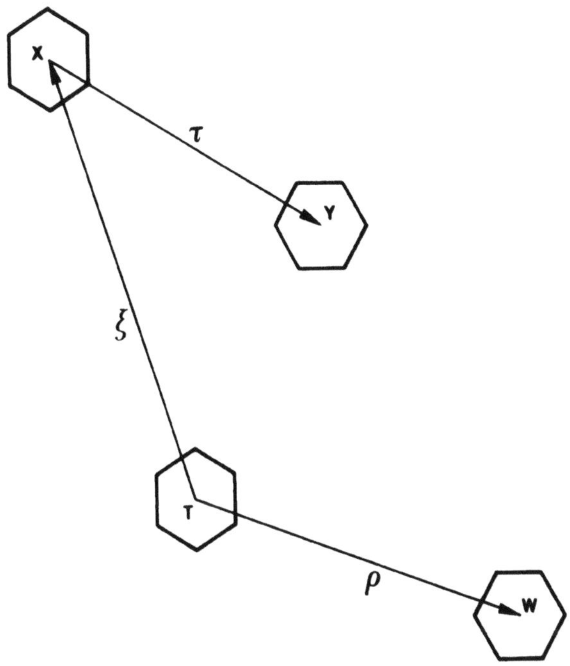

Figure 1.

Here the prime omits the origin term, $\zeta = 0$ if $\rho = \tau = 0$. The nth order ($n > 0$) matrix equations based on equation (31) for the electron repulsion contribution are partially decoupled since it is no longer necessary to calculate all density submatrices in order to set-up a given Fock submatrix ; a complete set of translationally inequivalent matrices is all that is required. In this notation, the first order Fock submatrix may be written

$$
{}^{\rho}F^{(1)}_{\mu\nu} = (1-\delta_{d,\rho})(\mu_T|-\tfrac{1}{2}\nabla^2|\nu_{T+\rho}) - {}^{\rho}V_{\mu\nu} + \sum_{\sigma\lambda}\sum {}^{d}P^{(0)}_{\sigma\lambda}\, {}^{\rho}\Gamma^{(d)}_{\mu\nu:\sigma\lambda}
$$

$$
+ \delta_{d,\rho}\sum_{\sigma\lambda}\sum {}^{d}P^{(1)}_{\sigma\lambda}\, {}^{d}M^{(d,d)}_{\mu\nu:\sigma\lambda} + \sum_{\tau\sigma\lambda}\sum\sum {}^{\tau}P^{(1)}_{\sigma\lambda}\, {}^{\rho}\Gamma^{(\tau)}_{\mu\nu:\sigma\lambda} \qquad (42)
$$

where d, for diagonal, now denotes the null vector labels are those shown in Fig. 1. Similarly, for the complete n^{th}, $n > 1$, Fock submatrix

$$
{}^{\rho}F^{(n)}_{\mu\nu} = \delta_{d,\rho}\sum_{\sigma\lambda}\sum {}^{d}P^{(n)}_{\sigma\lambda}\, {}^{\rho}M^{(d,d)}_{\mu\nu:\sigma\lambda} + \sum_{\tau\sigma\lambda}\sum\sum {}^{\tau}P^{(n)}_{\sigma\lambda}\, {}^{\rho}\Gamma^{(\tau)}_{\mu\nu:\sigma\lambda} \qquad (43)
$$

The method of solution for an ideal crystal should now be apparent. A complete set of translationally inequivalent matrices are calculated at each order. At the zeroth level, only the intramolecular submatrices, ${}^{d}P^{(0)}$, are non-zero. A complete set of inequivalent submatrices strictly involves the intramolecular submatrix together with all intermolecular submatrices involving the chosen reference and all the remaining molecules in the lattice. However, in practice relatively few, at most up to nearest neighbour, intermolecular submatrices need be calculated since ${}^{\rho}P^{(n)}$ decreases strongly as $|\rho|$ increases. Any elements of rotational symmetry present in the lattice could be used to reduce this number still further. Thus, in practice the calculation of the density matrix for a molecular crystal requires the solution of about eight to twelve, depending on the crystal structure, coupled matrix equations of the same dimensionality as the SCF equation for an isolated molecule of the same species as the lattice molecules. These matrix equations are of the density submatrices given by (15) and (17). For the purpose of the ideal crystal calculation these, and their associated matrix equations would be expressed in the relative vector notation. For example, the first order density submatrices are given by

$$
{}^{\rho}P^{(1)}_{\mu\nu} = 2\sum_{i}^{occ}\sum_{k}^{all}\{{}^{\bar{\rho}}A_{ki}\, C^{(0)}_{\mu i}\, C^{(0)}_{\nu k} + {}^{\rho}A_{ki}\, C^{(0)}_{\nu i}\, C^{(0)}_{\mu k}\} \qquad (44)
$$

${}^{\rho}A$ is the ρth submatrix of A and the summations are over the molecular orbitals of an arbitrarily chosen lattice molecule. Since all the molecules in the lattice are the same, it is unnecessary to identify with which molecule the molecular orbitals and the corresponding coefficients are associated. This has to be modified, of course, if there is more than one molecule per unit cell.

The energy per mole, W, of the crystal may readily be calculated from the various submatrices as follows[4]

$$W' = \frac{1}{2}\sum_{\mu\nu}\sum P_{\mu\nu}(H_{\mu\nu} + F_{\mu\nu}) + \text{nuclear repulsion} \quad (45)$$

$$= \frac{1}{2}\sum_{TW\mu}\sum\sum^{T} \sum_{\nu}^{W} {}^{TW}P_{\mu\nu}({}^{TW}H_{\mu\nu} + {}^{TW}F_{\mu\nu}) + N.R.' \quad (46)$$

N.R.' is the total nuclear repulsion energy for the crystal. If N is the number of unit cells, then applying the tranlational symmetry argument,

$$W' = \frac{1}{2}N\sum_{\rho\mu\nu}\sum\sum {}^{\rho}P_{\mu\nu}({}^{\rho}H_{\mu\nu} + {}^{\rho}F_{\mu\nu}) + N.R.' \quad (47)$$

Dividing by N to obtain the energy per unit cell, we find

$$W = \frac{1}{2}\sum_{\rho\mu\nu}\sum\sum {}^{\rho}P_{\mu\nu}({}^{\rho}H_{\mu\nu} + {}^{\rho}F_{\mu\nu}) + N.R. \quad (48)$$

N.R. now refers to the nuclear repulsion energy per unit cell. To estimate the crystal binding energy per molecule, we calculate the average energy per molecule in the crystal and subtract from it the energy of an isolated molecule, given here by

$$\frac{1}{2}\sum_{\mu\nu}\sum {}^{d}P^{(0)}_{\mu\nu}({}^{d}H^{(0)}_{\mu\nu} + {}^{d}F^{(0)}_{\mu\nu}) + n.r. \quad (49)$$

n.r. is the nuclear repulsion energy for a single molecule. The perturbation series for the average energy per molecule in the crystal is simplified by the block diagonal form of the zero order matrices and by the fact that

$$\sum_{\mu\nu}\sum {}^{\rho}P^{(1)}_{\mu\nu}\,{}^{\rho}F^{(0)}_{\mu\nu} = 0 \quad (50)$$

The final result for the binding energy per molecule, ΔW, is

$$\Delta W = \frac{1}{2}\sum_{\mu\nu}\sum\{{}^{d}P^{(0)}_{\mu\nu}\{{}^{d}H^{(1)}_{\mu\nu} + {}^{d}F^{(1)}_{\mu\nu}\} + {}^{d}P^{(1)}_{\mu\nu}\,{}^{d}H^{(0)}_{\mu\nu}$$

$$+ {}^{d}P^{(1)}_{\mu\nu}\{{}^{d}H^{(1)}_{\mu\nu} + {}^{d}F^{(1)}_{\mu\nu}\} + {}^{d}P^{(0)}_{\mu\nu}\,{}^{d}F^{(2)}_{\mu\nu} + {}^{d}P^{(2)}_{\mu\nu}({}^{d}H^{(0)}_{\mu\nu} + {}^{d}F^{(0)}_{\mu\nu})$$

$$+ \frac{1}{2}\sum_{\rho}{}' \sum_{\mu\nu}\sum [\rho P^{(1)}_{\mu\nu}(\rho H^{(1)}_{\mu\nu} + \rho F^{(1)}_{\mu\nu}) + \rho P^{(2)}_{\mu\nu}(\rho H^{(1)}_{\mu\nu} + \rho F^{(1)}_{\mu\nu})$$

$$+ \rho P^{(1)}_{\mu\nu}\, \rho F^{(2)}_{\mu\nu}] + \text{N.R.} \qquad (51)$$

N.R. is now the nuclear repulsion energy per unit cell minus the nuclear repulsion energy for a single molecule.

Although the calculations discussed here have been taken to the second order, the last terms in equation (51) combine first and second order contributions to yield terms that are overall of the third order. This procedure would be inconsistent in the calculation of molecular interaction constants, for example, but it is acceptable in the present application since the choice of perturbation is, to some extent, arbitrary. The inclusion of the last terms is recommended since studies with perturbation calculations based on semi-empirical MO theories showed that their inclusion improved the performance of the theory to the point where it could be confidently employed for the calculation of molecular configurations with dimers.

As calculation in eq. (51), the binding energy contains all of the interactions included in the single determinant SCF theory. Thus, the present method includes contributions from intermolecular electron exchange, intermolecular charge transfer, together with the major terms which can be derived from classical electrostatic theory. The latter include direct multipole-multipole interactions together with the corresponding polarization forces. Dipersion forces, which arise from Coulomb electron correlation, are not included and must be considered separately. The theory can confidently be applied to hydrogen bonded and charge transfer crystals. Useful insights can also be obtained for molecular crystals where dispersion energies make a substantial contribution to the binding energy.

A NOTE ON INTERMOLECULAR CHARGE TRANSFER

The transfer of electron density between sets of molecules in a crystal is symmetry allowed if the molecules are crystallographically independent. That is, if the crystal in question has several molecules in its unit cell some of which are not related by elements of the unit cell group. An obvious example of this is a crystal, such as oxalic acid dihydrate, composed of molecules of different chemical species. Unit cell molecules of the same species can also be independent. Ice-II provides an example of this situation. Here the twelve unit cell molecules fall into two independent sets each of six molecules.

Intermolecular transfer of electron density is clearly important to the electronic properties of molecular crystals. We show, as an example of an algebraic application of the perturbation approach, that it is probably quite small for the majority of molecular crystals, even when symmetry allowed.

If N_e is the total number of electrons in the crystal, then

$$N_e = \sum_{\mu\nu}\sum P_{\mu\nu} S_{\mu\nu} \tag{52}$$

The summations over μ and ν extend over all functions in the crystals basis set. Partitioning this expression into submatrices we find

$$N_e = N\sum_{\zeta\mu\nu}\sum\sum {}^{\zeta}P_{\mu\nu}\, {}^{\zeta}S_{\mu\nu} \tag{53}$$

where μ and ν are now summed over the basis functions for the two molecules connected by the vector ζ. N, as before, is the number of unit cells in the crystal. Expanding this expression into a perturbation series, remembering that $P^{(0)}$ and $S^{(0)}$ are block diagonal and that ${}^{d}S^{(1)} = 0$, we find for the charge density per unit cell :

Zero order :

$$\sum_{\mu\nu}\sum {}^{d}P^{(0)}_{\mu\nu}\, {}^{d}S^{(0)}_{\mu\nu} \tag{54}$$

First order :

$$\sum_{\mu\nu}\sum {}^{d}P^{(1)}_{\mu\nu}\, {}^{d}S^{(0)}_{\mu\nu} \tag{55}$$

Second order :

$$\sum_{\mu\nu}\sum {}^{d}P^{(2)}_{\mu\nu}\, {}^{d}S^{(0)}_{\mu\nu} + \sum_{\zeta\mu\nu}\sum\sum {}^{\zeta}P^{(1)}_{\mu\nu}\, {}^{\zeta}S^{(1)}_{\mu\nu} \tag{56}$$

The investigation of intermolecular charge transfer requires the notation used here to be generalized to handle the many-molecule unit cell. Two labels are required to identify a given lattice molecule in a crystal with more than one molecule per unit cell. These labels, T and u, identify the unit cell and site, respectively, occupied by a given molecule. Accordingly, the density and all other submatrices, must now be written

$${}^{TuWv}P \equiv {}^{u\rho v}P \tag{57}$$

The above equations are now rewritten in this notation :

Zero order :

$$\sum_{u\mu}\sum^{u}\sum_{\nu}^{u}\ {}^{u}du_{P_{\mu\nu}}(0)\ {}^{u}du_{S_{\mu\nu}}(0) \tag{58}$$

Since there are now several different species of molecule present, it is necessary to identify to which molecular list of basis functions the summations over μ and ν refer.

First order :

$$\sum_{u\mu}\sum^{u}\sum_{\nu}^{u}\ {}^{u}du_{P_{\mu\nu}}(1)\ {}^{u}du_{S_{\mu\nu}}(0) \tag{59}$$

Second order :

$$\sum_{u\mu}\sum^{u}\sum_{\nu}^{u}\ {}^{u}du_{P_{\mu\nu}}(2)\ {}^{u}du_{S_{\mu\nu}}(0) + \sum_{\zeta u\nu\mu}\sum\sum\sum^{u}\sum_{\nu}^{v}\ {}^{u\zeta v}P_{\mu\nu}(1)\ {}^{u\zeta v}S_{\mu\nu}(1) \tag{60}$$

The zero order contribution, equation (58), equals the number of electrons in the unit cell. Thus, in order to conserve charge, the first and second, in fact all, orders must separately equal zero. This does not imply that charge transfer must be zero but simply that, for each order, the electron density gained by one set of molecules must equal that lost by the other molecules. The fraction of the first order charge, ${}^{u}Q^{(1)}$, associated with the uth molecule is clearly

$${}^{u}Q(1) = \sum_{\mu}^{u}\sum_{\nu}^{u}\ {}^{u}du_{P_{\mu\nu}}(1)\ {}^{u}du_{S_{\mu\nu}}(0) \tag{61}$$

If this number is positive, it shows the uth molecule has accumulated excess electron density as a result of the crystal interactions. We now show equation (61) vanishes for all u so that the unit cell molecules neither gain nor lose electron density to the first order. First, substitute for $P^{(1)}$ in equation (61) using a generalized form of equation (44) :

$${}^{u}Q(1) = 2\sum_{i}^{occ\,u}\sum_{k}^{all\,u}\sum_{\mu}^{u}\sum_{\nu}^{u}\ {}^{u}du_{A_{ki}}\left({}^{u}C_{\mu i}(0)\ {}^{u}C_{\nu k}(0) + {}^{u}C_{\mu k}(0)\ {}^{u}C_{\nu i}(0)\right)\ {}^{u}du_{S_{\mu\nu}}(0) \tag{62}$$

$$= 2\sum_{i}^{occ\,u}\sum_{k}^{all\,u}\ {}^{u}du_{A_{ki}}\sum_{\mu}^{u}\sum_{\nu}^{u}\ {}^{u}du_{S_{\mu\nu}}(0)\left({}^{u}C_{\mu i}(0)\ {}^{u}C_{\nu k}(0) + {}^{u}C_{\mu k}(0)\ {}^{u}C_{\nu i}(0)\right) \tag{63}$$

$$= 4\sum_{i}^{occ\,u}\ {}^{u}du_{A_{ii}} = -2\sum_{i}^{occ\,u}\ {}^{u}du_{S_{ii}}(1) = 0 \tag{64}$$

It is a relatively straightforward matter to show that the second order term, $u_Q^{(2)}$, does not vanish. Intermolecular charge transfer is, therefore, a second and higher order effect. In view of the relatively weak intermolecular forces in molecular crystals this implies that this effect will be small for the majority of molecular crystals. Possible exceptions to this may be the π-complex crystals such as the trinitrobenzene-skatole system.

Analysis of the second-order term yields considerable insight into the processes responsible for intermolecular charge transfer in crystals. However, as this is a rather lengthy procedure, it will be discussed separately elsewhere.

CONVERGENCE PROPERTIES OF THE PERTURBATION APPROACH

There are two questions to be considered under this heading. The first is how convergent is the perturbation series with respect to order ? The second is how convergent are the $^{\rho}P$ with respect to ρ? In other words, what is the minimum molecular separation beyond which all remaining intermolecular submatrices can be considered negligible ? The non-empirical crystal theory has not, as yet, been implemented so that there are no non-empirical data available for the investigation of these two questions.
The theory has, however, been implemented at the semi-empirical level and applied to a wide range of crystals. The results from these calculations give some indication of how well the method works in practice.

The convergence of the SCF perturbation theory of intermolecular forces was studied by Bacon and Santry[11] These workers undertook a series of direct SCF calculations for $(HF)_2$, with various intermolecular separations, using Fock operators with a range of values for the perturbation scaling parameter, λ. The set of density matrices obtained by this procedure were fitted to matrix polynomials in λ. The matrix coefficients, $P^{(n)}$, of these polynomials are by definition the various order of the density matrix. It was found that these polynomials could not be taken beyond the fifth order, λ^5, because of severe rounding errors. Nonetheless, the calculations did provide some interesting results and, incidentally, a check on our perturbation algebra. It was found that $P^{(2)}$ was smaller than $P^{(1)}$ but nonetheless, comparable. The reason for this has been discussed by Middlemiss and Santry. [12] $P^{(3)}$ was found to be always more than an order of magnitude smaller than $P^{(2)}$. The perturbation series in P appeared to converge very strongly beyond the third order so that $P^{(4)}$ and $P^{(5)}$ are almost certainly negligible.

The convergence of the perturbation series in P for a crystal is illustrated by the data in Table 1.

Table 1. Zero-, first-, second- and third-order intramolecular density submatrices, $d_p(n)$, for a water molecule in the ice-lh lattice. The 2p orbitals point along the a, b' and c axes of the hexagonal lattice.

	1s	1s	2s	$2p_x$	$2p_y$	$2p_z$
$d_p(0)$	.867890 -.003916 .392738 -.059549 -.034385 .907502	-.003916 .867890 .392738 -.760821 -.439263 -.237672	.392738 .392738 1.728451 .283612 .163746 -.231569	-.059549 -.760821 .283612 1.485837 -.296854 -.114058	-.034385 -.439263 .163746 -.296854 1.828610 -.065848	.907502 -.237672 -.231569 -.114058 -.065848 1.221321
$d_p(1)$	-.014018 .001847 .002015 .002060 .001189 -.002725	.001847 -.027372 .004533 .005409 .003123 -.000412	.002015 .004533 -.000023 -.005576 -.003219 .001931	.002060 .005409 -.005576 .019742 .011398 .003406	.001189 .003123 -.003219 .011398 .006581 .001966	-.002725 -.000412 .001931 .003406 .001966 .015089
$d_p(2)$	-.018998 .000132 -.003008 .005638 .003255 -.018057	.000132 -.020816 -.002764 .017238 .009953 .000091	-.003008 -.002764 -.005907 -.022550 -.013019 .017919	.005638 .017238 -.022550 .020002 .027280 .004022	.003255 .009953 -.013019 .027280 -.011499 .002322	-.018057 .000091 .017919 .004022 .002322 .036996
$d_p(3)$	.000060 .000160 -.000093 .000023 .000013 -.000522	.000160 .000517 -.000383 .000882 .000509 .000304	-.000093 -.000882 -.000064 -.000257 -.000148 .000086	.000023 .000882 -.000257 -.000497 .000066 .000301	.000013 .000509 -.000148 .000066 -.000573 .000174	-.000522 .000304 .000086 .000301 .000174 .000160

Table 2. First-order intermolecular submatrices, $\rho_P^{(1)}$, for various intermolecular separations R for water molecules in ice-Ih. E_q gives the number of coupled equations that must be solved for each $|\rho|_{max}$ assuming all unit cell molecules are symmetry related and using the full unit cell symmetry to decouple the equation.

RÅ	E_q	1s	1s	2s	$2p_x$	$2p_y$	$2p_z$
1.70	3	.011164	.011165	.081649	-.003194	-.001844	-.149641
		-.000813	-.000813	.006399	.001080	.000624	-.008894
		.006620	.006620	-.037496	-.006607	-.003814	.048816
		-.002512	-.002000	.009976	.002040	.001546	-.012165
		-.000859	-.001746	.005760	.001546	.000254	-.007024
		.021882	.021882	-.085607	-.019520	-.011270	.099953
2.78	4	-.000217	.012547	-.000243	.001491	.000861	-.001696
		-.000129	.001873	.000188	.000292	.000168	-.000584
		.000548	-.000864	-.001554	-.003407	-.001967	.001321
		-.000510	-.000827	.001196	.002117	.001222	-.000724
		-.000294	-.000477	.000690	.001222	.000706	-.000418
		.001379	.001956	-.004210	-.009007	-.005200	.002720
3.67	7	.000479	-.000073	.000192	.000123	.000332	.000053
		.001321	.000067	.001400	.000848	.002548	-.000355
		-.000073	.000367	-.000866	-.000063	-.000838	-.000258
		-.000210	-.000954	.001682	-.000149	.001310	.000452
		-.000316	-.000607	.001058	-.000133	.000729	.000406
		.000035	-.000069	.000145	-.000006	.000225	-.000259

These data are for a water molecule in an ice crystal where the intermolecular interactions are relatively strong. They show that if taken to third order most elements of P would be accurate to five decimal places. The results from the polynomials studies suggest that they are probably accurate to six decimal places. This would be more than sufficient for most purposes and certainly beyond the accuracy of any foreseeable experimental data. In fact, for most purposes expansion up to $P^{(2)}$ should prove adequate.

The decrease in $^{\rho}P$ with increase in intermolecular separation is illustrated by the data in Table 2 for water molecules in the ice lattice. The situation is slightly complicated here by the fact that successive layers of water molecules are not all at the same orientation. The number tabulated under E_q gives the number of coupled equations to be solved, at each order, to calculate all density submatrices between molecules which make interatomic contacts of up to RÅ. These numbers show that $^{\rho}P$ decreases strongly with intermolecular separation. Therefore, relatively few coupled equations need be solved for the density submatrices in order to obtain reasonably accurate results.

APPENDIX

We briefly derive here the SCF perturbation theory necessary for the calculation of the crystal density matrix discussed in the foregoing sections. We start with Roothaans matrix equation

$$FC = SCE \tag{1A}$$

where F and S are, respectively, the Fock and Overlap matrices for the system referred to a basis set of atomic orbitals. C is the coefficient matrix for the expansion of the molecular or crystal orbitals in terms of the basis function. E is the orbital energy matrix.

Equation (1A) is expanded in a perturbation series to the second order :

$$(F^{(0)}+F^{(1)}+F^{(2)}+\dots)(C^{(0)}+C^{(1)}+C^{(2)}+\dots) = (S^{(0)}+S^{(1)}+S^{(2)}+\dots)(C^{(0)}+C^{(1)}+C^{(2)}+\dots)(E^{(0)}+E^{(1)}+E^{(2)}+\dots) \tag{2A}$$

Collecting terms by order we find, up to the second order

$$F^{(0)}C^{(0)} = S^{(0)}C^{(0)}E^{(0)} \tag{3A}$$

$$F^{(1)}C^{(0)} + F^{(0)}C^{(1)} = S^{(0)}C^{(0)}E^{(1)}+S^{(0)}C^{(1)}E^{(0)}+S^{(1)}C^{(0)}E^{(0)} \tag{4A}$$

$$F^{(2)}C^{(0)}+F^{(1)}C^{(1)}+F^{(0)}C^{(2)} = S^{(0)}C^{(0)}E^{(2)}+S^{(0)}C^{(2)}E^{(0)} +$$
$$S^{(2)}C^{(0)}E^{(0)}+S^{(0)}C^{(1)}E^{(1)}+S^{(1)}C^{(0)}E^{(1)}+S^{(1)}C^{(1)}E^{(0)} \quad (5A)$$

If these equations are solved subject to the constraint,

$$\tilde{C}SC = 1, \quad (6A)$$

E can be considered diagonal to all orders. Expanding (6A) in a perturbation series we find

$$\tilde{C}^{(0)}S^{(0)}C^{(0)} = 1 \quad (7A)$$

$$\tilde{C}^{(0)}S^{(0)}C^{(1)} + \tilde{C}^{(0)}S^{(1)}C^{(0)} + \tilde{C}^{(1)}S^{(0)}C^{(0)} = 0 \quad (8A)$$

$$\tilde{C}^{(0)}S^{(0)}C^{(2)}+\tilde{C}^{(0)}S^{(2)}C^{(0)}+\tilde{C}^{(2)}S^{(0)}C^{(0)}+\tilde{C}^{(0)}S^{(1)}C^{(1)} +$$
$$\tilde{C}^{(1)}S^{(1)}C^{(0)}+\tilde{C}^{(1)}S^{(0)}C^{(1)} = 0 \quad (9A)$$

It is assumed that equation (3A) has been solved, so that $F^{(0)}$, $S^{(0)}$, $C^{(0)}$ and $E^{(0)}$ are all known. Equation (4A) is solved by expanding $C^{(1)}$ in terms of complete set $C^{(0)}$:

$$C^{(1)} = C^{(0)}A \quad (10A)$$

where A is a matrix of mixing coefficients. Substitute for $C^{(1)}$ in (4A) and multiply to the left by $\tilde{C}^{(0)}$:

$$F^{(1)} + E^{(0)}A = E^{(1)} + AE^{(0)} + S^{(1)}E^{(0)} \quad (11A)$$

where $F^{(1)}$ and $S^{(1)}$ are <u>now</u> referred to a basis or zero order molecular orbitals,

$$S^{(1)}_{ij} = \sum_{\mu\nu}\sum C^{(0)}_{\mu i} S^{(1)}_{\mu\nu} C^{(0)}_{\nu j} \quad (12A)$$

$$F^{(1)}_{ij} = \sum_{\mu\nu}\sum C^{(0)}_{\mu i} F^{(1)}_{\mu\nu} C^{(0)}_{\nu j} \quad (13A)$$

Define a new F matrix, equation (14A)

$$\mathcal{F}^{(n)} = F^{(n)} - S^{(n)}E^{(0)} \quad (14A)$$

and substitute into (11A)

$$AE^{(0)} - E^{(0)}A = -E^{(1)} + \mathcal{F}^{(1)} \quad (15A)$$

Also, if $E^{(1)}$ is to be diagonal, from equation (8A),

$$A + \tilde{A} = -S^{(1)} \quad (16A)$$

From (15A), if $i \neq j$

$$A_{ij} = \mathcal{F}^{(1)}_{ij}/(\varepsilon^{0}_{j} - \varepsilon^{0}_{i}) \quad (17A)$$

and from (16A)

$$A_{ii} = -\frac{1}{2}S^{(1)}_{ii} \tag{18A}$$

Now that A has been derived, $P^{(1)}$ can be calculated :

$$\begin{aligned} P^{(1)}_{\mu\nu} &= 2\sum_i^{occ} (C^{(0)}_{\mu i}C^{(1)}_{\nu i} + C^{(1)}_{\mu i}C^{(0)}_{\nu i}) \\ &= 2\sum_i^{occ}\sum_k^{all} A_{ki}(C^{(0)}_{\mu i}C^{(0)}_{\nu k} + C^{(0)}_{\mu k}C^{(0)}_{\nu i}) \end{aligned} \tag{19A}$$

This equation is singular if the zero order solution contains degenerate orbitals. However, by making use of equation (16A), it can easily be reduced to a non-singular form :

$$P^{(1)}_{\mu\nu} = 2\sum_i^{occ}\sum_k^{vac} A_{ki}(C^{(0)}_{\mu i}C^{(0)}_{\nu k} + C^{(0)}_{\mu k}C^{(0)}_{\nu i}) - 2\sum_i^{occ}\sum_k^{occ} S^{(1)}_{ki}C^{(0)}_{\mu i}C^{(0)}_{\nu k} \tag{20A}$$

This can be conveniently expressed, for later usage, as

$$P^{(1)}_{\mu\nu} = 2\sum_i^{occ}\sum_k^{all} \mathcal{A}_{ki}(C^{(0)}_{\mu i}C^{(0)}_{\nu k} + C^{(0)}_{\mu k}C^{(0)}_{\nu i}) \tag{21A}$$

where, if both i and k occupied

$$\mathcal{A}_{ki} = -\frac{1}{2}S^{(1)}_{ki} \tag{22A}$$

and if i is occupied and k vacant

$$\mathcal{A}_{ki} = \mathcal{F}^{(1)}_{ki}/(\varepsilon^{(0)}_i - \varepsilon^{(0)}_k) \tag{23A}$$

Note, the $\mathcal{A}$ matrix is neither symmetric nor skew symmetric. Equation (21A) is for all practical purposes the solution to the first order equation. It has to be calculated iteratively since the evaluation of $P^{(1)}$ requires $\mathcal{F}^{(1)}$ which itself requires $P^{(1)}$ through $F^{(1)}$. The second order equations are solved in much the same way. $C^{(2)}$ is expanded in terms of $C^{(0)}$:

$$C^{(2)} = C^{(0)}B \tag{24A}$$

This is substituted into equation (5A) which, on multiplying to the left by $\tilde{C}$, yields

$$E^{(0)}B - BE^{(0)} = -\mathcal{F}^{(2)} - \mathcal{F}^{(1)}A + E^{(2)} + AE^{(1)} + S^{(1)}\mathcal{F}^{(1)} \tag{25A}$$

The following equation, derived from (15A), has been used in the derivation of (25A).

$$S^{(1)}E^{(1)} + S^{(1)}AE^{(0)} = S^{(1)}F^{(1)} + S^{(1)}E^{(0)}A \qquad (26A)$$

The matrices in equation (25A) are, as for the first order equation, referred to a basis of zero order molecular orbitals. The second order ortho-normality equation, (9A), yields directly on substitution for $C^{(1)}$ and $C^{(2)}$:

$$B + \tilde{B} = -S^{(2)} - S^{(1)}A - \tilde{A}S^{(1)} - \tilde{A}A \qquad (27A)$$

Noting, from equation (16A), that

$$A\tilde{A} = -AA - AS^{(1)} \qquad (28A)$$

equation (27A) can be written in the more useful form

$$B + \tilde{B} = -S^{(2)} + S^{(1)}S^{(1)} - A\tilde{A} \qquad (29A)$$

If B is constrained to satisfy this orthogonality condition $E^{(2)}$ can be assumed to be diagonal. Thus, for $k \neq i$, from (25A)

$$B_{ki} = F^{(2)}_{ki}/(\varepsilon_i^{(0)} - \varepsilon_k^{(0)}) + [\sum_{\ell}^{all} (F^{(1)}_{k\ell}A_{\ell i} - S^{(1)}_{k\ell}F^{(1)}_{\ell i}) - A_{ki}E^{(1)}_{ii}]/(\varepsilon_i^{(0)} - \varepsilon_k^{(0)}) \qquad (30A)$$

and

$$B_{ii} = -\frac{1}{2}\{S^{(2)}_{ii} - \sum_{\ell}^{all} (S^{(1)}_{i\ell}S^{(1)}_{\ell i} - A_{i\ell}A_{i\ell})\} \qquad (31A)$$

Now that B has been derived, the next step is to calculate the second order change in the density matrix, $P^{(2)}$

$$P^{(2)}_{\mu\nu} = 2\sum_{i}^{occ}\{C^{(1)}_{\mu i}C^{(1)}_{\nu i} + C^{(0)}_{\mu i}C^{(2)}_{\nu i} + C^{(2)}_{\mu i}C^{(0)}_{\nu i}\} \qquad (32A)$$

which, on substituting for $C^{(1)}$ and $C^{(2)}$ becomes

$$P^{(2)}_{\mu\nu} = 2\sum_{i}^{occ}\sum_{k}^{all}\sum_{\ell}^{all} A_{ki}A_{\ell i}C^{(0)}_{\mu k}C^{(0)}_{\nu\ell} + 2\sum_{i}^{occ}\sum_{k}^{all} B_{ki}(C^{(0)}_{\mu i}C^{(0)}_{\nu k} + C^{(0)}_{\mu k}C^{(0)}_{\nu i}) \qquad (33A)$$

This expression is singular and thus cannot be used for calculations where the zero order orbitals are degenerate. These singularities, however, can be easily combed out by the following manipulations. Consider the two contributions to $P^{(2)}$ in equation (33A) separately, taking the first term first :

First term =

$$2[\sum_{i}^{occ}\sum_{k}^{occ}\sum_{\ell}^{occ} + \sum_{i}^{occ}\sum_{k}^{vac}\sum_{\ell}^{occ} + \sum_{i}^{occ}\sum_{k}^{occ}\sum_{\ell}^{vac} + \sum_{i}^{occ}\sum_{k}^{vac}\sum_{\ell}^{vac}]$$

$$A_{ki}A_{\ell i}C_{\mu k}C_{\nu\ell} \tag{34A}$$

Of all these terms only the fourth is non-singular. The objective is to find contributions from the second term of (33A) which will either cancel or combine with terms in (34A) to yield a final overall non-singular expression.
Second term =

$$2\sum_i^{occ}\sum_k^{vac} B_{ki}g_{ik} + 2\sum_i^{occ}\sum_k^{occ}(B_{ik}+B_{ki})C_{\mu i}C_{\nu k} \tag{35A}$$

where

$$g_{ik} = C_{\mu i}C_{\nu k} + C_{\mu k}C_{\nu i} \tag{36A}$$

Using equation (29A), equation (35A) may be written as

$$2\sum_i^{occ}\sum_k^{vac} B_{ki}g_{ik} + 2\sum_i^{occ}\sum_k^{occ} C_{\mu k}^{(0)}C_{\nu i}^{(0)}\left(\sum_\ell^{all} S_{i\ell}^{(1)}S_{\ell k}^{(1)} - S_{ki}^{(2)}\right) - 2\sum_i^{occ}\sum_k^{occ}\sum_\ell^{vac} A_{i\ell}A_{k\ell}C_{\mu k}^{(0)}C_{\nu i}^{(0)} - 2\sum_i^{occ}\sum_k^{occ}\sum_\ell^{occ} A_{i\ell}A_{k\ell}C_{\mu k}C_{\nu i} \tag{37A}$$

The last term of (37A), which is singular, cancels the first term of (34A). All the remaining terms of (37A), save the first, are non-singular. The second and third terms of equation (34A) can be combined and rearranged to give

$$2\sum_i^{occ}\sum_k^{vac}\sum_\ell^{occ} A_{k\ell}A_{i\ell}g_{ik} \tag{38A}$$

which is combined with the remaining singular term, the first, of equation (37A) :

$$2\sum_i^{occ}\sum_k^{vac} g_{ik}\left(B_{ki} + \sum_\ell^{occ} A_{\ell k}A_{i\ell}\right) \tag{39A}$$

Expand B_{ki} using equation (30A), equation (40A) :

$$2\sum_i^{occ}\sum_k^{vac} g_{ik}\left[\frac{F_{ki}^{(2)}}{\varepsilon_i^{(0)}-\varepsilon_k^{(0)}} + \left(\sum_\ell^{all} F_{k\ell}^{(1)}A_{\ell i} - A_{ki}E_i^{(1)} - \sum_\ell^{all} S_{k\ell}F_{\ell i}^{(1)}\right)\Big/(\varepsilon_i^{(0)}-\varepsilon_k^{(0)}) + \sum_\ell^{occ} A_{k\ell}A_{i\ell}\right] \tag{40A}$$

Combine the occupied component of the second term with the fifth term.

Expand all the A's using equation (17A).

$$\sum_{\ell}^{occ}\left\{\frac{F^{(1)}_{k\ell}A_{\ell i}}{\varepsilon_i-\varepsilon_k}+A_{k\ell}A_{i\ell}\right\}=\sum_{\ell}^{occ}{}'\left\{\frac{F^{(1)}_{k\ell}}{(\varepsilon_i-\varepsilon_\ell)}\left[\frac{F^{(1)}_{\ell i}-S_{\ell i}\varepsilon_i^{(0)}}{(\varepsilon_i-\varepsilon_k)}-\frac{F^{(1)}_{i\ell}-S_{i\ell}\varepsilon_\ell^{(0)}}{(\varepsilon_\ell-\varepsilon_k)}\right]\right\}$$

$$+\ (\ell = i)\ \text{terms} \qquad (41A)$$

The $\ell = i$ terms has to be treated separately, as may be seen from equation (18A). Collecting terms in (41A)

$$= -\sum_{\ell}^{occ}{}'\frac{A_{k\ell}F^{(1)}_{\ell i}}{(\varepsilon_i-\varepsilon_k)}-\sum_{\ell}^{occ}{}'\frac{F^{(1)}_{k\ell}S^{(1)}_{i\ell}}{(\varepsilon_\ell-\varepsilon_k)}+(\ell = i)\ \text{terms} \qquad (42A)$$

Explicitly including the $\ell = i$ term, and completing the summation in equation (42A) yield,

$$= -\sum_{\ell}^{occ}\frac{A_{k\ell}F^{(1)}_{\ell i}}{(\varepsilon_i-\varepsilon_k)}-\sum_{\ell}^{occ}\frac{F^{(1)}_{k\ell}S^{(1)}_{i\ell}}{(\varepsilon_\ell-\varepsilon_k)}+\frac{A_{ki}F^{(1)}_{ii}}{(\varepsilon_i-\varepsilon_k)}+\frac{F^{(1)}_{ik}S^{(1)}_{ii}}{(\varepsilon_i-\varepsilon_k)}-\frac{1}{2}\frac{F^{(1)}_{ki}S^{(1)}_{ii}}{(\varepsilon_i-\varepsilon_k)}-\frac{1}{2}\frac{F^{(1)}_{ki}S^{(1)}_{ii}}{(\varepsilon_i-\varepsilon_k)} \qquad (43A)$$

Noting that

$$E^{(1)}_{ii} = F^{(1)}_{ii} \qquad (44A)$$

the third term in (43A) cancels the third term in (40A). Finally, the fourth term is cancelled by the fifth and sixth terms. Thus, collecting terms, equation (40A) equals

$$2\sum_{i}^{occ}\sum_{k}^{vac}g_{ik}\left[\frac{F^{(2)}_{ki}}{\varepsilon_i^{(0)}-\varepsilon_k^{(0)}}-\left(\sum_{\ell}^{occ}A_{k\ell}F^{(1)}_{\ell i}+\sum_{\ell}^{all}S_{k\ell}F^{(1)}_{\ell i}-\sum_{\ell}^{vac}F^{(1)}_{k\ell}A_{\ell i}\right)/(\varepsilon_i^{(0)}-\varepsilon_k^{(0)})-\sum_{\ell}^{occ}A_{k\ell}S^{(1)}_{i\ell}\right] \qquad (45A)$$

All of the terms in this equation are non-singular. Thus, the second order change in the density matrix can be written in the following non-singular form which can be applied to problems where the zero order basis contains degenerate orbitals

$$P^{(2)}_{\mu\nu} = 2\sum_{i}^{occ}\sum_{k}^{vac}\sum_{\ell}^{vac}A_{ki}A_{\ell i}C^{(0)}_{\mu k}C^{(0)}_{\nu\ell}+2\sum_{i}^{occ}\sum_{k}^{all}g_{ik}B_{ki} \qquad (46A)$$

where, if both i and k occupied

$$B_{ki} = -\frac{1}{2}\left[\sum_{\ell}^{vac} A_{i\ell}A_{k\ell} - \sum_{\ell}^{all} S_{i\ell}^{(1)}S_{\ell k}^{(1)} + S_{ki}^{(2)}\right] \tag{47A}$$

and if i occupied and k vacant

$$B_{ki} = F_{ki}^{(2)}/(\varepsilon_i^{(0)} - \varepsilon_k^{(0)}) - \left(\sum_{\ell}^{occ} A_{k\ell}F_{\ell i}^{(1)} + \sum_{\ell}^{all} S_{k\ell}^{(1)}F_{\ell i}^{(1)} - \sum_{\ell}^{vac} F_{k\ell}^{(1)}A_{\ell i}\right)/(\varepsilon_i^{(0)} - \varepsilon_k^{(0)}) - \sum_{\ell}^{occ} A_{k\ell}S_{i\ell}^{(1)} \tag{48A}$$

The calculation of $P^{(2)}$ must also be iterative because of the presence of $F^{(2)}$ in equation (48A). However, it should be noted that the most complicated terms required for the evaluation of $P^{(2)}$ need be calculated only once, prior to the start of the iterative calculation. In fact, the non-iterative contribution to $P^{(2)}$ may be used as a first guess to initiate the second order calculation.

REFERENCES

1. J. Ladik and G. Biczo, J. Chem. Phys. 42, 1658 (1965).
2. JM.André, J. Chem. Phys. 50, 1536 (1969).
3. J. Bacon and D. P. Santry, J. Chem. Phys. 56, 2011 (1772).
4. R. W. Crowe and D. P. Santry, Chem. Phys. 2, 304 (1973).
5. K. M. Middlemiss and D. P. Santry, Chem. Phys. 1, 128 (1973).
6. K. M. Middlemiss and D. P. Santry, Chem. Phys. Lett., in press.
7. K. M. Middlemiss and D. P. Santry, Theor. Chim. Acta., in press.
8. S. F. O'Shea and D. P. Santry, submitted for publication.
9. R. McWeeny and G. Diercksen, J. Chem. Phys. 44, 3554 (1966).
10. C. C. J. Roothaan, Rev. Mod. Phys. 23, 69 (1951).
11. J. Bacon and D. P. Santry, J. Chem. Phys. 55, 3743 (1971).
12. K. M. Middlemiss and D. P. Santry, submitted for publication.

INTERACTION BETWEEN POLYMER CONSTITUENTS AND THE STRUCTURE OF BIOPOLYMERS

R. Rein

Roswell Park Memorial Institute
666 Elm St. Buffalo, N.Y. 14203

and

State University of New York
at Buffalo
4248 Ridge Lea, Amherst, N.Y. 14226

I. INTRODUCTION

Generally macromolecules may appear in a variety of three dimensional structures. A macromolecular chain, for example, may be linear or helical or in the form of a random coil. A few cross-links may further introduce a complicated folding pattern into the chain. This structural flexibility is the consequence of hindered rotations around the acyclic bonds comprising the polymer chain and the availability of several rotational isomeric states for each one of the acyclic bonds. Cross-links between the constituents can be formed by covalent bridges or a variety of weaker interactions such as hydrogen bonds, ion bridges and London Van der Waals forces.

The transition of a macromolecule from one of its stable structural states (configuration) to another may often proceed with relative ease due to the fact that the free energies of the respective configurations may not differ greatly. This is in particular, the case in biopolymers, for which not only are the rotational isomeric state energies spaced closely, but the cross-links are due to weak bonds and disrupted with ease. The physical properties of these macromolecules as well as the chemical behavior depends on their three dimensional structure ; more importantly, the biological activity is a function of these structures. Macrobiomolecules perform their functions generally only in the native ordered states. The DNA double helix, and sterically precisely organized enzyme molecules, are prominent examples of the important role the structural organization plays in the activity of these molecules.

A fundamental objective of macromolecular theory is then an analysis and quantitative description of the force fields governing macromolecular structure. Once these force fields are characterized to a sufficient degree of accuracy, theories of statistical physics can be implemented to predict stable structure and transitions for the more important biomacromolecules such as nucleic acids and proteins. With this accomplished, the more ambitious program of relating biological activity, to structural features and intermolecular forces, in a quantitative fashion, can be endeavored.

From what has been said up to this point, it is clear that of special interest in the context of macromolecular physics is
1. the study of molecular interactions between polymer constituents
2. the study of conformational energies of the polymer chains.
It has to be realized that for predicting macromolecular structure, it is not enough to characterize these interactions for the final configuration, as for example, to calculate the hydrogen bond energies and stacking interactions in the case of DNA for the Watson-Crick structure only. The stable structure of a macromolecule corresponds to a global minimum in a multidimensional configurational space. In this configurational space, forces arising from one type of molecular interaction may be opposed or reinforced by forces arising from other types of interactions. The balance of all the operative forces will determine the position of the minimum and the corresponding stable structure. This point is illustrated in the case of DNA for example, by the fact that the position of a base in the double helical stack will depend not only on hydrogen bonding energies, but also on the stacking energies and the conformational energy of the sugar phosphate strand joining to successive bases with the energy of each one of their components depending on a set of connected configurational variables. Putting this into the language of statistical mechanics, the energies of all the permissible configurations of the macromolecule should be known in order to form the partition function from which, in principle, all the desired structural and thermodynamic information should be obtainable. For an N-atomic system having 3N-6 variables with N in the range of hundreds, a rigorous evaluation of the energy of this hyper space is, of course, an impossible task. The approach taken in such cases is the consideration of a model in a drastically reduced space of configurational variables. The choice of the model is guided by recognizing the existing geometrical constraints and connections of variables as well as by distinguishing between so-called hard and soft variables. Where the former effect strong valence forces, i.e. bond lengths and valence angles, the latter effects torsional angles H-bond length or angles, stacking distances and orientations and other weaker forces of non-bonded type.

It follows that subject to the above simplifications the theoretical task is the characterization of the energy hyper space of the soft variables.

The present paper is designed to present a review and comprehensive discussion of methods for calculation of interactions and intermolecular potentials which are required for the above task. We will be concerned with the questions on the nature of the leading interaction in macromolecules for various ranges of molecular separation and the best approximations to calculate these interactions at the present state of application of quantum theory. To illustrate the application of the interaction results in elucidation of macromolecular structure, a statistical treatment of a simplified DNA model will be reviewed also.

Experimental results on DNA and DNA-like synthetic polymer melting are abundant. Melting curves due to disordering of the double helix at elevated temperatures can be interpreted in terms of the statistical model for the relevant intermolecular interactions. In this way one can deduce interaction energies from the melting experiments and compare it with values computed from quantum theory. Alternatively, theoretical melting curves can be computed and compared with the experimental one. In either case as it will be shown, the consistency of the statistical model and the interaction energies with experimental properties can be established.

II. INTERMOLECULAR INTERACTION THEORY

It would appear that the calculation of intermolecular interactions could be best accomplished in a super molecule approach, i.e. by treating the interacting molecules as a single system followed by a calculation of the energies of the separated constituents and taking the difference of these quantities as the interaction energy. A closer examination indicates that this approach is practical only in a more restricted domain of applications. These studies require accurate ab-initio procedures and are best applied in the short range domain of molecular interactions.

In the general context of calculating intermolecular interaction potentials, the super molecule approach is not very practical. First, because the interaction energy is a small quantity appearing as a difference of two large quantities, thus a meaningful calculation requires a high degree of accuracy in the calculation of the combined and separate molecule energies which exceeds the practical limits except for the most sophisticated ab-initio calculations. But an expensive molecular orbital calculation for every single geometrical configuration of the biomolecular complex would drastically reduce the number of configurations which can be considered, thus making it effectively very difficult to gather a sufficient number of points to construct a potential function of interest. However, there is also a more fundamental obstacle imposed by the nature of the wave function. A single determinatal SCF wave function may be a good description of a molecule at or near its equilibrium configuration. Whether such a description remains optional for a biomolecular complex particularly at longer separations, is not yet satisfactorily answered. Experience indicates that such situations could be handled only by wave functions constructed from a large number of electronic configurations. With present day computational thechnology such calculations are practical, only for the smallest systems such as two hydrogen atoms, and are thus not applicable in the biomolecular context.

A major alternative to the super molecular approach for treatment of intermolecular interactions is based on the assumption that the individual interacting molecules preserves their individuality. That is, the molecular orbitals of the individual molecules can be utilized for construction of good zero order basis for a description of the perturbation of the interacting system. When the interacting molecules are sufficiently apart their electronic systems are not significantly overlapping. The above assumption is indeed a very good approximation for the interacting system.

At close approach such as Van der Waals contacts, the assumption of the overlap between the constituents is not valid any more and the additional effect of exchange, non orthogonality[1] and charge transfer come into play.

Recently, studies concerned with small Hydrogen Bonded systems, notably the studies of Dreyfus and Pullman[2], of Morokuma[3] and of Kollman and Allen[4], have provided an insight into the nature of these close range interaction components, as well as how these components change with increasing separation distance. This work is of interest here partly because of the conceptual interest of these studies and also because of their practical value as guide lines to what components have to be included in an interaction scheme at a given separation distance ; and what empirical corrections can be made for the more difficult to calculate short range terms. Therefore we shall return in the next section to a more detailed discussion of the method and results of component analysis.

In the thirties, when London[5] developed his theory of intermolecular forces, wave functions for larger molecules were not yet available. To overcome this difficulty, the perturbation operation to the Hamiltonian ($1/r_{12}$) has been expanded in a Taylor series. Depending on the center of expansion two schemes have been derived. In the first one the expansion is over the entire molecule, and in the second the molecule is first divided into segments and then the expansion is made for each segment. For both cases the interaction energies can be expressed in terms of interactions between multipoles of the two molecules and transition multipoles of the two molecules for first and second order interactions. The expansions are terminated after the first non-zero term. That means a dipole-dipole interaction is calculated for the first method, and an atomic point charge-atomic point charge interaction for the second. The experimental knowledge of dipole moments and polarisabilities for the first and second order, respectively thus circumvents the need for explicit knowledge of wave function. However, it has been shown recently[(6-10)] that truncation of the interaction series to one term is generally not valid below separation distances of the order of tens of Å angstroms, thus excluding most biological problems of interest.

The approach taken in the present study is to carry the expansion to higher terms until convergence is assured.

While dipole and quadrupole moments for a collection of polyatomic molecules is now available[(11)], the segmental moments as well as the higher moments, are derivable only from wave functions. This introduces the requirement for reliable wave functions. The reliability of the wave function in the particular context can be best established by calculation of moments which have been experimentally observed.

The question of dispersion interactions is of interest, not only from the practical point of view, but also conceptually. There are at least three different ways to look on dispersion forces. In a static view they appear as the consequence of interaction of induced changes of one molecule with those of another molecule, where the magnitude of the changes is calculable from the first order perturbation correction to the respective wave function. Another way to look on dispersion forces is to consider them as arising from electron correlations[(12)]. Yet another way is to obtain the dispersion forces for an electrodynamic point of view (13-14-22). The interesting point is that whichever of these descriptions is adopted by a suitable set of approximations, it can be brought into an equal form in which the dispersion interactions will appear in terms of polarizabilities.

Thus, in summary, it appears that a comprehensive consideration of intermolecular interaction has to deal with the above outlined three major aspects of the theories :

(1) the short range effects
(2) the convergent representation of the electrostatic interactions
(3) the nature of dispersion forces.

In following, we consider each of these aspects in further detail.

III. COMPONENT ANALYSES

Among the SCF supermolecule energy decomposition schemes which have been proposed, the one due to Morokuma[3] seems to be the most comprehensive. Thus, it appears convenient to review first the basic concepts underlying the decomposition method.

For a biomolecular complex, one can consider the following wave functions of interest :

i. ψ_{AO} and ψ_{BO} are the ground state wave functions of molecules A and B respectively. These wave functions are generally a single configuration LCAO-MO-SCF type.

ii. $\psi^{o}_{AB} = \psi_{AO}\,\psi_{BO}$ is the zeroth order wave function for the complex taken as the Hartree product of the wave functions of the separated molecules.

iii. $\psi^{p}_{AB} = \psi^{p}_{A}\,\psi^{p}_{B}$, the polarized wave function of the complex considered as a Hartree product of the polarized wave functions of the separated molecules, ψ^{p}_{A} and ψ^{p}_{B}, respectively. These polarized wave functions are obtained by solving the SCF equations of the respective molecules in the Coulomb field of their partners. This achieved conveniently by adding to the Hamiltonian of, for example, molecule A the terms

$$V^{o}_{B}(i) = \sum_{B} \frac{Z_B}{r_{B_i}} - \sum_{K} J_K(i) \text{ where } V^{o}_{B} \text{ is the rigid Coulomb}$$

field of B, the rest of the symbols have their conventional meaning and the indices (i) are for electrons of molecule A, B, and J and K stand for the nuclei and electron operators of molecule B, respectively. The coulomb operator $J_K(i)$ is formed from the orbitals of ψ_{BO} . V^{o}_{A} is obtained by the same procedure by interchanging the rôle of the molecules.

iv. $\psi_{AB} = a\,\psi_{AO}\,\psi_{BO}$ is the antisymetrized wave function of the product found from the zeroth order wave functions of the separate molecules.

v. $\psi^{SCF}_{AB} = (a\,\psi_{A}\,\psi_{B})_{SCF}$ is the SCF wave function of the super molecule, that is to say, the SCF description of the complex as a single molecule.

With the wave functions i.-v, one can associate the following Hamiltonians :

i. F_A and F_B the Hamiltonians of the separate molecules

ii. $F^0_{AB} = F_A + F_B$ the Hamiltonian of the non-interacting complex

iii. $F^C_{AB} = F_A + F_B + V^0_A + V^0_B$ the Hamiltonian of the electrostatically interacting complex

iv. $F^e_{AB} = F'_A + F'_B + V'_A + V'_B$ The Hamiltonian of the electrostatically, and by exchange, interacting complex. The prime on the operators denotes the terms due to anti-symmetrization which are included.

v. F_{AB} the SCF Hamiltonian of the complexes.

It should be recognized that the difference between iv and v is that iv contains Coulomb fields V'_A and V'_B which are obtained from unoptimised orbitals of the partner molecules, while the Hamiltonian in v takes into consideration the presence of the partner molecules and all orbitals included in the Hamiltonian are fully optimised.

From the described wave functions and Hamiltonians, one can form the following eigenvalue equation for the complex :

I. $F_A \psi_{AO} = E_A \psi_{AO}$; $F_B \psi_{BO} = E_B \psi_{BO}$

II. $F^o_{AB} \psi^o_{AB} = E_0 \psi^o_{AB}$

III. $F^c_{AB} \psi^o_{AB} = E_1 \psi^o_{AB}$

IV. $F^c_{AB} \psi^p_{AB} = E_2 \psi^o_{AB}$

V. $F^e_{AB} \psi^a_{AB} = E_3 \psi^a_{AB}$

VI. $F^{SCF}_{AB} \psi^{SCF}_{AB} = E_4 \psi^{SCF}_{AB}$

A convenient way for solving these sets of equations is as follows :

One first solves equation I, which gives the separate molecule energies and wave functions. The sum of these energies gives also, E_0 the energy of the non-interacting complex. Using the wave functions from the first step and suppressing intermolecular exchange integrals, one performs the first iteration of eq.III which yields E_1. The first iteration of equation V (not suppressing exchange integrals), gives E_3, the SCF solution of eq.VI gives E_4. In obtaining E_2 from eq.IV, first the exchange integrals are suppressed and the equation is solved iteratively from the set of orbitals comprising A, while the orbitals of B are kept fixed to their first iterative value. To complete the procedure, the roles of A and B are reversed.

It is easy to recognize that the energies obtained in this way are comprised from the following terms :

$$E_0 = E_A^0 + E_B^0$$

$$E_1 = E_0 + E_{electrostatic}$$

$$E_2 = E_0 + E_{electrostatic} + E_{polarization}$$

$$E_3 = E_0 + E_{electrostatic} + E_{exchange\ repulsion}$$

$$E_4 = E_0 + E_{exchange\ repulsion} + E_{polarization} + E_{charge\ transfer} + E_{electrostatic}$$

Rearranging these expressions gives the energy components in an explicit form :

$$E_{electrostatic} = E_1 - E_0$$

$$E_{polarization} = E_2 - E_1$$

$$E_{exchange\ repulsion} = E_3 - E_1$$

$$E_{charge\ transfer} = E_4 + E_1 - E_3 - E_2$$

Morokuma has applied his decomposition scheme to study the components contributing to hydrogen bonding in the case of water dimers and in the case of formamide water dimers. The contribution to hydrogen bonding in the case of water dimers has been studied also by Coulson and Danielson[15] Further results are available on the contributing components to hydrogen bonding in the case of the linear dimer of formamide from the work of Dreyfus and Pullman[2]. These results are summarized in table I.

TABLE I

DECOMPOSITION OF HYDROGEN BOND ENERGIES

-kcal/mole dimer

System	E_{total}	$E_{el.stat.}$	$E_{exh.rep.}$	$E_{char.trans.}$	E_{pol}	$E_{disp.}$	References
$(H_2O)_2$	8.09	8	-9.86	8.16	0.2	1.54	(4)
$(H_2O)_2$	8.6	6	-8.4	8.		3.	(3)
$H_2CO..H_2O$	3.45	4.64	-6.71	5.34	0.18		(2)
$(H_2CO)_2$	8.	10.	-7.	5.			(15)

The most interesting feature which emerges from these results is that the electrostatic contribution to hydrogen bonding near the equilibrium geometry seems to be approximating fairly well, the total interaction energy. This appears to be a consequence of the fact that exchange repulsion in this domain is nearly compensated by the attractive terms contributing in this range, i.e. the charge transfer and polarization term.

Another point of great interest is to observe how the various components change with increasing separation between the interacting systems. Both Morokuma, and Dreyfus and Pullman have published curves representing the magnitude of the various components for a range of distances from the equilibrium configuration to about 10 Å. From these results, it is apparent that all but the electrostatic terms are negligible for distances larger than 5 Å.

IV. MULTIPOLE REPRESENTATIONS OF MOLECULAR CHARGE DISTRIBUTION

Before proceeding with the analysis of the interaction schemes suitable for calculation of intermolecular interaction energies, we are going to describe the alternative multipole representations of charge distribution on which these schemes are based.

1. Dipole Moment

We consider first, the molecular dipole moment which is defined as :

$$\mu = \sum_{\substack{A \\ \text{atoms}}} Z_A R_A - \sum_{\lambda\sigma} P_{\lambda\sigma}\mu_{\lambda\sigma} \qquad (1)$$

where Z_A and R_A are the charge and position vectors, respectively, of the Ath nucleus, in a space-fixed coordinate system, $P_{\lambda\sigma}$ is the element of the charge bond order matrix, and $\mu_{\lambda\sigma}$ is the moment arising from the electron charge density in the same coordinate system as R_A.

A displacement equation defines the moment component with respect to a new origin

$$\mu_{\lambda\sigma} = \langle\lambda|r|\sigma\rangle = \langle\lambda|r'|\sigma\rangle + R\langle\lambda|\sigma\rangle \qquad (2)$$

where R is the displacement to the new origin (R connects the space-fixed axial system with the new origin), and r is the position vector defined with respect to the original reference point, and r' is the position vector defined with respect to the new origin.

After partitioning the second term of Eq.(1) into integrals arising from one- and two-center densities, respectively, and introducing Eq.(2) to simplify the integration, the expression for the dipole moment becomes

$$\mu = \sum_A [R_A (Z_A - \sum_\lambda P_{\lambda\lambda}) - \sum_\lambda \sum_{\sigma\neq\lambda} P_{\lambda\sigma} \langle\lambda|r'|\sigma\rangle - \frac{1}{2} \sum_{B\neq A} \sum_\lambda \sum_\nu P_{\lambda\nu} (\langle\lambda|r'|\nu\rangle - R_{AB} \langle\lambda|\nu\rangle)] \quad (3)$$

where the sums A and B are over atoms, the sums λ and σ are over orbitals on center A, the sum ν is over orbitals on center B, and R_A and R_{AB} are the displacement vectors from the molecular origin to a local origin, the origin of r'.

The first three terms in Eq.(3) correspond to the one-center density contributions and the last two terms to the two-center density contributions to the dipole moment. Various semiempirical methods lead to somewhat different dipole moments. This is partly due to different assumptions involved but also due partly to different properties of such semiemperical orbitals. Thus dipole moments calculated from IEHT wave functions contain both types of contributions, while for CNDO wave functions the second contribution due to different centers vanishes.

There are several alternative ways to perform the calculation of the two-center density contributions. These alternatives depend on the choice of the local coordinate systems in which the integrations implied in Eq.(3) are performed. More specifically, there are alternative ways of defining the origin of the integration variable r' in the respective local coordinate systems in which the densities are specified. This amounts to choosing the points in the molecules, i.e., atomic or interatomic centers (R_{AB} specifies such a point), that the bicentric density moments are referred to in the intermediate stages of calculation. While the final result for the molecular dipole moment is independent of the choice, its selection is of substantial interest for atomic segmental multipole representation of molecular charge distributions. Hence we disgress from the topic of molecular dipole moments to introduce this alternative representation.

2. Atomic Multipole Representation

The molecular density matrix, ρ , can be partitioned

$$\rho(1) = \sum_i \rho_i(1), \tag{4}$$

where $\rho_i(1)$ are the densities assigned to the ith atom.

$$\rho_i(1) = \sum_\lambda [P_{\lambda\lambda}\chi_\lambda(1)\chi_\lambda(1) + \sum_{\sigma\neq\lambda} P_{\lambda\sigma}\chi_\lambda(1)\chi_\sigma(1) + \sum_\nu P_{\lambda\nu}f_{\lambda\nu} \chi_\lambda(1)\chi_\lambda(1)] + Z_i\delta(R_i - r_1) \tag{5}$$

where the sums λ and σ are over all orbitals on atom i and the sum ν is over all orbitals on all centers except atom i. The subscripts, λ , σ, and ν are indications of all the quantum numbers and parameters in the Slater orbitals (χ_λ , χ_σ , and χ_ν) and the center from which the orbital is defined ; $f_{\lambda\nu}$ is the fraction of the density in the case where the Slater orbitals are referred to two different centers and this is the part assigned to center χ_λ. Obviously

$$f_{\lambda\nu} + f_{\nu\lambda} = 1. \tag{6}$$

It is easy to recognize that the integral of ρ_i corresponds to formal atomic charge on atom i. Various definitions of formal atomic charges then correspond to various definitions of ρ_i. A choice of ($f_{\lambda\nu} = \frac{1}{2}$) corresponds, for example, to formal charge defined according to the Mulliken division. Alternatively, one could choose the f's to satisfy Lowdin's(1) definition of formal charge.

By a straightforward generalization, formal atomic multipoles can be defined as

$$<{}_m\chi^k> = \int {}_m\chi^k \rho_m \, d\tau \tag{7}$$

where the subscript on the left indicates that the operator component χ^k is defined relative to the atomic origin m.

It is important to recognize that any molecular multipole moment referred to an arbitrary single origin is defined uniquely in terms of the corresponding set of atomic multipole components of the type defined in Eq.(7) and the corresponding transformation equations that are generalizations of Eq.(2).

For quadrupole moment this transformation is

$$M_{ij} = R_i R_j q' + R_i \mu_j' + R_j \mu_i' + M_{ij}' \tag{8}$$

where M_{ij} is a quadrupole-like operator (an operator from which all the components of the quadrupole moment can be obtained), R_i and R_j are vector components from the unprimed origin to the primed one, and

$$M_{ij} = \langle 0|X_i X_j|0\rangle \tag{9}$$

It is of further importance to note that the electrostatic potential of the molecular charge distribution can be expressed as the superposition of potentials of an atomic multipole series. Such a representation in certain instances may be of substantial practical advantage relative to a single-center multipole representation. The most important advantage is the applicability in certain interaction situations which are not tractable in the single center representation because the geometrical conditions do not justify the underlying expansion. We will return to discuss this aspect in greater length in a later section.

3. Calculation of moments

The method of calculation of moments described in this section follows closely the description presented in our earlier papers[6-7]

The following are the main steps involved in calculation of the moments :

i. Partitioning of the density matrix according to Eq.(5) leads to a set of atomic and diatomic densities with orbital components defined relative to atomic centers A and A,B, respectively. These centers are designated by Cartesian coordinates X_A, Y_A, and Z_A relative to a fixed coordinate system which is the same for all orbitals.

   Slater orbitals adopted have the form

$$\chi_\mu^A = B_n r^{n-1} e^{-\delta r} Y_{\ell m}(\theta,\phi) \tag{10}$$

and are specified in spherical coordinate systems centered at atom A with the polar axis in the +z direction and +x direction defining the reference meridian. The subscript μ stands for (n,ℓ, m,δ), and B_n is the normalization factor

$$B_n = \left[\frac{(2\delta)^{2n+1}}{(2n)\,!} \right]^{1/2} \tag{11}$$

It is to be noted that the orbitals in a bicentric density are not aligned properly at this stage.

ii. Assignment of local operator components and specification of coordinate systems in which they are defined. These are

X_{iA} , Cartesian component of the dipole operator defined relative to origin on atomic center A ;

X_{iAB}, Cartesian component of the dipole operator defined in an axis system whose polar direction is along the directed line from center A toward center B and whose origin is at the midpoint between A and B,

$X_{iA}X_{jA}$, Cartesian component of a type of quadrupole operator relative to origin A.

$X_{iAB}X_{jAB}$, Cartesian component of the quadrupole operator in the A, B midpoint coordinate system.

iii. Application of the operators defined in (ii) to their corresponding densities defined in (i) yields the following integrals :

$$\mu_{iA} = \int X_{iA}\rho_A(r)\, dr \tag{12}$$

$$\mu_{iAB} = \int X_{iAB}\rho_{AB}(r)\, dr \tag{13}$$

$$m_{iAjA} = \int X_{iA}X_{jA}\rho_A(r)\, dr \tag{14}$$

$$m_{iAjB} = \int X_{iAB}X_{jAB}\rho_{AB}(r)\, dr \tag{15}$$

These integrals are the components of the local one-center and two-center electronic moments and their set, i.e., all A and all A,B, in fact corresponds to a unique moment representation of the molecular electron distribution (see below), where $\rho_A(r)$ is a local one-center charge density and $\rho_{AB}(r)$ is a local two-center charge density.

iv. The one-center dipole moment integral (Eq.(12)) after expansion of ρ_A is

$$\mu_{iA} = \sum_{\mu\nu}^{\text{on atom } A} P_{\mu\nu} \int \chi_\mu^A X_{iA} \chi_\nu^A \, dr. \quad (16)$$

The only nonzero contributions to μ_{iA} are from Slater orbitals for which $\ell_\mu = \ell_\nu \pm 1$. Formulas for these cases in terms of one-center overlap integrals have been presented in an earlier paper (Rein et al., 1970a). The one-center quadrupole integral (Eq.(14)) after expansion is

$$M_{iAjA} = \sum_{\mu\nu}^{\text{on atom } A} P_{\mu\nu} \int \chi_\mu^A X_{iA} X_{jA} \chi_\nu^A \, dr \quad (17)$$

These integrals have been discussed elsewhere. Their calculation proceeds by expressing $X_{iA} X_{jA}$ in terms of r^2 multiplied by a linear combination of even spherical harmonics. By virtue of this decomposition and the orthogonality of spherical harmonics, Equations (13) and (15), after expansion, are

$$\mu_{iAB} = \sum_{\mu}^{A} \sum_{\nu}^{B} P_{\mu\nu} \int \chi_\mu^A X_{iAB} \chi_\nu^B \, dr \quad (18)$$

(on atom A, B)

$$m_{iAjB} = \sum_{\mu}^{A} \sum_{\nu}^{B} P_{\mu\nu} \int \chi_\mu^A X_{iAB} X_{jAB} \chi_\nu^B \, dr. \quad (19)$$

(on atom A, B)

We observe that (18) and (19) are not yet in a form convenient for integration. In order to bring these expressions to a more convenient form for integration the orbitals have to be expressed in the coordinate system in which the operators are referred, i.e., in a system where their polar axes are along the line joining the centers and their reference meridian coincides. The coordinate rotation affects only the spherical harmonics. The unrotated spherical harmonics can be expressed as a linear combination of harmonics $Y_{\ell m}$ in the rotated system

$$Y_{\ell m} = \sum_{\sigma=-\ell}^{\ell} D_\ell^m (\phi, t) Y_{\ell\sigma} \quad (20)$$

where t stands for cos θ ; φ and θ are angular coordinates of the new orientation from the position coordinates of the centers ; and $D_\ell^{m\sigma}(\phi, t)$ are the well-known rotational coefficients.

The integrals of (18) and (19) after inserting (20) are

$$\int X^{A}_{n\ell m\delta} \begin{matrix} X_{iAB} \\ \text{or} \\ X_{iAB}X_{jAB} \end{matrix} X^{B}_{n'\ell' m'\delta'}\, dr = \sum_{\sigma=-\ell}^{\ell} \sum_{\sigma=-\ell'}^{\ell'} D^{mA\sigma}_{\ell A}(\phi,t) \tag{21}$$

$$\times\ D^{m'B\sigma'}_{\ell' B}(\phi,t) < n\ell\sigma\delta \left| \begin{matrix} X_{iAB} \\ \\ X_{iAB}X_{jAB} \end{matrix} \right| n'\ell'\sigma'\delta' >$$

The integrals in Eq.(21) are calculated by expressing the respective operators in terms of cylindrical operators. The further details in the specification of the local coordinate origin will be given in the computational section. The dipole operators are $z, \rho e^{i\phi}$ $\rho e^{-i\phi}$, and the quadrupole components can be expressed as a linear combination of the five cylindrical operators

$$zz\ ,\quad z\rho e^{i\phi}\ ,\quad \rho^2 e^{+2i\phi}\ ,\quad \rho^2 e^{-2i\phi}\ ,\quad \rho^2\ .$$

These cylindrical operators have ellipsoidal representation and the integrals are reduced to ellipsoidal expression, details of this procedure and a list of ellipsoidal expressions are given in Ref. 19.

In order to complete the calculation of the atomic moments, two further steps are required. First, the moments have to be rotated into a new local coordinate system with axes parallel to the fixed axis system. This is accomplished by the standard rotational matrices.

Second, they have to be translated to the respective atomic origin to which they are referred according to the partitioning of the densities. For this partitioning we have to consider the cases, which corresponds to $f^{AB}_{\lambda\sigma} = \frac{1}{2}$, and effects a translation of one-half of each moment component to center A and one-half to center B. As we have already indicated for the monopoles, this just corresponds to the Mulliken division of overlap charge. The translation is effected by Eq.(2) and Eq.(8) for the dipole and quadrupole components respectively.

The final step of the calculation of the local moments is transformation of the atomic moments into the traceless definition

$$Q^{A}_{ij} = \frac{1}{2} \int \rho_A(r)(3X_iX_j - r^2\delta_{ij})\, dr. \tag{22}$$

This is the form in which the calculated atomic moment values appear in the tables in Ref.7.

The molecular origin related moments which are in comparable form with experimentally measured quantities are

$$\mu = \sum_A Z_A R_A - \sum_{\mu\nu} P_{\mu\nu}\mu_{\mu\nu} \tag{23}$$

where the components are connected with the local moments through the translational and rotational relations discussed before. Or more explicitly, for a dipole component

$$(\mu_x)^{AB}_{\mu\nu} = \frac{1}{2}(X_A + X_B)S_{\mu\nu} + \frac{1}{2} R(\cos\theta \cos\phi \; X_{\mu\nu} - \sin\phi \; Y_{\mu\nu} + \sin\theta \cos\phi Z_{\mu\nu}) \tag{24}$$

where $(\mu_x)^{AB}_{\mu\nu}$ is the x component of the dipole moment $\mu_{\mu\nu}$ due to the overlap density $\chi^A_\mu \chi^B_\nu$ and expressed relative to the fixed axial system. X_A, X_B, ϕ, θ are as defined previously, R is the inter-center distance ; $S_{\mu\nu}$ is the overlap integral between orbitals ϕ_n and ϕ_ν ; and $X_{\mu\nu}$, $Y_{\mu\nu}$, and $Z_{\mu\nu}$ are the bicentric moment components in the AB axis system in which system the integration has been performed.

$$Q_{ij} = \frac{1}{2} |e| \int \rho(X)(3X_iX_j - r^2\delta_{ij}) \, dX + \frac{1}{2} \sum_A^N |e| \; Z_A(3\bar{X}_{Ai}\bar{x}_{Aj} - R_A^2 \; \delta_{ij}) \tag{25}$$

where $\rho(X)$ is the density matrix ; X_iX_j are the Cartesian components of the position operator r relative to the fixed axial system ; Z_A is the core or nuclear charge on atom A ; and $\bar{X}_{Ai}$, $\bar{X}_{Aj}$ are the components of the position vector R_A of the Ath nucleus in the fixed axis system and the sum of over all atoms in the molecule.

The connection between Q_{ij} and the density moments which have been calculated in the local coördinate system is established via the following relations :

The electronic contribution to Q_{ij} is

$$M_{ij} = \int \rho(r)X_iX_j \, dr = \sum_{\mu\nu} P_{\mu\nu}\langle\mu|X_iX_j|\nu\rangle \tag{26}$$

The integral in (26) is related to the local moments by the translational Eq. (8).

$$\langle\mu|X_iX_j|\nu\rangle = (m_{ij})^{rot}_{\mu\nu} + X_{\overline{iAB}}(\mu_j)^{rot}_{\mu\nu} + X_{\overline{jAB}}(\mu_i)^{rot}_{\mu\nu} - |e|$$

$$X_{\overline{iAB}}\, X_{\overline{jAB}}\, S_{\mu\nu} \qquad (27)$$

where the first term is the local quadrupole integral of Eq.(27) after rotational alignment with the fixed axis system. Similarly, the $(\mu_i)^{rot}$ are the rotated dipole components of Eq.(24). The X_{iAB}'s are the Cartesian components of the midpoint between centers A and B on which orbitals μ and ν's are defined, respectively.

4. Interaction Schemes

In this section the interaction energy between non-overlapping molecules is considered following our earlier descriptions(19). Expansion techniques are used to express the interaction in a series of terms between the multipoles of molecule 1 with those of molecule 2. Part of the analysis is similar to that of Buckingham(20) and McLean and Yoshimine(21) in their discussion of the interaction of a molecular charge distribution with an external electric field. These considerations are extended to deal with interaction formulas in terms of the atomic multipoles which have been treated in the previous section.

These two interaction schemes are related to earlier formulations of London and subsequent workers in such a way that the two London methods appear as first approximations to the more general expansions of the present work. More specifically, the molecular dipole approximation(5) and the replacement of the molecule by point charges at the atomic centers (called the point charge or monopole approximation), the two alternative interaction schemes which were in general use for electrostatic (first order) interactions, are both equivalent to terminating the expansion after one nonzero term. In the latter case the molecule is first divided in segments and a separate expansion is made for each segment.

For the molecular dipole approximation, it had been previously suggested that this termination was not valid in many instances, particularly in biological systems. Subsequent studies used the monopole method. Based on Lowdin's(1) criterion of spherically disconnected charges it had been shown(23) that for segments the expansion is valid for distances of interest, whereas for the first scheme it is not valid. The assumption had been that the first term of each segment would dominate, but this was never shown explicitly.

The immediate purpose of this analysis is, therefore, to investigate the convergence of the electrostatic term in the two interaction schemes for larger molecules. Toward this objective we presented above the definition and computational method of the molecular and atomic moments. This, together with the formulas which will be presented here, constitutes a complete computational scheme. Starting from molecular LCAO wave functions, the electrostatic interaction between the two molecules up to and including octopole - octopole terms can be calculated in either of the two approximations. The computational methods can be extended to higher moments as well and can be used in a perturbation formulation to obtain higher-order interaction terms when suitable excited-state wave functions are available.

V. DERIVATION OF MULTIPOLE INTERACTION SERIES

If a charge distribution $\rho(r)$ is placed in an external field $\Phi(r)$ than the electrostatic energy of the system is

$$W = \int \rho(r)\Phi(r)\, dr \tag{28}$$

r is the position vector over the charge distribution and the origin is within the distribution. If the external potential Φ is due to another continuous charge distribution (an atom or a molecule) that does not overlap with the charge distribution $\rho(r_i)$ then

$$\Phi(r) = \int \frac{\rho'(r')}{|r-r'|}\, dr' \tag{29}$$

If $\rho(r)$ and $\rho'(r')$ are composed of segmental or atomic densities as defined above

$$\rho(r) = \sum_a \rho_a(r) \tag{30}$$

(28) and (29) can be rewritten as

$$W = \sum_a \int \rho_a(r)\Phi(r)\, dr \tag{31}$$

$$\Phi(r) = \sum_\beta \Phi_\beta(r) = \sum_\beta \int \frac{\rho_{\beta'}(r')}{|r-r'|}\, dr' \tag{32}$$

where r and r' are position vectors over the segmental densities with origins chosen at the respective atomic centers. There are two Taylor expansions which are utilized to express W in terms of multipole series. In considering these expansions we will use them alternatively in the sense of Eqs. (28)and (31) and Eqs. (29) and (32), respectively, to derive the two alternative interaction schemes.

First, we begin with a Taylor expansion of the potential Φ about a suitable origin (by convention the center of mass is chosen as this origin for the molecular scheme, and the set of atomic origins in the segmental scheme). For this expansion it is assumed the ϕ is slowly varying about the origin, so that higher derivatives can be neglected.

$$\Phi(r) = \Phi(o) + r\cdot\nabla\Phi\ (o) + \frac{1}{2} \sum_{i,j} r_i r_j \frac{\partial^2 \Phi}{\partial r_i \partial r_j} (o) + \ldots, \quad (33)$$

where Φ and all its derivatives are evaluated at the origin. Since Φ is assumed to be due to another molecule (or its segments) that does not overlap with the one represented by the charge distribution $\rho(r)$, $\rho\nabla^2\Phi = 0$, and we can subtract

$$\frac{1}{6} r^2\ \nabla^2\Phi = 0. \quad (34)$$

Substituting (33) and (34) into Eq. (28) and identifying the multipole moments as a in Section 2

$$W = q\,\Phi\,(o) + \mu\cdot\nabla\Phi\ (o) + \frac{1}{2} \sum_{i,j} Q_{ij} \frac{\partial\Phi}{\partial r_i \partial r_j} (o) + \cdots. \quad (35)$$

The same steps but with $\rho(r)$ first partitioned according to Eq. (30) leads to

$$W = \sum_{\alpha\beta}\sum W_{\alpha\beta}$$

$$W_{\alpha\beta} = \left[q_\alpha\ \Phi_\beta(r) + \mu_\alpha\ .\nabla\Phi(r) + \frac{1}{2} \sum_{i,j} (Q_{ij}) \frac{\partial\Phi_\beta}{\partial r_i \partial r_j} (r) \right]_{r=r_\alpha} \quad (37)$$

where q_α, μ_α, and $(Q_{ij})_\alpha$ are the atomic charge dipole and quadrupole components as previoulsy defined.

The external potential we are considering arises from the non-overlapping charge distribution of the second molecule according to Eq. (37) or Eq. (35). The operators can be expanded in a second Taylor expression which, after the interactions implied either by Eq.(37) or by Eq. (35), leads to the two alternative expressions for the potential

$$\Phi(R) = \frac{q'}{|R|} + \frac{\mu' \cdot R}{|R|^3} + \frac{R \cdot Q' \cdot R}{|R|^5} + \ldots \qquad (38)$$

$$\Phi(r_\beta) = \sum_\beta \frac{q'_\beta}{|r_\beta|} + \frac{\mu'_\beta \cdot r_\beta}{|r_\beta|^3} + \frac{r_\beta \cdot Q'_\beta \cdot r_\beta}{|r_\beta|^5} + \ldots, \qquad (39)$$

where R is the vector from the origin of the primed charge distribution ; the r'_β s are similarly the vectors from the respective atomic origins in the primed charge distribution ; q', μ', Q' are the multipole moments of the primed charge distribution referred to the molecular origin ; and q'_β , μ'_β , and Q'_β are the respective atomic multipole moments of the primed charge distribution.

Substituting Eq.(38) and its respective derivatives into Eq.(36) and repeating the whole procedure by interchanging the roles of molecules 1 and 2, then the interaction energy appears as a series of interactions of successive multipoles of the two molecules

$$W = \sum_{i=0} \sum_{j=0} W_{ij}, \qquad (40)$$

where the successive interaction terms up to octupole-octupole interactions are

Monopole-monopole

$$W_{00} = \frac{qq'}{|R|} \qquad (41)$$

Monopole-dipole

$$W_{01} = (q'\mu \cdot R)|R|^{-3} \qquad (42)$$

Dipole-monopole

$$W_{10} = -(q\mu' \cdot R)|R|^{-3} \qquad (43)$$

Dipole-dipole

$$W_{11} = (\vec{\mu} \cdot \vec{\mu}')|R|^{-3} - 3(\vec{\mu} \cdot \vec{R})(\vec{\mu}' \cdot \vec{R})|R|^{-5} \qquad (44)$$

Monopole-quadrupole

$$W_{02} = (q'\vec{R} \cdot \overleftrightarrow{Q} \cdot \vec{R})|R|^{-5} \qquad (45)$$

Quadrupole-monopole

$$W_{20} = (q\vec{R}\cdot\overleftrightarrow{Q}'\cdot\vec{R})|R|^{-5} \quad (46)$$

Dipole-quadrupole

$$W_{12} = 2(\vec{\mu}' \cdot \overleftrightarrow{Q} \cdot \vec{R})|R|^{-5} - [5(\vec{\mu}'\cdot\vec{R})(\vec{R}\cdot\overleftrightarrow{Q}\cdot\vec{R})]\,|R|^{-7} \quad (47)$$

Quadrupole-dipole

$$W_{21} = -2(\vec{\mu}\cdot\overleftrightarrow{Q}'\cdot\vec{R})|R|^{-5} + [5(\vec{R}\cdot\overleftrightarrow{Q}'\cdot\vec{R})(\vec{\mu}\cdot\vec{R})]\,|R|^{-7} \quad (48)$$

Quadrupole-quadrupole

$$W_{22} = \frac{2}{3}\,\frac{1}{6}\,[\overleftrightarrow{Q} : \overleftrightarrow{Q}']\,|R|^{-5} - \frac{20}{3}\,[\vec{R}\cdot\overleftrightarrow{Q}'\cdot\overleftrightarrow{Q}\cdot\vec{R})][R]^{-7}$$

$$+ \frac{35}{3}\,(\vec{R}\cdot\overleftrightarrow{Q}\cdot\vec{R})(\vec{R}\cdot\overleftrightarrow{Q}'\cdot\vec{R})\;[R]^{-9}, \quad (49)$$

Monopole-octupole

$$W_{03} = q\vec{R}\cdot\;(\overleftrightarrow{\Omega}'\cdot\vec{R})\;\cdot\;\vec{R}/R^{7} \quad (50)$$

Octupole-monopole

$$W_{30} = -q'\vec{R}\cdot(\overleftrightarrow{\Omega}\cdot\vec{R})\cdot\vec{R}/R^{7} \quad (51)$$

Dipole-octupole

$$W_{13} = 3\vec{\mu}\;.\;(\overleftrightarrow{\Omega}'\cdot\vec{R})\cdot\vec{R}/R^{7} - 7(\vec{\mu}\cdot\vec{R})\;(\vec{R}\cdot(\overleftrightarrow{\Omega}'\cdot\vec{R})\cdot\vec{R})/R^{9} \quad (52)$$

Octupole-dipole

$$W_{31} = 3\vec{\mu}'.(\overleftrightarrow{\Omega}.\vec{R}).\vec{R}/R^{7} - 7(\vec{\mu}'.\vec{R})(\vec{R}.(\overleftrightarrow{\Omega}.\vec{R}).\vec{R})/R^{9} \quad (53)$$

Quadrupole-octupole

$$W_{23} = (\overleftrightarrow{Q}:\overleftrightarrow{\Omega}').\vec{R}/R^{7} - 7(\vec{R}.\overleftrightarrow{Q}).((\overleftrightarrow{\Omega}'.\vec{R}).\vec{R})/R^{9}$$

$$+ 10.5(\vec{R}.\overleftrightarrow{Q}.\vec{R})(\vec{R}.(\overleftrightarrow{\Omega}'.\vec{R}).\vec{R})/R^{11} \quad (54)$$

Octupole-quadrupole

$$W_{32} = -\vec{R}.(\overleftrightarrow{\Omega}:\overleftrightarrow{Q}')/R^7 + 7(\vec{R}.(\vec{R}.\overleftrightarrow{\Omega}))\ .\ (\overleftrightarrow{Q}'.\vec{R})/R^9$$

$$- 10.5(\vec{R}.(\overleftrightarrow{\Omega}.\vec{R}).\vec{R})(\vec{R}.\overleftrightarrow{Q}.\vec{R})/R^{11} \qquad (55)$$

Octupole-octupole

$$W_{33} = 0.4\ T_r[(\overleftrightarrow{\Omega})(\overleftrightarrow{\Omega}')]/R^7 - 8.4(\vec{R}.\overleftrightarrow{\Omega}):(\vec{R}.\overleftrightarrow{\Omega}')/R^9$$

$$+ 37.8(\vec{R}.(\vec{R}.\overleftrightarrow{\Omega})).((\overleftrightarrow{\Omega}'.\vec{R}).\vec{R})/R^{11}$$

$$- 46.2(\vec{R}.(\vec{R}.\overleftrightarrow{\Omega}).\vec{R})(\vec{R}.(\vec{R}.\overleftrightarrow{\Omega}').\vec{R})/R^{13} \qquad (56)$$

where $\vec{R}$ is the vector from α atom in A to β atom in B and $R = |R|$

$\vec{\mu}$ is a dipole vector

$\overleftrightarrow{Q}$ is a quadrupole tensor of rank 2

$\overleftrightarrow{\Omega}$ is an octopole tensor of rank 3

The notation is such that

. designates dot product

: designates a trace between two dyadics

$$A.(\overleftrightarrow{\Omega}.\vec{B}).\vec{C} = \Omega_{ijk}\ A_i B_j C_k$$

$$\vec{A}.(\overleftrightarrow{\Omega}:\overleftrightarrow{Q}) = \Omega_{ijk}\ A_i Q_{jk}$$

$$\overleftrightarrow{Q}.\vec{A} = \begin{pmatrix} Q_{1j}A_j \\ Q_{2j}A_j \\ Q_{3j}A_j \end{pmatrix} \qquad (\overleftrightarrow{\Omega}.B).C = \begin{pmatrix} \Omega_{1jk}B_jC_k \\ \Omega_{2jk}B_jC_k \\ \Omega_{2jk}B_jC_k \end{pmatrix}$$

where repeated subscript in the above implies summation over that index.

The expression for the interaction energy in terms of a series of interactions between successive atomic multipoles of molecule 1 with those of molecule 2 is obtained by a straightforward generalization of the procedure by which Eq.(40) has been obtained, i.e., the potential defined by Eq.(10) for successive atomic positions is substituted into the interaction equation

$$W = \sum_{\alpha} \sum_{\beta'} \sum_{i=0} \sum_{j=0} W_{ij}^{\alpha\beta'} , \qquad (57)$$

where the interaction W is decomposed as in Eqs.(41) - (56) for each atomic pair α', β' . Thus, for example, a typical term $W_{12}^{\alpha\beta}$ has the following meaning,

$$W_{20}^{\alpha\beta'} = (\frac{1}{2} q^{\alpha} R_{\alpha\beta'} Q^{\beta'} . R_{\alpha\beta'}) |R_{\alpha\beta'}|^{-5}, \qquad (58)$$

so that this term represents the interaction between the monopole of atom α of molecule 1 with the quadrupole tensor of atom β' of molecule 2 with $R_{\alpha\beta'}$, the vector pointing from atom α to atom β'.

This completes the description of the two interaction schemes which are used in this study.

VI DISPERSION INTERACTIONS

1. The Approach of London

In terms of a visualization the dispersion interaction can be considered as interactions between instantaneous dipole moments of any two interacting molecules. In view of the electronic motions these dynamic interactions are always present and contribute to lower the energy of the system of two molecules in a vacuum.

London[5] used the dipole-dipole approximate of the interaction Hamiltonian

$$V = (\mu.\mu')|R|^{-3} - 3\,(\mu.R)(\mu'.R)|R|^{-5} \qquad (59)$$

in a second order perturbation expression to calculate dispersion energies

$$E_{disp} = \sum_{\alpha'} \sum_{\beta'} \frac{|\langle \Psi_{\alpha}\Psi_{\alpha'}|(\mu.\overset{=}{T}.\mu')|R|^{-3}\, \Psi_{\beta}\Psi_{\beta'}\rangle|^{2}}{E_{\alpha}+E_{\beta} - E_{\beta'}-E_{\alpha'}} \qquad (60)$$

where α and β designate the ground states and $E_{\alpha'}$, $E_{\beta'}$, are the energies of the states $\psi_{\alpha'}$ and $\psi_{\beta'}$ respectively and where T has the form

$$\bar{\bar{T}} = 3\left(\frac{R}{|R|} \otimes \frac{R}{|R|}\right) - \bar{\bar{1}} \qquad (61)$$

Where R is the vector joining the mid-points of two bonds

Because excited state wave functions of sufficient accuracy have not been available in the thirties and in fact are even presently not available for large molecules, equation (60) can be applied only with further approximations.

The approximations used by London were to replace the denominator by a mean value and evaluate the sum in the nominator by the corresponding molecular polarizabilities.

With these approximations, London's formula for the dispersion energies is

$$E_{disp} = -\frac{3}{2}\frac{I\,I'}{I+I'}\frac{\alpha\alpha'}{|R|^6} \qquad (62)$$

where I and I' are the molecular ionization potentials. More recently this expression has been further refined by improving the convergence and by taking into account also the anisotropy of the polarizabilities.

The modified formula suggested by Rein, Claverie and Pollak (16,17) is

$$E_{disp} = -\frac{1}{4}\frac{I_1 I_2}{I_1+I_2}\sum_{i=1}^{b_1}\sum_{j=1}^{b_2}\frac{1}{r_{ij}}\Big(6\alpha_1^T\alpha_2^T + \alpha_1^T\delta_2(3(\vec{\alpha}_2^L.\vec{r})+1)$$

$$+\ \alpha_2^T\delta_1(3(\vec{\alpha}_1.\vec{r})^2+1) + \delta_1\delta_2[3(\vec{\alpha}_1^L.\vec{r})(\vec{\alpha}_2^L.\vec{r})-(\vec{\alpha}_1^L.\vec{\alpha}_2^L)]^2\Big) \qquad (63)$$

where α_1^L, α_1^T, α_2^L, α_2^T are the longitudinal and transverse bond polarizabilities for bonds 1 and 2 ; r is a unit vector in the direction of the line joining the midpoints of the two bonds, $\delta = \alpha^L - \alpha^T$. The distance between the midpoint of the ith bond on molecule 1 to midpoint of the jth bond on molecule 2. I_1 and I_2 are molecular ionization potentials.

2. Polarization Energies

The polarization energy is obtained similarly from second order perturbation theory[18]. It involves the interaction of transition charges with static charge and the effect of this has been approximated by the electric field created by the charge distribution of molecule one with the transition charges of two and vice versa at the point of location of the transition dipoles associated with the bonds of the other molecule. The dipole location was taken to be the midpoint of the bonds on the other molecule. Bond polarizabilities for specific types of bonds were employed as an approximation to the localized transition dipoles.

A refinement of these methods is to perform the calculations by expanding the electric field to full multipole representation and to observe the convergence of the centers.

The following equation due to Rein, Claverie and Pollak[16] was used as a starting point in determining this interaction.

$$E_{pol} = -\frac{1}{2} \sum_{k}^{b_2} \alpha_k^T (\vec{E}_K . \vec{E}_K) + \delta_k (\vec{E}_k . \vec{\alpha}^L)^2 \qquad (64)$$

where $\delta_k = \alpha_k^T - \alpha_k^L$

and where α_K^T, α_K^L are the transverse and longitudinal bond polarizability on the other molecule, and $\vec{E}_k$ is the full multipole representation of the electric field felt at the midpoint of the kth bond.

The electrostatic field $\vec{E}$ has been derived by Egan, Swissler and Rein[10] as follows

$$\vec{E} = -\text{grad}\,\Phi \quad \text{where grad}\,\phi = \sum_m \frac{\partial \Phi}{\partial X_m} \vec{e}_m \quad m=1,2,3 \qquad (65)$$

where letting $R_{\beta k}$ be the distance from the βth

where $\beta = 1,2,\ldots a_1$

atom on one molecule to the midpoint of the kth bond we obtain for the potential

$$\Phi = \sum_{\beta}^{a_1} \frac{q_\beta}{|\vec{R}_{\beta k}|} + \frac{\vec{\mu}_\beta . \vec{R}_{\beta K}}{|\vec{R}_{\beta K}|^3} + \frac{\vec{R}_{\beta k} . Q_\beta . \vec{R}_{\beta k}}{|\vec{R}_{\beta k}|^5} + \vec{R}_{\beta k} . \frac{(\Omega_\beta : \vec{R}_{\beta K} \vec{R}_{\beta k})}{|\vec{R}_{\beta K}|^7} \qquad (66)$$

and for the field

$$\vec{E}_k = \sum_{\beta}^{a_1} \frac{q_\beta \vec{R}_{\beta K}}{|\vec{R}_{\beta K}|^3} - \frac{\vec{\mu}_\beta}{|\vec{R}_{\beta K}|^3} + 3\frac{(\vec{\mu}_\beta . \vec{R}_{\beta K})\, \vec{R}_{\beta K}}{|\vec{R}_{\beta K}|^5}$$

$$- \sum_{m,j} \frac{Q_{mj} X_j}{|R_{\beta\kappa}|^5} \vec{e}_m + \frac{5}{2} \vec{R}_{\beta\kappa} \sum_{i,j} Q_{IJ} \frac{X_i X_j}{|R_{\beta\kappa}|^7}$$

$$- \sum_{j,k,\ell} \frac{3X_k X_\ell}{|R_{\beta\kappa}|^7} \Omega_{jk\ell} \vec{e}_j + \sum_{jk\ell} 7\Omega_{jk\ell} \vec{R}_{\beta\kappa} \frac{(X_j X_k X_\ell)}{|R_{\beta\kappa}|^9} \quad (67)$$

VII SHORT RANGE REPULSION ENERGY

The importance of overlap and exchange energies in systems of finite overlap has been well recognized. Short range overlap energies play an important role in influencing bond strengths and molecular geometries. Quantum theory applying to this area was developed by Löwdin(1) and others(9,30). The importance of short range effects in DNA interactions has been discussed by Rein and Pollak(18) and more recently by Claverie(31), who used an empirical formula to obtain estimates of this energy for different DNA base complexes.

The expression used by Claverie to calculate the overlap repulsion is :

$$E_{rep.} = C \sum_{\mu} \sum_{\mu'} \exp(-\alpha \frac{r_{\mu\mu'}}{r_\mu + r_{\mu'}}) \quad (68)$$

where $r_{\mu\mu'}$ is the distance between atom pair μ and μ' ; r_μ and $r_{\mu'}$ are van der Wall's radii of the respective atoms; α and c' are adjustable parameters. Kitaigorodskii(39) fitted the constants in an energy expression containing also the attractive terms in order to reproduce the crystal structure of certain hydrocarbons, and found that $\alpha = 13$, $c = 3 \times 10^4$ k cal/mol reproduces the equilibrium geometries. Claverie performed the fitting in the case of stacked planar conjugated systems and proposed a value $c = 8.1 \times 10^4$.

VIII CONFORMATIONAL ENERGIES

The interest to predict the preferred conformations of polypeptides and polynucleotides has prompted the development of several methods for calculation of conformational energies. Broadly speaking, these methods fall in one of the following two categories:

a) Molecular orbital methods(24,25)

b) Empirical potential functions(26,27).

We would consider here the latter one with some refinements introduced recently.

It would in fact be more appropriate to refer to these potential functions as semiempirical since they are comprised from terms which arise from perturbation theory in a semiempirical form, complemented by empirical functions to account for short range effect.

Thus, a typical potential function is of the form

$$V(\phi_i,\psi_i) = V\ (\phi_i) + V_\psi(\psi_i) + V_{vdw}(\phi_i,\psi_i) + V_e(\phi_i,\psi_i) \qquad (69)$$

where $V\ (\phi_i)$ and $V_\psi(\psi_i)$ are intrinsic three-fold torsional potentials ascribed to the successive single bonds. The intrinsic three-fold torsional potentials, following the suggestion of Ramachandran[27] have the form:

$$V\ (\phi_i) = V_\phi^o/2\ (1 + \cos 3\phi_i) \qquad \text{and} \qquad (70)$$

$$V_\psi(\psi_i) = V_\psi^o/2\ (1 + \cos 3\psi_i) \qquad (71)$$

where values of V_ϕ^o and V_ψ^o are assigned semiempiricaly. Non bonded interactions $V_{vdw}(\phi_i\psi_i)$ are calculated using Buckingham potential[20]

$$V_{ij} = - A_{ij}/\gamma_{ij}^6 + B_{ij}\ \exp(-\mu\gamma_{ij}) \qquad (72)$$

It can be recognized that the first term corresponds to dispersion energies between atom pairs with Aij corresponding to $\frac{I_iI_j}{I_iI_j}\alpha_i\alpha_j$ while the second term corresponds to short range repulsion between atoms penetrating van der Walls distances. Values of Aij and Bij have been tabulated for example, by Ramachandran and Sasisekharan[27]. In the defined treatment V_e the electrostatic interaction between non bonded atoms is calculated by the segmental multipole approximations described in a previous section.

IX CONFIGURATIONAL STATES

1. General Considerations

It was stressed in the introduction that the general aim underlying the study of the force laws governing the interactions between polymer units is to arrive at a description of the macroscopic properties of the polymer. Once these force laws have been elucidated, methods of statistical mechanics can be applied for calculation of the desired macroscopic properties. How elaborate a model one can choose for the description of a polymer depends on the extent to which interaction potentials have been characterized. That is, it depends on the range of configurational variables for which the interaction energies are known.

In the case of DNA, with the methods reviewed in this paper for calculation of interaction and conformational potentials, it seems that a comprehensive statistical model is becoming feasible. That is to say, a model which is explicit in the complete set of soft variables, i.e., torsional angles, defines the state of the polymer strands. Nonetheless, such a study has not yet been performed for the double helix, although recent work by Olson and Flory(32) on single strands is along this direction.

There is abundant literature(34) on statistical mechanical studies of the DNA double helix on a much more approximate level. The principle approximation leading to the simplification of the treatment is the reduction of the configurational variables to only two states per base pair, i.e., in viewing DNA as a ladder each pair (base pair) can be intact or broken, which is equivalent with the pair being hydrogen bonded or not.

To understand the degree of approximation involved in these models it is worthwhile to consider the structure of the sugar phosphate strand. A repeat unit in this strand contains five skeletal torsional angles, five torsional angles defining the deoxyribose conformation and the torsional angle around the glycosydine bond. Thus in replacing the description of the configurational states of a repeat unit by two states is tantamount to choosing only two states out of a configuration space spanned by 22 angle variables, indeed a very drastic reduction.

It is interesting to examine how such a simplified model accounts for experimentally observed properties of the double helix, as well as what agreement one can obtain between the interaction parameter deducible from such a model and experimental data with the values calculated by the quantum mechanical methods described in the previous sections.

In order to perform such a comparison we will review first the statistical model underlying such a comparison.

2. Statistical Model

We will now briefly, review the modified Ising model[35,36,37] for the statistical treatment of DNA. The DNA molecule is considered as a ladder, with its rungs as the bonding between the bases on the complementary strands, the simplest statistical model one can choose is the Ising model for one-dimensional systems, with only the nearest neighbor correlations. Let us define a parameter, σ, for each bond so that

$$\sigma = \begin{cases} +1 \text{ if the bond is intact} \\ -1 \text{ if the bond is broken} \end{cases}$$

Let $F(\alpha)$ be the bond energy of bond type α (α= A-T or G-C), taking the energy of the broken bond as zero ; and let $J(\alpha\beta)$ be the stacking energy of two nearest neighbor bonds, α and β, which will depend generally on the relative orientation of the α and β bonds. We assume that the probability that the α base pair will be in state σ, and its nearest neighbor pair β in state σ', is given by the Boltzman factor : i.e.,

$$f_{\alpha\beta}(\sigma,\sigma') = s(\alpha)^{\frac{1+\sigma}{4}} s(\beta)^{\frac{1+\sigma'}{4}} \left[t(\alpha\beta)\right]^{\frac{(1+\sigma)(1+\sigma')}{4}} \tag{73}$$

where

$$s(\alpha) \equiv \exp\left(\frac{-F(\alpha)}{kT}\right)$$

$$s(\beta) \equiv \exp\left(\frac{-F(\beta)}{kT}\right)$$

$$t(\alpha\beta) \equiv \exp\left(\frac{-J(\alpha\beta)}{kT}\right)$$

k is the Boltzman factor, and T is the temperature. It is obvious that both s and t together take into account the probability of a bond being broken directly and indirectly. It may be noted that if we consider only one pair of bonds, then in (73) we should have $s(\alpha)$ and $s(\beta)$ raised to powers $\frac{1+\sigma}{2}$ and $\frac{1+\sigma'}{2}$ respectively, instead of $\frac{1+\sigma}{4}$ and $\frac{1+\sigma'}{4}$. Since we are concerned here with a sequence of bonds, and each bond energy is counted twice, we have chosen the latter powers. The probability $P(\sigma_1, \sigma_2 \ldots \sigma_N)$ that for LNA $\alpha\beta \ldots \beta$ the state of bonds will be $\sigma_1, \ldots, \sigma_N$ is clearly the product of pair probabilities (1): i.e.,

$$P(\sigma_1,\ldots, \sigma_N) = \frac{1}{Z} f_{\alpha\beta}(\sigma_1,\sigma_2)\, f_{\beta\alpha}(\sigma_2, \sigma_3) \ldots f_{\beta\alpha}(\sigma_N,\sigma_1) \quad (74)$$

where

$$Z = \sum_{\sigma_1=\pm 1} \cdots \sum_{\sigma_N=\pm 1} f_{\beta\alpha}(\sigma_1,\sigma_2) \ldots f_{\beta\alpha}(\sigma_N,\sigma_1) \quad (75)$$

is the normalizing function. If we substitute for f's from (73), we get the partition function for the Ising model of ferromagnetism,[17] and we can, therefore, evaluate it explicitly as a function of $F(\alpha)$, T and $J(\alpha\beta)$. We can then anticipate that all the average physical properties are derivable for Z. For the present discussion, the property we are interested in is the fraction of bonds intact (as a function of temperature). If Θ denotes the fraction of bonds intact, from (4)

$$\Theta = \frac{1}{N}\left(\frac{\partial \ln Z}{\partial\, \ell ns(\alpha)} + \frac{\partial \ln Z}{\partial\, ins(\beta)}\right) \quad (76)$$

intact as a function of $H(\alpha)$, $H(\beta)$, $S(\alpha)$, $S(\beta)$, $J(\alpha\beta)$, and $J(\alpha\beta)$. In addition to the fraction of bonds intact, the other experimentally measurable quantity is the slope of the melting curve near the melting point or, equivalently, the transition width of the melting curve defined by

$$W = -\frac{1}{T_m}\left(\frac{\partial\Theta}{\partial T}\right)^{-1}_{T_m} \quad (77)$$

Where T_m is the melting point. We could write w in terms of $H(\alpha)$, $H(\beta)$, $S(\alpha)$, $S(\beta)$, $J(\alpha\beta)$, and $J(\alpha\beta)$.

Thus if data on melting curves for polymers with known sequences is available, the stacking energies and entropies can be evaluated.

Goel, et al[36], using the available data on melting temperature and widths of the melting curves for dA: dT, dAT:dAT, dG:dC, dAC:dTG and dAG:dTC, have obtained the stacking energies for various base combinations.

An inspection of the results reviewed in tables V and VI shows that the balance of forces of these complexes is indeed different than in the case of hydrogen bonded complexes. That is to say, that while in the latter case the electrostatic interaction represents within a good approximation the total interaction energy, this is definitely not the case for the former. Second, the super molecule scheme does not include correlation energies. Therefore, based on these studies, one cannot arrive even at a qualitative conclusion regarding the status of the dispersion forces as calculated from modified London theory.

The second approach for testing the theory is by way of comparing its macroscopic predictions with experiments. We have discussed in detail the complexity and limitations of the approach in view of the simplifying assumptions used in the present form of statistical theory.

It is encouraging to find (table V and VI), that in spite of the limitations imposed by the Ising statistics, there is a fair agreement between the values of the stacking energies obtained from melting data and those predicted from quantum theory. This agreement lends some support to the notion, that the empirical treatment of the repulsion terms and the calculation of dispersion terms from polarizabilities are justified.

The status of the theory in the range of separations larger than 5 Å appears to be somewhat better clarified. The overlap terms in this region are definitely negligible. The convergence of the representation of the electrostatic interactions by the segmental multipole method as well as the convergence of the representation of the second order terms by the bond polarizabilities seems to be assured as it can be recognized from tables II and III.

In summary, it appears that the methodology of the interacion energy calculation has now been sufficiently advanced. With this methodology a more detailed mapping of the interaction potentials seems to be feasible. And, the time is ripe for introducing more elaborate statistical descriptions for the macroscopic properties of interest for biopolymers, especially DNA. Further correlations of these predicted macroscopic properties with experiments will provide further and more stringent tests to the interaction theory as well as guide lines for possible further refinements.

It seems appropriate to conclude the discussion by stressing the fact that with the developed methodology approaching a phase where prediction of macroscopic properties appears as a reasonable task, one looks forward to fruitful application of these methods for explanation of fundamental biological phenomena from a molecular and physical point of view.

TABLE II CONTRIBUTIONS TO ELECTROSTATIC ENERGIES IN INVERSE POWERS OF R(kcal)

Pair	R^{-1}	R^{-2}	R^{-3}	R^{-4}	R^{-5}	R^{-6}	R^{-7}	Mean
G-C	- 18.055	-20.200	-23.567	-24.028	-23.359	-23.405	-22.483	-22.517
G ↘ C	- 3.551	- 2.052	- 2.382	- 2.359	- 2.4605	- 2.435	- 2.398	- 2.520
G ↗ C	- 5.234	- 5.073	- 5.188	- 5.027	- 5.096	- 5.133	- 5.073	- 5.117
↑G G	+ 8.601	+ 3.364	+ 9.397	+ 6.378	+ 6.826	+ 7.097	+ 7.080	+ 6.856
C C↑	+ 7.563	+ .830	+ 5.227	+ 4.243	+ 3.535	+ 3.791	+ 3.851	+ 4.234
A-T	- 6.081	- 7.218	-10.116	-10.960	- 9.882	-10.010	- 8.993	- 9.037

TABLE III CONTRIBUTIONS TO POLARIZATION ENERGIES IN INVERSE POWERS OF R (kcal)

(Chart illustrates the effect of convergence)

Pair	R^{-2}	R^{-4}	R^{-6}	R^{-8}	Total
G–C	-3.422	-3.786	-3.841	-3.841	-3.841
G $\searrow$ C	- .553	- .5910	- .5917	- .5917	- .5917
G $\nearrow$ C	- .554	- .5820	- .5823	- .5823	- .5823
$\uparrow$ G G	-1.625	-2.221	-2.252	-2.252	-2.252
C C $\uparrow$	-1.153	-1.596	-1.604	-1.604	-1.604
A–T	- .763	-1.028	-1.080	-1.090	-1.090

TABLE IV ENERGY CONTRIBUTIONS IN HYDROGEN BOUNDED BASE PAIRS

(ALL ENERGIES IN kcal/mole)

Pairs	E_{el}	E_{pol}	E_{disp}	E_{rep}	Total
G-C	-22.483	-3.841	-2.490	+5.28	-23.534
A-T	- 8.993	-1.09	-1.77	+3.81	- 8.04

TABLE V ENERGY CONTRIBUTIONS TO STACKING FOR G-C (kcal/mole) COMPLEX

Pairs	E_{el}	E_{pol}	E_{disp}	E_{rep}	Stacking Total	Stacking[37] Experimental
G ↘ C	-2.398	- .591	- .647			
G ↗ C	-5.073	- .582	- .650			
G ↑ G	+7.080	-2.252	-4.541	+3.740		
C ↓ C	+3.851	-1.604	-3.161			
Sub-Total	+3.460	-5.029	-8.998	+3.740	-6.827	-5.89

TABLE VI ENERGY CONTRIBUTIONS TO STACKING FOR A-T (kcal/mole) COMPLEX

Pairs	E_{el}	E_{pol}	E_{disp}	E_{rep}	Stacking Total	Stacking[37] Experimental
A ↗ T	- .233	- .192	- .686			
A ↘ T	- .231	- .057	-. 453			
A ↑ A	+ .6641	-1.780	-5.350	+4.08 †		
↑ T T	+1.208	-1.004	-5.354			
Sub-Total	+1.408	-3.033	-11.822	+4.08	-9.367	-5.40

† Value pertains to entire stacking interactions

X DISCUSSION AND RESULTS

The purpose of this paper has been to review the current status of methods for calculating intermolecular interactions between biopolymer units. A further objective was to clarify the nature of forces contributing to the various domains of intermolecular separations as well as to examine the status of approximations applicable in the respective regions. Yet another aim was to test the predictive value of the theory by establishing a connection with macroscopic properties and comparing the theoretically predicted values with those derived from experimental data. For this reason, a statistical model describing DNA has been introduced.

Among these questions perhaps the most fundamental is the one concerned with the nature of contributing forces. One would like to ascertain that the theoretical formalism used accounts fully for all the interaction terms significant for the investigated case. Considering the form of perturbation theory generally used for treatment of interactions and elaborated in this paper, this is a non-trivial problem. First, because there is no a priori argument to say that a second order non-overlapping perturbation theory should yield the leading interaction terms in the whole range of interest. Second, many further approximations are introduced in the treatment, i.e., replacement of the interaction operator by a truncated expandion, use of approximate wave functions for the ground state charge distribution, and replacement of certain matrix elements by empirical quantities. The status and quantitative balance of these approximations is an open question.

There are two alternative approaches which one may take under these circumstances, unfortunately neither of them is simple or definitive in an absolute sense.

One of these approaches is to compare with results obtained from other theoretical schemes, preferably with more complete and refined ones. We have reviewed the component analysis of Morokuma[3], Dreyfus and Pullman[2] and Kollman and Allen[4] with this goal in mind. The results obtained by these workers provide support to the following two important conclusions.

1) The forces dominant in the range exceeding 5 Å separation are the same which one obtains from non-overlapping perturbation theory.
2) At separation below 5 Å forces arising from overlap are coming into play, but the repulsive and attractive terms arising from overlap are often balancing each other at equilibrium separation.

There are two serious shortcomings for a broader generalization of the last conclusion. First, these are specific to hydrogen bonded systems, thus the behaviour of different types of molecular complexes, such as for example, stacked bases do not have necessarily to behave in the same way at their equilibrium configuration.

REFERENCES

1. Löwdin,P.-O. (1956). Advances in Physics 5, No. 17, 153 ; J. Chem. Phys. 18, 365, (1950).

2. Dreyfus, M. and A. Pullman (1970). Theor. Chim. Acta 17, 109.

3. Morokuma, K. (1971). J. Chem. Phys. 55, 1236.

4. Kollman, P.A. and L.C. Allen (1970). J. Chem. Phys. 52, 5085.

5. London, F. (1937). Trans. Faraday Soc. 33, 8.

6. Rein, R., J. R. Rabinowitz, and T.J. Swissler (1972). J. Theor. Biol. 34, 215.

7. Rein, R. (1973). Adv. Quant. Chem. 7, 335.

8. Stamatiadou, M., T. Swissler, J. Rabinowitz and R. Rein (1972). Biopolymers 11, 1217.

9. Rabinowitz, J., T. Swissler and R. Rein (1972). Intern. J. Quant. Chem. 6, 353.

10.Egan, J.T., T. Swissler and R. Rein (1974). Intern. J. Quant. Chem. (in press).

11.Rabinowitz, J.R., and R. Rein (1972). Intern. J. Quant. Chem. 6, 669.

12.Sinanoglu, O., (1964). Adv. Chem. Phys. 6, 315.

13.Lifshitz, E.M. (1955). Zh. Eksperim. Theor. Fiz. 29, 94.

14.Nir, S., R. Rein and L.Weiss (1972). J. Theor. Biol. 34, 135.

15.Coulson, C.A. and V. Danielson (1954). Arkiv Fysik, 8, 239.

16.Rein, R., P. Claverie and M. Pollak (1968). Intern. J. Quant. Chem. 2, 129.

17.Claverie, P. and R.Rein (1969). Intern. J. Quant. Chem. 3, 537.

18.Rein, R., and M. Pollak (1967). J. Chem. Phys. 47, 2039.

19. Rein, R., Clarke, G.E., and Harris, F.E., (1970). "Quantum Aspects of Heterocyclic Compounds in Chemistry and Biochemistry" (E.D. Bergmann and B. Pullman eds.) p.86, Israel Acad. Sci. Humanities, Jerusalem.

20. Buckingham, A.D. (1954). Quart. Rev. Chem. Soc. (London) 13, 183.

21. McLean, A.D. and M. Yoshimine (1967). J. Chem. Phys. 47, 1927.

22. Nir. S. (1974). J. Chem. Phys. (in press).

23. Pollak, M. and R. Rein (1967). J. Chem. Phys. 47, 2045.

24. Perahia, D., A. Saran and B. Pullman (1973). In "Conformation of Biological Molecules and Polymers" (E.D. Bergmann and B. Pullman, eds.) p. 225. The Israeli Academy of Sciences and Humanities, Jerusalem.

25. Pullman, B., and H. Berthold (1973). In "Conformation of Biological Molecules and Polymers" (E.D. Bergmann and B. Pullman, eds.) p. 209. The Israeli Academy of Sciences and Humanities, Jerusalem.

26. Scheraga, H.A. (1968). Adv. Phys. Org. Chem., 6, 103.

27. Ramachandran, G.N., and V. Sasisekharan (1968). Adv. in Prot. Chem. 23, 283.

28. Egan, J.T., (1974). Private Communication.

29. McWeeny, R. (1959). Proc. Roy. Soc. A. (London) 253, 242.

30. Murrel, J.N., M. Randic and D.R. Williams (1965). Proc. Roy. Soc. A. (London) 284, 566.

31. Claverie, P., (1968). "Molecular Associations in Biology" B. Pullman, ed., Academic Press., New York and London.

32. Olson, W.K. and P.J. Flory (1972). Biopolymers 11, 1-66.

33. Brant , D.A., W.G. Miller and P.J. Flory (1967). J. Mol. Biol. 23, 47.

34. Scheraga, H.A. and D. Poland (1970). "Theory of Helix Coil Transition in Biopolymers." Academic Press, New York and London.

35. Goel, N.S. and E. P. Montroll (1968). Biopolymers 6, 731.

36. Rein, R., N.S.Goel, N. Fukuda, M. Pollak, and P. Claverie (1969). Ann. N. Y. Acad. Sci. 153, 805.

37. Goel, N.S., N. Fukuda and R. Rein (1968). J. Theor. Biol. 18, 350.

38. Egan, J.T. (1974). Research Report, SUNYAB.

39. Kitaigorodskii, A.I. (1961). Tetrahedron 14, 230.

AKNOWLEDGEMENT.

I would like to acknowledge the support of N.A.S.A. Grant NGR 33-015-002.

I am much indebted to Dr. Shlomo Nir for reading the manuscript and for useful suggestions.

I would also like to thank Ruth Harvey, Randy Wheeler, Janice Fleischauer and Ruth Kuhfahl for their technical help and their ability to decipher illegible handwriting.

CONFORMATION OF CONSTITUENTS IN MOLECULAR CRYSTALS

Massimo Simonetta

Universita di Milano

Instituto di Chimica Fisica

Via Saldini, 50

30133 Milano (Italy)

LECTURE I - GENERAL FORMULATION OF THE PROBLEM

THE CHOICE OF POTENTIAL FUNCTIONS

Molecular Mechanics (also known as empirical force field calculations) is a terms which covers empirical methods for the calculation of molecular energies (or transition states, or molecular crystals) for given geometries. The earliest application of the method was due to Westheimer (1) who studied the influence of steric effects on the rate of racemisation of substitued biphenyls.

Several years later Hendrickson took advantage of the availability of electronic computers and performed more comprehensive calculations of this kind for cycloalkanes (with 5 to 7 carbon atoms) in different conformations (2) and for medium size rings (3). A different computational scheme was suggested by Wiberg (4), who used Cartesian instead of internal coordinates, which included an automatic search for the minimum energy geometry using the method of the steepest descent.

The method was further developed to more sophisticated force fields and in extending the areas of applications : a comprehensive survey has been published.

The method is an inductive method in which analytical functions are built to fit a large set of experimental data from a class of molecules : the same functions are then used to calculate observables for molecules in the same or similar class.

The general outline of the method is the following. For a molecule in a given geometry a strain-energy is calculated as the difference in energy between the molecule in the assumed geometry and an hypothetical strain-free molecule. The strain energy can be expressed as the sum of four components :

$$S.E. = \Delta E_c + \Delta E_b + \Delta E_t + \Delta E_{nb} \qquad (1)$$

where ΔE_c and ΔE_b are the energies for bond-length and bond-angle deformations, ΔE_t is the torsional energy and ΔE_{nb} takes care of non-bonded interactions. The various terms in the right hand side of eq. (1), can be calculated for any assumed geometry if a force field model is available. Such force fields have been derived many times, and by many workers in the field : they have been based on different experimental data, on different assumptions and are in general dependent on the aim of the calculations, that is on the class of molecules and the properties dealt with. In the following a number of force fields, chosen among the most widely and most successfully used, are described.

1. Allinger force field

This force field model has been initially constructed for the calculation of the structures and energies of saturated hydrocarbons at 25°C (6). The term ΔE_c is expressed

$$\Delta E_c = \sum_{\text{bonds}} \frac{1}{2} k_1 \, (1 - 1^o)^2 \qquad (2)$$

where the sum is over all bond lengths, 1° are natural bond lengths and k_1 are empirical stretching force constants.

The values of these parameters, as well as the other parameters needed to calibrate the force field, were chosen in such a way as to optimize the results for the geometries of a set of hydrocarbons. The optimized parameters are,

$$\begin{aligned} k_1(\text{C-C}) &= 4.4 \text{ mdyn/Å} \; ; \; 1^o(\text{C-C}) = 1.512 \text{ Å} \; ; \\ k_1(\text{C-H}) &= 4.6 \text{ mdyn/Å} \; ; \; 1^o(\text{C-H}) = 1.094 \text{ Å} \end{aligned} \qquad (3)$$

The bending function is

$$\Delta E_b = \sum_a \frac{k_\theta}{2} \, (\Delta\theta^2 + k_\theta \Delta\theta^3) \qquad (4)$$

where the sum is over all bond angles and $\Delta\theta = \theta - \theta_o$. The natural bond angles and bending force constants are as follows :

H–C(H)–C (C–C–H with two H):
$\theta_o(C\text{-}C\text{-}H) = 107.8°$
$\theta_o(H\text{-}C\text{-}H) = 111.2°$

H–C(H)–C with one H and two C:
$\theta_o(C\text{-}C\text{-}H) = 112.8°$
$\theta_o(H\text{-}C\text{-}H) = 108.5°$
$\theta_o(C\text{-}C\text{-}C) = 110.2°$

C–C(H)–C with three C:
$\theta_o(C\text{-}C\text{-}H) = 108.4°$
$\theta_o(C\text{-}C\text{-}C) = 110.6°$

C–C(C)–C with four C:
$$\theta_o(C\text{-}C\text{-}C) = 109.4° \qquad (5)$$

$k_\theta(C\text{-}C\text{-}C) = 0.40$ mdynÅ/rad^2 $\quad k_\theta(C\text{-}C\text{-}H) = 0.29$ mdynÅ/rad^2

$k_\theta(H\text{-}C\text{-}H) = 0.20$ mdynÅ/rad^2 $\quad k'_\theta = 0.007$

A stretch-bend interaction term has been included also

$$E_{str\text{-}bend} = k_{1\theta}(\Delta 1_{abc})\,(\Delta\theta_{abc}) \qquad (6)$$

where $k_{1\theta}$ is the stretch-bend force constant,

$\Delta 1_{abc} = |(1_{ab} - 1^{o}_{ab})| + |(1_{bc} - 1^{o}_{bc})|$, $\Delta\theta_{abc} = |\theta_{abc} - \theta^{o}_{abc}|$

the assumed values being

$$k_{1\theta}(C\text{-}C\text{-}C) = -\ 0.09 \text{ mdyn/rad} ;$$
$$k_{1\theta}(C\text{-}C\text{-}H) = -\ 0.04 \text{ mdyn/rad}; \qquad (7)$$
$$k_{1\theta}(H\text{-}C\text{-}H) = 0$$

The torsional term ΔE_t is a sum over all dihedral angles about all C-C bonds :

$$\Delta E_t = \frac{1}{2} \sum_{C\text{-}C} V_o\,(1 + \cos 3\phi) \qquad (8)$$

where V_o is 0.50 Kcal/mole when the dihedral angle is between $0 \pm 60°$ and zero outside that range. Exceptionally the value $V_o = 1.0$ kcal/mole is used for cyclobutane rings. Non-bonded interactions were calculated by means of the Hill equation :

$$\Delta E_{nb} = \sum_{i \neq j} E_{ij}$$

$$E_{ij} = -\ 2.25\ \varepsilon\left(\frac{r^*}{r}\right)^6 + 8.28\ \frac{10^5}{\varepsilon}\ e^{-\frac{r}{0.0736\ r^*}} \qquad (9)$$

where $r^*_C = 1.50$ Å $\varepsilon_C = 0.116$ kcal/mol; $r^*_H = 1.50$ Å $\varepsilon_H = 0.060$ kcal/mole.

The heats formation at 25°C were calculated by means of the formula

$$\Delta H^o_+ = S.E. + \Delta H_{conf} + \Delta H_{bond} + \Delta H_{1^o} + \Delta H_{3^o} + \Delta H_{4^o} \qquad (10)$$

where S.E. is given by eq. (1), ΔH_{conf} represents the contribution from configurations other than the most stable one coexisting at 25°, ΔH_{bond} is the sum of contributions from C-C and C-H bonds in the molecule ;ΔH_{1^o}, ΔH_{3^o}, ΔH_{4^o} are terms effectively correcting branching effects and refer repsectively to primary, tertiary and quaternary carbons.

The parameters were obtained by means of an optimization process in which experimental and calculated heats of formation for a set of 32 selected compounds were compared. The value of these parameters are shown below :

	C-C	C- H	H-C(H)(H)... C-C-H (with H above, H below)	C-C-C (with H above, C below)	C-C-C (with C above, C below)	
ΔH(kcal/mol)	2.97	-4.47	0.79	-0.74	-1.75	(11)

Allinger's force field was extended to allow the calculation of structures and energies of molecules containing the carbonyl group (7). More interesting to us was the extension to include compounds containing non conjugated olefinic linkages (8), and to hydrocarbons containing delocalized electronic systems (9). To include alkenes other parameters have to be added in the force field model. They are van der Waals constants for

$$C_{sp^2} \quad r^* = 1.85\ \text{Å} \quad \varepsilon = 0.033\ \text{kcal/mole}$$

Stretching constants

$$C_{sp^2} - C_{sp^2} \quad l_o\ 1.496\ \text{Å} \quad k_1 = 4.4\ \text{mdyn/Å}$$

$C_{sp^2} - C_{sp^2}$	1.332 Å	k_1 = 9.6 mdyn/Å	(12)
$C_{sp^2} - H$	1.090 Å	4.6 mdyn/å	

Bending constants

Angle	Type	θ_o(degrees)	k_θ(mdynÅ/rad^2)
$H-C_{sp^2}-H$		122.2	0.20
$C_{sp^3}-C_{sp^2}-H$		117.9	0.24
$C_{sp^2}-C_{sp^2}-H$	1°	118.9	0.24
$C_{sp^2}-C_{sp^2}-H$	2°	119.9	0.24
$C_{sp^3}-C_{sp^2}-C_{sp^3}$		116.6	0.40
$C_{sp^3}-C_{sp^2}-C_{sp^2}$	2°	122.2	0.40
$C_{sp^3}-C_{sp^2}-C_{sp^2}$	3°	121.7	0.40
			$k_\theta3$=0.007

Torsional constants (kcal/mole)

Angle	k_ω	k'_ω
$C-C_{sp^3}-C_{sp^3}-C$	0.50	
$C-C_{sp^3}-C_{sp^3}-H$	0.50	
$C-C_{sp^3}-C_{sp^2}-C_{sp^3}$	2.10	
$H-C_{sp^3}-C_{sp^2}-C_{sp^3}$	2.10	
$C-C_{sp^3}-C_{sp^2}-H$	1.90	
$H-C_{sp^3}-C_{sp^2}-H$	1.90	
$C-C_{sp^3}-C_{sp^2}=C_{sp^2}$	0.0	
$H-C_{sp^3}-C_{sp^2}=C_{sp^2}$	0.0	
$C_{sp^3}-C_{sp^2}=C_{sp^2}-C_{sp^3}$	13.55	-0.68
$H-C_{sp^2}=C_{sp^2}-C_{sp^3}$	13.55	
$H-C_{sp^2}=C_{sp^2}-H$	13.55	

The stretch-bend constants are,

$$k_{1\theta}(C\text{-}C_{sp}3\text{-}C) = -\ 0.09\ \text{mdyn/rad};\ k_{1\theta}(C\text{-}C_{sp}3\text{-}H) = -0.04\ \text{mdyn/rad}.$$

The potential function for the two-fold barriers is

$$E_{\omega} = \frac{1}{2}k_{\omega}\ (1-\cos 2\omega) + \frac{1}{2}k'_{\omega}\ (1+\cos\omega) \tag{13}$$

while bond enthalpy increments (kcal/mole^{-1}) for the calculation of heat of formation at 25°C are :

$C_{sp}3\text{-}C_{sp}3$	2.97	Iso($C_{sp}3$)	- 0.74	
$C_{sp}3\text{-}C_{sp}2$	-0.13	Iso($C_{sp}2$)	- 0.17	(14)
$C_{sp}2\text{=}C_{sp}2$	29.33	Neo	- 1.75	
C-H	-4.47	Methyl groups	0.74	

For compounds containing delocalized systems the problem of correlating bond orders, bond lengths and force contants arises. The prescription suggested by Allinger starts with the assumption of an approximate geometry. Then a π-system calculation is performed to find the bond orders and, hence, force constants. Standard methods are used then to find the minimum energy geometry. If this geometry differs substantially from the assumed one, the calculation is started again, until self consistency is obtained.

The π-electron calculation is made by the VESCF (variable electro negativity self consistent field) method (10), a modification of the Pariser and Parr method in which the effective nuclear charge for each π-orbital is varied according to the electron density around the corresponding nucleus. Accordingly valence state ionization potentials, one- and two-centre electron repulsion integrals and penetration integrals became functions of the electron density, through their dependence on the effective nuclear charge. The natural bond lengths and bond order are connected by the relation

$$l_o = 1.511 - 0.179\ P_{ij} \tag{15}$$

while bond orders and force constants are related according to

$$k_s = 5.0 + 4.6\ P_{ij} \tag{16}$$

where P_{ij} is the bond order between atoms i and j and 5.0×10^{-5} dyn/cm is the $C_{sp}2\text{-}C_{sp}2$ single bond force constant. The torsional function accross a double bond has been established by means of the following definitions :

$$E\ (\omega) = E\ (0°)\ \cos^2\omega \qquad E(0°) = 54.2\ \rho_{ij}^2\ \text{kcal/mole}$$

2. Schleyer's force field (5)

This force field applies only to alkanes. Parametrisation has been accomplished starting with values obtained from experimental data such as stretching and bending constants, torsional constants for rotation barriers and equilibrium hydrocarbon crystalline structures for non-bonded interaction constants. All the parameters are optimized to fit the experimental gas phase enthalpies of formation for a balanced set of different structural types. Natural bond lengths and angles were chosen in order to fit the experimental structures of propane and isobutane. The resulting molecular force field is shown below

Bond	k_r	r_o
C-H	4.6	1.100
C-C	4.4	1.520

$$\Delta E_c = 0.5\ k_r\ (r-r_o)^2 \qquad (18)$$

Angle	R	R'	R"	θ_θ	k_θ
C-C-R (R' R")	C	C	C	109.5	0.57
	C	C	H	110.1	0.57
	C	H	H	110.4	0.57
	H	C	C	109.2	0.40
	H	C	H	109.0	0.40
	H	H	H	109.5	0.40
H-C-R (R" R')	H	C	C	109.1	0.33
	H	C	H	109.2	0.33

$$\Delta E_b = 0.5 k_\theta \left|\Delta\theta^2 - k'_\theta\ \Delta\theta^3\right|$$

$$\Delta\theta \leqslant 25°$$

$$\Delta\theta = |\theta - \theta_o|$$

$$k'_\theta = 0.55$$

Dihedral angle	$k\phi \times 10^3$
H-C-C-H	4.85
H-C-C-C	4.85
C-C-C-C	3.11

$$\Delta E_t = 0.5\ k_\phi (1 + \cos 3\ \phi)$$

Non bonded distance	α	$\varepsilon \times 10^4$	r_m
H ... H	12.0	2.78	3.20
C ... H	12.0	2.08	3.35
C ... C	12.0	6.599	3.85

$$\Delta E_{nb} = \frac{\varepsilon}{1 - \frac{6}{\alpha}} \left| \frac{6}{\alpha} e^{\left(1 - \frac{r}{r_m}\right)} - \left(\frac{r}{r_m}\right)^6 \right|$$

The parameters needed to calculate heats of formation are given as group increments, the following values having been chosen :

CH_3	CH_2	CH	C	
- 10.82	-5.88	-2.82	-0.82	(19)

3. Bartell force field (11)

This force field was again designed for hydrocarbons, with the aim of calculating geometries. The force field can be defined as a modified Urey-Bradley force field (12), in which all non bonded interactions are included and are expressed in the form :

$$E_{nb} = \sum_i a_i \, r^{b_i} \, e^{-c_i r} \qquad (20)$$

r being the non bonded distance. The total potential energy for a molecule is then given by

$$E = \sum_1 \frac{1}{2} k_1 \, (1 - 1_o)^2 + \sum_n E^n_{nb} \qquad (21)$$

where the first sum is over all internal coordinates (bond lengths, bond angles, dihedral angles) and the second sum is over all non bonded intermolecular distances.

The following potential functions were chosen for non bonded interactions

$$\begin{aligned} E_{CC} &= 2.993 \; 10^5 \, r^{-12} - 3.252 \; 10^2 \, r^{-6} \\ E_{CH} &= 4.471 \; 10^4 \, e^{-2.04r} \, r^{-6} - 1.249 \; 10^2 \, r^{-6} \qquad (22) \\ E_{HH} &= 6.591 \; 10^3 \, e^{-4.08r} - 49.2 \, r^{-6} \end{aligned}$$

For force constants k_1 the following values were used :

$$k_{CC} \quad 2.2 \text{ mdyn/}\mathring{A} \qquad k_{CH} \quad 4.1$$

$$k_{CCC} = 0.687 \text{ mdyn}\mathring{A}/\text{rad}^2 \quad k_{CCH} = 0.320 \quad k_{HCH} = 0.520 \qquad (23)$$

$k_r = 0.0853$ mdyn $\mathring{A}/\text{rad}^2$ (rotation around C-C bond)

Reference values of bond angles and bond lengths were taken as

$$l^o_{C-C} = 1.24\ \mathring{A} \quad l^o_{C-H} = 1.05\underline{6} \qquad (24)$$

120° and 109.47° at trigonal and tetrahedral carbon respectively.

4. Lifson force field (13-14)

Again this isopotential function fitted to alkanes and cycloalkanes. It has been defined as a consistent force field (CFF) since it has been tailored to perform conformational and vibrational analysis. The parameters included in the potential functions have been chosen at first in such a way as to minimize differences between calculated and observed values of anthalpies, equilibrium conformations and vibrational frequencies for a set of alkane molecules. Later a study of the crystalline phase was included and other observables have been added in the parameter calibration procedure, namely : unit-cell parameters, heat of sublimation, lattice vibrations and thermal expansion of the unit cell. Molecular vibrations in the crystal as well as in isolated molecules have been considered. The optimized parameters to be used in the potential functions are shown below

	1/2 k_b(kcal/mol $\mathring{A}^2$)	b_o($\mathring{A}$)		1/2 K_θ	1/2 F_{ij}	r_{ij} ($\mathring{A}$)
C-H	286.4	1.099	H-C-H	39.5	1.7	1.8
C'-H	310.6	1.102	C-C-H	25.3	42.9	2.2
C-C	110.3	1.490	C-C-C	15.5	55.0	2.5
C'-C'	110.3	1.444	C-C'-H	18.3	51.7	2.2
C-C'	110.3	1.467				

(C' denotes a methyl carbon) $\theta^o = 109.47°$

For C-C-C and C-C-C' angles a linear term is added, with K_θ=-6.2 kcal/mol rad (25)

K_ϕ(C-C-C-C) = 1.161 kcal/mol $K_{\theta\theta'}$(C-C-C-C)=2.3 $K_{\theta\theta'}$(H-C-C-H)=9.5

$K_{\theta\theta'}$(H-C-C-C) is the geometric mean of C-C-C-C and H-C-C-H values. The torsional potential of the C-CH_3 group is equally distributed over the nine pairs across the C-C bond, giving each a value of $\frac{1.161}{9}$.

$$\text{H...H} \quad 1/2\ r^* = 1.774\ \text{Å} \quad \varepsilon^{1/2} = 0.0508 \quad e_{eff} = 0.11$$

$$\text{C...C} \quad 1/2\ r^* = 1.808 \quad \varepsilon^{1/2} = 0.4297$$

The potential is

$$V = 1/2 \sum_{lst} \sum_{l's't} V_{nb}\ (r_{sts't'}, ll') + \sum_{l\ s} V_{intra,\ s} \qquad (26)$$

t labels atoms, s labels molecules, l labels cells. The double sum appears only in crystal energy calculations. Terms with l=l' are included provided s≠s'. The second term represents the molecular energy. The non-bonded energy is given by

$$V_{nb} = 2\varepsilon \left[\left(\frac{r^*}{r}\right)^9 - 1.5 \left(\frac{r^*}{r}\right)^6 \right] + \frac{e_{eff}\ e'_{eff}}{r} \qquad (27)$$

e is the residual charge in unit of electronic charge.
The intramolecular energy is :

$$V_{intra} = \frac{1}{2} \sum_i k_b(b_i-b_o)^2 + \frac{1}{2} \sum_g K_\theta(\theta_g-\theta_o)^2 + \sum_m K'_\theta(\theta_m-\theta_o)$$
$$+ \sum_d \frac{1}{2} F_{ij}(r_{ij}-r^o_{ij})^2 + \sum_n \frac{1}{2} K_\phi(1+\cos 3\phi) - \sum_n K_{\theta\theta'}(\theta-\theta_o)$$
$$(\theta'-\theta_o) \cos\phi + \sum_{nb} V_{nb} \qquad (28)$$

The first sum is over all bonds, the second over bond angles, the third over bond angles involving three carbon atoms and the fourth is over non bonded pairs of atoms that are connected to a common atom; n sums over carbon-carbon bonds and V_{nb} is summed over all pairs of atoms that are not bonded to each other non bonded to a common atom.

5. Williams force field (15-17)

These are non bonded potentials and were derived from the crystal structures and thermodynamics of hydrocarbons.
The potentials have the form

$$E = Ad^{-6} + B\ e^{-Cd} \qquad (29)$$

where A, B, C are parameters and d is a non-bonded interatomic distance. The values for the C parameters were assumed :

$$C_{C\ldots C} = 3.60\ \text{Å}^{-1} \qquad C_{C\ldots H} = 3.67\ \text{Å}^{-1} \qquad C_{H\ldots H} = 3.74\ \text{Å}^{-1} \tag{30}$$

The six parameters A_{CC} A_{CH} A_{HH} B_{CC} B_{CH} B_{HH} were first calibrated by fitting 72 experimental observations on nine aromatic hydrocarbons. The following values (kcal mol^{-1}) represent the best compromise between goodness of fit and amount of computational work:

$$\begin{array}{llll} A_{CC} - 535 & B_{CC}\ 74460 & A_{CH} - 139 & B_{CH}\ 9411 \\ A_{HH} - 36 & B_{HH}\ 4000 & & \end{array} \tag{31}$$

The experimental data used were heat of sublimation, lattice constants, elastic moduli and geometrical structures. The lattice sums were carried out with the following limits : 6.0, 5.5, Å for C...C, C...H, H...H interactions, respectively. It was found that this summation limits lead to about 80 % of the total lattice energy.

Later aliphatic hydrocarbons were included and observational equations for aromatic and non aromatic structures were combined. The parameters values shifted as shown below :

$$\begin{array}{llllll} A_{CC} & B_{CC} & A_{CH} & B_{CH} & A_{HH} & B_{HH}\ \text{(kcal/mol)} \\ -568 & 83630 & -125 & 8766 & -27.3 & 2654 \end{array} \tag{32}$$

Coulombic intermolecular interactions in crystalline hydrocarbons have been also considered. In the expression for the lattice energy the following summation has been included :

$$S = \frac{1}{2} W \Sigma \frac{q_j q_k}{r_j k} \tag{33}$$

where r_{ij} are distances between atoms on different molecules, q_j and q_k are (point) charges on atoms j and k and W is a calibration factor. It was found that with the assumption

$A_{CH} = \sqrt{A_{CC} A_{HH}}$, $B_{CH} = \sqrt{B_{CC} B_{HH}}$, the best fit for 118 observational equations involving structural parameters for 18 compounds plus two equations in the lattice energy for benzene and n-hexane leads to the following parameters values :

$$A_{CC} = 1880 \quad A_{HH} = 168 \quad B_{CC} = 299000 \quad B_{HH} = 12000\ \text{(kJ/mol)}$$

$$\Delta e = 0.358 \tag{34}$$

where Δe is the point charge on each hydrogen atom (in electron charge unit). The lattice sums in this work were evaluated using a convergence acceleration technique (18).

The Lifson potential has been complemented with a quantum mechanical part in order to obtain potential surfaces for conjugated molecules in ground and excited states (19). Molecular mechanics methods have been detailed for studying polymer (20, 21) and polypeptide conformations (22, 23).

It is worthwhile mentioning that in molecular mechanics calculations minimisation procedures play a fundamental role. Many techniques are available to this aim, and some of them have been critically reviewed (24). Here only two of such methods will be mentioned. The first one is the steepest descent method, introduced by Wiberg (4). The starting point is an assumed geometry; a small increment δ is then chosen and each coordinate changed positively or negatively by δ. The energy change is proportional, through δ, to the potential derivative of energy with respect to this variable. All variables are tested this way, one at a time, all being changed by an amount proportional to its partial derivative and the energy at this new geometry is computed. The process is repeated until the energy change in a cycle is less than a predetermined amount. In the second method (25) the potential energy is expanded as a function of internal coordinates, in a power series up to quadratic terms about a set of trial coordinates. The potential energy is transformed then to Cartesian coordinates and differentiated with respect to these new coordinates. The necessary conditions for a minimum lead to a set of linear algebraic equations whose solution gives the displacements of the coordinates. These are used to modify the initial coordinates and a new geometry is obtained; the whole procedure is repeated until all coordinate displacements are less than a predetermined limit.

BIBLIOGRAPHY

1. F.H. Westheimer in "Steric effects in organic chemistry", M.S. Newman Ed., J. Wiley - New York (1956) ch. 12.

2. J.B. Hendrickson, J. Am. Chem. Soc. 83, 4537 (1961).

3. J.B. Hendrickson, ibid. 89, 7036, 7043, 7047 (1967).

4. K.B. Wiberg, ibid. 87, 1070 (1965).

5. E.M. Engler, J.D. Andose, and P.v.R. Schleyer, ibid. 95, 8005

(1973).

6. N.L. Allinger, M.T. Tribble, M.A. Miller, and D.H. Werts, ibid. 93, 1637 (1971).

7. N.L. Allinger, M.T. Tribble and M.A. Miller, Tetrahedron 28, 1173 (1972).

8. N.L. Allinger and J.T. Sprague, J. Am. Chem. Soc. 94, 5734 (1972)

9. N.L. Allinger and J.T. Sprague, ibid. 95, 3893 (1973).

10. R.D. Brown and N.L. Hefferman, Australian J. Chem. 12, 319 (1959).

11. E.J. Jacob, H.B. Thompson and L.S. Bartell. J. Chem. Phys. 47, 3736 (1967).

12. M. Simonetta and P. Beltrame in "The chemistry of aryl halides" S. Patai Ed., Interscience Publ., London (1972) ch. 1 p. 15.

13. S. Lifson, and A. Warshel. J. chem. Phys., 49, 5116 (1968).

14. A. Warshel and S. Lifson ibid. 53, 582 (1970).

15. D.E. Williams, J. Chem. Phys. 45, 3770 (1966).

16. D.E. Williams, ibid. 47, 4680 (1967).

17. D.E. Williams, Acta Cryst. A 30, 71 (1974).

18. D.E. Williams, Acta Cryst. A 27, 452 (1971).

19. A. Warshel and M. Karplus, J. Am. Chem. Soc. 94, 5612 (1972).

20. P.J. Flory, J.E. Mark and A. Abe, ibid. 88, 639 (1966).

21. A. Abe, R.L. Jermigen and P.J. Flory, ibid. 88, 631 (1966).

22. H.A. Sheraga, Adv. Phys. Org. Chem. 6, 103 (1968).

23. H.A. Sheraga, Chem. Rev. 71, 195 (1971).

24. J.E. Williams, P.J. Stang and P.v.R. Schleyer, Ann. Rev. Phys. Chem. 19, 531 (1968).

25. R.H. Boyd, J. chem. Phys. 49, 2574 (1968).

LECTURE II - APPLICATION TO ISOLATED MOLECULES, INTERMEDIATES AND TRANSITION STATES

The potentials described in Lecture I as well as a number of potentials derived from or similar to them have been used to attack a variety of problems in chemistry. The number of applications is such that a complete review would be outside the scope of this lecture, an extensive list being given in ref. 1. Here we shall mention a number of representative examples of which the main ones relate to conformational analysis, that is the calculation of structures and energies for a range of molecules. Allinger has extensively applied his force field to hydrocarbons and carbonyl compounds (2-5). He mainly concentrated his effort on the calculation of (i) rotational barriers (see Table 1) :

Table I : Energy as a function of torsional angle θ° in n-butane

θ°	E (kcal/mol)
0°	4.55
60°	0.73
120°	2.94
180°	0.00

θ = 0 for eclipsing methyl groups,
(ii) the strain energy of carbocyclic rings (Table II) :

Table II : Strain energy of carbocyclic rings relative to cyclohexane (kcal/CH_2 group)

N (number of CH_2 groups in the ring)	S.E. calc.	S.E. exp.
4	6.5	6.5
5	1.4	1.2
6	0.0	0.0
7	0.9	0.9
8	1.4	1.2
9	1.6	1.4
10	1.4	1.3
12	0.6	0.7

and (iii) heats of formations and strain energy for alkanes of the following series (chain, ring, forced ring, polycyclic) and for straight and branched chain ketones, monocyclic ketones, polyketones, aldehydes.

The heats of formations and strain energies, as well as the heats of hydrogenation for cyclic olefins were calculated and showed excellent agreement with data from calorimetric experiments. The method was extended to hydrocarbons containing delocalized systems and, for example the barrier to rotation in butadiene (fig. 1) estimated

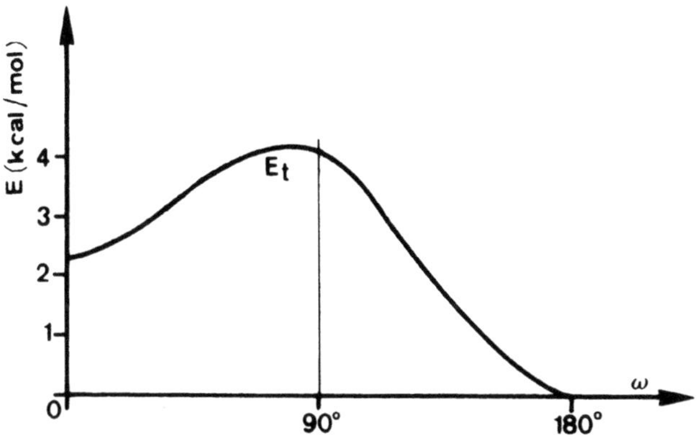

Fig. 1 The rotational barrier in butadiene

The equilibrium geometries of a series of planar aromatic hydrocarbons (naphtalene, anthracene, phenantrene, triphenylene, pyrene, perylene) were calculated and are, for the most part, correct within the experimental error. Other interesting results are contained in the calculated geometries of bicyclo |5.5.0| dodecahexane, o-di-t-butylbenzene, 1,3 cyclopentadiene, 1,3 cyclohexadiene : it was found that in bicyclo |9.9.1| undecapentaene the outer ring bond lengths are constant (in agreement with experiment (6)) while the isomer bicyclo |5.3.1| undecapentaene is predicted to have complete bond alternation. The geometries of |12|, |14| , |16| and |18| annulenes were also calculated : the skeleton of the minimum energy is shown below

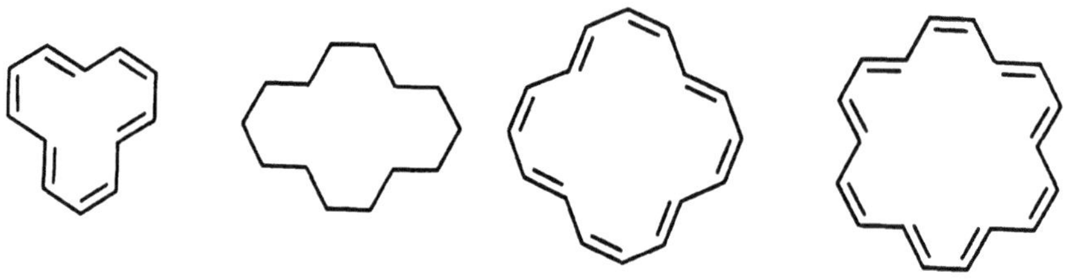

| 12 | annulene non planar bond alternations
| 14 | annulene ring distorted aromatic
| 16 | annulene non planar bond alternation
| 18 | annulene non planar bond alternation

Particular attention to the evaluation of heats of formation of alkanes has been given by Schleyer (1), who claims high accuracy in the determination of geometries; bond lengths are generally reproduced within 0.01 Å, bon angles within 1-2°. For a balanced set (39 compounds) of acyclic, cyclic and polycyclic alkanes, used in the parametrisation of the force field, the standard deviation between calculated and experimental heats of formation is 0.83 kcal/mole. It is claimed that molecular mechanics at this level of sophistication is competitive with experimental methods for the determination of geometries and enthalpies of molecules. A useful interpretation of the sources of strain in adamantane has been offered (7).

A specially interesting test-ground is the molecule tri-t-butyl-methane. The structure of this molecule has been investigated by a synchronous attack using electron diffraction vibrational spectroscopy and molecular mechanics (8). Both the Lifson (9) and Bartell (10) force fields have been tried, the former leading to the most useful results. In Table III the observed and calculated average structure parameters (with assumed local C_{3v} symmetry of the carbon skeleton of the t-butyl groups (II)) are shown. The experimental geometry is between the calculated ones, but closer to that obtained from Bartell force field.

The infrared C-H stretching frequencies in this molecule are distributed over a wider range ($\sim$170 cm^{-1}) than the corresponding frequencies in isobutane ($\sim$70 cm^{-1}) suggesting that the force constants differ significantly from site to site, in a way related to the different C-H bond lengths. An empirical relationship between bond lengths and vibrational frequencies was established and bond lengths calculated from the experimental spectrum were compared with the bond lengths obtained through the Bartell force field (Table IV). The results from electron diffraction, vibrational spectroscopy and molecular mechanics are in good agreement, indicating the site of strain as the small H_t-C_t-C_q bond angle and as the long C_t - C_q distance.

Table III[a] Structural data for tri-t-butylmethane

	Experiment	Lifson	Bartell[b]
$\bar{r}(C-H)$	1.111	1.100	1.112
$\bar{r}(C_t-C_q)$	1.611	1.569	1.595
$\bar{r}(C_q-C_m)$	1.548	1.528	1.552
$\bar{\alpha}(H_t-C_t-C_q)$	101.6	100.9	102.4
$\bar{\delta}(C_q-C_t-C_{q'})$	116.6	116.5	115.5
$\bar{\beta}(C_t-C_q-C_m)$	113.0	114.7	112.5
$\bar{\varepsilon}(C_m-C_q-C_m)$	105.8	103.8	106.2
$\bar{\gamma}(C_q-C_m-H_m)$	114.2	111.4	112.4
$\Delta\bar{\tau}(H_t-C_t-C_q-C_m)$	10.8	16.4	15.8
$\lvert\Delta\bar{\tau}\rvert$ (methyl)[c]	18.0	7.4	13.5

a) t = tertiary, q = quaternary, m = methyl ; τ = torsion angles; α, β, γ, δ bond angles; distances in Å, angles in degrees

b) calculated C-C distances decreased by 0.006 Å

c) average of absolute magnitude.

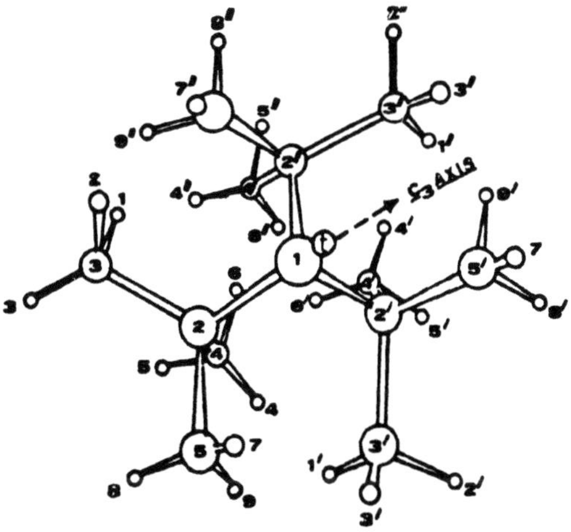

Figure 2.

Table IV: C-H bond lengths (Å) from the observed frequency and calculated from Bartell force field

CH[a]	$r_{Bartell}$	r_v	CH	r_B	r_v(Å)
1	1.104	1.105	7	1.112	
9	1.105	1.107	8	1.116	
4	1.106	1.09	5	1.118	1.122
6	1.108	-	2	1.118	
3	1.111	1.112	t	1.137	1.132

a) identifying number of hydrogen (fig. 2)

Molecular mechanics was used also to determine the geometry of the following set of molecules : mono-, di-, tri-, tetra-phenyl ethylene and all their possible methyl derivatives, a total of 20 hydrocarbons[12] (see Table V).

Most of the internal geometrical parameters were kept constant, the energy being minimized with respect to twist angles of the phenyl groups,the C=C ethylenic bond length and the C-C length in the bonds between the phenyl groups and the ethylenic fragment.

The results are reasonable : rotation angles of the phenyl rings were found in the range 26-62°; double bond lengths in the range 1.33-1.35 Å, quasi-single bonds in the range 1.47-1.50 Å. In the styrene derivatives it was found that the phenyl group shows a stronger interaction with the methyl group in the cis β position, a lesser one with an α methyl group and practically no interaction with the trans β methyl group. The steric interaction between two phenyl groups separated by a double bond is smaller than that between a phenyl and a methyl group, while such interactions are almost of the same magnitude when the two groups are bonded to the same carbon atom. Spectroscopic and thermochemical data, when available, were well rationalized in terms of the geometries.

A Bartell potential was used in the investigation of a set of five phenyl- and naphthyl-naphthalenes (13). It was found that 2-phenyl-naphthalene and 2- (2-naphthyl) naphthalene are almost planar, while 1-phenyl-naphthalene, 1- (1-naphthyl) naphthalene and 1- (2-naphthyl) naphthalene have a flat region of low energy between 60° and 120°. The energy and oscillator strengths of electronic transitions of these molecules in different conformations were calculated by the "molecules in molecules" method (14) and compared with the experimental spectra.

Table V(a) Equilibrium geometry of methyl, phenyl ethylenes :

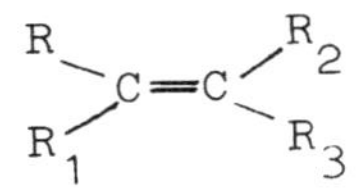

Compound	methyl groups	phenyl groups	C=C	Bond lengths (Å) C-R	C-R_1	C-R_2	C-R_3	θ(°)
styrene	-	R	1.336	1.474				26
α-methyl-styrene	R_1	R	1.339	1.479				32
cis-β-methylstyrene	R_2	R	1.343	1.494				55
trans-β-methylstyrene	R_3	R	1.340	1.474				26
cis-2-phenyl-2-butene	R_1R_3	R	1.341	1.478				33
trans-2-phenyl-2-butene	R_1R_2	R	1.327	1.500				59
ββ-dimethylstyrene	R_2R_3	R	1.340	1.493				55
αββ-trimethylstyrene	$R_1R_2R_3$	R	1.340	1.501				60

Table V(b)

trans-stilbene	–	RR_3	1.346	1.466			1.466	17
cis-stilbene	–	RR_2	1.340	1.480		1.480		43
trans-α-methylstilbene	R_1	R	1.342	1.490				40
		R_3					1.488	50
cis-α-methylstilbene	R_1	R	1.342	1.490				50
		R_2				1.479		40
trans-αα'-dimethyl-stilbene	R_1R_2	RR_3	1.326	1.500			1.500	62
cis-αα'-dimethyl-stilbene	R_1R_3	RR_2	1.326	1.484		1.484		46

Table V(c)

1,1-diphenylethylene	-	RR_1 R	1.341	1.481 1.502	1.481			33 60
2-methyl-1,1-diphenyl ethylene	R_2	R_1	1.341		1.476			30
2,2-dimethyl-1,1-di-phenylethylene	R_2R_3	RR_1 R R_1 R_2	1.341	1.501 1.484	1.501			60 45 30 40
triphenylethylene	-		1.346		1.473	1.477		
methyl-triphenyl-ethylene	R_3	R R_1 R_2	1.344	1.470	1.505	1.499		30 60 60
tetraphenylethy-lene	-	$R_1R_2R_3$	1.330	1.489	1.489	1.489	1.489	42

The agreement between calculated and measured spectra obtains only for the conformations of minimal energy. The conformations and spectra of some flexible molecules, such as

$$CH_2=CH-(CH_2)_n-CH=CH_2 \quad \text{and} \quad CH_2=CH-CH=CH(CH_2)_n-CH=CH-CH=CH_2$$

with n=1-6 were also investigated (15). In conformation calculations of this type only torsional and non-bonded effects need to be considered. These molecules show different conformations of similar energy, separated by low energy barriers, and the distribution of the molecules among the conformations was calculated by means of the usual Boltzmann expression. Again calculated and experimental U.V. spectra were compared: the general features of the spectra are in good agreement, and in particular the fact that one CH_2 group has a poor isolating effect between two chromophores is reproduced by the calculations.

Conformational analysis has been extended to treat the interesting problem of bridgehead carbonium ions in that force field applicable to carbonium ions has been produced (16). Sovolysis rates of bridgehead substituted polycyclic bromides are determined by the difference in free energy between the solvated transition state and the solvated molecule. Assuming solvation and entropy effects to be the same in the transition state and reactants, solvolysis rates were correlated with differences in energy and hence with differences in strain energy between the isolated systems. Also, the parent hydrocarbon was assumed as a model for the reactant and the bridgehead carbonium ion as a model for the transition state. Then molecular mechanics can be used to calculate rate constants. Fig. 3 shows the correlation between the calculated differences in strain energies and the experimental relative rate constants at 25°C in 80% ethanol for the compounds :

$(CH_3)_3CBr$

I

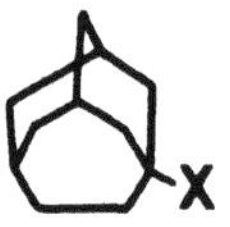

II

III

IV

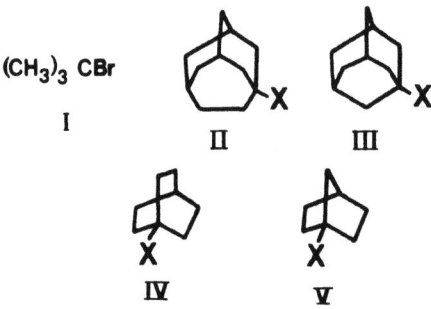

V

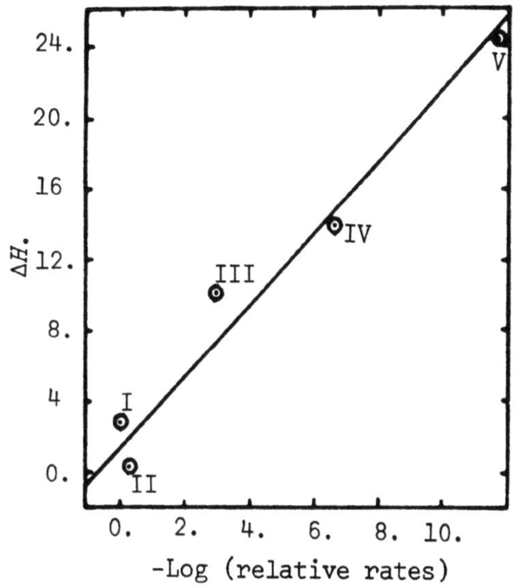

Fig. 3

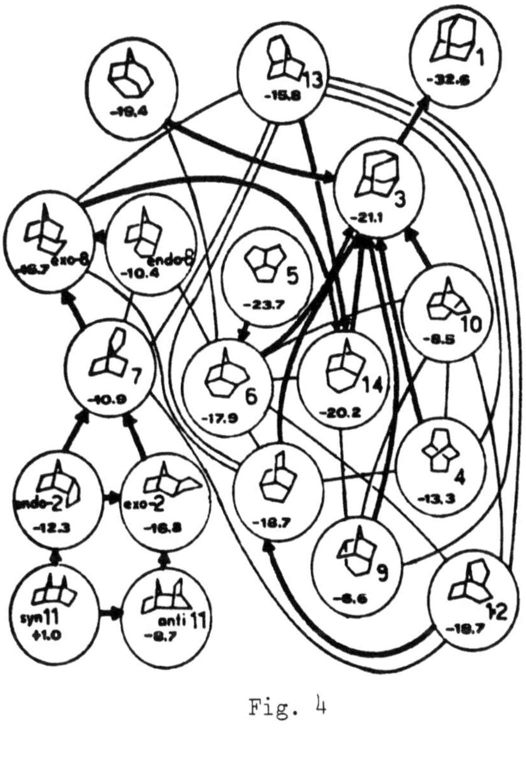

Fig. 4

Later the calculation of bridgehead reactivities has been extended to a wider variety of systems, and radical as well as carbonium ion reactivities have been correlated (17). The mechanism of adamantane rearrangements has also been investigated with the aid of molecular mechanics (18). Tricyclodecanes can rearrange to adamantane following many different paths. From the graph of fig. 4 it has been deduced that at least 2897 pathways for the conversion of tetrahydrobicyclopentadiene (2) to adamantane (1) are possible. Heats of formations of the tricyclodecanes were calculated using the Schleyer force field: the results are reported in fig. 4, indicating the relative stabilities of the various isomers. Starting from 2 the first step (to compound 7) is endothermic, but all subsequent steps can be exothermic. Since the isomerisation is a Lewis acid catalyzed reaction the stability of carbonium ions was also investigated; the results for ions formed from isomers 7, exo-8, 14 and 3 are shown in fig. 5. From the data of fig. 5 it may be inferred that the most favourable pathway from endo-2 to 1 should be 2 → 7 → exo-8 → 14 → 3 → 1.

This route was tested experimentally. Starting from exo-8, the rearrangement to 1 in the presence of $AlBr_3$ 1:1 in CS_2 is fast, even at low temperature, faster than the rearrangement starting from 2. The presence of two intermediates was revealed, and they were identified as compounds 14 and 3. The results of the kinetic experiments are visualized in fig. 6. From the synergistic application of molecular mechanics and kinetic experiments the pathways shown in dark arrows in fig. 4 can be predicted to be the most favourable routes leading from tricyclodecane to adamantane. Another example of reaction paths investigated by means of molecular mechanics is the Cope rearrangement, a thermal reaction that obtains in diallylic systems :

$$A=B-C-C'-B'=A \longrightarrow C=B-A-A'-B'=C'$$

Three molecules have been considered, namely exa-1,5-diene, cis-1,2-divinylcyclopropane and cis-1,2-divinylcyclobutane (19). For the first molecule two reaction paths can be envisaged, through a chair-like or boat-like activated complex.

The calculation predicts the activation energy in the first case to be lower by about 5.0 kcal/mole, in nice agreement with experimental findings (20). For the other two molecules, the opening of the small ring during reaction to cis-cis-cyclohepta-1, 4-diene or cis-cis-cycloocta-1,5-diene favours the reaction. The calculated activation energy for the cyclobutane derivative is about 18.5 kcal/mole, to be compared with an experimental value of 23.8 kcal/mole (21). The situation concerning the cyclopropane derivative was less clear: an energy barrier to the rearrangement of 17.2 kcal/mole was considered, which could not be easily recon-

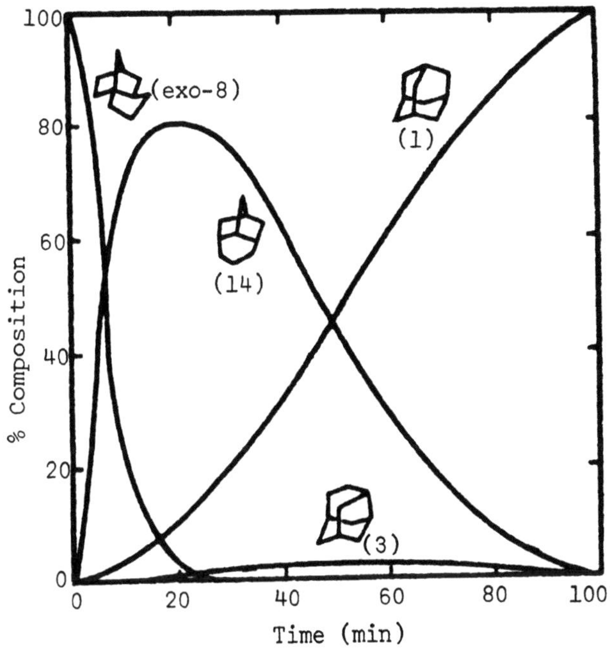

100
80
60
40
20
0
% Composition
(exo-8)
(1)
(14)
(3)
0
20
40
60
80
100
Time (min)

Fig. 6

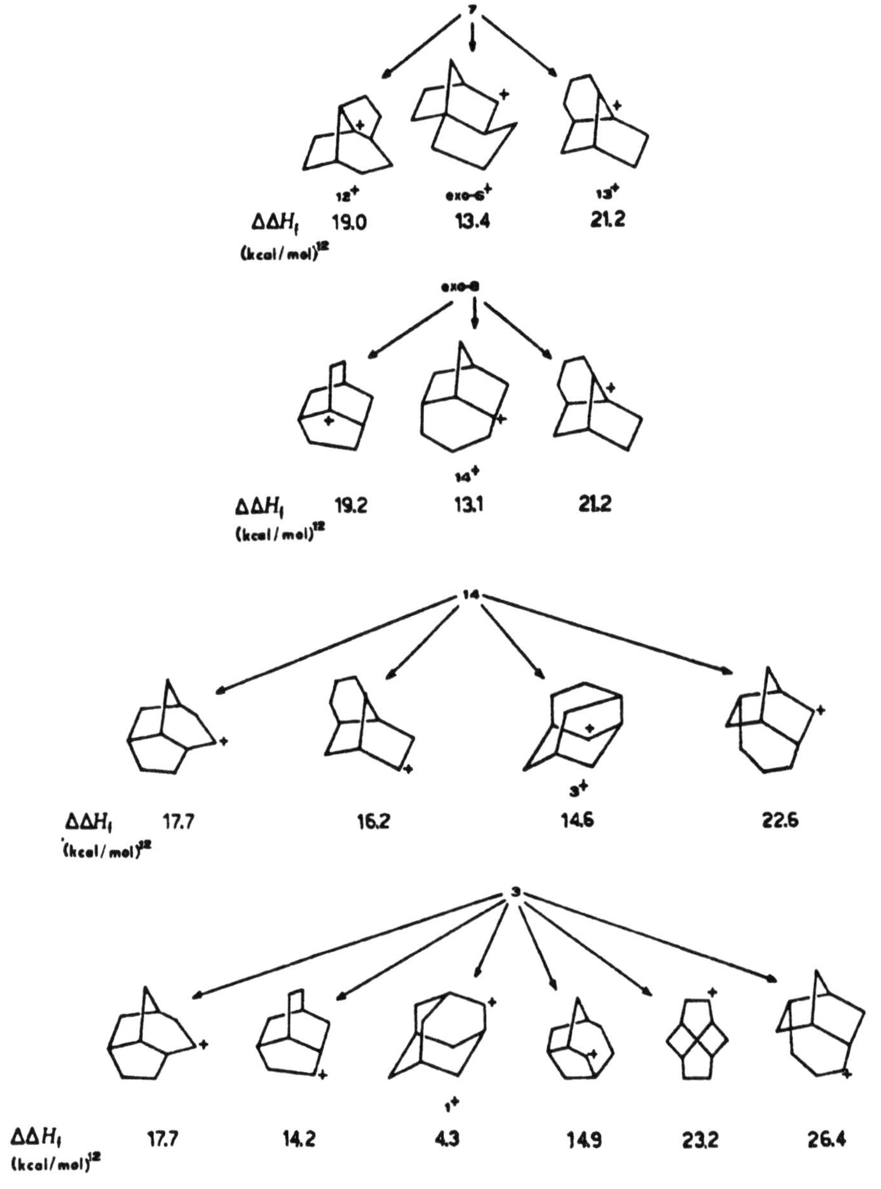
7
12+
exo-6+
13+
ΔΔH_f 19.0 13.4 21.2
(kcal/mol)12
exo-6
14+
ΔΔH_f 19.2 13.1 21.2
(kcal/mol)12
14
3+
ΔΔH_f 17.7 16.2 14.6 22.6
(kcal/mol)12
3
1+
ΔΔH_f 17.7 14.2 4.3 14.9 23.2 26.4
(kcal/mol)12

Fig. 5

ciled with the widely accepted conclusion that cis-divynylcyclopropane rearranges so fast that it cannot be isolated, even at - 40°C. However a recent experiment for isolating the compound was reported (22) and the measured activation energy for its thermolysis is 19.4 kcal/mole; calculated (-5.5 e.u.) and experimental (-5.3 e.u.) activation entropies are in excellent agreement. The activation energies for the cis-trans isomerisation reaction of 2-butene, 2-pentene, β-methylstyrene and stilbene, as well as the enthalpy differences and the geometry of the most stable conformation of the isomers have been calculated by the same method, satisfactory agreement with experimental data being obtained (23).

BIBLIOGRAPHY

1. E.M. Engler, J.D. Andose and P.v.R. Schleyer - J. Am. Chem. Soc. 95, 8005 (1973).
2. N.L.Allinger, M.T. Tribble, M.A. Miller and D.H. Wurtz - ibid. 93, 1637 (1971).
3. N.L.Allinger, M.T. Tribble and M.A. Miller - Tetrahedron 28, 1173 (1972).
4. N.L.Allinger and J.T. Sprague - J. Am. Chem. Soc. 94, 5734 (1972).
5. N.L. Allinger and J.T. Sprague - ibid. 95, 3893 (1973).
6. M. Dobler and J. Dunitz - Helv. Chim. Acta 48, 1429 (1965).
7. P.v.R. Schleyer, J.E. Williams and R.K. Blanchard - J. Am. Chem. Soc. 92, 2377 (1970).
8. L.S. Bartell and H.B. Burgi - ibid. 94, 5239 (1972).
9. E.J. Jacob, H.B. Thompson and L.S. Bartell - J. Chem. Phys. 47, 3736 (1967).
10. S. Lifson and A. Warshel - ibid. 49, 5116 (1968).
11. H.B. Burgi and L.S. Bartell - J. Am. Chem. Soc. 94, 5236 (1972).
12. G. Favini and M. Simonetta - Theor. Chim. Acta 1, 294 (1963).
13. A. Gamba, E. Rusconi and M. Simonetta - Tetrahedron 26, 871, (1970).
14. G. Favini, A. Gamba and M. Simonetta - Theor. Chim. Acta 13, 175 (1969).

15. I.R. Bellobono, A. Gamba and M. Simonetta - Rend. Acc. Naz. Lincei VIII 49, 102 (1970).

16. G. J. Gleicher and P.v.R. Schleyer - J. Am. Chem. Soc. 79, 582 (1967).

17. R.C. Bingham and P.v.R. Schleyer - ibid. 93, 3189 (1971).

18. E.M. Engler, M. Farcasin, A. Sevin, J.M. Cense and P.v.R. Schleyer - ibid. 95, 5769 (1973).

19. M. Simonetta, G. Favini, C. Mariani and P. Gramaccioni - ibid. 90, 1280 (1968).

20. W. von Doering and W. Roth - Tetrahedron 18, 67 (1962).

21. G.S. Hammond and C.D. de Boer - J. Am. Chem. Soc. 86, 899 (1964).

22. J.M. Brown, B.T. Golding and J.J. Stofko jr. - J. Chem. Soc. Chem. Comm. 319 (1973).

23. T. Beringhelli, A. Gavezzotti and M. Simonetta - J. Mol. Structure 12, 333 (1972).

LECTURE III - APPLICATIONS TO THE CRYSTALLINE STATE

We have seen that molecular mechanics can be used to find the geometry of organic molecules of medium size, that is molecules which pose the problems in the application of spectroscopic and electron diffraction techniques for the structure determination. Of course, their crystal structures can be studied by X-ray diffraction but the conformation in the crystal phase may be different from that of the gas phase geometry, due to the action of crystal field packing forces. However, we have seen that intermolecular interactions can be taken care of in force field calculations so that the geometry of the molecule can be obtained both in the gas and in the crystalline phase. This last result can be taken as a probe for the reliability of the geometry calculated for the isolated molecule. Let us take as an example biphenyl. The twist angle between the phenyl ring is about 40° in the gas phase (1) and 0° in the crystalline phase (2). Calculations for the isolated molecule included as variable parameters the inter-ring distance r, the twist angle θ between the rings and the CCH angle $\hat{X}$ for hydrogen atoms in ortho position (3). The minimum of energy is found for r=1.51 Å (exp. 1.49), θ =35° (exp. 42°), $\hat{X}$=121° (exp. 120° (assumed)). The calculation for the crystal was made by locating the center of mass of molecules in the experimental position, so that only 6 variables vere needed : the internal variables r, θ, $\hat{X}$ and three Euler angles $\phi_1\phi_2\phi_2$. At the minimum the molecule was planar, with r=1.518 (exp. 1.507) and the orientation was such to give a discrepancy index R=0.15 to be compared to the values 0.17 for the h01 zone and 0.15 for the 0kl zone obtained by the experimentalist at this stage of refinement.

The crystal structure determination of p-p'-bitolyl (4) has shown that the monoclinic crystals contain two molecules in the asymmetric unit with the long molecular axes parallel to the b axis of the cell and with twist angles of 36 and 40°. Conformational energy calculations were carried out with the center of mass and axes in the experimental position and assuming regular carbon and hydrogen hexagons and four degrees of freedom to the system viz the angles of rotation ϕ_i (i=1,2,3,4) around the molecular axes of the four phenyl rings.

The minimum energy was found for ϕ_1=20°, ϕ_2=55°, ϕ_3= -18°, ϕ_4=58° (exp. values ϕ_1=20°, ϕ_2=55°, ϕ_3=-18°, ϕ_4=-58°). It is gratifying that the orientation of the rings correspond within 1°. With the <u>para</u> substituents the packing is such that the two rings are twisted by about the same angle as in gaseous biphenyl, while biphenyl itself is planar in the solid state due to packing forces.

p-Nitrobiphenyl is an example of a biphenyl derivative with only one substituent. Crystals of this compound are orthorombic with eight molecules in the unit cell (5). In the isolated molecule calculation the variables were the inter ring distance r, the twist angle θ° between the phenyl rings and the rotation angle ψ° of the nitro group with respect to the phenyl ring. The energy minimum was found at $r=1.51$ Å, $\theta=35^\circ$, $\psi=0^\circ$. In the crystal packing calculations the reference molecule was given two degrees of freedom: namely the dihedral angles ϕ' and ϕ'' of the first and second phenyl ring with respect to the same ring in the experimental conformation. The minimum obtains for $\phi'=3^\circ$, $\phi''=2^\circ$. The dihedral angle between the phenyl groups is 32° (exp. 33°) and the molecule as a whole is only slightly rotated by 2.5° with respect to the experimental conformation. For this molecule some new atom-atom potentials are needed. These were taken from Kitaigorodsky's work (6), except for the N...N, N...C, N...H interactions, where the potentials proposed by Mirskaja and Nantchitel, as quoted by Ahmed and Kitaigorodsky (7), were used.

The crystal structure determination of 2-bromo-1, 1-di-p-tolylethylene (8) has shown that phenyl ring cis to bromine atom is turned 68° out of the ethylene plane, while the second phenyl is 24° out of the same plane. In conformational calculations for the isolated molecule the minimum energy is obtained at angles of 55° and 25° respectively. But if the optimum conformation is searched for the molecule in the crystal the calculated twist angles become 69° and 25°.

Another example is 1,1-di-(p-nitrophenyl)ethylene, a monoclinic crystal with four molecules in the unit cell (9). The calculated geometry for the isolated molecule shows the following features : $\theta_1=\theta_2=35^\circ$, $\phi_1=\phi_2=0^\circ$, $r_{C=C}=1.336$, $r_{C_{ethylene}-C_{1st\ phenyl}} = r_{C_{ethylene}-C_{2nd\ phenyl}} = 1.508$ Å, where θ's are the twist angles between phenyl rings and ethylenic fragment, and ϕ's are twist angles for the nitrogroups, with respect to the rings. The C_2 axis along the double bond was not imposed, but was an outcome from the calculations.

For the model crystal the results are: $\theta_1=36^\circ$, $\theta_2=35^\circ$, $\phi_1=0^\circ$, $\phi_2=5^\circ$, $r_{C=C}=1.336$, $r_{C-C}=1.508$ Å. The experimental values are $\theta_1=38^\circ$, $\theta_2=40^\circ$, $\phi_1=12^\circ$, $\phi_2=14^\circ$, $r_{C=C}=1.33$, $r_{C-C}=1.49$. The results are less good than usual, but a better refinement analysis should be carried on the X-ray data, since the direction of highest thermal libration for the oxygen atoms is the one normal to the N-O bonds.

In nucleophilic substitution reactions at aromatic carbons a mechanism had been suggested long ago involving an activated complex or an intermediate of the following structure :

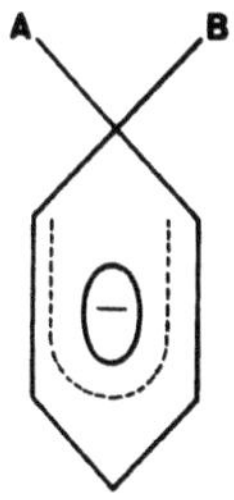

The mechanism is not of interest here, nor is the long lasting controversy on the number of reaction steps. However in some special cases it is possible to isolate the intermediate. One example is the reaction of symmetric trinitrophenetole with alkali ethoxide.

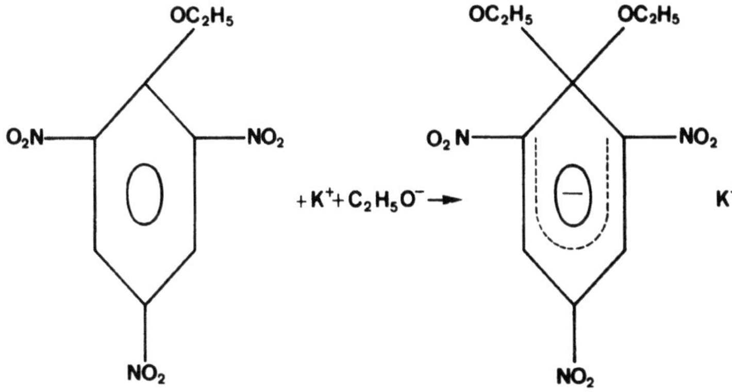

The structure of the Meisenheimer salt shows two molecules in the asymmetric unit (10). Both were located, but the position of only three of the four ethoxy groups was determined : the fourth group was apparently disordered. On a difference Fourier synthesis a large positive region appeared instead of definite peaks. Again packing calculations were the remedy. Assuming normal bond lengths and angles in the ethoxy group and after locating all the other atoms in the asymmetric unit in their known position rotation of the C_2H_5 group around the C-O bond, and of CH_3 group around the O-C bond were allowed and the repulsion energy of all pairs of non bonded atoms whose distance was less than the touching distance was calculated. This energy surface has shown two minima of about the same value. Assuming that each ethoxy group of that kind had a 50/50 chance to be in one or the other of these two positions half C and O atoms were put in the position corresponding to the energy minima. Least squares refinement led to an R value of 0.64 and to reasonable bond angles and lengths for all atoms, including the disordered ethoxy chain. The 3-nitrogroups are roughly in the plane of the ring, in both molecules.

The method was also applied to the study of the conformation of crystalline polyethylene (11). The calculation was carried on in two steps. First by a quantum mechanical computation, using the SFC-MO method in its CNDO approximation (12), the most stable geometry of a single chain was determined. At the beginning all bond angles were assumed to be tetrahedral while variation of bond lengths was allowed. The following order of stability was found for the configurations suggested by spectroscopic evidence:

$$(T)_\infty > (TG)_\infty > (TGTG')_\infty > (G)_\infty \qquad \text{(see fig. 1)}$$

When the energy for the most stable conformation was calculated including variation of bond angles in the minimisation procedure the following geometry was found :

C-C 1.52 Å C-C 1.13 Å $\widehat{CCC} = 114°$ $\widehat{HCH} = 105°$

Bond angles satisfy orthogonality conditions.

Figure 2 shows energy levels as a function of wave number k, in a.u. per CH_2 group. Filled and vacant orbitals are well separated as expected for a good insulator. In the second step the energy for the crystal energy was minimized for $(T)_\infty$ chains of different geometry as a functions of the orthorhombic unit cell parameters a and b and of θ, the angle between axis a and the projection on the ab plane of the C-C bond through the origin. As expected the most stable conformation of the chain in the crystal turned out to be the same as for the isolated chain : the calculated lattice parameters were a = 7.078 Å, b = 4.97 Å, and θ = 42°, to be compared with experimental value a = 7.478 Å, b = 4.97 Å and θ = 42°. The packing energy per CH_2 unit is 1.95 kcal/mole while the experimental heat of fusion was found to be of the order of 1 kcal/mole.

We come now to an example of solution of a crystal structure by means of molecular mechanics : 1,6:8,13-propane-1,3-diylidene-|14|-annulene. The geometry of the isolated molecule was calculated first.

Bond distances and angles involving H atoms were assumed as well as one value (1.397 Å) for all the C-C bond lengths in the annulene ring and mm2 symmetry for the molecule. The carbon skeleton has then 9 degrees of freedom. Energy has been minimized with respect to all of them.

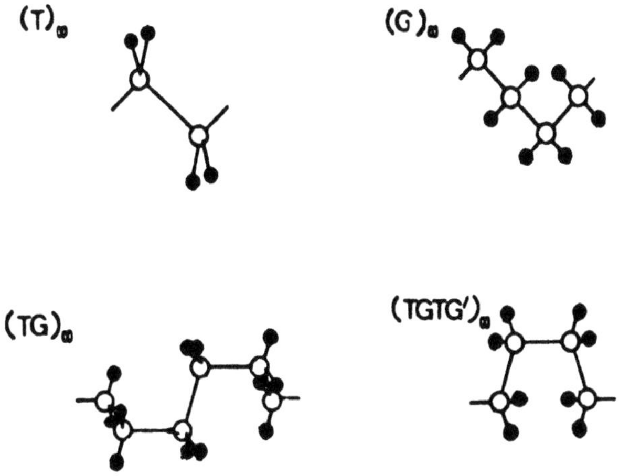

Fig. 1

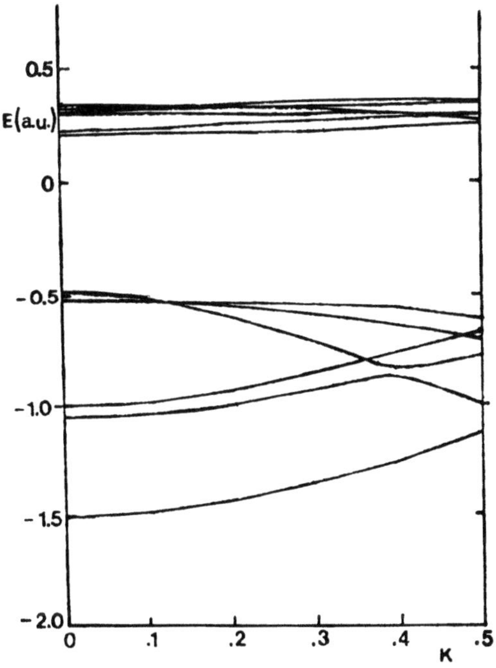

Fig. 2

The orientation of the molecular model in the orthorhombic unit cell was determined by comparing overlap of Patterson functions calculated for different orientations with the same function as obtained by X-ray experimental data.

The position of the reference molecule in the unit cell was determined by shifting the oriented molecule along the axes x and y (z is a polar axis) and calculating at each point the sum of non-bonded interactions between each atom of the model and each atom of molecules generated by symmetry around the reference molecule. One very clear minimum was found. From this point the structure was refined by least squares technique, to a final R = 0.06 (13). The annulene ring is aromatic, in accordance with its chemical and spectroscopic behaviour. Bond lengths in the ring are between 1.38 - 1.41 Å (mean value 1.395 uncorrected, 1.401 corrected), and molecular mm2 symmetry is preserved within experimental errors.

The ring is not planar, but the distance from the best plane is < 0.6 Å. Torsional angles around C-C bond in the ring reach the maximum value of 28°.

A second example of the solution of a crystal structure by means of the atom-atom potential approach was 11,11-difluoro-1,6-methano|10|annulene (14).

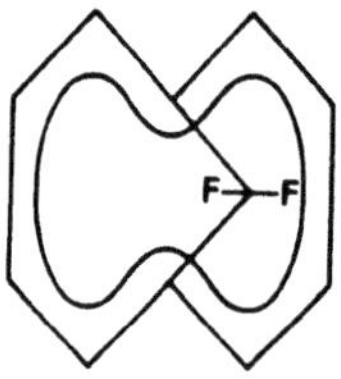

The method of solution followed the same route as for the preceeding compounds : i.e. determination of the orientation by means of Patterson functions and of the position in the orthorhombic unit cell by minimisation of the packing energy.

A well defined minimum was found. The molecule shows chemical and spectroscopic aromatic behaviour. C-C distances in the ring are spread in the range 1.33-1.47. The compound could valence-isomerize to a bis-norcaradiene structure. However no evidence to support this was found, since one of the "double bond" is 1.44 Å and the C_1-C_6 distance is 2.25 Å

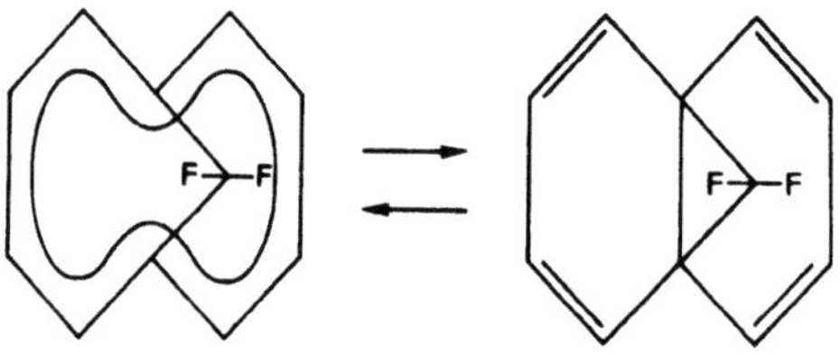

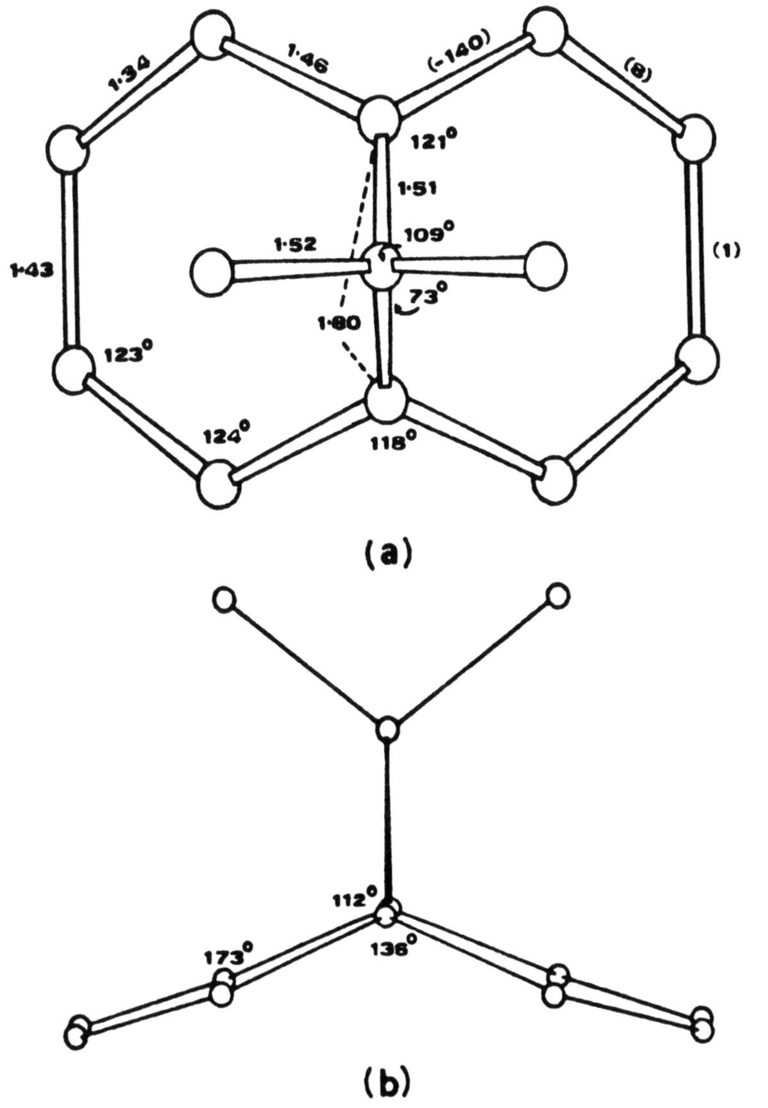
1·34
1·46
(-140)
(8)
121°
1·51
1·52
109°
1·43
(1)
73°
1·80
123°
124°
118°
(a)
112°
173°
136°
(b)

Fig. 3

In this connection a comparison can be made with 11,11-dimethyltricyclo|4,4,2,0^{1,6}|undeca-2,4,7,9-tetraene, a compound in which the bisnorcaradiene form obtains. The crystal is triclinic, space group P1 with two molecules in the asymmetric unit (15).

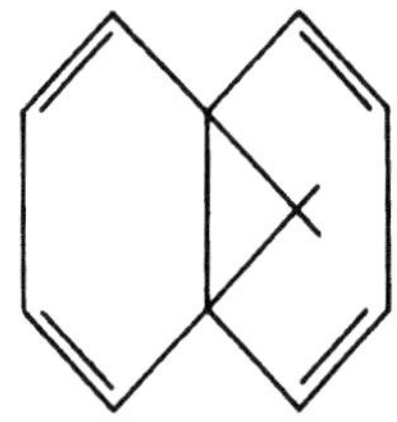

The average molecular structure is shown in figure 3. Chemical and spectroscopic behaviour support this finding (16). We may try to answer the question why is the difluoro compound an annulene unlike the dimethyl compound. In the last compound we have a cyclopropane ring, showing a very long C-C bond, perhaps the longest C-C bond ever found; in the difluoro derivative the bond has disappeared. In both cases there is a strong internal strain acting on this bond. The difference in resisting power is due to an electron substituent effect. Using the Walsh model for describing orbitals in a cyclopropane ring, R. Hoffmann has shown through a molecular orbital analysis that π-electron attracting substituents weaken the adjacent bonds and strengthen the opposite one, while π-electron donors weaken all bonds in the ring (17). To substantiate such qualitative arguments we made an "ab initio" M.O. calculation, using the SCF-GTO-3G technique proposed by Pople (18).

Table I - Structural data for annulenes

Compd	σ_s	D_{max} Å	ϕ_{max} deg	Molecular symmetry
1	0.006	0.49	24	mm2
2	0.013	0.57	28	mm2
3	0.016	0.46	34	mm2
4	0.025	0.75	35	mm2
5	0.047	0.50	41	mm2
6	0.062	0.69	74	m

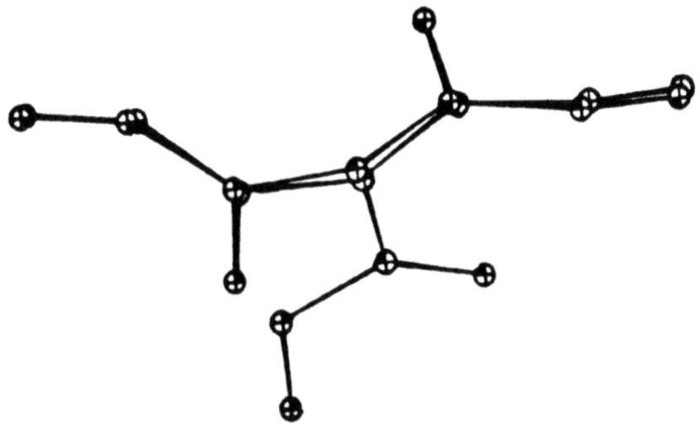

Fig. 4

This is a complete SCF calculation including all valence electrons. The valence shell orbitals are expressed as linear combination of 3 gaussians, with coefficients and exponents that make the orbital as close as possible to a Slater orbital. The geometry of cyclopropane, 1,1-difluoro-and 1,1-dimethyl-cyclopropane were optimized. The results, though the effects are weak, are in complete agreement with qualitative predictions.

Another example of the use of molecular mechanics in the study of conformation of organic crystals is 1,6 : 8,13 butane 1,4 diylidene|14|annulence (19). Again the geometry of the isolated molecule was calculated first. The molecule was assumed to have point group symmetry $2(C_2)$ and all bond lengths in the ring were taken as 1.397 Å. The energy was minimized with respect to the eleven internal degrees of freedom of the C-atom skeleton. Standard values were assumed for C-H bond length and the direction of these bonds were determined by the orthogonality conditions for the hydrid carbon orbitals. The minimum energy was found to correspond to mm2 symmetry. In view of this symmetry and of the fact that the space group was Fmm2 with 4 molecules in the unit cell all that was needed was to find the polar axis and to choose among two possible orientations of the molecule, rotated by 90° with respect to each other. It turned out that for only one position was the packing energy reasonable (-2.5 kcal/mole) : for all others some contacts were too short to be admissible. The R index was R=0.24, a good starting point for least-squares refinement.

I would like to mention now the work on 7-methoxycarbonyl-anti-1,6 : 8,13-bismethano |14| annulene (20) :

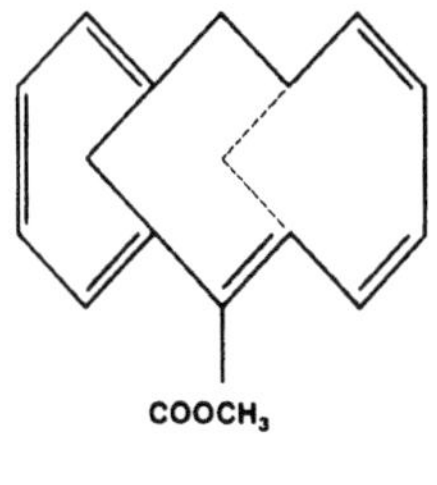

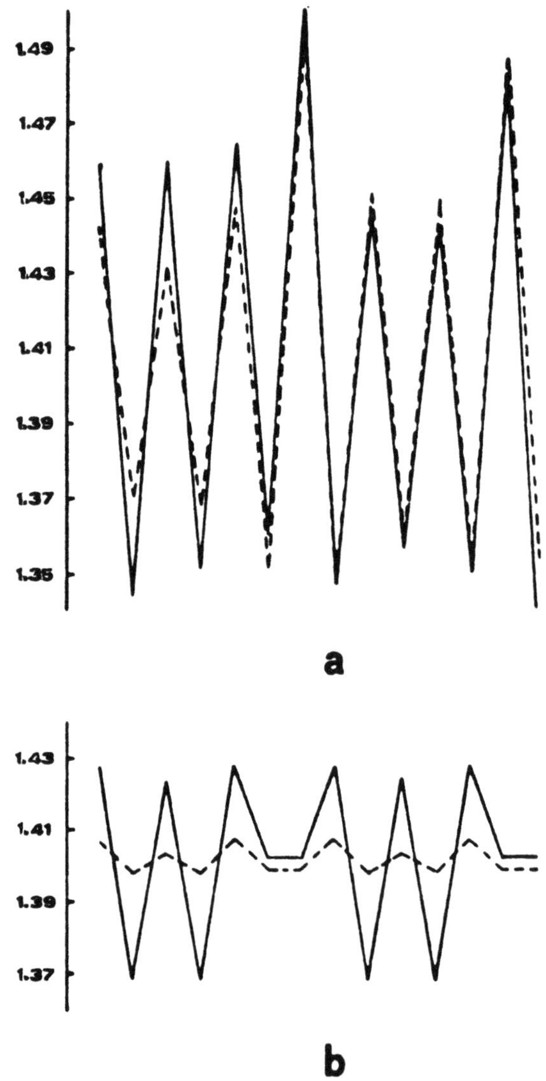

Fig. 5

The hydrocarbon has a low melting point but the methoxycarbonyl derivative was suitable for X-ray studies. Spectral data and chemical behaviour predict a non aromatic structure. The crystal structure was solved by direct methods to a final R=0.051(21) and the final geometry for the molecule is shown in figure 4.

The striking feature is the systematic succession of long and short bond distances in the annulene perimeter at variance with the results found for similar annulenes. Table I compares the structural data for six annulenes. The difference in bond length along the perimeter cannot be attributed to a more pronounced deviation from planarity of the perimeters. Butano|14|annulene shows the highest deviation from planarity but does not show the highest deviation in bond lengths. The olefinic or aromatic behaviour can instead be referred back to the misalignement angles, ϕ_{ij}, between the directions of adjacent $2p_z$ orbitals on carbon atoms i and j in the ring. The maximum value is shown in the anti derivative. The reliability of our assumption was substantiated by means of Hückel molecular orbital calculations for the π-electrons in the ring in which the angles ϕ_{ij} are kept constant and equal to the experimental value, the C-C distances are assumed all equal to 1.40 Å at the beginning of the calculation and the exchange integral β is given by $\beta = \beta_0 \frac{S}{S_0}$, where S is the overlap integral in the considered geometry and S_0 the overlap integral for the Slater $2p_z$ carbon orbitals with parallel axes at a distance of 1.4 Å.

At the end of the calculation C-C bond orders and hence bond distances are evaluated and new β integrals are calculated. The calculation is repeated (always with the experimental ϕ_{ij}) till consistency obtains. The results shown in figure 5 for the anti compound and butano|14|annulene are in very satisfactory agreement with experiment, suggesting the adeguacy of the proposed explanation.

It is interesting that the same conclusion was reached by Allinger (22) who determined by means of his force field the geometry of bicyclo|4,4,1|undecapentaene and bicyclo|5,3,1|undecapentaene, finding for the former constant values for the C-C distances in the ring in agreement with X-ray results (23), but bond alternation in the latter, even if the degree of planarity in both rings is very close.

We have seen some applications of molecular mechanics in the study of molecular packing. Other problems in this area can be tackled by the same method, such as the determination of the equilibrium orientation of molecules in a crystal lattice, the determination of lattice parameters, the study of crystals with disordered elements, of hydrogen bonds in crystals, of solid solutions and of binary crystals. An excellent review of such possibilities can be found in a work by A. Kitaigorodsky (24).

BIBLIOGRAPHY

1. A. Almenningen, and O. Bastiansen - Kgl. Norsk Vid. Selsk. Skr. n° 4 (1958).

2. J. Trotter - Acta Cryst. 14, 1135 (1961).

3. G. Casalone, C. Mariani, A. Mugnoli, and M. Simonetta - Mol. Phys. 15, 339 (1968).

4. G. Casalone, C. Mariani, A. Mugnoli, and M. Simonetta - Acta Cryst. B 25, 1741 (1969).

5. G. Casalone, A. Gavezzotti, and M. Simonetta - J. Chem. Soc. Perkin II, 342 (1973).

6. A.I. Kitaigorodsky - Tetrahedron 14, 230 (1961).

7. N.A. Ahmed, and A.I. Kitaigorodsky - Acta Cryst. B 28, 739 (1972).

8. G. Casalone, C. Mariani, A. Mugnoli, and M. Simonetta - Acta Cryst. 22, 228 (1967).

9. G. Casalone, and M. Simonetta - J. Chem. Soc. B, 1180 (1971).

10. R. Destro, C.M. Gramaccioli, and M. Simonetta - Acta Cryst. B 24, 1369 (1968).

11. G. Morosi, and M. Simonetta - Chem. Phys. Letters 8, 358 (1971).

12. J.A. Pople, and G.A. Segal - J. Chem. Phys. 44, 3289 (1966).

13. A. Gavezzotti, A. Mugnoli, M. Raimondi, and M. Simonetta - J. Chem. Soc. Perkin II, 425 (1972).

14. C.M. Gramaccioli, and M. Simonetta - Acta Cryst. B 28, 2231 (1971).

15. R. Bianchi, G. Morosi, A. Mugnoli, and M. Simonetta - Acta Cryst. B 29, 1196 (1973).

16. E. Vogel - Pure and Applied Chem. 20, 237 (1969).

17. R. Hoffmann, and R.B. Davidson - J. Am. Chem. Soc. 93, 5699 (1971).

18. W.J. Hehre, R.F. Stewart, and J.A. Pople - J. Chem. Phys. 51, 2657 (1969).

19. C.M. Gramaccioli, A. Mugnoli, T. Pilati, M. Raimondi, and M. Simonetta, Acta Cryst. B 28, 2365 (1972).

20. E. Vogel, U. Haberland, and H. Günther - Angew. Chem. Int. Ed. 9, 513 (1970).

21. C.M. Gramaccioli, A. Mimun, A. Mugnoli, and M. Simonetta - J. Am. Chem. Soc. 95, 3149 (1973).

22. N.L. Allinger, and J.T. Sprague - ibid. 95, 3893 (1973).

23. M. Dobler, and J.D. Dunitz - Helw. Chim. Acta 48, 1429 (1965).

24. A.I. Kitaigorodsky - Adv. in Str. Res. by Diffr. Meth. 3, 173 (1970).

LECTURE IV - LATTICE DYNAMICS

Molecular mechanics have been used for the study of properties of molecular crystals, the basic assumption making this application possible being the pairwise additivity of atom-atom interaction to provide the total intermolecular non-valence energy. The same non-bonded interaction potentials that have been used to calculate the geometry and heat of formation of molecules in the gas phase and to solve and refine crystal structures -that is to calculate the molecular packing that leads to an energy minimum - can be used also to investigate several properties of organic crystals, mostly related to the rigid body molecular motions and hence to lattice dynamics.

With the assumption of rigid-body motions, the thermal vibrations of each molecule can be described by means of an angular and translational displacement, expressed in the form (2) :

$$\underline{u}(kl) = \underline{U}(k\underline{q})\exp i|\underline{q}.\underline{r}(kl) - \omega(\underline{q})t| \qquad (1)$$

where $\underline{u}$ is a vector in a six-dimensional space (there are three components for the translation and three rotational ones around the principal axes of inertia of each molecule), k is an index referring to the asymmetric unit in the cell to which each molecule belongs (assuming one molecule in each asymmetric unit), l is an index referring to cell, $\underline{q}$ is the wave vector (expressed in fractions of the reciprocal unit cell), $\underline{U}$ is a six-dimensional vector that gives the amplitude of each displacement, $\underline{r}$ is the vector from the origin to the center of mass of each molecule and ω is the frequency of the vibrational motion.

The potential energy of a molecule can be expressed as a function of its displacement from the equilibrium position :

$$E = E_o + \sum_i \left(\frac{\partial E}{\partial u_i(kl)}\right)_o u_i(kl) + \frac{1}{2}\sum_{ij}\left(\frac{\partial^2 E}{\partial u_i(kl)\partial u_j(k'l')}\right)_o u_i(kl)u_j(kT) + \ldots + \qquad (2)$$

If we ignore terms after the second derivatives(harmonic approximation) and we recall that at equilibrium the first derivatives are zero

$$\left(\frac{\partial E}{\partial u_i(kl)}\right)_o = 0 \qquad (2)$$

The potential energy is given by :

$$E = E_o + \frac{1}{2} \sum_{ij} \left(\frac{\partial^2 E}{\partial u_i(kl)\partial u_j(k'l')}\right)_o u_i(kl)u_j(k'l') \quad (3)$$

or

$$E = E_o + \frac{1}{2} \sum_{ij} \Phi_{ij} u_i(kl)u_j(k'l') \quad (4)$$

where Φ_{ij}'s are force constants, defined as

$$\Phi_{ij}(kl,k'l') = \left(\frac{\partial^2 E}{\partial u_i(kl)\partial u_j(k'l')}\right)_o \quad (5)$$

The classical equations of motion for the molecule (kl) are:

$$m_i \ddot{u}_i(kl) = - \frac{\partial E}{\partial u_i(kl)} = - \sum_{k'l'j} \Phi_{ij} (kl,k'l')u_j(k'l') \quad (6)$$

where m_i is the mass of the molecule for i=1,2,3, and the moment of inertia about the principal axes for i=4,5,6. By means of eq. (1), the equations of motions assume the form :

$$-m_i\omega^2(\underline{q})U_i(k\underline{q}) = - \sum_{l'k'j} \Phi_{ij}(kl,k'l')U_j(k'\underline{q})e^{i|\underline{q}.\underline{r}(k'l')-\underline{q}.\underline{r}(kl)|} \quad (7)$$

where i=1,2... 6, k,k'=1...S(number of molecules in the unit cell) l'=1...N (number of cells in the crystal). Eq. (7) can be written as :

$$m_i \omega^2(\underline{q})U_i(k\underline{q}) = \sum_{k'j} U_j(k'\underline{q}) \sum_{l'} \Phi_{ij} (kl,k'l')\exp i\underline{q}|\underline{r}(k'l')-\underline{r}(kl)| \quad (8)$$

Let us define the new variables :

$$\xi_i = \sqrt{m_i}\ u_i \quad (9)$$

Then we have :

$$\Phi'_{ij} = \frac{\partial^2 E}{\partial\xi_i\partial\xi_j} = \frac{1}{\sqrt{m_i}} \frac{1}{\sqrt{m_j}} \Phi_{ij} \quad (10)$$

After defining

$$U'_i(k\underline{q}) = \sqrt{m_i}\ U_i(k\underline{q}) \quad (11)$$

the equations of motion become :

$$\omega^2(\underline{q})U'_i(k\underline{q}) = \sum_{k'j} U'_j(k'\underline{q}) \sum_{l'} \Phi'_{ij}(kl,k'l')\exp i\, \underline{q}|\underline{r}(k'l)-\underline{r}(kl)| \tag{12}$$

The equations (in number of 6S) can be written in matrix form, by means of the so called modified dynamical matrix (3), with elements

$$M_{ij}(k,k',\underline{q}) = \sum_{l'} \Phi'_{ij}(kl,k'l')\ \exp i\, \underline{q}|\underline{r}(k'l') - \underline{r}(kl)| \tag{13}$$

Then we have :

$$\omega^2(\underline{q})U'_i(k\underline{q}) = \sum_{k'j} U'_j(k'\underline{q})\ M_{ij}(k,k'q) \tag{14}$$

or

$$\omega^2(\underline{q})\underline{U}'(k\underline{q}) = \underline{M}(k\underline{q})\underline{U}'(k\underline{q}) \tag{15}$$

This means that the eigenvalues of the matrix $\underline{M}$ are the squares of the frequencies, while U' (k$\underline{q}$) are the corresponding eigenvectors.

The first problem in setting up eqs (15) is the choice of the potentials from which force constants are calculated. Since the interaction between the two molecules is expressed as the sum of the interactions between all atoms in one molecule with all the atoms in the second molecule, all we need is the atom-atom potential for atom pairs of different kind. If we limit consideration to hydrocarbons we need only carbon-carbon, carbon-hydrogen and hydrogen-hydrogen functions. Many of the functions used to this purpose have the form

$$V(r) = A\exp(-Br) - Cr^{-6} \tag{16}$$

Some of the values for parameters A, B, C given by different authors are listed in Table I.

TABLE I

Ref.	Contact	A(kcal/mol)	B(Å^{-1})	C(kcal Å^{6}/mole)
4	H...H	2654.0	3.74	27.3
	H...C	8766.0	3.67	125.0
	C...C	83630.0	3.60	568.0
5	H...H	4000.0	3.74	36.0
	H...C	9411.0	3.67	139.0
	C...C	74460.0	3.60	535.0
6	H...H	42000.0	4.86	57.0
	H...C	42000.0	4.12	154.0
	C...C	42000.0	3.58	358.0

With atom-atom potential function it is possible to calculate, for each given wave vector $\underline{q}$, the force constants that appear in the matrix elements. However caremust be taken when calculating derivatives, since the displacements are related to the principal axes of inertia of each molecule. We have :

$$\frac{\partial^2 E}{\partial u_i \partial u'_j} = \frac{\partial r}{\partial u_i} \frac{\partial^2 E}{\partial r^2} \frac{\partial r}{\partial u'_j} + \frac{\partial E}{\partial r} \frac{\partial^2 r}{\partial u_i \partial u'_j} \tag{17}$$

Also each derivative is obtained by summing up contributions over all packing distances below an assigned value (usually 5.5 Å).

Before starting the calculation of force constants there is another problem. We have assumed that the molecules are in the equilibrium position but the equilibrium position found experimentally is not identical to the position of minimum energy for the potential we have assumed : initially all molecules must be shifted to the equilibrium position on the basis of the assigned potential functions, and the amount of the rotational and translational shifts can be used to give an idea of the goodness of the functions used. For this, the unit cell parameters are taken from experiment and left unchanged. Once the force constants are at hand, building the matrix $\underline{M}$ is not particularly difficult, and its diagonalisation is carried on by routine techniques. Frequencies calculated at zero wave vector can be compared with experimental results from Raman or infrared spectroscopy. As an example the results for naphthalene are shown in Table II (7)

TABLE II - RAMAN FREQUENCIES FOR NAPHTHALENE

	Exp	Calc
Symmetric Modes	146	139
	90	92
	71	62
Antisymmetric Modes	124	129
	86	77
	59	49

Of course frequencies can be calculated for wave vectors spanning the first Brillouin zone in any direction. In this way dispersion curves are obtained. Fig. 1 shows dispersion curves along crystallographic axes for 1,6 : 8,13 butane-1,4-diylidene |14|annulene (8). Another example is given in Fig. 2, where the calculated dispersion curves for deuteroanthracene along the 001 axes are compared with result from coherent inelastic neutron scattering (9).

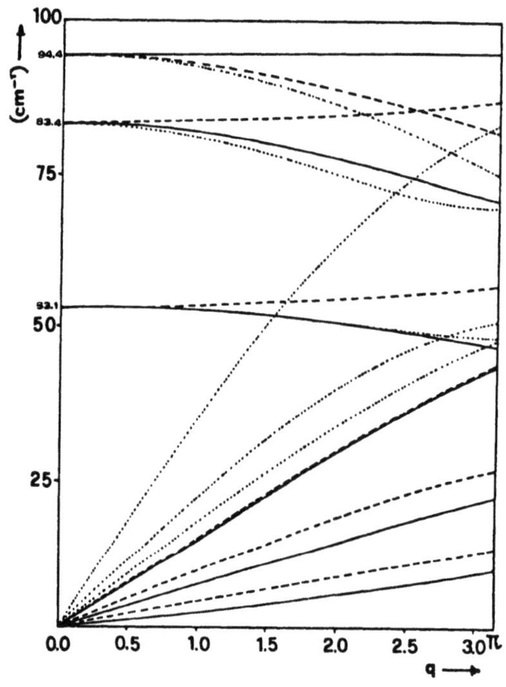

Fig. 1

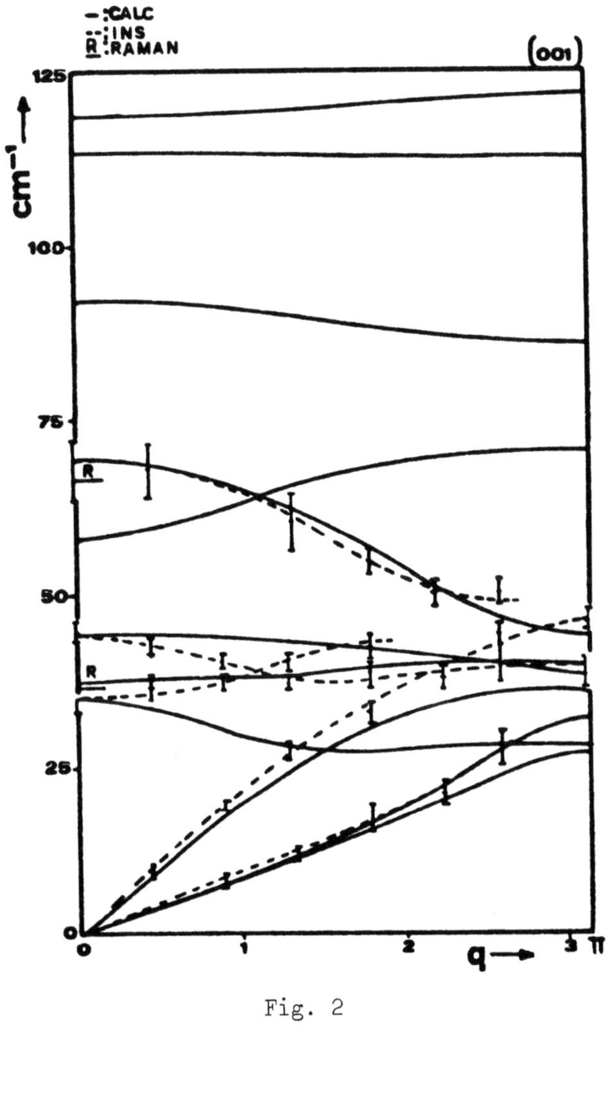

Fig. 2

It is well known that the rigid-body motion of molecules in crystals can be described by means of three tensors, $\underline{T}$ for translation, $\underline{L}$ for libration and $\underline{S}$ for correlation of libration and translation (10). Anisotropic temperature factors B_{ij}, as usually obtained from X-ray data refinement, (when referred to a Cartesian axis system and devided by $2\pi^2$) produce the mean square displacements tensor $\underline{U}$. If the molecule behaves like a rigid body the components of $\underline{U}$ can be expressed in the following form :

$$U_{ij} = \sum_{\kappa l} G_{ij\kappa l} L_{\kappa l} + \sum_{\kappa l} M_{ij\kappa l} S_{\kappa l} + T_{ij} \tag{18}$$

Since the coefficients G_{ijkl}, H_{ijkl} are known functions of atomic coordinates it is possible, if $\underline{T}$, $\underline{L}$, $\underline{S}$ are known, to calculate $\underline{U}$ and compare with the tensor derived from $\underline{B}$, that is from experiment. In fact the components of $\underline{T}$, $\underline{L}$, $\underline{S}$ are obtained through a least squares procedure, minimizing the function $\Sigma(U_{ij}$ calc-U_{ij} obs$)^2$ where the unknown are six components of $\underline{T}$ six components of $\underline{L}$ (both are symmetric tensors) and eight components of $\underline{S}$ (one of the diagonal elements is undetermined).

The tensor $\underline{T}$, $\underline{L}$, $\underline{S}$ can be obtained, quite independently, by means of a lattice-dynamical treatment. In a crystal made of N cells, each containing S molecules,there are 6SN normal modes. The mean energy of each normal mode will be :

$$\varepsilon(\omega) = \hbar\omega \left|\frac{1}{2} + (\exp\{\frac{\hbar\omega}{\kappa T}\} - 1)^{-1}\right| \tag{19}$$

giving a mean energy per cell of $\frac{\varepsilon}{N}$. it is also well known that the mean square displacement in harmonic motion as frequency ω is related to the energy by the formula :

$$s^2 = \frac{E}{m\omega^2} \tag{20}$$

By solution of eqs. (15) we obtain, for each wave vector, the eigenvalues ω and the corresponding normalized eigenvectors $\underline{U}'$ (in the 6S dimensional space). The mean-square displacement of, say, molecule 1 in the i direction will be

$$\frac{U_i'(1,\underline{q})^2 \; \varepsilon(\omega)}{Nm_i \; \omega^2} \tag{21}$$

for the mode of frequency ω. Summing over the N points in the Brillouin zone, we obtain :

$$T_{ij} = \sum_{\substack{6SN \\ modes}} \frac{U_i^! U_j^! \varepsilon(\omega)}{Nm\omega^2} \tag{22}$$

$$L_{kl} = \sum_{6SN} \frac{U'_k U'_l \varepsilon(\omega)}{N\omega^2 \sqrt{I_k I_l}} \qquad S_{ik} = \sum_{6SN} \frac{U_i^! U_k^! \varepsilon(\omega)}{N\sqrt{mI_k}\, \omega^2}$$

Where I_k are moments of inertia, i and j label translational components (i,j = 1, 2, 3), k and l rotational components (k, l = 4, 5, 6). For the actual calculations it is necessary to sample throughout the Brillouin zone, and it has been found that uneven sampling, with maximum density of points in proximity to the origin, and with a weight proportional to the extension of the interval, is the most effective. Once tensors $\underline{T}$, $\underline{L}$, $\underline{S}$ are obtained, it is a trivial matter to calculate the B_{ij} for the different atoms in the molecule and to compare with data from X-ray experiments. A good fit of calculated and experimental B_{ij} has been found for a number of hydrocarbons (11), even when "crystallographic" and "dynamical" $\underline{T}$, $\underline{L}$, $\underline{S}$ tensors are in noticeable disagreement.

From frequencies obtained for all sampled points a density of states can be easily built, as

$$g(\nu_k,\nu'_k) = N \sum_i \eta_i(\nu_k,\nu'_k)\, W_i\, \Delta\nu_k \qquad \nu_k < \nu_{max} \tag{23}$$

where $\eta_i(\nu_k,\nu'_k)$ is the number of the frequencies within the interval $\Delta\nu_k$ between ν_k and ν'_k, W_i is a weight proportional to the extension of the sampling interval in the Brillouin zone, and N is a normalizing factor. The summation at the numerator is extended to all sampling points in the Brillouin zone. In fig. 3 is shown, as an example, the density of states calculated for 1,6 : 8,13-butane-diylidene|14|annulene.

From a density of states, a number of thermodynamical functions can easily be evaluated. The vibrational energy for external modes is given by :

$$E_{vib(ext)} = 6R \sum_j g(\nu_j) \frac{\hbar\nu_j}{k} \frac{1}{e^{\frac{\hbar\nu_j}{kT}} - 1} \Delta\nu_j + E_o(ext) \tag{24}$$

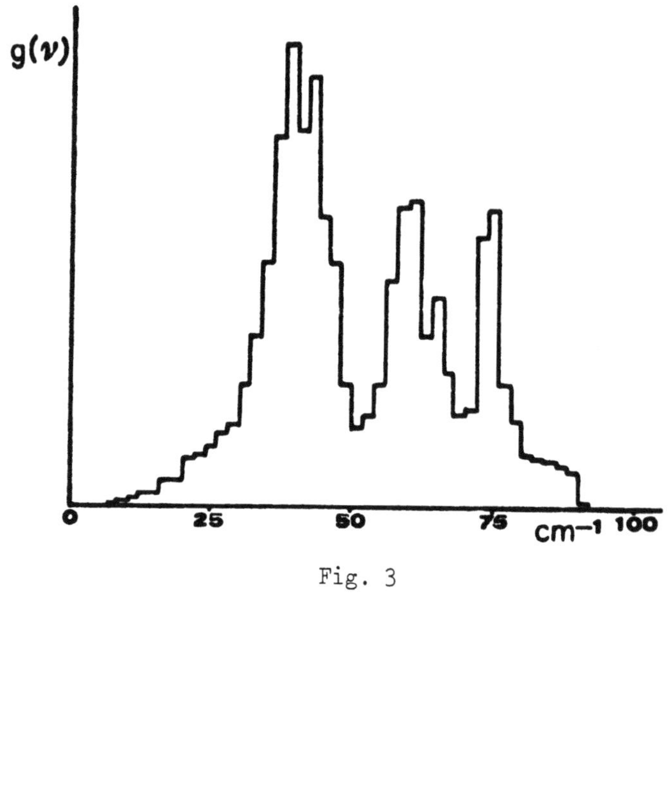

Fig. 3

where $$E_{o(ext)} = 6R \sum_j g(\nu_j) \frac{\hbar\nu_j}{2k} \Delta\nu_j \tag{25}$$

in the zero point energy.

Similarly contributions of external modes to the specific heat c_v, entropy and free energy can be obtained according to standard formulae of statistical thermodynamics :

$$c_v(ext) = 6R \sum_j g(\nu_j) \left(\frac{\hbar\nu_j}{kT}\right)^2 \exp|\hbar\nu_j/kT| / |\exp(\frac{h\nu_j}{kT})-1|^2 . \Delta\nu_j \tag{26}$$

$$S_{(ext)} = E_{vib(ext)}/T - R \sum_j g(\nu_j) \ln|1-\exp(\frac{-h\nu_i}{kT})| \Delta\nu_j \tag{27}$$

$$F_{(ext)}-F_{o(ext)} = -RT \sum_j g(\nu_j) \ln|1-\exp\frac{-h\nu_j}{kT}| \Delta\nu_j \tag{28}$$

In order to obtain E, c_v, S, $F-F_o$ for the crystal the external modes contribution has to be added to the corresponding contribution of internal modes; the latter can be evaluated according to standard formulae of statistical thermodynamics, such as :

$$E_{vib(int)} + R \sum_i \frac{h\nu_i}{k} \frac{1}{e^{\frac{h\nu_i}{kT}}-1} + E_{o\ int} \tag{29}$$

with $$E_{o\ int} = R \sum_i \frac{h\nu_i}{2k} \tag{30}$$

$$c_{v(int)} = \sum_i R \left(\frac{h\nu_i}{kT}\right)^2 \exp\frac{h\nu_i}{kT} \frac{1}{|\exp(\frac{h\nu_i}{kT})-1|^2} \tag{31}$$

$$S_{int} = \frac{E_{vib(int)}}{T} - R \sum_i \ln|1-\exp(-h\nu_i/kT)| \tag{32}$$

$$F_{(int)}-F_{o(int)} = \sum_i RT \ln|1-\exp(h\nu_i/kT)| \tag{33}$$

Where the summations $\sum_i$ are extended to all internal modes. For internal vibration frequencies experimental values can be used. The heat of sublimation and vapor pressure can be calculated as well.

In Table 3 calculated and experimental results for naphtalene are compared (12).

TABLE III - Thermodynamic functions for Naphtalene, at 298.1°K

	ΔH (kcal/mol)	c_p cal/mol°K	c_v cal/mol °K	S cal/mol °K	$-F_T - F_0$ cal/mol
calc.	16.537	39.45	36.56	40.99	6376.0
exp.	15.4-17.4	39.6	36.0	39.9	6004.9

If, then, we limit our attention to molecular crystals of rigid hydrocarbons, which we can consider an ideal crystal in that the potential functions we have chosen are effective, we can calculate the equilibrium structure, the external vibrational frequency dispersion curves, the anisotropic temperature factors, and thermodynamical functions. The calculated values for all these properties, all derived from the assumed potential, are in quite good agreement with the values obtained from experiments on the real crystal : it appears that molecular mechanics can be used to predict, with a considerable amount of accuracy and confidence, the behaviour of molecular crystals of the same kind, for which experimental data are not available.

BIBLIOGRAPHY

1. A.I. Ktaigorodski - Adv. Struct. Res. by Diff. Meth. 3, 173 (1970).

2. M. Born, and K. Huang - Dynamical Theory of Crystal Lattices Oxford University Press - London (1954).

3. G.S. Pawley - Phys. Stat. Sol. 49 b, 475 (1972).

4. D.E. Williams - J. Chem. Phys. 47, 4680 (1967).

5. D.E. Williams - ibid. 45, 3770 (1965).

6. A.I. Kitaigorodski - J. Chim. Phys. 63, 6 (1966).

7. G.S. Pawley - Phys. Stat. Sol. 20, 347 (1967).

8. C.M. Gramaccioli, M. Simonetta, and G.B. Suffritti - Chem. Phys. Letters 20, 23 (1973).

9. G. Filippini, C.M. Gramaccioli, M. Simonetta, and G.B. Suffritti - J. Chem. Phys. 59, 5088 (1973).

10. V. Schomaker, and K.N. Trueblood - Acta Cryst. B 24, 63 (1968).

11. G. Filippini, C.M. Gramaccioli, M. Simonetta, and G.B. Suffritti - Acta Cryst. A 30, 189 (1974).

12. G. Filippini, C.M. Gramaccioli, M. Simonetta, and G.B. Suffritti - unpublished results.

APPLICATIONS OF THE SCF-Xα SCATTERED-WAVE METHOD TO MOLECULAR CRYSTALS AND POLYMERS

Keith H. Johnson*

Department of Metallurgy and Materials Science

Massachusetts Institute of Technology

Cambridge, Massachusetts 02139, U. S. A.

Frank Herman

IBM Research Laboratory

San Jose, California 95193, U. S. A.

Roland Kjellander†

Department of Physical Chemistry

The Royal Institute of Technology

S-100 44 Stockholm 70, Sweden

ABSTRACT

To facilitate the application of the self-consistent-field X scattered-wave (SCF-Xα-SW) method to certain types of molecular crystals and polymers, it is possible to introduce two modifications of the theory and computational procedure : (1) the use of overlapping atomic spheres and (2) the partitioning of the system into component interacting molecular units. The theoretical assumptions un-

*Research at M.I.T. sponsored in part by the General Electric Fondation.

†Visiting Scientist, Department of Metallurgy and Materials Science, M.I.T. Visit sponsored in part by Kungafonden (H.M. King Gustav VI Adolf's Foundation), the Sweden-America Foundation, the American-Scandinavian Foundation, and the Swedish Natural Science Research Council.

derlying the use of overlapping spheres are discussed first, and the theory is illustrated for molecular TCNQ, a component of highly conducting organic molecular crystals of the TTF-TCNQ type. The results obtained for TCNQ suggest that the SCF-Xα-SW method, in conjunction with overlapping atomic spheres, can be used to produce realistic electronic energy levels and charge distributions for large planar organic molecules. To calculate the electronic structure of a crystalline or polymeric array of interacting polyatomic molecules, it is possible to make use of some unique features of the SCF-Xα-SW method. The calculations are carried out on one part of the system at a time using overlapping atomic spheres and a local intersphere potential, iterating between interacting parts until self consistency is attained. By exploiting the local molecular symmetry of each component part, the calculations can be further simplified, and only moderate amounts of computer time are required. This method is not dependent on the assumption of molecular periodicity or Bloch's theorem and therefore is applicable to disordered or amorphous polymers and molecular solids, where there is at most only short-range molecular order. Results for a water trimer are presented as a simple illustration of the computational procedure for a weakly coupled polymer. Extensions of this procedure to more complex systems, such as the molecular crystal TTF-TCNQ, are also discussed.

I. THE USE OF OVERLAPPING ATOMIC SPHERES IN THE SCF-Xα SCATTERED-WAVE METHOD

The self-consistent-field Xα scattered-wave (SCF-Xα-SW) method[1-3] of calculating the electronic structures of polyatomic molecules and clusters in solids has been successfully applied to a wide range of systems.[4,5] The method is based on the division of molecular space into nonoverlapping atomic, interatomic, and extramolecular regions of spherically averaged and volume averaged potential, including Slater's Xα approximation to exchange correlation.[6,7] The one-electron Schrödinger equation is solved numerically in each region in the partial-wave representation, the wavefunctions and their first derivatives being joined continuously across the boundaries of each region via multiple-scattered-wave formalism.[1,3] The calculations are iterated until self-consistency is attained.

This procedure leads to an accurate description of electronic structure for a large variety of polyatomic systems in only moderate computation times. However, there are molecular systems where the geometry results in an outer sphere which is poorly packed with nonoverlapping atomic spheres, producing a large intersphere region which contains a disproportionate amount of electronic charge. Examples of such systems where "non-muffin-tin" corrections are required include π-electron molecules such as ethylene and benzene and larger planar organic molecules like TCNQ, which are constituents of important typed of molecular crystals. For such systems the SCF-Xα-SW method, in its conventional muffin-tin representation, leads to ionization energies which are considerably overestimated and to poor total energies.

Realistic corrections for the muffin-tin approximation can be introduced in the SCF-Xα-SW method through the straightforward use of overlapping atomic spheres.[8,9] To show that the theory and computational procedure can be extended to overlapping atomic spheres, to good approximation (subject to the restrictions described below), is relatively simple and requires only the multiple-scattered-wave formalism presented in Refs. 1-3 . Details will be given below. Systematic investigations of the use of overlapping spheres in the SCF-Xα-SW technique have already been presented,[9,10] and recently Slater[11,12] has discussed the overlapping-sphere approximation in relation to the Wigner-Seitz cellular theory of solids and the cellular multiple-scattering approach of Williams.[13]

Consider a cluster of N overlapping atomic spheres of radii b_α ($\alpha = 1,2,\ldots N$) subject only to the restriction that each sphere does not overlap a nearest neighboring sphere beyond its nucleus i.e., $b_\alpha < R_{\alpha\beta}$ where $R_{\alpha\beta}$ is the distance between the nuclei of neighboring atoms α and β. Inside each atomic sphere, we represent the one-electron wavefunction in terms of the single-center partial-wave expansion

$$\Psi_I(\vec{r}_\alpha) = \sum_L C_L^\alpha R_\ell(\varepsilon;r_\alpha)Y_L(\vec{r}_\alpha) \quad ; r_\alpha \leq b_\alpha < R_{\alpha\beta} \tag{1}$$

where $L = (\ell,m)$ is the partial-wave (angular-momentum) index. The functions $R_\ell(\varepsilon;r_\alpha)$ are solutions of the radial Schrödinger equation

$$\left[-\frac{1}{r_\alpha^2}\frac{d}{dr_\alpha} r_\alpha^2 \frac{d}{dr_\alpha} + V(r_\alpha) + \frac{\ell(\ell+1)}{r_\alpha^2} - \varepsilon \right] \times R_\ell(\varepsilon;r_\alpha) = 0 \tag{2}$$

in which $V(r_\alpha)$ is the spherical average of the Coulomb and X_α exchange-correlation potentials, and ε is the trial one-electron energy. The functions $Y_L(\vec{r}_\alpha)$ are real spherical harmonics and the expansion coefficients C_L^α are to be determined. In the region of overlap between two neighboring atomic spheres α and β, we have the choice of using either the single-center expansion (1) or a similar expansion

$$\Psi_I(\vec{r}_\beta) = \sum_L C_L^\beta R_\ell(\varepsilon; r_\beta) Y_L(\vec{r}_\beta) \quad ; r_\beta \leq b_\beta < R_{\alpha\beta} \tag{3}$$

taken with respect to nucleus β.

In the intersphere region, II, Schrödinger's equation reduces to the ordinary scalar wave equation

$$(\nabla^2 + \varepsilon - \bar{V})\Psi_{II}(\vec{r}) = 0 \tag{4}$$

where $\bar{V}$ is the constant intersphere potential, which is usually chosen as the average of the Coulomb and Xα exchange-correlation potentials over the intersphere volume

$$\Omega_{II} = \frac{4}{3}\pi(b_o{}^3 - \sum_{\beta=1}^{N} b_\beta{}^2) \tag{5}$$

where b_o is the radius of the outer sphere surrounding the molecule or cluster. The volume $\frac{4}{3}\pi b_\beta{}^3$ of the overlapping atomic spheres are assumed to be additive, so that the effective intersphere volume is considerably less than it would be in the limit of nonoverlapping spheres.

The solutions of the differential equation (4) can be written in the multicenter partial-wave representation

$$\Psi_{II}(\vec{r}) = \sum_L A_L^\alpha \, k_\ell{}^{(1)}(\kappa r_\alpha) Y_L(\vec{r}_\alpha) + \sum_{\beta\neq\alpha} \sum_L A_L^\beta \, k_\ell{}^{(1)}(\kappa r_\beta) Y_L(\vec{r}_\beta) \tag{6}$$

in which $(\bar{V} - \varepsilon)^{1/2} \quad ; \varepsilon < \bar{V}$ (7)

and where $k_\ell{}^{(1)}(\kappa r_\alpha)$ and $k_\ell{}^{(1)}(\kappa r_\beta)$ are modified spherical Hankel functions of the first kind centered on nuclei α and β respectively. We now wish to match the wavefunction (6) to the wavefunction (1) at each atomic sphere radius $r_\alpha = b_\alpha$, and then to do the same with their respective first derivatives. To facilitate this, we can use the multipole expansion theorem[3]

$$k_\ell{}^{(1)}(\kappa r_\beta) Y_L(\vec{r}_\beta) = k_\ell{}^{(1)}(\kappa|\vec{r}_\alpha - \vec{R}_{\alpha\beta}|) Y_L(\vec{r}_\alpha - \vec{R}_{\alpha\beta}) \tag{8}$$

$$= 4\pi \sum_{L'} (-1)^{\ell+\ell'} \sum_{L''} I_{L''}(L;L') k_{\ell''}{}^{(1)}(\kappa R_{\alpha\beta}) Y_{L''}(\vec{R}_{\alpha\beta}) i_{\ell'}(\kappa r_\alpha) Y_{L'}(\vec{r}_\alpha)$$

where $I_{L''}(L;L') \equiv \int_0^{2\pi} d\phi \int_0^{\pi} d\theta \; \sin\theta Y_{L''}(\theta,\phi) Y_L(\theta,\phi) Y_{L'}(\theta,\phi)$ (9)

and where $i_{\ell'}(\kappa r_\alpha)$ is a modified spherical Bessel function. The expansion theorem (8) is valid where $r_\alpha < R_{\alpha\beta}$. Since the atomic sphere overlap condition is $b_\alpha < R_{\alpha\beta}$, expression (8) is valid at the sphere radius $r_\alpha = b_\alpha$, where the wavefunction matching is to be carried out.

Introducing the notation

$$G_{LL'}^{\alpha\beta}(\varepsilon) \equiv (1-\delta_{\alpha\beta})4\pi(-1)^{\ell+\ell'} \sum_{L''} I_{L''}(L;L')k_{\ell''}^{(1)}(\kappa R_{\alpha\beta})Y_{L''}(\vec{R}_{\alpha\beta}) \quad (10)$$

in Eq. (8), substituting (8) into (6), and then matching (6) to (1) at $r_\alpha = b_\alpha$, we obtain for each partial wave the expression

$$C_L^\alpha R_\ell(\varepsilon;b_\alpha) = A_L^\alpha k_\ell^{(1)}(\kappa b_\alpha) + \sum_{\beta L'}\sum G_{LL'}^{\alpha\beta}(\varepsilon)A_{L'}^\beta i_{\ell'}(\kappa b_\alpha) \quad (11)$$

matching first derivatives at $r_\alpha = b_\alpha$ yields

$$C_L^\alpha \frac{dR_\ell(\varepsilon;r_\alpha)}{dr_\alpha}\Big|_{r_\alpha=b_\alpha} = A_L^\alpha \frac{dk_\ell^{(1)}(\kappa r_\alpha)}{dr_\alpha}\Big|_{r_\alpha=b_\alpha} + \sum_{\beta L'}\sum G_{LL'}^{\alpha\beta}(\varepsilon)A_{L'}^\beta \times \frac{di_{\ell'}(\kappa r_\alpha)}{dr_\alpha}\Big|_{r_\alpha=b_\alpha} \quad (12)$$

If we divide (11) by (12) we can eliminate the coefficients C_L^α and obtain a set of linear homogeneous equations in the coefficients A_L^α and $A_{L'}^\beta$. Using the notation

$$[f(x),g(x)] \equiv f(x)\frac{dg(x)}{dx} - g(x)\frac{df(x)}{dx} \quad (13)$$

and

$$t_\ell^\alpha(\varepsilon) \equiv -\frac{[R_\ell(\varepsilon;b_\alpha),\ i_\ell(\kappa b_\alpha)]}{[R_\ell(\varepsilon;b_\alpha),\ k^{(1)}(\kappa b_\alpha)]} \quad (14)$$

these equations can be written in the form

$$\sum_{\beta L'}\sum \{\delta_{\alpha\beta}\delta_{LL'}[t_\ell^\alpha(\varepsilon)]^{-1} - G_{LL'}^{\alpha\beta}(\varepsilon)\}A_{L'}^\beta = 0 \quad (15)$$

These are precisely the secular equations of the conventional multiple-scattering method,[1-3] except that the quantities $t_\ell^\alpha(\varepsilon)$ (which describe the atomic partial-wave scattering amplitudes) are evaluated at the overlapping sphere radii $r_\alpha = b_\alpha < R_{\alpha\beta}$, rather than at the nonoverlapping muffin-tin radii.

The one-electron energy eigenvalues ε_i correspond to zeros of the determinant of the secular equation (15), and the partial-wave coefficients A_L^α can be obtained by solving these equations at the eigenvalues. From Eqs. (11) and (12) we can derive the following relation between the coefficients A_L^α and C_L^α :

$$A_L^\alpha = (-1)^\ell \kappa b_\alpha^{\ 2} [R_\ell(\varepsilon;b_\alpha), i_\ell(\kappa b_\alpha)]\ C_L^\alpha \tag{16}$$

Thus, having computed the coefficients A_L^α for each energy eigenvalue $\varepsilon=\varepsilon_i$, we can determine the C_L^α coefficients from the formula (16). From the latter coefficients and the radial wavefunctions $R_\ell(\varepsilon_i;r_\alpha)$, one can construct the spherically averaged orbital charge density within each atomic sphere α for each orbital eigenvalue ε_i, namely

$$\rho(\varepsilon_i;r_\alpha) = \sum_L |C_L^\alpha|^2 |R_\ell(\varepsilon_i;r_\alpha)|^2 \quad ; \ r_\alpha \leq b_\alpha \tag{17}$$

The total orbital charge in each atomic sphere is

$$q_I(\varepsilon_i;b_\alpha) = 4\pi \int_0^{b\alpha} \rho(\varepsilon_i;r_\alpha) r_\alpha^{\ 2} dr_\alpha \tag{18}$$

The normalization of each orbital wavefunction (excluding orbital and spin degeneracy) is defined so that the integral of the orbital charge density over the whole space of the molecule or cluster is equal to unity, i.e.,

$$\sum_{\alpha=1}^{N} q_I(\varepsilon_i;b_\alpha) + q_{II}(\varepsilon_i) + q_{III}(\varepsilon_i;b_o) = 1 \tag{19}$$

where $q_{II}(\varepsilon_i)$ is the effective orbital charge in the intersphere region and $q_{III}(\varepsilon_i;b_o)$ is the orbital charge in the extramolecular region (the region exterior to the outer sphere of radius b_o). The intersphere orbital charge density is taken to be a constant obeying the relation

$$\rho_{II}(\varepsilon_i) = q_{II}(\varepsilon_i)/\Omega_{II} \tag{20}$$

where Ω_{II} is the effective intersphere volume defined in expression (5). Just as Ω_{II} for overlapping atomic spheres is less than it would be in the nonoverlapping muffin-tin limit, $q_{II}(\varepsilon_i)$ is likewise reduced because of the additivity of the overlapping spherical charges $q_I(\varepsilon_i;b_\alpha)$. Thus $\rho_{II}(\varepsilon_i)$ is a very reasonable approximation to the volume averaged intersphere charge density in the overlapping-sphere model.

For relatively large amounts of overlap (but within the limit $b_\alpha < R_{\alpha\beta}$) and highly localized orbitals, it is possible that the condition

$$\sum_{\alpha=1}^{N} q_I(\varepsilon_i;b_\alpha) > 1 \tag{21}$$

will be encountered. Because of the normalization condition (19), the effective intersphere charge $q_{II}(\varepsilon_i)$ and charge density $\rho_{II}(\varepsilon_i)$ will be negative for such an orbital. This is not a serious problem and can be rectified by renormalizing the orbital charge $q_I(\varepsilon_i;b_\alpha)$ in each atomic sphere (and $q_{III}(\varepsilon_i;b_o)$, if nonzero) so that $q_{II}(\varepsilon_i)$ is identically zero.

The total electronic charge density in each atomic sphere is obtained by summing the individual orbital contributions (17) over all occupied orbitals, i.e.,

$$\rho_I(r_\alpha) = \sum_{\substack{i \\ \text{occup.}}} n_i\, \rho_I(\varepsilon_i;r_\alpha) \tag{22}$$

when n_i is the orbital occupancy (including degeneracies). Likewise, one can define the total charge densities ρ_{II} and $\rho_{III}(b_o)$ in the intersphere and extramolecular regions, respectively. On the basis of expression (19), we can write the following condition for dividing up the total electronic charge :

$$q_{tot} = \sum_{\alpha=1}^{N} q_I(b_\alpha) + q_{II} + q_{III}(b_o) = n \tag{23}$$

where n is the total number of electrons in the molecule or cluster.

From the total electronic charge density we can solve Poisson's equation of classical electrostatics to determine the Coulomb part of the potential to be used in carrying out the SCF procedure. This potential is spherically averaged in each atomic sphere, taking the form

$$V_C(r\ ) = \frac{2q_\alpha(r_\alpha)}{r_\alpha} + 2p_\alpha(r_\alpha) + 2\sum_{\beta\neq\alpha} \frac{q_\beta(b_\beta)}{R_{\alpha\beta}} + 2p_o(b_o) + V_C(\rho_{II}) \quad \text{(in Rydberg units)} \tag{24}$$

in which the following quantities are introduced:

$$q_\alpha(r_\alpha) \equiv -Z_\alpha + 4\pi\int_0^{r_\alpha} \rho_I(r_\alpha')\, r_\alpha'^2\, dr_\alpha' \tag{25}$$

$$p_\alpha(r_\alpha) \equiv 4\pi\int_{r_\alpha}^{b_\alpha} \rho_I(r_\alpha')r_\alpha' \, dr_\alpha' \tag{26}$$

$$q_\beta(b_\beta) \equiv -Z_\beta + 4\pi\int_0^{b_\beta} \rho_I(r_\beta)r_\beta^2 dr_\beta \tag{27}$$

$$p_o(b_o) \equiv 4\pi\int_{b_o}^{\infty} \rho_{III}(r_o)r_o dr_o \tag{28}$$

The first term on the right side of expression (24) [using Eq. (25)] clearly describes the Coulomb attractive potential due to the positive charge Z_α on the atomic nucleus at the origin plus the repulsive potential or "inner shielding" arising from the spherical distribution of electronic charge located between the nucleus and the field point r_α. Using Eq. (26), the second term on the right side of Eq. (24) describes the "outer-shielding" potential due to the spherical shell of charge lying between radii r_α and b_α. Using (27), the third term of (24) represents an approximate spherical average (with respect to nucleus α) of the attractive Coulomb potentials associated with the other nuclei $Z_\beta(\beta \neq \alpha)$ in the molecule plus the outer-shielding potential arising from their spherically symmetrical electronic charge distributions. This term is written simply as if the effective charges $q_\beta(b_\beta)$ were concentrated at their respective nuclei. The term $2p_o(b_o)$ in Eq. (24) describes the outer shielding potential arising from the spherically averaged charge in the extramolecular region outside the boundary of the molecule or cluster. Finally the term $V_C(\rho_{II})$ accounts for the outer shielding of each atomic sphere by the uniform intersphere charge density.

Analogous formulae can be derived for the Coulomb contributions to the potential in the intersphere and extramolecular regions, respectively. The $X\alpha$ exchange-correlation contribution to the potential is constructed in each region by substituting the corresponding charge density in the formula

$$V_{X\alpha}(\vec{r}) = -6\alpha \, [(3/8\pi)\rho(\vec{r})]^{1/3}. \tag{29}$$

Expression (24) is rigorously correct only in the muffin-tin limit of nonoverlapping spheres. Nevertheless, it appears to be a very realistic approximation to the solution of Poisson's equation in the SCF procedure for overlapping atome spheres and it is perfectly consistent with the approximations introduced in solving Schrödinger's equation for overlapping spheres. This is confirmed by the empirical observation that the use of overlapping spheres leads to realistic electronic energy levels and charge distributions for the TCNQ molecule[9] (see following section) and for ethylene, benzene, and carbon monoxyde,[8] phosphine,[10] ferrocene,[14] platinum-olefin complexes,[15] and many other unpublished examples.

II. ELECTRONIC STRUCTURE OF THE TCNQ MOLECULE

A. Preliminary Remarks Concerning Organic Crystals

In this section we will discuss the application of the overlapping atomic sphere version of the scattered-wave method to the TCNQ molecule. Before doing so, however, it is desirable to place this work in proper perspective. Organic crystals, which are our ultimate concern, can be classified into three broad categories depending on the shape of their constituent molecules.[16] The first category is composed of crystals formed from small or symmetrically shaped molecules, such as methane and hexamethylenetetramine. The crystal structures are characteristically close-packed arrangements reflecting the simple form of these molecules and the symmetrical, nondirected distribution of bonds between them.

The second and third categories consist of crystals formed from long molecules, such as long-chain paraffins, and flat molecules, such as anthracene and TCNQ. The long molecules are usually arranged in stacks with their axes parallel, and the flat molecules are usually stacked together with their planes parallel to one another, in both cases leading to anisotropic crystal structures.

In order to determine the electronic structure of an organic crystal, it is expedient to take advantage of the fact that such a crystal is usually composed of more or less well-defined molecules which are individually stable and which interact only weakly with one another. Thus, one can begin by studying the individual molecules, and then one can take their interactions into account in a self-consistent fashion. In this section we will be concerned with the TCNQ molecule, which is the constituent of a large number of charge-transfer salts[17-19], the most famous of which is TTF-TCNQ, a highly conducting organic solid.[20,21] How to deal with the interactions between adjacent TCNQ (or TTF) molecules in TTF-TCNQ, and between adjacent TCNQ and TTF molecules, is considered briefly in the final section.

B. Theoretical Alternatives

As has already been indicated in the previous section, a molecule is represented in the scattered-wave method by a collection of atomic spheres surrounded by a bounding outer sphere. A representative (nonoverlapping) atomic sphere model for the TCNQ molecule is shown in Fig. 1. The potentials and charge densities are spherically averaged inside the atomic spheres and outside the outer sphere, and volume averaged in the intersphere region. In most cases, it is possible to achieve a higher degree of physical realism by spherically averaging in atomic-like regions than by volume avera-

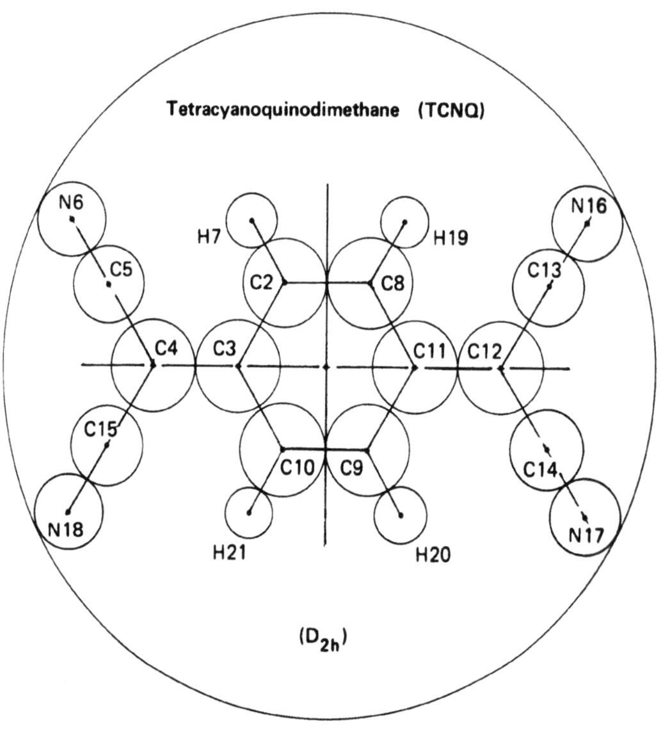

Figure 1. Nonoverlapping atomic sphere model for the TCNQ molecule. The outer sphere is assigned serial number 1 (OUT1), and the atomic spheres serial numbers 2 through 21. The molecule lies in the x-y plane, with the x and y axes pointing along the long and short molecular axes, respectively. Numerical values for the atomic positions and atomic sphere radii are given in Ref. 9. After Ref. 25.

ging outside such regions. Accordingly, it is desirable to make the atomic spheres as large as possible, and the intersphere region as small as possible. (It is not necessary for the outer sphere to lie outside all the atomic spheres. However, for the scattered-wave formalism to apply, the outer sphere must encompass the centers of all the atomic spheres.)

The original form of the scattered-wave method based on nonoverlapping atomic spheres is ideally suited for symmetrically shaped (globular) molecules since for such molecules the atomic spheres account for a considerable fraction of the outer sphere volume. One runs into difficulties as soon as one considers rod-like or dish-like molecules for which the atomic spheres make up only a small fraction of the outer sphere volume (for TCNQ, less than 3 percent!!) For rod-like and dish-like molecules, there are a number of theoretical alternatives:

1. The outer sphere can be replaced by a prolate or oblate spheroid which surrounds the atomic spheres as snugly as possible. The formalism for this geometrical arrangement has been worked out by Scheire and Phariseau,[22] but the necessary computer programs have not yet been developed. There is no question that this approach leads to a considerable reduction in intersphere volume.

2. The outer sphere can be replaced by a right cylinder which just fits around the set of atomic spheres. The appropriate formalism has been developed by Liberman,[23] but again the computer programs have not been written. Although the cylinder is less efficient than the spheroid from the standpoint of minimizing the intersphere volume, the right cylinder can be used, in conjunction with cyclic boundary conditions imposed on its bases, to deal with one-dimensional periodic structures (polymers, dislocations, etc.).

3. Whether the outer boundary is a sphere, a spheroid, or a cylinder, the atomic spheres can be supplemented by (nonoverlapping) interstitial spheres. We will have more to say about this later.

4. Whatever the form of the outer boundary, the interior of the molecule can be partitioned into space-filling atomic (and possibly interstitial) polyhedra, as is done in the Wigner-Seitz-Slater cellular method for solids.[24] The formalism for this cellular version of the multiple-scattering method has been developed by Williams,[13] and the computer programs are currently being written by Williams and his collaborators. One of the questions concerning this version is whether it will be necessary to surround the atomic polyhedra by a "coat" of superficial polyhedra to provide a suitable interface between the molecule proper and the outside world. For example, in dealing with a diatomic molecule such as O_2, each O atom can be represented by

a truncated octahedron which is surrounded by eight others, seven superficial polyhedre and an eighth containing the other O atom. Thus, the two oxygen polyhedra are fully encapsulated by 14 superficial polyhedra. If this type of coating proves to be an essential feature of the cellular version, then one will have to deal with very large basis sets, since each polyhedron requires its own spherical harmonic expansion.

5. The most direct approach is simply to increase the size of the atomic spheres so that they include more volume. By using overlapping atomic spheres, one improves the physical realism in the immediate vicinity of the nonoverlapping atomic spheres, including the bonding regions between adjacent atoms, where such improvement can be expected to do the most good. Moreover, one can make immediate use of existing computer programs, which is an important practical consideration. Of course, one can also incorporate interstitial spheres, but this would represent a refinement, the principal improvement will come from the overlapping atomic spheres.

C. Scattered-wave Calculations for the TCNQ Molecule

A number of different geometrical models of the TCNQ molecule have been used by Herman and Batra[25,26] in their studies of this molecule by the multiple scattering method. These include a non-overlapping atomic sphere model (Model I) and several overlapping atomic sphere models, two of which have already been published (Model III, Ref. 25, ModelIV, Ref. 26). Supplementary numerical information bearing on these and other overlapping atomic sphere models (e.g., Model II) are reported by Herman, Williams, and Johnson.[9] We will discuss Models I through IV briefly in what follows, together with new results which complement the earlier ones.

All the studies of the TCNQ molecules were based on a statistical exchange approximation for which α was set equal to 0.75 for all regions of space. The TCNQ molecule was assumed to be planar and to have D_{2h} symmetry. The atomic coordinates were obtained from the x-ray crystallographic data for the TCNQ crystal[27] by appropriate averaging. (The numerical values for the atomic positions are listed in Ref.9.) The symmetry notation of Cotton[28] will be used to describe the D_{2h} point group.

As a compromise between numerical accuracy and computational effort, spherical harmonics up to $l_{max} = 4$ and 2 were used outside the outer sphere and inside all the atomic spheres, respectively. All the numerical work was carried out using computer programs originally developed at M.I.T. by Johnson and Smith[29] and subsequently modified at IBM San Jose Research Laboratory by Liberman and Batra[30] and Herman. All excitation and ionization energies were determined by the transition-state method,[7] so that relaxation effects were taken into account.

D. Nonoverlapping Atomic Sphere Model

The nonoverlapping atomic sphere model (Model I) shown in Fig. 1 and also in Fig. 2 (dashed circles) was constructed as follows :

The atomic radii for spheres C2,C3, C4, and H7 were determined by the requirement that adjacent spheres should touch. In determining the radii of C5 and N6, the additional requirement was imposed that their radii should be proportional to the atomic radii appropriate to carbon-carbon and carbon-nitrogen double bonds. The outer sphere (not shown in Fig. 1) was drawn externally tangent to the outermost (N6) atomic spheres. Although these choices are all clearly arbitrary, they serve to maximize the volume enclosed by the set of atomic spheres and to minimize the intersphere volume. This is consistent with the idea of representing as large a fraction of the molecule as possible by (relatively more realistic) spherically averaged atomic regions, and as small a fraction by (relatively less realistic) volume averaged intersphere region.

The results obtained for Model I (cf. Refs. 9, 25, and 26) were disappointing in many respects : (1) The calculated value of the virial ratio (total potential energy divided by total kinetic energy) was -1.858, as opposed to the theoretical value of -2, indicating a serious imbalance of potential and kinetic energies, at least within the present approximational framework.[31] (2) The calculated value of the first ionization energy was 16.4 eV, which is considerably larger than the experimental value of 9.6 eV obtained from photoemission measurements for TCNQ vapor.[32] (3) the higher-lying occupied energy levels were found to be much closer togheter than is indicated by photoemission measurements for TCNQ vapor and solid TCNQ.[33] (4) The lowest excitation energy (weighted mean of singlet and triplet) was found to be about 1 eV, which is too small by a factor of 2. (The main absorption peak for TCNQ molecules in solution is 3.1 eV.[34] This is usually attributed to the lowest singlet-singlet transition. If we assume a singlet-triplet splitting of about 1 eV, then the weighted mean of the singlet-singlet and singlet-triplet transition energies should be about 2.3 eV.)

In brief, Model I leads to occupied levels which have excessively large ionization energies. Moreover, the higher-lying occupied levels and the lower-lying unoccupied levels are bunched too close together, so that optical excitation energies are too low. These shortcomings must be attributed to the fact that the unphysical intersphere region plays too large a role in determining the electronic energy levels and charge distribution, and the more physically realistic atomic sphere regions too small a role. In this connection;

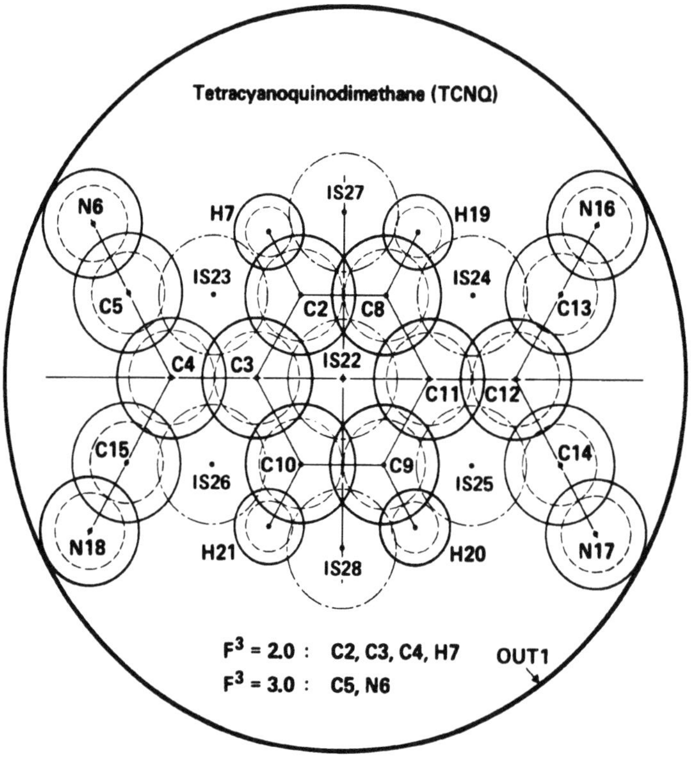

Figure 2. Nonoverlapping atomic sphere Model I (dashed circles) and overlapping atomic sphere Model IV (solid circles) for the TCNQ molecule. The outer sphere for Model I is not shown in the interest of clarity. The ratio of the volumes of corresponding overlapping and nonoverlapping atomic spheres is denoted by F^3. Interstitial spheres are not used in Model I (Ref. 25) or Model IV (Ref. 26). These spheres, denoted by serial numbers IS22 through IS28, are incorparated in Models VII through X, as described in the text. All interstitial spheres have the same radius as C2 in these particular models (VII through X). Adapted from Ref. 26.

it should be noted that the TCNQ molecule has 32 core electrons and 72 valence electrons, making a total of 104 electrons. The ground state solution for Model I places 0.3 electrons outside the outer sphere, 26.4 electrons in the intersphere region, and 45.3 valence plus 32.0 core electrons in the atomic sphere regions. Examining the valence orbitals individually, we find that 30 to 40 percent of the one-electron charge lies outside the atomic spheres for each of these. While it may not be unreasonable to find 30 to 40 percent of a delocalized orbital lying outside the atomic spheres, it is certainly unreasonable for 30 to 40 percent of the more localized orbitals to lie outside the atomic spheres.

E. Overlapping Atomic Sphere Models : General Considerations

It is reasonable to conclude that Model I is unsatisfactory because too much of the valence electronic charge lies outside the atomic spheres in the physically unrealistic intersphere region. In order to remove charge from this region and transfer it to more realistic atomic-like regions, we will enlarge the size of our atomic spheres, causing them to overlap one another. As is clear from any picture of the electronic charge distribution of a molecule, most of the charge in the intersphe region lies in the immediate vicinity of (i.e., just outside) the nonoverlapping atomic spheres. Thus, from a purely geometrical point of view, overlapping atomic spheres provide a highly efficient means for capturing a large fraction of the electronic charge lying outside nonoverlapping atomic spheres.

To make this point more concrete, we show in Fig. 3 the total valence electron charge distribution for the TCNQ molecule in a plane parallel to the plane of the molecule and lying 0.5 Bohr units away from the latter. The circles represent the intersection of the 20 overlapping atomic spheres and the outer sphere with the plane of the molecule. The contour of maximum charge density just happens to coincide, very nearly, with the outer portions of the carbon and nitrogen circles. There is a shallow minimum at the center of the quinone ring. As one moves away from the molecule on the outside, the charge density falls rapidly. Neighboring contours differ by about a factor of 2, so that the outermost contour repredents a charge density about 0.001 as large as the maximum charge density contour. For the overlapping atomic sphere model shown in Fig. 3, about 9 electron charges lie outside the atomic spheres (as opposed to about 27 for nonoverlapping atomic spheres). Thus about 2/3 of the intersphere charge of Model I (nonoverlapping spheres) can be shifted into atomic-like regions by enlarging the atomic spheres to the extent shown in Fig. 3.

For modest amounts of overlap, the increase in physical realism will undoubtedly overshadow the errors introduced by spherically

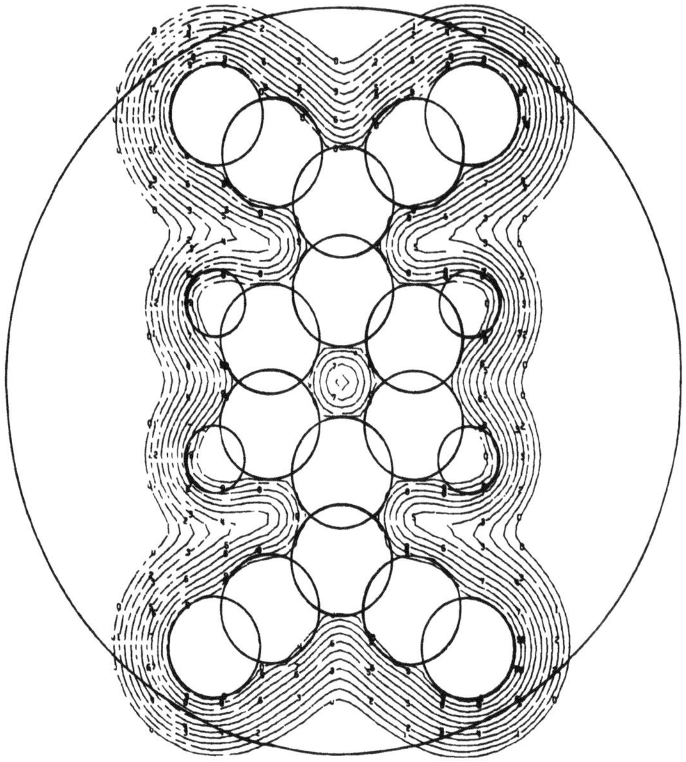

Figure 3. Computer generated contour map of the total valence electron charge distribution for the TCNQ molecule. Based on overlapping atomic sphere Model VI (cf. Refs. 9 and 35 for details). The contours describe the charge distribution in a plane parallel to the molecular plane but displaced from the latter by 0.5 Bohr units. (The radius of C2 is 1.6 Bohr units.) The intersections of the atomic and outer spheres of Model VI with the molecular plane are described by the circles. Note that for Model VI the outer sphere lies somewhat beyond the outermost atomic spheres. The plots were made at IBM San Jose using a computer program developed at G.E. and M.I.T. by R.P. Messmer, S.K. Knudson, and K.H. Johnson.

averaging potentials and charge distributions inside the overlapping atomic spheres. These errors will become progressively more important and will lead to difficulties at sufficiently large overlap (if one does not take suitable precautions). Accordingly, one should introduce no more overlap than necessary.

F. Overlapping Atomic Sphere Models. Uniform Scalling

Since overlapping atomic sphere models are still in an early stage of development, it is appropriate to deal with the choice of atomic radii pragmatically. As can be seen from Fig. 1, the TCNQ molecule has six different types of atoms (C2, C3, C4, C5, N6 and H7). If one assigns a different radius to each type of atom, and does not insist on an externally tangent outer sphere, the general overlapping atomic sphere model for TCNQ would have seven degress of freedom. This is too many to deal with directly.

In Ref. 25, Herman and Batra proceeded on the assumption that the principal degree of freedom was a uniform scaling of all the nonoverlapping atomic spheres. In their preliminary studies, the outer sphere was drawn externally tangent to the outermost atomic spheres. In later studies, the outer sphere radius was modified as an added refinement.

We will denote by F and F^3 the ratios of the overlapping and nonoverlapping atomic sphere radii and volumes, respectively. We will refer to F^3 as the scale factor. As F^3 is increaded from 1 (corresponding to Model I, no overlap) to 2, the first ionization energy is reduced from 16.4 to 10.3 eV, the first excitation energy is increased from 0.9 to 1.9 eV, and, generally speaking, the theoretical energy level spectrum comes considerably closer to experiment than before. Moreover, the intersphere charge (QINT) falls from 26.4 to 12.5. Each of the valence orbitals becomes more concentrated in the atomic sphere regions, the localized orbitals more so than the delocalized orbitals. Some of the delocalized orbitals continue to have as much as 35 percent of their charge in the intersphere region, but a few (the four lowest-lying sigma orbitals which are strongly localized on the cyano groups) actually end up with slightly negatibe intersphere charges (this is an artifact of the present set of approximations). As is indicated in greater detail in the previous section and in Ref. 9 whenever the intersphere charge for a particular orbital comes out negative, this quantity is set equal to zero and the remaining charge is suitably renormalized.

As F^3 is increased from 1 to 2, the calculated virial ratio changed from -1.858 to -1.954, a value considerably closer to the theoretical value of -2. One way of fixing F^3 is to require the calculated first ionization energy to agree with the experimental value. This empirical approach leads to the choice $F^3 = 2.1$, corresponding to Model II (cf. Ref. 9 for additional details). The calculated energy level structure is compared with the experimental photoemission spectrum for TCNQ vapor in Fig. 4. (In this figure, Model II is represented by the long dashes.) It will be seen that the agreement between theory and experiment is satisfactory.

Another way of fixing F^3 is to require the calculated virial ratio to become equal to -2. This "theoretical" approach leads to the choice $F^3 = 2.6$. In Ref. 25, Herman and Batra report results for $F^3 = 2.5$, for which the virial ratio is -1.993 (close enough to -2 for all practical purposes). The calculated energy level spectrum for $F^3 = 2.5$ (Model III) agrees quite nicely with the experimental spectrum on a relative energy scale, but not on an absolute energy scale. In particular, all the theoretical levels are about 1.2 to 1.5 eV higher (less negative) than they should be. This deficiency is easily remedied by suitably changing the outer sphere radius, which here plays the role of a second adjustable parameter. When the outer sphere radius is increased, all the energy levels drop, more or less by the same amount. Thus, the calculated and experimental energy level spectra can be brought into close registry.

For purposes of comparison, we note that for Model III (F^3= 2.5, externally tangent outer sphere), the intersphere charge (QINT) is 8.6. While only 4 of the 36 valence orbitals had negative intersphere charges for $F^3 = 2$, now 16 have negative intersphere charges, though most of these are just slightly negative. The first excitation energy for $F^3 = 2.5$ is 2.2 eV, and the first ionization energy is 8.1 eV. The former is not changed when the outer sphere radius is increased, while the latter is readily adjusted by this means to the experimental first ionization energy (9.6 eV).

Model III clearly distinguished three classes of molecular orbitals. There are 4 strongly localized sigma orbitals (concentrated on the cyano groups) ; these have moderately negative intersphere charges which we set equal to zero. There are 16 delocalized orbitals, including 8π orbitals, 4 nitrogen lone pairs, and 4π' orbitals belonging mostly to the cyano groups (these are π orbitals in the plane of the molecule). The intersphere charge for these 16 delocalized orbitals ranges from 0.2 to 0.3. The remaining 16 orbitals are intermediate, having slightly positive or slightly negative intersphere charges (plus or minus a few hundredths). The electronic charge distributions for some of these orbitals are depicted in Refs. 9 and 35.

As the overlap is increased still further, the total intersphere charge (QINT) continues to fall, but less rapidly than initially. By the time we reach $F^3 = 3$, the intersphere charge is only 6.15. The decrease from $F^3 = 1$ to about 2.5 is quite rapid since this is determined primarily by the localized orbitals being drawn into the atomic spheres. Above about 2.5, the decrease in QINT is slower, since the localized orbitals are now as localized as they can be, and it is only the delocalized orbitals that can still be brought into the atomic spheres by further increase in their size.

The energy level structure varies rapidly with increasing overlap in the range between F^3 = 1 and 2, and progressively less rapidly thereafter, with certain marked exceptions. The exceptions are four highly localized sigma bonding orbitals localized on the cyano groups which progressively acquire large negative intersphere charges and as a consequence experience anomalously large downward energy shifts. These shifts are artifacts of the present set of approximations and must be discounted as unphysical. We will return to these shifts after discussing Model IV.

G. Overlapping Atomic Sphere Models : Nonuniform Scaling

At this point we wish to remind the reader that there are single, double, and triple bonds in the TCNQ molecule (cf. Fig.4) and that the corresponding bond lengths are significantly different from one another. In particular, the nonoverlapping atomic sphere for C5 (triple bond on cyano group) is considerably smaller than the nonoverlapping atomic spheres for C2 (double bond on quinone ring) and C3, C4 (double bond on bridging carbons). Of course, uniform scaling maintains these proportions.

While uniform scaling makes some allowance for differences in local environment, it is not obvious that the optimum set of overlapping atomic spheres should have the same proportions as the original set of nonoverlapping atomic spheres. It is conceivable, for example, that uniform scaling exaggerates the difference in size between C2, C3, and C4 on the one hand, and C5 on the other. In order to explore alternatives to uniform scaling, Herman and Batra (in Ref. 26) scaled the double and triple bond regions separately, thereby allowing C5 in particular to adjust itself relative to C2, C3, C4.

Since there are now two degrees of freedom, it is necessary to impose two conditions in order to determine the two scaling factors (F^3 for C2, C3, C4, and H7; and F^3 for C5 and N6). (The outer sphere was drawn externally tangent to the outermost atomic spheres.) One condition was that the calculated first ionization energy should agree with the experimental value. The second was that the calculated virial ratio should be equal to -2. Imposing these two conditions leads to Model IV, which is defined by F^3 = 2.1 for C2, C3, C4, and H7; and F^3 = 3.0 for C5 and N6. The energy level structure for Model IV (cf. Ref. 26) is shown in Fig. 4 by the small dashes, where it is compared with the energy level structure for Model II (F^3 = 2.1 for all atoms) and with the experimental photoemission spectrum for TCNQ vapor.

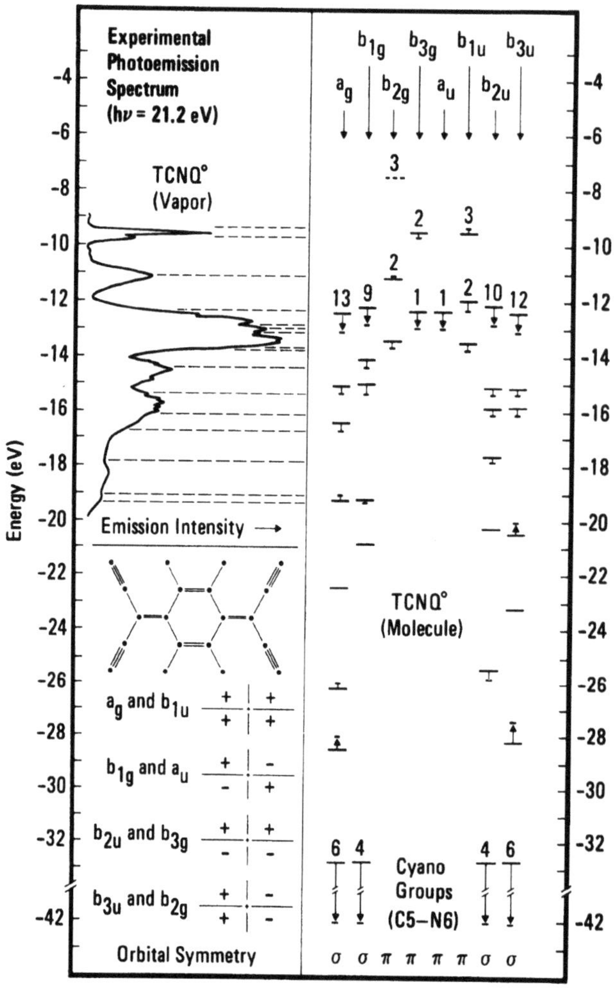

Experimental Photoemission Spectrum (hν = 21.2 eV)
TCNQ° (Vapor)
Emission Intensity →
Energy (eV)
a_g and b_1u
b_1g and a_u
b_2u and b_3g
b_3u and b_2g
Orbital Symmetry
a_g b_1g b_2g b_3g a_u b_1u b_2u b_3u
TCNQ° (Molecule)
Cyano Groups (C5–N6)
σ σ π π π π σ σ

It will be seen form Fig. 4 that the energy level structures for Models II and IV are quite similar, except for some small shifts among the higher-lying occupied levels, and some large shifts of the four lowest-lying occupied levels ($6a_g$, $4b_{1g}$, $4b_{2u}$, and $6b_{3u}$). All of these shifts are produced by the increase in size of the C5 and N6 atomic spheres. The agreement between theory and experiment appears to be somewhat better for Model IV than for Model II, at least so far as the higher-lying occupied levels are concerned. Judging from ESCA measurements for TCNQ[36] and theoretical studies of HCN,[37] the four lowest-lying occupied levels should lie at -30 eV or slightly below. The results given by Model II for these levels are realistic; those given by Model IV are anomalously low.

For Model II, these levels have slightly negative intersphere charges, while for Model IV these charges are strongly negative. Even though the overall charge distribution continues to be realistic after these charges are set equal to zero and the orbital char-

Figure 4. Comparison of calculated energy level structure of the TCNQ molecule and experimental photoemission spectrum of TCNQ vapor (Ref. 32). The long dashes refer to Model II (cf. Ref. 9) and the short dashes to Model IV (cf. Ref. 26). Where short dashes do not occur, they coincide with the long dashes. In some instances the shift from Model II to Model IV is indicated by an arrow. The dashed lines in the upper left corner show the connection between the experimental spectrum and Model IV. The symmetry notation for the D_{2h} point group is that of Ref. 28. The symbols σ and π refer to reflection with respect to the x-y plane of the molecule. The remaining orbital symmetry properties are indicated schematically in the lower left corner. For example, fully symmetric linear combinations of s and p_z orbitals located on the nitrogen atoms would be labeled a_g and b_{1u}, respectively. For each symmetry species, the levels are labeled by serial numbers in order of increasing energy. The highest occupied orbitals are $3b_{1u}$ and $2b_{3g}$. The lowest unoccupied orbital is $3b_{2g}$ (shown dashed). The experimental spectrum was obtained by Dr. J. N. A. Ridyard using a Perkin-Elmer Photoemission Spectrometer Model PS 18. After Ref. 9.

ge distributions in the remainder of the molecule are renormalized, the energy levels (calculated before renormalization) are not to be trusted for such large negative intersphere charges. It would certainly be desirable to introduce modifications in the existing formalism that would eliminate negative intersphere charges and related difficulties, such as the large energy shifts just mentioned. For example, one could use smaller atomic spheres for the most highly localized molecular orbitals, thereby avoiding excessive overlap for such orbitals and at the same time using acceptably large overlap for the less localized ones.

Within the existing theoretical framework, the most sensible approach is to tescale the various overlapping atomic sphere radii so as to eliminate as many (large) negative orbital interstitial charges as possible. As long as these charges are very close to zero, it hardly matters whether they are positive or negative, but when they become strongly negative, say more negative than -0.1, a serious attempt should be made to reduce their negativity by suitable rescaling.

In spite of minor difficulties, such as negative orbital intersphere charges introduced by excessively large overlap, the overlapping atomic sphere model in general represents a considerable improvement over the nonoverlapping atomic sphere model for TCNQ, as we have demonstrated. The same should be true for other large disk-like molecules as well as for long rod-like molecules. Future improvements will involve the incorporation of nonspherical terms in the atomic regions and nonconstancy terms in the intersphere region. Even in its present form, however, the overlapping sphere model appears capable of dealing effectively with large planar organic molecules and with arrays formed from such molecules.

H. Effects of Outer Sphere Radius and Size of Basis Set

Model I through IV have all made use of spherical harmonic expansions for which $l_{max} = 4$ for the outer sphere, and $l_{max} = 2$ for all the atomic spheres. Moreover, the outer sphere was made externally tangent to the outermost atomic spheres. We have also carried out studies using fewer spherical harmonics and outer spheres that were not externally tangent (Cf. Fig. 3). We have found that the number of spherical harmonics for the hydrogen spheres and for the outer sphere can be reduced without signicicantly affecting the calculated energy level structure or charge distribution. To illustrate this point, and to set the stage for our subsequent discussion of interstitial spheres, we compare selected results for several models in Table I to IV. For the moment, we will concentrate on Models V and VII.

Table I. Comparison of charge distributions for various overlapping atomic sphere models for the TCNQ molecule. The outer sphere and intersphere charges, Q(OUT1) and Q(INT), appear in the rows denoted by OUT1 and INT. The total charge in the C2 sphere, Q(C2), appears in the row denoted by C2, etc. The sum over all regions should equal 104 for each model, where 104 is the total number of electrons.

Repesentative Region and Number of Equivalent Regions		Model and Total Inner Spheres				
		V	VII	VIII	IX	X
		21	21	22	26	28
OUT1	1	0.26	0.05	0.05	0.05	0.05
C2	4	5.59	5.60	5.54	5.35	5.27
C3	2	5.65	5.67	5.51	5.30	5.29
C4	2	5.58	5.59	5.61	5.37	5.37
C5	4	5.54	5.54	5.54	5.46	5.46
N6	4	6.25	6.25	6.26	6.24	6.26
H7	4	0.63	0.63	0.63	0.59	0.56
IS22	1	0	0	1.05	1.05	1.06
IS23	4	0	0	0	0.71	0.71
IS27	2	0	0	0	0	0.44
INT	1	9.25	9.37	9.12	8.17	7.66

Table II. Comparison of potentials in the intersphere region (INT) and at the centers of representative interstitial spheres (IS22, IS23, IS27) for various models of the TCNQ molecule. All entries are in Rydberg units and have multiplied by minus 1.

Region	Model and Total Inner Spheres				
	V	VII	VIII	IX	X
	21	21	22	26	28
INT	0.412	0.386	0.373	0.335	0.317
IS22			0.886	0.911	0.928
IS23				0.947	0.947
IS27					0.738

Table III. Comparison of energy eigenvalues for various overlapping atomic sphere models for the TCNQ molecule. All entries are in Rydberg units and have been multiplied by minus 1. The lower-lying unoccupied levels are shown. Closely spaced levels are presented together.

Orbitals		Model and Total Inner Spheres				
		V	VII	VIII	IX	X
		21	21	22	26	28
$4b_{2g}$	π^*	0.08	0.08	0.04		
$3a_u$	π^*	0.17	0.16	0.12	0.07	0.05
$3b_{3g}$	π^*	0.18	0.17	0.15	0.10	0.10
$10b_{1g}$	σ^*	0.12	0.16	0.14	0.15	0.15
$11b_{2u}$	σ^*	0.17	0.16	0.14	0.16	0.16
$2a_u$	π^*	0.22	0.22	0.17	0.13	0.12
$14a_g$ } $13b_{3u}$	σ^*	0.24	0.23	0.21	0.24	0.24
$4b_{1u}$	π^*	0.28	0.28	0.23	0.14	0.13
$3b_{2g}$	π^*	0.45	0.45	0.40	0.33	0.32

Table IV. Comparison of energy eigenvalues for various overlapping atomic sphere models for the TCNQ molecule. All entries are in Rydberg units and have been multipled by minus 1. The higher-lying occupied levels are shown. Closely spaced levels are presented together.

		Model and Total Inner Spheres				
		V	VII	VIII	IX	X
Orbitals		21	21	22	26	28
$3b_{1u}$	π	0.59	0.59	0.55	0.48	0.47
$2b_{3g}$	π	0.62	0.62	0.54	0.49	0.47
$2b_{2g}$	π	0.73	0.72	0.67	0.61	0.60
$2b_{1u}$	π	0.81	0.80	0.76	0.70	0.69
$9b_{1g}$ } $10b_{2u}$	σ	0.86	0.86	0.84	0.79	0.79
$1b_{3g}$ } $1a_u$	π	0.87	0.87	0.85	0.80	0.79
$12b_{3u}$ } $13a_g$	σ	0.88	0.88	0.86	0.83	0.83
$1b_{2g}$ } $1b_{1u}$	π	0.92	0.92	0.90	0.84	0.83
$8b_{1g}$	σ	0.94	0.94	0.87	0.82	0.81
Next 7 levels	σ	1.06 ±0.02	1.06 ±0.02	1.03 ±0.01	0.99 ±0.01	0.98 ±0.01

Model V describes an overlapping atomic sphere model similar to Model IV except that $F^3 = 2.0$ rather than 2.1 for C2, C3, C4, C5, and H7. Model VII is identical to Model V except that a smaller basis set is used. For H7 and the outer sphere (OUT1), $l_{max} = 1$ and 2 are used instead if $l_{max} = 2$ and 4, respectively. For the remaining atomic spheres, $l_{max} = 2$ as before.

As can be seen from Table I, reducing the basis set for H7 has no effect on the total charge in the hydrogen spheres, suggesting that $l_{max} = 1$ is adequate for hydrogen. Reducing the basis set for OUT1 leads to a reduction in the total charge outside the outer sphere, Q(OUT1), from 0.26 to 0.05 electronic charge. At the same time, the total intersphere charge, Q(INT), increases by 0.12 from 9.25 to 9.37, partially compensating for the reduction of 0.21 in Q(OUT1). The remaining 0.09 is distributed among the various atomic spheres, affecting each of these to a negligible extent. According to Tables II and III, the energy eigenvalues for Models V and VII are nearly identical to one another. The difference between corresponding levels is typically less than 0.01 Ryd.

Accordingly, we can use smaller basis sets for H7 and OUT1 without affecting the electronic structure to any significant degree, and we will do so when to incorporate interstitial spheres in Models VII through X. We have also investigated[25,26] the consequences of reducing l_{max} from 2 to 1 for the various carbon and nitrogen spheres. This reduction leads to an increase in Q(INT) by about 10 percent, and to compensating decreases in the various atomic sphere charges, Q(ATOM), by 1 percent or less (each). There are detailed changes in the energy level structure, but most of these can be described by a rigid shift of one set of levels relative to the other set. The differential changes are usually small (of the order of a few tenths of an eV). In our earlier work[25,26] we used the larger basis set as a compromise between accuracy and computational effort. This was appropriate as long as we did not include interstitial spheres. When we do, the compromise shifts to the basis set we have already described for Model VII.

We have also begun with Model V and moved the outer sphere outward, as discussed in Refs. 9 and 35 (cf. Fig. 3). This leads to Model VI, which is the same as Model V except that the outer sphere is as far away from the nitrogen nuclei as the neighboring carbon nuclei are. Shifting the outer sphere does not have any significant effect on the separation of adjacent energy levels. The principal effect is a rigid shift in energy of all the levels. Thus, one can use the radius of the outer sphere as an adjustable parameter, as we have done in Ref. 26, to bring the calculated and experimental first ionization energies into registry (for example).

Generally speaking, overlapping atomic sphere models lead to more definite results for relative energies than for absolute energies, in the sense that the former are less sensitive to the details of the model than are the latter. Accordingly, it makes good sense to fix the absolute energy scale by adjusting at least one energy level.

I. Inclusion of Interstitial Spheres

The volume of the intersphere region can be reduced by contracting the radius of the outer sphere (within limits), by increasing the radii of the various atomic spheres (again within limits), and by introducing interstitial spheres. The latter may overlap neighboring atomic spheres as well as each other. Of course, the larger the interstitial spheres, the more volume they remove from the intersphere region. The present formalism applies equally to atomic and interstitial spheres.

The advantage of including interstitial spheres is that the intersphere region can be partitioned into regions that can be dealt with individually. One can obtain self-consistent charge densities and potentials for each of these regions, though according to the present formalism these quantities would be spherically averaged. Even so, these spherical averages may, under many conditions, represent improvements over the original volume averages of the intersphere region as a whole. This point should not be taken for granted, however, and each particular situation should be cardfully examined to see whether interstitial spheres improve or worsen the overall molecular model.

The principal disadvantage of incorporating interstitial spheres is that additional spherical harmonics have to be used for each of these spheres, thereby increasing the size of the basis set and hence the length and cost of the computation. There is also the danger that the introduction of an interstitial sphere in a particular region will improve the effective molecular potential for certain orbitals but not for others, thereby exaggerating the differences between them. Accordingly, one should introduce the interstitial spheres in as uniform and consistent a manner as possible.

To illustrate the consequences of introducing interstitial spheres, we will now consider three additional models for the TCNQ molecule.

Model VIII differs from Model VII only by having an interstitial sphere (IS22) at the center of the quinone ring (cf. Fig. 2). Model IX has in addition to IS22 four more interstitial spheres, IS23 through IS26. Model X has 7 interstitial spheres in all; IS22 through IS28. All the interstitial spheres have the same radius as C2, and all are described by spherical harmonic expansions with

ℓ_{max} = 1. The remaining portion of the basis set (atomic spheres and outer sphere) is identical to that previously described for Model VII. Thus, Model VII (no interstitial spheres) is the reference against which Models VIII, IX, and X should be compared.

As can be seen from Fig. 2, IS22 is fully surrounded by neighboring atomic spheres in the plane of the molecule, IS23 is almost fully surrounded, and IS27 is only half surrounded. It is easier to justify placing an interstitial sphere at IS22 than at IS23 or IS27, because potentials and charge densities are expected to be more nearly spherically symmetrical inside the quinone ring than outside. Nevertheless, it is instructive to examine the consequences of placing interstitial spheres at the three different types of positions typified by IS22, IS23, and IS27.

Turning to Table I, we see that the intersphere charge, Q(INT), gradually becomes smaller as more and more of this charge is reassigned to interstitial spheres as we proceed from Model VII to Model X. Note also that the exterior charge, Q(OUT1), remains fixed for these four models. The clearest way to determine the redistribution of (intersphere) charge is to take the differences between successive values of Q(INT). Going from Model VII to VIII, Q(INT) is reduced from 9.37 to 9.12, indicating that 0.25 electron charge has been removed from the intersphere region by IS22. The total charge inside IS22, Q(IS22), is actually 1.05 for Model VIII, but this includes 0.80 charge that has been reassigned from the four C2-type and the two C3-type atomic spheres. (Bear in mind that IS22 overlaps its six neighboring atomic spheres.)

By similar arguments, we find that 0.24 charge is removed from the intersphere region and placed in each of the spheres IS23 through IS26, and 0.26 charge in each of the spheres IS27 and IS28. Thus, even though IS22 through IS28 are located in three different types of environments, each of these interstitial spheres ends up removing roughly 1/4 of an electron charge from the original intersphere region. Actually, IS27 and possibly IS23 are more properly described as <u>peripheral</u> or <u>superficial</u> spheres. In what follows we will refer to them as <u>interstitial</u> or <u>peripheral</u> spheres depending on the context.

According to the charge density contour map shown in Fig. 3, virtually all of the electronic charge in the intersphere region bordering the plane of the molecule can be included in a set of overlapping <u>peripheral</u> spheres which surround the molecule. (All of these are centered in the plane of the molecule and, and for the purposes of discussion, can be assumed to have the same radius as C2.) The six peripheral spheres that we have already considered, IS23 through IS28, are amoung the 32 that together can encompass the molecule, very nearly, in its plane. The remaining 26 include:

two located next to C4 and C12; four pairs of two straddling the cyano groups (one pair overlaps C5 and N6, for example); four pairs of two lying adjacent to the previous pairs but further out; four located near the hydrogen spheres; and four located on the far side of the nitrogen spheres. (Inclusion of these last four requires the use of a larger outer sphere.)

On the basis of the information contained in Table I, it is reasonable to expect each of the 28 peripheral spheres next to carbon and nitrogen spheres, as well as IS22, to capture 1/4 electron charge from the intersphere region. The four peripheral spheres next to the hydrogen atoms would each capture perhaps 1/8 electron charge. If we were to augment Model VII with IS22 and the 32 peripherical spheres just described, we estimate that Q(INT) would fall from 9.37 to about 1.62, where the difference is taken simply as 29 x 0.25 + 4 x 0.125 = 7.75. Most of the residual intersphere charge (I.62 electron charges) would most likely lie above and below the atomic spheres, where the interstitial and peripheral spheres dc not penetrate. This residual charge could be reduced still further by increasing the radii of the various atomic spheres, but this should not be done if it leads to excessive atomic sphere overlap. One could also increase the radii of the interstitial and peripheral spheres, but this would probably be less effective than increasing the atomic sphere radii.

Alternatively, we could do away with the outer sphere entirely, and with it the intersphere region. Each molecular orbital would then be normalized to the region enclosed by the sum of all the atomic, interstitial, and peripheral spheres. With only 1 or 2 electron charges (our rough estimate was 1.62) to be redistributed among 20 atomic spheres and 33 interstitial / peripheral spheres, we would expect 0.1 or less electron charge to be added to each of the carbon and nitrogen spheres, a quantity negligible compared with the 5 or more electron charges already in each of these spheres. We would expect the energy level structure to be changed hardly at all by eliminating the intersphere region in this manner.

To recapitulate, it is clearly possible to remove virtually all the intersphere charge by increasing the radii of the overlapping atomic spheres and introducing 33 interstitial/peripheral spheres. Altough the size of the basis set is increased, the calculations remain tractable. By surrounding the atomic spheres by interstitial/peripheral spheres in the plane of the molecule, one could probably reduce l_{max} from 2 to 1 for the carbon and nitrogen spheres, thereby compensating in part for the interstitial/peripheral sphere augmentation. If one were to study a TCNQ dimer by this technique, one would also consider inserting a sheet of interstitial spheres between the neighboring TCNQ molecules. One might then render the calculations tractable by employing partitioning techniques, as described elsewhere in this paper.

It would obviously be more efficient to leave out the interstitial/peripheral spheres entirely and deal only with even larger overlapping atomic spheres. However, there are difficulties associated with excessively large overlap that should be overcome before this course is attempted. Within the existing framework, it is feasible to treat the TCNQ molecule as a set of 20 overlapping atomic spheres augmented by 33 overlapping interstitial/peripheral atomic spheres. It would be of considerable interest to see how the electronic structure determined on the basis of this augmented model compares with the present results.

It should be emphasized that a molecule such as TCNQ readily lends itself to interstitial sphere augmentation because the principal atoms, carbon and nitrogen, have nearly the same atomic size. In a case such as TTF, where the sulfur atom is considerably larger than the carbon atom, it is less convenient, though not impossible, to surround the molecule in its plane with a set of overlapping interstitial spheres (each having the same radius as one of the carbon atoms). In the neighborhood of the sulfur atoms, it might even be referable to use pairs of overlapping interstitial spheres (one above the other) to make up for the difference in size between the sulfur sphere and the carbon-like interstitial spheres.

In Table II we compare the potentials in selected regions for the various models. The volume averaged intersphere potential changes from -0.386 to -0.317 Rydberg as we progress from Model VII to Model X. In each of the interstitial/peripheral spheres, the spherically averaged potential has a radial dependence which looks like an upside down parabola. The potential is flat over most of the interior, and then plummets downward close to the sphere boundary due to the overlap with the neighboring atomic spheres, which have much deeper potentials. The flat portions for IS22, IS23, and IS27 are all considerably lower (-0.928, -0.947, -0.738) than the intersphere value, -0.317 (Model X), indicating how significant a correction the interstitial/peripheral spheres make to the intersphere potential.

It might be thought at first glance that spherically averaged peripheral sphere potentials for IS27 and IS28 would be highly inaccurate in the hemispheres lying away from the molecule. However, since the molecular orbitals that we are considering are extremely small in these regions, such inaccuracies have very little if any practical effect on the energy levels. The principal effects come from the hemispheres which are contiguous to the molecule.

The energy levels for the lower-lying unoccupied orbitals and the higher-lying occupied orbitals are compared for the various models in Table III and IV. Considering the fact that the molecule space is partitioned in a significantly different way for the two extremes, Model VII and X, it is indeed remarkable that the eigenvalues for these two models are so similar. The most striking feature is the similarity in the eigenvalues for the unoccupied orbitals, which are highly delocalized.

The differences between the eigenvalues for Models VII and X can be resolved into a rigid shift of one set of levels relative to the other (by about 1 eV), and into differential shifts that are sometimes as large as 1 eV but are usually smaller. The rigid shift is due to the charge in the value of the intersphere potential, which in turn is brought about by changes in the intersphere charge and volume. Analogous comparisons of excitation and ionization energies determined by the transition state method would reveal smaller rigid shifts because of cancellation effects.

A close examination of the differential shifts indicates that these are due primarily to "preferential treatment." That is to say, some orbitals benefit more from the seven interstitial spheres than do others. Since we have only partially surrounded the molecule with peripheral spheres, we have not dealt with all the orbitals on an equal footing, and that is why differential shifts occur. In particular, we have given preference to the quinone ring and largely ignored the cyano groups. Clearly, if interstitial spheres are to be introduced at all, they should be distributed as uniformly and consistently as possible.

It is of considerable interest to learn that the differential effects produced by "preferential treatment" are indeed quite small, usually amounting to a fraction of an eV. This result supports the view that the electronic energy level structure is determined overwhelmingly by the electronic energy charge distribution within the overlapping atomic spheres, provided these are made sufficiently large, and that this structure is affected only slightly by the residual charge outside. In a sense, this accounts for the success of the overlapping atomic sphere model in its present form. At the same time, this result suggests that further improvements in this model should be sought, particularly improvements that make possible the inclusion of as much of the total molecular charge in overlapping atomic spheres as possible.

Information regarding the relative merits of the molecular orbitals and derived quantities (e.g., dipole moments, oscillator strengths) obtained from Models VII through X is not yet available. The principal result to date is that the calculated electronic structure of the TCNQ molecule is affected only slightly by adding seven overlapping interstitial spheres to the twenty overlapping atomic spheres. This gives us added confidence in our earlier results.[9,25,26]

J. Discussion

For large planar organic molecules such as TCNQ, overlapping atomic sphere provide considerably greater physical and chemical realism than do nonoverlapping atomic sphere models. Our experience has been that the calculated energy level spectrum assumes a certain "natural" form that is relatively insensitive to the geometrical details of the model, including the presence of interstitial spheres, provided the atomic spheres overlap sufficiently ($F^3 \gtrsim 2$) but not excessively ($F^3 \lesssim 3$), where F^3 is the ratio of corresponding overlapping and nonoverlapping atomic spheres. Although the detailed form of the energy level spectrum can be modified by adjusting the various atomic sphere radii, as well as the outer sphere radius, the gross features remain the same for a wide range of chemically reasonable choices of these radii.

III. GENERALIZED MOLECULAR PARTITIONING IN THE SCF-X SCATTERED-WAVE METHOD

In the standard SCF-Xα-SW method, we treat all atoms of the system on an equal basis and solve Schrödinger's equation accordingly. However, there are many large polyatomic systems where this approach, even with its relatively efficient computational procedure, becomes unnecessarily complicated. Examples include a central molecule with polyatomic ligands, a molecule interacting with its environment (e.g., a molecule in a solvant), a large molecule with interacting groups of atoms of entirely different kinds and properties, and a macromolecular system composed of recognizable molecular units (e.g., a molecular crystal or polymer, not necessarily ordered). It is clear that for such systems some generalizations and simplifications of the SCF-Xα-SW method would be useful.

In the original SCF-Xα-SW technique, we have already partitioned the system into different regions: atomic, interatomic, and extramoleuclar. However, we will now go a step further and focus attention on a particular local subgroup on neighboring atomic region. Let us enclose this cluster of atoms, called set $\underline{a}$, in a new sphere ($\underline{s}$). All of the other atoms of the system, called set $\underline{b}$, lie in the extramolecular region of set $\underline{a}$.

The secular equation in the SCF-Xα-SW method can formally be written as

$$MA = 0 \tag{30}$$

where M is the secular matrix and A is a column vector made up of the coefficients of the wavefunction in a multicenter partial-wave representation.[1-3] With the above molecular partitioning, the secular equation can be put into the form

$$\begin{pmatrix} M_{aa} & M_{ab} & M_{ao} \\ M_{ba} & M_{bb} & M_{bo} \\ M_{oa} & M_{ob} & M_{oo} \end{pmatrix} \begin{pmatrix} A_a \\ A_b \\ A_o \end{pmatrix} = 0 \qquad (31)$$

where the submatrix M_{ij} contains all matrix elements with the row index of an atom in the set i and the column index of an atom in the set j (index o stands for the outer sphere which, in principle, surrounds the entire system), and where A_j is defined analogously. By inspection of the standard formulae for the matrix elements,[2,3], we can write (cf. Eq. 15)

$$(M_{aa})^{\alpha\beta}_{LL'} = \{\delta_{\alpha\beta}\delta_{LL'}[t^{\alpha}_{\ell}(\varepsilon)]^{-1} - G^{\alpha\beta}_{LL'}(\varepsilon)\}\delta^{a}_{\alpha}\delta^{a}_{\beta} \qquad (32)$$

in which

$$\delta^{a}_{\alpha} = \begin{cases} 1 & \text{if } \alpha\varepsilon a \\ 0 & \text{otherwise} \end{cases} \qquad (33)$$

with the matrix elements M_{bb} analogously represented. From the standard formulae,[2,3] we can also write

$$(M_{oo})_{LL'} = \delta_{LL'}[t^{o}_{\ell}(\varepsilon)]^{-1} \qquad (34)$$

$$(M_{ab})^{\alpha\beta}_{LL'} = [-G^{\alpha\beta}_{LL'}(\varepsilon)]\delta^{a}_{\alpha}\delta^{b}_{\beta} \qquad (35)$$

$$(M_{ao})^{\alpha}_{LL'} = [-S^{\alpha o}_{LL'}(\varepsilon)]\delta^{a}_{\alpha} \qquad (36)$$

with the remaining elements of the matrix (31) similarly represented. Only the diagonal submatrices of (31) are necessarily square.

We shall next fix our attention on the surface of sphere s, investigate the scattered-wave representation of the wave-function there, and then rewrite the secular equations so that they explicitly involve s. In the spirit of the overlapping-sphere approximation (see Section I), sphere s, which encloses all the atoms of set a, may possibly overlap some of the atoms of set b. To simplify writing, we assume that the outer sphere has the same center as sphere s. We will later remove the outer sphere, so that this is not a serious restriction. In the following, we will also assume that $\varepsilon < \bar{V}$. By using standard partial-wave expansions of the scattered-wave method,[2,3] one can represent the intersphere wavefunction near the surface of s as

$$\Psi_{II}(\vec{r}) = \sum_{L} [(B_s)_L k^{(1)}_{\ell}(\kappa r_s) + (B'_s)_L i_{\ell}(\kappa r_s)] Y_L(\vec{r}_s) \qquad (37)$$

where we have used

$$B_s = -M_{oa}A_a \tag{38}$$

$$B'_s = A_o - M_{sb}A_b \tag{39}$$

and where

$$(M_{sb})^{\beta}_{LL'} = [-G^{s\beta}_{LL'}(\varepsilon)]\ \delta^b_\beta \tag{40}$$

By introducing the coefficients

$$A_s = B'_s \tag{41}$$

$$A'_s = B_s \tag{42}$$

we may interpret expression (37) in two alternative ways. When observing from inside sphere s (region a), we may interpret the two components of (37), respectively, as an incoming wave with coefficients B_s directed toward the "inner" surface of s and an outgoing wave with coefficients A_s. When observing from outside sphere s (region b), we may inerpret the two components of (37), respectively, as an outgoing wave with coefficients A'_s directed away from the "outer" surface of s and an incoming wave with coefficients B'_s. We will use this notation even when relations (41) and (42) are not applicable (see below). Thus we will always write

$$\Psi_{II}(\vec{r}) = \sum_L [(B_s)_L k^{(1)}_\ell(\kappa_a r_s) + (A_s)_L i_\ell(\kappa_a r_s)]\, Y_L(\vec{r}_s) \tag{43}$$

$$(r_s \leqslant b_s)$$

$$\Psi_{II}(\vec{r}) = \sum_L [(A'_s)_L k^{(1)}_\ell(\kappa_b r_s) + (B'_s) i_\ell(\kappa_b r_s)]\, Y_L(\vec{r}_s) \tag{44}$$

$$(r_s \geqslant b_s)$$

where b_s is the radius of sphere s. In the present cas

$$\kappa_a = \kappa_b = \kappa = \sqrt{V-\varepsilon} \tag{45}$$

The main point of this exercise is that the continuity of (43) and (44) (and that of their respective first derivatives) at $r_s = b_s$ is equivalent to the coupling of the solutions of Schrödinger's equation for regions a and b, since the intersphere wavefunction is written entirely with respect to s. When observing from region b, we would like to view s, containing a, as if it were simply an "atomic" sphere, ignoring the detailed molecular structure of a.

Alternatively, when observing from region a, we would like to view s as if it were simply an "outer" sphere, ignoring the detailed molecular structure of b. To implement this formally, let us introduce two nonsingular diagonal matrices M_{ss} and M'_{ss} through the relations

$$B_s = M_{ss} A_s \tag{46}$$

$$B'_s = M'_{ss} A'_s \tag{47}$$

In the present case, where expressions (45), (41), and (42) are valid, these matrices are related by

$$M'_{ss} = (M_{ss})^{-1} \tag{48}$$

M'_{ss} and M_{ss}, respectively, play the same roles at s as the matrices $[t^{\alpha}_{\ell}(\varepsilon)]^{-1}$ and $[t^{o}_{\ell}(\varepsilon)]^{-1}$ do at each atomic sphere α and at the outer sphere, respectively, in the standard formulation of the scattered-wave method.[2,3] However, the solution of the problem must generally be known before M_{ss} and M'_{ss} can be explicitly determined, hence their major role is in the formulation of the theoretical model and computational procedure.

our goal is to partition the original secular equation (31) for the entire system into smaller coupled secular equations which describe the individual interacting parts a and b separated by the spherical boundary s. To accomplish this, it is necessary to introduce some formal mathematical relations between the pertinent matrix elements. For example, we can write

$$M_{ao} M_{sb} = -M_{ab} \tag{49}$$

a relation which can be proven in the partial-wave representation by using standard multipole expansion theorems of the scattered-wave method.[2,3] We can also write

$$M_{os} = -1 \tag{50}$$

$$M_{sa} = M_{oa} \tag{51}$$

where we have used the fact that the sphere s and the outer sphere (to be removed later) have the same center. Using the above expressions and Eds. (38) and (39), one can derive the following relations:

$$M_{oa}A_a = M_{os}(-M_{oa}A_a) = M_{os}B_s = M_{os}A'_s \tag{52}$$

$$M_{so}A_o + M'_{ss}A'_s + M_{sb}A_b = 0 \tag{53}$$

$$M_{ba}A_a = M_{bs}(-M_{oa}A_a) = M_{bs}B_s = M_{bs}A'_s \tag{54}$$

$$M_{ss}A_s + M_{sa}A_a = 0 \tag{55}$$

$$M_{ao}A_o + M_{ab}A_b = M_{ao}(A_o - M_{sb}A_b) = M_{ao}B'_s = M_{as}A_s \tag{56}$$

With the above expressions and Eqs. (41), (42), (46), and (47), we can write the matrix equations

$$\begin{pmatrix} M'_{ss} & M_{sb} & M_{so} \\ M_{bs} & M_{bb} & M_{bo} \\ M_{os} & M_{ob} & M_{oo} \end{pmatrix} \begin{pmatrix} A'_s \\ A_b \\ A_o \end{pmatrix} = 0 \tag{57}$$

$$\begin{pmatrix} M_{aa} & M_{as} \\ M_{sa} & M_{ss} \end{pmatrix} \begin{pmatrix} A_a \\ A_s \end{pmatrix} = 0 \tag{58}$$

Substituting (41) into (47) and (42) into (46), we obtain

$$A_s = M'_{ss}A'_s \tag{59}$$

$$A'_s = M_{ss}A_s$$

which is consistent with (48).

Eqs. (57) through (59) now replace the original secular equation (31), with (57) describing the electronic structure of region b, (58) describing the electronic structure of region a, and (59) describing the coupling of the wavefunctions across the sphericall boundary s separating regions a and b. It should be recalled that these equations have been derived under the condition (45) that the constant intersphere potential $\bar{V}$ is the same in both regions a and b.

We may actually introduce different constant potentials $\bar{V}_a$ and $\bar{V}_b$ representing the local interatomic volume averages of the molecular potential in regions a and b, respectively. The coupling equations at s are more complicated than (59), but Eqs. (57) and (58) remain essentially the same in form, with the matrix elements of (57) depending on

$$\kappa_b = \sqrt{\bar{V}_b - \varepsilon} \tag{60}$$

and those in (58) depending on

$$\kappa_a = \sqrt{\bar{V}_a - \varepsilon} \tag{61}$$

It is necessary only to find the new coupling conditions at s. We want Ψ_{II} (as given by Eqs. (43) and (44)) to be continuous and to have continuous first derivative at s. The matching of (43) and (44) at $r_s = b_s$ yields

$$(B_s)_L k_\ell^{(1)}(\kappa_a b_s) + (A_s)_L i_\ell(\kappa_a b_s) = (A'_s)_L k_\ell^{(1)}(\kappa_b b_s) + (B'_s)_L i_\ell(\kappa_b b_s) \tag{62}$$

$$(B_s)_L \left.\frac{dk_\ell^{(1)}(\kappa_a r_s)}{dr_s}\right|_{r_s=b_s} + (A_s)_L \left.\frac{di_\ell(\kappa_a r_s)}{dr_s}\right|_{r_s=b_s} = (A'_s)_L \left.\frac{dk_\ell^{(1)}(\kappa_b r_s)}{dr_s}\right|_{r_s=b_s} + (B'_s)_L \left.\frac{di_\ell(\kappa_b r_s)}{dr_s}\right|_{r_s=b_s} \tag{63}$$

By eliminating one coefficient at a time from Eqs. (62) and (63), we can write

$$(A_s)_L = (-1)^{\ell+1}\kappa_a b_s^{\,2}\ \{(B'_s)_L\,[\,i_\ell(\kappa_b b_s), k_\ell^{(1)}(\kappa_a b_s)\,]\} + (A'_s)_L\,[\,k_\ell^{(1)}(\kappa_b b_s),\ k_\ell^{(1)}(\kappa_a b_s)\,]\,\} \tag{64}$$

$$(A'_s)_L = (-1)^{\ell+1}\kappa_b b_s^{\,2}\ \{(B_s)_L[\,i_\ell(\kappa_b b_s),\ k_\ell^{(1)}(\kappa_a b_s)\,] + (A_s)_L\,[\,i_\ell(\kappa_b b_s),\ i_\ell(\kappa_a b_s)\,]\,\} \tag{65}$$

where we have used Eq. (13) and

$$[\,i_\ell(\kappa b),\ k_\ell^{(1)}(\kappa b)\,] = (-1)^{\ell+1}/\kappa b^2 \tag{66}$$

It is obvious that when $\kappa_a = \kappa_b = \kappa$ (cf. Eq. (45)), expressions (64) and (65) reduce respectively to (41) and (42).

Let us now introduce the notation

$$N_\ell = [\,i_\ell(\kappa_b b_s),\ k_\ell^{(1)}(\kappa_a b_s)\,] \tag{67}$$

$$(M'_r)_{LL'} = -\delta_{LL'}\ \frac{[\,k_\ell^{(1)}(\kappa_b b_s),\ k_\ell^{(1)}(\kappa_a b_s)]}{N_\ell} \tag{68}$$

$$(M'_t)_{LL'} = \delta_{LL'} \frac{(-1)^{\ell+1}}{\kappa_a b_s^2 N_\ell} \tag{69}$$

$$(M_r)_{LL'} = -\delta_{LL'} \frac{[i_\ell(\kappa_b b_s), i_\ell(\kappa_a b_s)]}{N_\ell} \tag{70}$$

$$(M_t)_{LL'} = \delta_{LL'} \frac{(-1)^{\ell+1}}{\kappa_b b_s^2 N_\ell} \tag{71}$$

By using the definitions (46) and (47), the coupling equations at s can be written (cf. (59))

$$B'_s = M'_{ss}A'_s = M'_t A_s + M'_r A'_s$$

$$B_s = M_{ss}A_s = M_t A'_s + M_r A_s \tag{72}$$

and can be interpreted as the conditions by which an incoming spherical wave (from regions b and a, respectively) is split into a reflected wave and a transmitted wave at the surface of s.

Eqs. (57) and (58), together with (72), constitute the coupled secular equations for our partitionable molecular system. Theses equations, in principle, allow one to focus on a particular part a of the system, with the rest of the system, part b, interacting with a but treated as a boundary condition at s. Alternatively, one can focus on part b, treating the interacting part a as a boundary condition at s. The coupling of the secular equations (57) and (58) for parts b and a via Eqs. (72) lead to a self-consistent solution of the Schrödinger'equation for the entire system.

The above secular equations can be used in their exact form, or they can be simplified to various degrees, depending on the relative strengths of coupling between the interacting parts and on the size of the system. For example, if the molecule is fairly large, then the outer sphere surrounding the entire system can be effectively removed without significantly affecting the solution of Schrödinger's equation by increasing the outer-sphere radius b_0 to infinity, giving

$$\lim_{b_0 \to \infty} M_{00}^{-1} = 0 \text{ (cf. Eq. (34))} \tag{73}$$

$$\lim_{b_0 \to \infty} \bar{V}_b = 0 \tag{74}$$

In this limit, partitioning theory and contraction of the secular matrix equation (51) lead to $A_0 = 0$ and

$$\begin{pmatrix} M'_{ss} & M_{sb} \\ M_{bs} & M_{bb} \end{pmatrix} \begin{pmatrix} A'_s \\ A_b \end{pmatrix} = 0 \qquad (75)$$

Afurther simplification can be effected by partially decoupling the secular equations (75), (72), and (58). In (75), $-M_{sb}A_b$ describes the coupling of region b to region a at the surface s, whereas $-M_{bs}A'_s$ describes the reverse couplinb of a to b via s. If we are interested initially in determining the local electronic structure of region a, but including the effective coupling of region b, then, to first approximation, we can set $M_{bs}=0$ and keep $M_{sb} \neq 0$. Under these conditions, (75) reduces to

$$B'_s = M'_{ss}A'_s = -M_{sb}A_b \quad \text{(cf. (39))} \qquad (76)$$

$$M_{bb}A_b = 0 \qquad (77)$$

Expression (77) is the secular equation for region b as if it were decoupled from region a. If we know the eigenvalues of (77) for some appropriate starting molecular potential in region b, then we can obtain the coefficients A_b. The latter, when substituted into Eq. (76), give us the coefficients B'_s representing "incoming spherical waves" directes from b toward the surface of sphere s surrounding region a. These coefficients are then used as a constraint when solving the secular equation, (58), for the electronic structure of region a, thus effectively including the coupling between regions b and a.

Under this constraint, the secular equation, (58), can be transformed into the inhomogeneous form

$$\begin{pmatrix} M_{aa} & M_{as} \\ M_{sa} & T^{-1}_{ss} \end{pmatrix} \begin{pmatrix} A_a \\ A_s \end{pmatrix} = \begin{pmatrix} 0 \\ M_i B'_s \end{pmatrix} \qquad (78)$$

where

$$(T_{ss})_{LL'} = \delta_{LL'} t^s_\ell(\varepsilon) = -\delta_{LL'} \frac{[k^{(1)}_\ell(\kappa_a b_s),\ k^{(1)}_\ell(\kappa_b b_s)]}{[i_\ell(\kappa_a b_s),\ k^{(1)}_\ell(\kappa_b b_s)]} \qquad (79)$$

and

$$(M_i)_{LL'} = \delta_{LL'} \frac{(-1)^{\ell+1}}{\kappa_b b_s{}^2 [k^{(1)}_\ell(\kappa_a b_s),\ k^{(1)}_\ell(\kappa_b b_s)]} \qquad (80)$$

The left side of (78) is similar to the secular matrix in the standard SCF-Xα-SW method,[2,3] with $t_\ell^s(\varepsilon)$ now playing the role of the outer-sphere scattering t-matrix and $k_\ell^{(1)}(\kappa_b r_s)$ representing the solution of the radial Schrödinger equation in the extramolecular region.

The partial decoupling carried out above can, of course, be reversed so that one can focus on the electronic structure of region b with the effects of a described in terms of an inhomogeneity in the secular equation for b. This procedure can be carried out iteratively between regions a and b until self-consistency is attained for the entire system.

Another level of approximation in calculating the electronic structure of a under the influence of b is to set $M_{sb} = 0$ in Eq. (76), leading to $B_s' = 0$, while retaining the spherical averages of the Coulomb and X exchange contributions of b in the atomic and extramolecular regions of a. This is a type of "molecular-field" approximation, in which there is a complete decoupling of regions a and b, except for spherically averaged Coulomb and exchange contributions which affect the diagonal elements (the scattering t-matrices) of the secular matrices. It is justified whenever the molecular subgroups constituting a and b are far enough apart so that the off-diagonal matrix elements M_{sb}, which have a radial dependence of the type $\frac{e^{-\kappa_b R_{sb}}}{R_{sb}}$ (see Refs. (2) and (3)), are vanishingly small. This is precisely the situation encountered in weakly coupled polymers and certain types of molecular crystals.

A more complete and detailed exposition of the theory underlying generalized molecular partitioning in the SCF-Xα-SW method will be published elsewhere.[38]

IV APPLICATION OF GENERALIZED MOLECULAR PARTITIONING TO WEAKLY COUPLED POLYMERS[39]

Let us now turn to the case of a polymeric system that can be divided into interacting units numbered 1,2,...., n (not necessarily equal), which we for simplicity assume are linearly arranged in numerical order. We now focus our attention on two consecutive units, j, j+1. Call this set a_j and the rest of the system b_j. Because of the radial dependence of the off-diagonal matrix elements, it is a good approximation (if the partitioned units are not too small) to put the matrix elements coupling a_j with the units in b_j, except for the nearest neighbors, equal to zero. By noticing that the coupling to the nearest neighbors is contained in an analogous treatment of the sets a_{j+1} and a_{j-1}, and that these elements are also rather small if the distance between two units is not too short, we may put all matrix elements coupling a_j and b_j equal to zero. This can be done without losing too much accuracy if we go through the sequence $\{a_j\}_{j+1}^{n-1}$ several times. Since we may assume that the outer sphere for the total system is very large for the cases of interest here, we can skip its influence of the coupling between a_j and b_j as well.

Having uncoupled a_j from b_j, we can solve the secular equation for a_j separately. This, however, does not mean that a_j is unaffected by the presence of b_j. One must still account for any Coulomb and exchange effects. How this can be accomplished is described below.

Since we treat a_j as if b_j were not present (except through the molecular potential), we can surround a_j with its own outer sphere (thus decreasing the effective interatomic region), introduce a local volume averaged intersphere potential and construct a spherically averaged potential outside this sphere. This means that we can use the standard SCF-X -SW computer programs for the computations, if they are changed to include the Coulomb and exchange interactions with b_j.

By carrying out the calcualtions repeatedly for all a_j until self-consistency is obtained, we will solve the problem for the entire system. How well orbitals which are delocalized through several units are represented generally in this model remains to be answered. However, it is clear from the above analysis that most of the occupied molecular orbitals should be reasonably well represented in the case of weakly coupled units. The way a delocalized orbital is recognized is that approximately the same energy level will be obtained in two consecutive calculations (for pair j-1, j and pair j, j+1) and that the electron distribution for this state in the unit in common (j) is essentially the same in both calculations.

The way the system is normally partitioned in the SCF-Xα-SW method, namely, into atomic spheres, gives us a convenient method of including the Coulomb and exchange effects between the neighboring units. When performing a calculation on the pair j, j+1 one simply mixes the potential for each atom in the unit j+1 with the potential obtained for the corresponding atom in a calculation on the pair j+1, j+2. The corresponding procedure is carried out for unit j. We will call the units where the mixing is perfomed "mix zones".

Although the different contributions to the potential V should generally be mixed with different weights, we may simply perform the mixing by taking a weighted average value of the total potential in the mix zones[40]

$$V_I^{used} = (1-a)\, V_I + a\, V_{II} \qquad (81)$$

where index I stands for unit j+1 in the pair j, j+1, index II for unit j+1 in the pair j+1, j+2 and a is the weighting factor. This leads to a correct mixing of the main contribution to V if a is chosen close to 0.5.

We also have the possibility of mixing the spherically averaged atomic electronic charge densities ρ in the same way.[40] This may be important since the Xα exchange potential is proportional to $\rho^{1/3}$.

If a pair of units have some symmetry, it may be used to simplify the secular equations even further. If, however, the asymmetry of the total system is of major importance to the problem, these local symmetries cannot be used, of course.

A picture illustrating the procedure and an obvious generalization if it is given in Fig. 5. Of course, the method may also be applied to certain nonlinear systems.

The standard SCF-Xα-SW computer program has been modified in order to perform these MIX-calculations in the case of three units. This new version, SCF-Xα-SW-MIX, differs from the original by also being more automatic. With the new program, it is possible to mix the potential and/or electron density in the way described above. By economical sharing of variables, it does not need a much larger core space than the ordinary SCF-Xα-SW program, and the time needed per iteration is not much greater than that for a similar nonMIX run.

The SCF-Xα-SW-MIX method has been applied to the sequential water trimer (Fig. 6a). By partitioning the trimer into three water molecules, the middle molecule serving as mix zone, one reduces the computation to that on two water dimers (Fig. 6b), which happen to be geometrically equivalent in this case. The atomic distances, angles, sphere radii, and α-parameters used are given in TableV.

We can see that each dimer has a plane of symmetry. This means that we can simplify the calculations by using these local symmetries as an approximation. However, in the present calculation only the symmetry of dimer II was used. This does not make any difference in the results, since the SCF-Xα-SW-MIX method treats H_{31} and H_{32} as equivalent anyway.

The starting orbitals of the two dimers in the iterative procedure are equal, which means that any difference in the final result is due to the mixing. There are 10 two-electron states in each

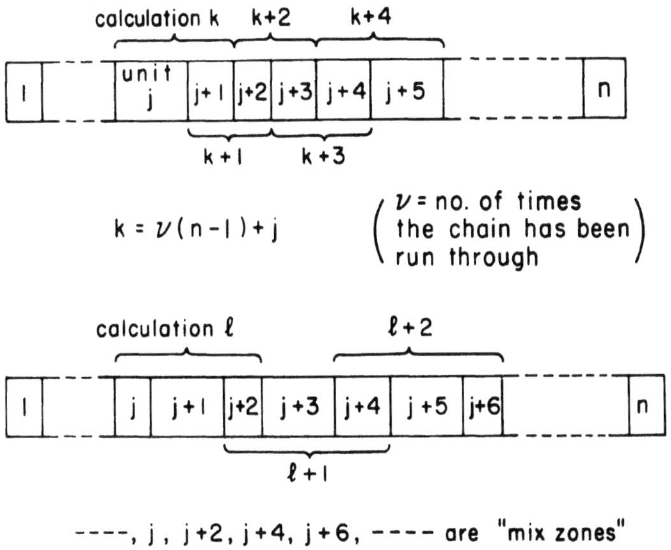

Figure 5. Schematic picture illustrating the MIX procedure as applied to a linear chain of atomic clusters in two different cases. Each rectangle represents a cluster of atoms.

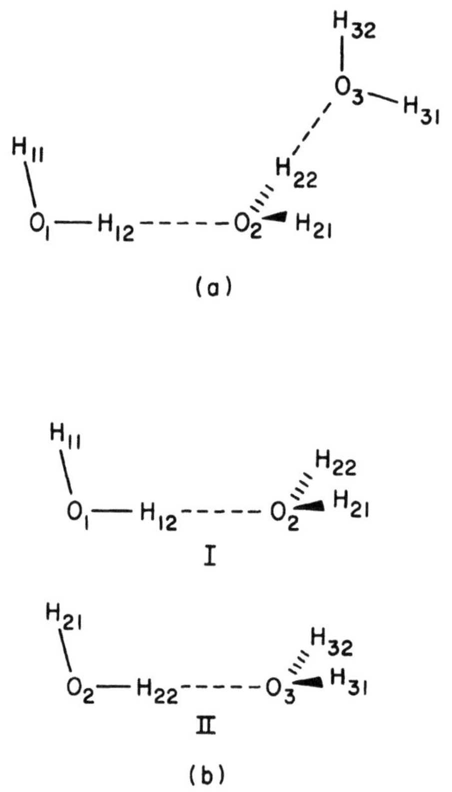

Figure 6. a) Sequential water trimer.

b) The trimer is partitioned according to the MIX scheme into two dimers. Molecule 2 is the MIX zone in this case.

Table V

$r_{OH} = 0.9572\ \mathring{A}$	$R_O^{sphere} = 0.808\ \mathring{A}$
$\Theta\ HOH = 104.54°$	$R_H^{sphere} = 0.414\ \mathring{A}$
$rO_1O_2 = 2.85\ \mathring{A}$	$R_{outer}^{sphere} = 2.392\ \mathring{A}$
$\Theta\ O_1O_2(HO_2H)_{plane} = 125.3°$	
	$\alpha_O = 0.74447$
	$\alpha_H = 0.77725$
	$\alpha_{outer} = \alpha_{inter} = 0.75103$

dimer, but only 15 in the trimer, indicating that 5 of each must coincide in the final result. These are, of course, the states that are centered at the middle water molecule.

In this calculation, both the potential and charge density were were mixed with a weighting factor of 0.5 in the mix zone. Already after 3-4 iterations, the trends in electron distribution and energy levels were very clear. At that point, each pair of energy levels (one in each dimer) of the four valence states centered at the middle water molecule differed by only 0.2-0.4 %, except for the highest (the one closest to $\bar{V}$) which differed by 2 %. These figures should be compared with the difference between the energies ot two similar states of the different molecules in the trimer, which is 2-8 %. This means that at least each of the three lowest pair of levels practically coincide. All these values were actually kept until full convergence was reached. At convergence, the maximal relative difference of the total electron density of the central molecule for the two dimers was 0.2-0.3 %.

Table VI shows some of the final results of the computations for the dimers. One should note the following when interpreting the figures :

1. The atomic populations (Q) are the integrated electrons densities inside each atomic sphere.
2. The electron density is normalized to unity in <u>each</u> dimer.

3. The quantities are calculated from the mixed potential, but are not mixed themselves.

To determine the actual population in the trimer, we let the population in each dimer determine the relative distribution of electrons in that dimer. By taking mean values of Q in the mix zone and renormalizing to 2.0 in the trimer (because of spin degeneracy) we get the final results shown in Table VII. Figure 7 relates the energy levels of the trimer to those of a single water molecule with the same geometry (computed by the SCF-Xα-SW method), and Table VIII shows the electron distribution in that molecule. (R_{outer}^{sphere} = 1.196 Å for a single water molecule.) By inspection of the symmetry representations and charge distributions for the states of the single water molecule in Table VIII, we can assign the type of the different valence orbitals for the monomer. In the order of increasing energy they are "symmetric H-O-H bond," "antisymmetric H-O-H bond," "symmetric lone pair," and "antisymmetric lone pair" (orthogonal to the H-O-H plane). We will keep this classification for the trimer as well, since each of its orbitals has a main contribution from one of the water molecules. We can see from Tables VII and VIII that the orbitals which are most changed due to the hydrogen bonding are the symmetric lone pairs, which become significantly delocalized in the trimer. In spite of this, the MIX procedure seems to represent them pretty well. We can point out that the energy levels obtained for the orbital of this type centered on the middle molecule are -0.773 and -0.770 for the two dimers, which means, as we have said before, that they practically coincide. Another interesting observation valid for all states centered on molecule 2 is that the electron distribution on the middle molecule is essentially the same for dimers I and II in spite of the fact that H_{22} is in a hydrogen bond for dimer II, but not for I. All this talks in favor of the MIX procedure.

By comparing the atomic total electron distributions for the monomer and the trimer, we can see that we generally have a larger fraction of the electrons inside the atomic spheres in the trimer than in the monomer. That this in fact may be a real rearrangement of electrons is seen for the asymmetric H-O-H bond for molecule 3 (ε = -0.980), by comparison with the corresponding state for the monomer. For all other states except the asymmetric lone pairs, the electron density of the oxygen atom, on which the orbital is centered, is decreased in the trimer, reflecting the delocalization of the orbitals. It is interesting to note that molecule 3 always has the lowest energies and molecule 1 the highest. Another observation is that the total electron density for the hydrogen in a hydrogen bond is less than that for the other hydrogen atoms.

It is difficult to come further in an interpretation of the results without electron density contour maps, but the existing SCF-Xα-SW electron density mapping program can, after some modifications, be used for this purpose.

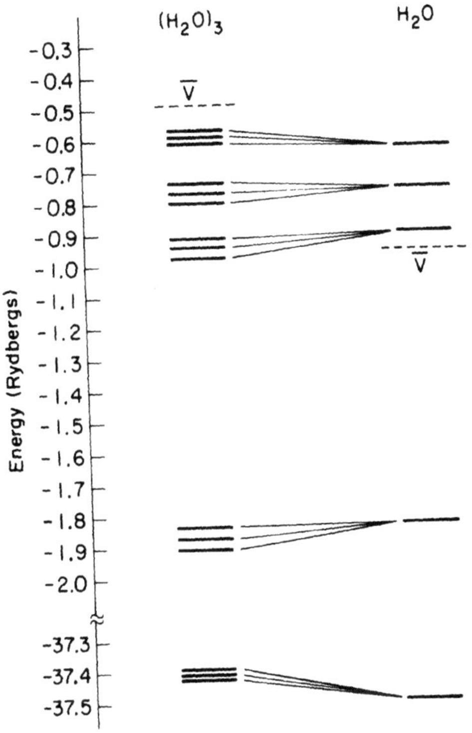

Figure 7. The electronic energy levels of the sequential water trimer obtained by the SCF-Xα-SW-MIX procedure and those of a single water molecule computed with the standard SCF-Xα-SW program. The molecules have the same geometry in both cases. The constant potential ($\bar{V}$) in the interatomic region is also shown.

Table VI. Energies and electron distributions (Q) for the lowest and highest valence states in the two coupled dimers. (Charge counted positive.)

Dimer No.	Energy of State (Ry)	Molecule 1		Molecule 2		Molecule 3		Inter-sphere Charge	Outer Sphere Charge
		$Q_{H_{11}}$ Q_{O_1}	$Q_{H_{12}}$	$Q_{H_{21}}$ Q_{O_2}	$Q_{H_{22}}$	$Q_{H_{31}}$ Q_{O_3}	$Q_{H_{32}}$		
II	-1.907			0.0020 0.036	0.0033	0.0421 0.718	0.0421	0.142	0.0144
I	-1.872	0.0021 0.037	0.0033	0.0421 0.716	0.0404			0.145	0.0147
II	-1.869			0.0427 0.718	0.0403	0.0025 0.039	0.0025	0.146	0.0087
I	-1.835	0.0428 0.716	0.0386	0.0025 0.040	0.0024			0.148	0.0091
II	-0.614			0.0001 0.035	0.0018	0.0001 0.707	0.0001	0.233	0.0231
II	-0.602			0.0 0.713	0.0	0.0001 0.001	0.0001	0.269	0.0178
I	-0.588	0.0001 0.040	0.0021	0.0001 0.697	0.0001			0.237	0.0244
I	-0.577	0.0 0.704	0.0	0.0001 0.001	0.001			0.275	0.0195

Table VII. Energies and electron distribution (Q) for the states in the water trimer. The total electron distribution is also given (Counted positive.)

Energy (Ry)	Molecule 1			Molecule 2			Molecule 3			Inter-sphere Charge	Outer Sphere Charge
	$Q_{H_{11}}$	Q_{O_1}	$Q_{H_{12}}$	$Q_{H_{21}}$	Q_{O_2}	$Q_{H_{22}}$	$Q_{H_{31}}$	Q_{O_3}	$Q_{H_{32}}$		
-1.907	0.0	0.0	0.0	0.0040	0.073	0.0066	0.0841	1.436	0.0841	0.284	0.0287
-1.871	0.0039	0.071	0.0062	0.0807	1.365	0.0767	0.0047	0.075	0.0047	0.290	0.0227
-1.835	0.0856	1.433	0.0773	0.0051	0.080	0.0048	0.0	0.0	0.0	0.296	0.0182
-0.980	0.0	0.0	0.0	0.0	0.0	0.0	0.1781	1.199	0.1781	0.414	0.0309
-0.946	0.0001	0.001	0.0001	0.1754	1.148	0.1642	0.0029	0.045	0.0029	0.423	0.0376
-0.913	0.1774	1.133	0.1566	0.0021	0.049	0.0041	0.0	0.0	0.0	0.433	0.0443
-0.803	0.0	0.0	0.0	0.0325	0.265	0.0037	0.0616	1.086	0.0616	0.445	0.0456
-0.772	0.0283	0.236	0.0038	0.0422	0.889	0.0567	0.0096	0.220	0.0096	0.464	0.0404
-0.740	0.0400	1.051	0.0758	0.0118	0.288	0.0128	0.0	0.0	0.0	0.485	0.0356

Table VII (Continued)

Energy (Ry)	Molecule 1		Molecule 2		Molecule 3		Inter-sphere Charge	Outer Sphere Charge
	$Q_{H_{11}}$ Q_{O_1}	$Q_{H_{12}}$	$Q_{H_{21}}$ Q_{O_2}	$Q_{H_{22}}$	$Q_{H_{31}}$ Q_{O_3}	$Q_{H_{32}}$		
-0.614	0.0 0.0	0.0	0.0002 0.071	0.0037	0.0002 1.414	0.0002	0.466	0.0462
-0.595	0.0002 0.078	0.0041	0.0 1.371	0.0	0.0001 0.001	0.0001	0.503	0.0421
-0.577	0.0 1.409	0.0	0.0002 0.001	0.0002	0.0 0.0	0.0	0.551	0.0391
-37.427	0.0 0.0	0.0	0.0 0.0	0.0	0.0 2.0	0.0	0.0	0.0
-37.410	0.0 0.0	0.0	0.0 2.0	0.0	0.0 0.0	0.0	0.0	0.0
-37.395	0.0 2.0	0.0	0.0 0.0	0.0	0.0 0.0	0.0	0.0	0.0
Total Electron Charge	0.336 7.41	0.324	0.354 7.60	0.334	0.341 7.48	0.341	5.05	0.431

Table VIII. Energies and electron distribution (Q) for a single water molecule. (Charge counted positive.)

Energy (Ry)	Symmetry	Q_{H_1} / Q_O	Q_{H_2}	Inter-sphere Charge	Outer Sphere Charge
-1.835	A_1	0.0791 / 1.471	0.0791	0.244	0.126
-0.904	B_1	0.1604 / 1.153	0.1604	0.266	0.260
-0.761	A_1	0.0683 / 1.279	0.0683	0.266	0.319
-0.631	B_2	0.0 / 1.355	0.0	0.362	0.283
-37.499		0.0 / 2.0	0.0	0.0	0.0
Total Electron Charge		0.308 / 7.26	0.308	1.137	0.988

Summing up, we see that the effect of the MIX procedure is to include the effects of the third water molecule when performing a computation on one of the dimers, and that in the final result the properties of the middle water molecule are very alike in the two dimers, allowing the self-consistent joining of the dimers to a trimer.

A full treatment of the SCF-Xα-SW-MIX scheme will be published elsewhere.[41]

V. APPLICATIONS OF GENERALIZED MOLECULAR PARTITIONING TO COMPLEX MOLECULAR CRYSTALS AND POLYMERS

The SCF-Xα-SW-MIX approximation to generalized molecular partitioning can be straightforwardly extended to more complex polymeric systems and molecular crystals where the interactions between component molecules are relatively weak. For example, in Figs. 8 and 9

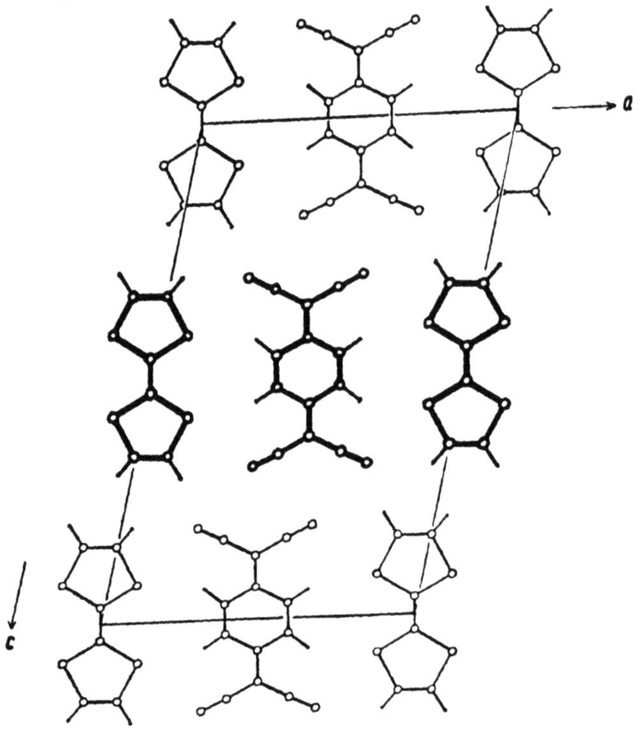

Figure 8. Crystal structure of TTF-TCNQ as determined by Kistenmacher et al. (Ref. 27).

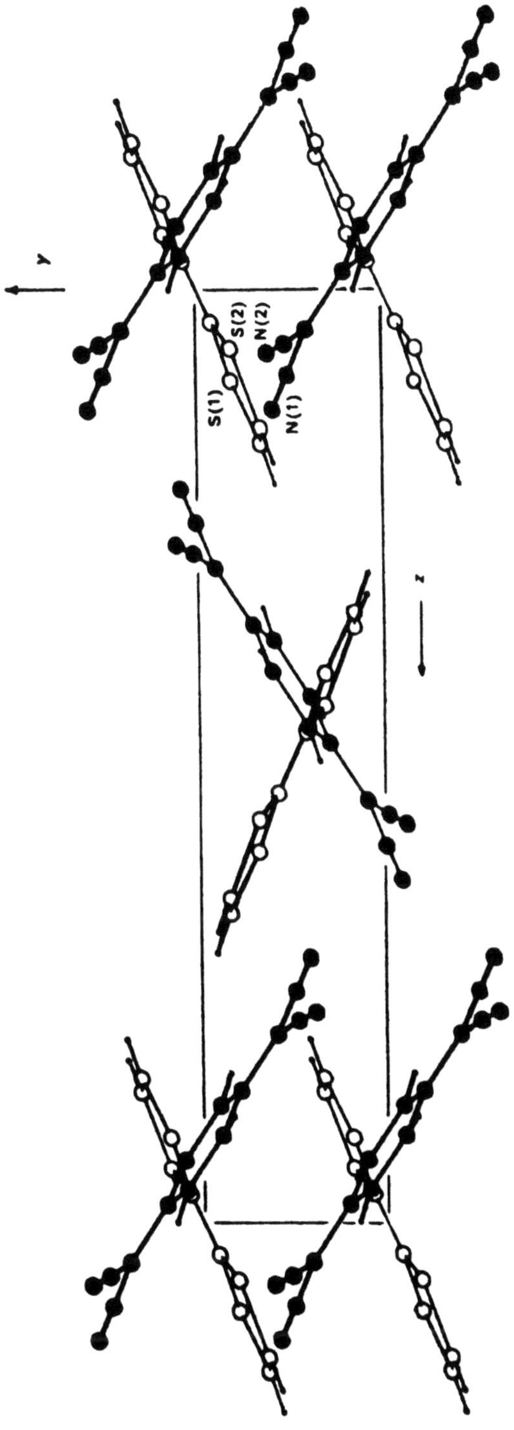

Figure 9. Crystal structure of TTF-TCNQ as determined by Kistenmacher et al. (Ref. 27).

are two views of the structure of a TTF-TCNQ molecular crystal, taken from the published crystallographic work of Kistenmacher et al.[27] The MIX procedure applied to a water trimer in the preceding section can be applied in a similar fashion to TTF-TCNQ-TTF and $(TCNQ)_3$ trimers (characteristic of the local crystalline structure of Figs. 8 and 9) to determine the first-order effects of neighboring TTF and TCNQ molecules on the electronic structure of a single TCNQ molecule. This would be a first step beyond the work on an isolated TCNQ molecule described in Section II, aimed at an understanding of the electronic structure and related physical properties of TTF-TCNQ crystals.

A more accurate treatment of solid-state effects in such crystals will require further extension of the MIX computational procedure and possibly use of the more general version of the partitioning technique described in Section III. The generalized partitioning theory of Section III can also be used, in principle, to determine the electronic structures of strongly coupled polymers and molecular crystals, where interactions between component monomers, dimers, etc. are too large for the MIX procedure to be justified. Modifications of the SCF-Xα-SW computer programs to implement the generalized partitioning theory are currently in progress.

Finally, for the sake of completeness, it should be noted that the SCF-Xα-SW method can be used as a basis for band-structure calculations on periodic molecular crystals and polymers, by applying Bloch's theorem as a boundary condition on the component monomers, dimers, etc. A theoretical formalism for implementing this concept, leading to a band-structure method analogous to the Korringa-Kohn-Rostoker technique, has already been described by Johnson and Smith.[1]

It should be emphasized, however, that the SCF-Xα-SW-MIX technique and generalized molecular partitioning approach to molecular crystals and polymers have the advantage over band theory that they are also applicable to disordered or amorphous systems where there is at most only short-range order and Bloch's theorem is not valid. A further advantage of the partitioning method over band theory is that it is naturally suited to the localized nature of the chemical bonding and electronic excitations in most molecular crystals and polymers. Localized excitations (e.g., excitons), including relaxation effects, can be described easily and accurately in the SCF-Xα-SW molecular partitioning approach by using Slater's transition-state concept,[1,6,7] whereas such excitations are difficult to describe within the framework of band theory.

REFERENCES

1. J.C. Slater and K.H. Johnson, Phys. Rev. B5, 844 (1972); K.H. Johnson and F.C. Smith, Jr., Phys. Rev. B5, 831 (1972).

2. K.H. Johnson, J. Chem. Phys. 45, 3085 (1966).

3. K.H. Johnson, in Advances in Quantum Chemistry, Vol. 7, edited by P.-O. Löwdin (Academic Press, New York, 1973), p. 143.

4. K.H. Johnson, J.G. Norman, Jr., and J.W.D. Connolly, in Computa tional Methods for Large Molecules and Localized States in Solids, edited by F. Herman, A.D. McLean, and R.K. Nesbet (Plenum Press, New York, 1973), p. 161.

5. K.H. Johnson in Annual Review of Physical Chemistry, Vol. 26, edited by H. Eyring (Annual Reviews, Inc., Palo Alto, California, 1975), to be published.

6. J.C. Slater, in Advances in Quantum Chemistry, Vol. 6, edited by P.-O. Löwdin (Academic Press, New York, 1972), p. 1.

7. J.C. Slater, The Self-Consistent-Field for Molecules and Solids, Volume 4 of Quantum Theory of Molecules and Solids (McGraw-Hill Book Company, New York, 1974).

8. N. Rösch, W.G. Klemperer, and K.H. Johnson, Chem. Phys. Lett. 23, 149 (1973).

9. F. Herman, A.R. Williams, and K.H. Johnson, J. Chem. Phys., October 15, 1974 issue, in press (provides a theoretical justification for the use of overlapping spheres and a sketch of Models I through IV for molecular TCNQ).

10. J.G. Norman, Jr., J. Chem. Phys. (submitted for publication).

11. J.C. Slater, Intern. J. Quantum Chem. (in press).

12. J.C. Slater, J. Chem. Phys. (in press).

13. A.R. Williams, Intern. J. Quantum Chem. (in press).

14. N. Rösch and K.H. Johnson, Chem. Phys. Lett. 24, 179 (1974).

15. N. Rösch, R.P. Messmer, and K.H. Johnson, J. Amer. Chem. Soc. 96, 3855 (1974).

16. R.C. Evans, An Introduction to Crystal Chemistry (Cambridge University Press, 1966), Second Edition.

17. R. Foster, Organic Charge-Transfer Complexes (Academic Press, London, 1969). F.H. Herbstein, in Perspectives in Structural Chemistry, eds. J.D. Dunitz and J.A. Ibers (John Wiley and Sons, New York, 1971), Vol. 4, p. 166.

18. I.F. Shchegolev, Phys. Status Solidi 12 (a) 9 (1972).

19. H.R. Zeller, in Festkörperprobleme, ed. H.J. Queisser (Pergamon, New York, 1973), Vol. XIII.

20. A.J. Epstein, S. Etemad, A.F. Garito, and A.J. Heeger, Phys. Rev. B5, 952 (1972); L.B. Coleman, J.J. Cohen, D.J. Sandman, F.G. Yamagishi, A.F. Garito, and A.J. Heeger, Solid State Commun. 12, 1125 (1973).

21. J. Ferraris, D.O. Cowan, V. Walatka, Jr., and J.H. Perlstein, J. Amer. Chem. Soc. 95, 948 (1973).

22. L. Scheire and P. Phariseau, Chem. Phys. Letters 26, 149 (1974); Intern. J. Quantum Chem., in press; Physica, in press.

23. D.A. Liberman, unpublished.

24. J.C. Slater, Phys. Rev. 45, 794 (1934).

25. F. Herman and I.P. Batra, Nuovo Cimento B, June, 1974 issue (Models I and III).

26. F. Herman and I.P. Batra, Phys. Rev. Letters 33, 94 (1974) (Models I and IV).

27. R.E. Long, R.A. Sparks, and K.N. Trueblood, Acta Cryst. 18, 932 (1965) (TCNQ crystal). The C-H bond distance was taken as 1.09 Å, a representative neutron scattering value. See also: A. Hoekstra, T. Spoelder, and A. Vos, Acta Cryst. B28, 14 (1972); T.J. Kistenmacher, T.E. Phillips, and D.O. Cowan, Acta Cryst. B30, 763 (1974).

28. F.A. Cotton, Group Theory with Chemical Applications (Wiley-Interscience, New York, 1971). Second edition.

29. K.H. Johnson and F.C. Smith, Jr., M.I.T. Scattered Wave Computer Programs, unpublished.

30. D.A. Liberman and I.P. Batra, IBM Research Report RJ 1224 (May, 1973), unpublished. This report documents a modified version of Ref. 29. The formalisms underlying Refs. 29 and 30 are identical.

31. The expressions used for the potential and kinetic energies reflect the approximations we are making, namely, spherically

averaging potentials and charge densities in the atomic spheres and outside the outer sphere, and volume averaging these quantities in the intersphere region. These simplifications also affect the calculated virial ratio. Accordingly, this quantity should be regarded as a rough figure of merit rather than an exact one. For a detailed discussion of the virial theorem within the context if the statistical exchange approximation, see Ref. 7.

32. J.N.A. Ridyard, Perkin-Elmer Limited, private communication. Spectrum obtained with Perkin Elmer Model PS18. The authors are grateful to Dr. Ridyard for permission to reproduce his experimental spectrum for TCNQ vapor.

33. W.D. Grobman, R.A. Pollak, D.E. Eastman, E.T. Maas, Jr., and B.A. Scott, Phys. Rev. Letters 32, 534 (1974).

34. H.T. Jonkman and J. Kommandeur, Chem. Phys. Letters 15, 496 (1972).

35. F. Herman, W.E. Rudge, I.P. Batra, B.I. Bennett, and K.H. Johnson, to be published.

36. R.A. Pollak, private communication.

37. D.W. Turner, C. Baker, A.D. Baker, and C.R. Brundle, Molecular Photoelectron Spectroscopy (Willey-Interscience, London, 1970). Since the 2σ orbital for HCN is analogous to the lowest-lying bonding σ orbitals for TCNQ, the following information is relevant: The 2σ orbital energy for HCN has been found to lie between 33 and 37 eV, according to several orbital energy calculations which are summarized in Table 13.1 (p. 345) of this reference. Of course, these calculations neglect relaxation effects, which can amount to few eV.

38. R. Kjellander, to be published.

39. This section describes preliminary work published elsewhere. See R. Kjellander, Chem. Phys. Lett., submitted for publication.

40. This simple mixing formula probably suppresses some of the effects due to the neighboring units and is regarded as a preliminary one, which can be used to examine the legitimacy of the partial decoupling. A more elaborate mixing scheme, which is under development, includes the full contributions to the potential from the nearest neighboring units and takes care of teh main effects due to the delocalization of the orbitals over neighboring pairs of units when computing the total electron density.

41. R. Kjellander, to be published.

ELECTRON STATES OF DISORDERED CHAINS+

W.L. McCUBBIN

School of Mathematics and Physics

University of East Anglia

Norwich. U.K.

I. INTRODUCTION

In the small space remaining it will be possible to do little more than summarise some of the approaches that have been made to this problem.

It should be emphasised at the outset that one-dimensional arrays of potential wells V $(x-x_a)$ are very much more tractable mathematically than linear chains, which are three-dimensional objects extensive in only one dimension. The reason for this is, briefly, that the matching of functions (wave-functions) at a point is very much simpler than matching over a surface. It is therefore tempting to try to map the linear chain problem on to a one-dimensional one. An example of this approach(1) will be discussed in some detail in the lectures.

However persuasive one's argument regarding the validity of the one-dimensional mapping (and the argument is essentially about the separability of the one-electron potential) it will always be very desirable to treat the disordered polymer as a linear chain, i.e. recognising fully its 3D quality. An approach which attempts to do so has been initiated by Tang Kai and McCubbin(2).

+ *An extended version of this lecture is available from the author on request.*

For the benefit of those who have not made contact with the theory of disordered systems, we shall summarise our reasons for believing that the disordered chain problem has not only mathematical interest but also quite concrete physical consequences.

II. EFFECT OF STRUCTURAL DISORDER ON THE EIGENSTATES OF ONE-DIMENSIONAL ARRAYS

The energy eigenvalue spectra of random one-dimensional arrays show certain characteristic differences from those of periodic chains. A discussion of the methods by which these differences were confirmed has been given elsewhere(3); for the present purposes we need simply state the results themselves.

1. The allowed bands increase in width as the degree of short-range disorder (SRDO) increases, irrespective of whether there is any change in long-range order (LRO)*.
2. When LRDO exists in addition to SRDO, the bands acquire tails instead of the precise cut-off characteristic of periodic (one-dimensional) systems. In other words, $dN(E)/dE \rightarrow 0$ instead of ∞ at the band edges.
   There may in addition be a slight upward displacement of the band system, but this effect is marginal and will not have important practical sequences.
   Apart from numerical shifts in the eigenvalues, the *character* of the eigenfunctions may change :
3. The eigenfunctions may become localised, especially in the tail regions.

III. EFFECT OF STRUCTURAL DISORDER ON THE EIGENSTATES OF LINEAR POLYMERS

The possibility of extending some of the theory for one-dimensional arrays to linear chains was investigated by McCubbin and Teemull(4). The circumstances in which this can be done are in fact quite limited and the notion of replacing the linear chain by a one-dimensional one must be employed for certain purposes.

---

* *We note that LRO is broken if the possibility of finding the nth atom outside the nth cell becomes finite. Clearly, perfect periodicity of nearest-neighbour distances (i.e. SRO) can be surrendered without impairing the LRO. The converse is not true.*

In the work described in ref.(1), some attempt to mitigate the errors of this approach was made by choosing paths through the disordered chain that might be expected to give, respectively, upper and lower bounds to the range of localisation (effect (3) above). The amplitude of ψ was calculated at each maximum of the one-electron potential by the method described in ref. (4). If we denote by A_n the amplitude at the nth cell, then $\ell n|A_n|$ versus n becomes a straight line for sufficiently large n and the gradient gives the range of localisation, R. By repeating this chalculation for a number of energies, the function R(E) could be found.

Explicit calculations for polyethylene showed

a) that the range of localisation decreased significantly (but not abruptly) near the band edges,

b) that the lower bound to R could be as low as $\sim$ 6 Å, i.e. that very strong localisation of electrons could occur at certain energies even without severe disorder.

In the work described in ref.(2), the actual geometry of the system was taken into account. A fully *trans* polyethylene chain was disordered by (rigid) rotations about successive carbon-carbon bonds. The sacrifice one had to make for this degree of realism was the assumption that after M such disordered CH_2 groups the chain was cyclically continued. By studying the band structure as a function of M, the asymptotic behaviour could be observed.

The preliminary results do indeed show a significant band broadening. In the case of the lowest unoccupied band the broadening actually pushes the lower band edge below the zero of energy, which has interesting consequences for our model of carrier trapping in these systems.

At the outset of this work we were quite sceptical that the wellknown *qualitative* effects of disorder on the eigenvalue spectrum would prove to be important in polymer chains. The results obtained so far certainly prove otherwise.

References

1. L.L. Moseley and W.L. McCubbin, unpublished.
2. A.H. Tang Kai and W.L. McCubbin, unpublished.
3. W.L. McCubbin in "Transfer and storage of energy by molecules- Vol. IV : The Solid State" Chapter IV, edited by A.M. North and G.M. Burnett, Wiley, 1974.
4. W.L. McCubbin and F.A. Teemull, Phys. Rev. A6, 2478 (1972).

SEMIEMPIRICAL ENERGY BAND STRUCTURES OF PERIODIC DNA AND PROTEIN MODELS

J. Ladik

Lehrstuhl für Theoretische Chemie

Technische Universität München, München, BRD

ABSTRACT

After the formulation of the PPP CO method for one-dimensional and the CNDO/2 and MINDO/2 CO methods for one- and two-dimensional periodic systems, they have been applied for different periodic DNA models and to the polyformamide and polyglycine chains and two-dimensional networks. The paper discusses the main features of the band structures obtained for these systems.

I. INTRODUCTION

Biopolymers like DNA and proteins play a central role in molecular biology. DNA is the carrier of genetic information, whereas the different proteins are the catalysts (enzymes) of the different biochemical reactions and they are very important building stones (structural proteins) of the living cell.

The stereo structure of DNA is fairly well known today. On the other hand we know its chemical structure only partially, because the sequence of the four nucleotide bases in the different DNA molecules is still unknown. By the proteins we find a reversed situation : the conformation of only about 20 smaller protein molecules have been determined with the aid of X-ray diffraction, but we know the sequences of the amino acids in a great number of proteins.

Having the stereo structure of DNA and at least of the main polypeptide chain of the proteins, in other words knowing the positions of the atomic nuclei in these chains, using the Born-Oppenheimer approximation it was possible to start the quantum mechanical calculation of the electronic structure of these macromolecules. In these calculations step by step always larger segments of DNA and proteins have been treated. First the single constituent molecules (nucleotide bases or amino acids) were treated which was followed by the computation of interacting subunits (base pairs, dinucleotides and oligopeptides, respectively).

As next step to approximate the electronic structure of the aperiodic DNA and protein macromolecules the energy band structures of periodic DNA and protein models have been computed. In the case of DNA a single base (homopolynucleotides), a base pair (poly(base pairs)) and two interacting base pairs have been taken as elementary cell. In all these periodic DNA models the stacked base pair have been taken in the same relative position as in the Watson-Crick stereomodel of the in vivo B form of DNA (1). Recently the calculations have been supplemented also by an investigation of the sugar-phosphate chain of DNA and of poly(base pairs) containing a H_2O molecule or a Mg^{2+} ion in the unit cell.

In the case of the proteins, the first calculations consisted of the treatment of the infinite π orbitals in the H-N-C = O ... H-N-C = O ... system followed by the calculation of the polypeptide chain

$$- \overset{O}{\underset{\parallel}{C}} - \underset{H}{\overset{|}{N}} - C - \overset{H}{\underset{|}{N}} - \text{(polyformamide)}$$

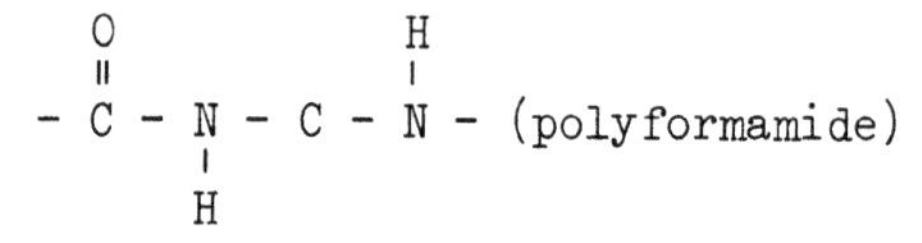

and the polyglycine chain

$$- \overset{O}{\underset{\parallel}{C}} - \underset{H}{\overset{|}{N}} - CH_2 - \overset{|}{\underset{O}{\overset{\parallel}{C}}} - \overset{H}{\underset{|}{N}} - CH_2 -$$

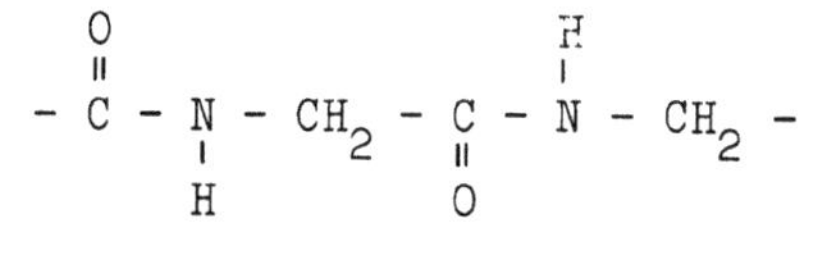

Most recently these calculations have been extended also to two-dimensional polyformamide and polyglycine networks.

On the basis of the band structures obtained it was possible to perform approximate calculations for the different physical and chemical properties of these systems. This made the interpretation of the UV absorption spectrum and the collective excitations in DNA possible and has given a deeper insight into the mechanism of charge transportation in these macromolecules.

The interpretation of the physical and chemical properties of more and more realistic DNA and protein models is not the last aim of their quantum mechanical investigation. One expects that interpreting the different physical and chemical properties of these systems it will become possible to establish the microphysical mechanisms which underlie those important biological processes in which they play a very essential role. Thus, the final goal of these rather cumbersome calculations is to find correlations between the electronic structure of DNA and of the proteins, respectively, and their biological functions.

II. METHODS

For the calculation of the energy band structures of the mentioned one-dimensional periodic DNA models and one- and two-dimensional periodic protein models different semiempirical crystal orbital (CO) methods have been used. The calculation of the π electron structure of the different periodic DNA models and of the H-N-C = O ... H-N-C = O ... chain have been performed with the aid of the Pariser-Parr-Pople (PPP) CO method for one-dimensional systems, while for the sugar-phosphate chain of DNA (poly(SP)), for the homopolynucleotides and for the one-dimensional polypeptide and polyglycine chains the CNDO/2 and MINDO/2 CO method for one-dimensional systems have been applied. Finally in the case of the two-dimensional polyformamide and polyglycine systems the two-dimensional versions of the mentioned all-valence-electron CO methods have been used. Here we shall show only how one obtains the expressions of the PPP CO methods for linear chains starting from the corresponding _ab initio_ equations (2). In the case of the other semiempirical CO methods we shall give only the final expressions.

We have seen previously (2) that in the _ab initio_ case we can get the band structure from the matrix equation

$$\underline{\underline{H}}(k)\underline{d}(k)_h = \varepsilon(k)_h \, \underline{\underline{S}}(k) \, \underline{d}(k)_h \tag{1}$$

where

$$\underline{\underline{H}}(k) = \sum_{q=-\infty}^{+\infty} \underline{\underline{H}}(q)e^{ikqa} \tag{2a}$$

$$\underline{\underline{S}}(k) = \sum_{q=-\infty}^{+\infty} \underline{\underline{S}}(q)e^{ikqa} \tag{2b}$$

The elements of the matrices $\underline{\underline{H}}(q)$ are defined as (2)

$$[\,H(q)\,]_{r,s} = \langle \chi_r^{o}(1)\,|\hat{H}^N|\,\chi_s^{q}(1)\rangle + \sum_{u,v=1}^{m} \sum_{q_1,q_2=-\infty}^{+\infty} P(q_1-q_2)_{u,v} \times$$

$$[\,\langle \chi_r^{o}(1)\,\chi_u^{q_1}(2)\,|\,\chi_s^{q}(1)\,\chi_v^{q_2}(2)\rangle -$$

$$\frac{1}{2}\langle \chi_r^{o}(1)\,\chi_u^{q_1}(2)\,|\,\chi_v^{q_2}(1)\,\chi_s^{q}(2)\rangle\,] \tag{3}$$

with

$$\hat{H}^N = -\frac{1}{2}\Delta_1 - \sum_{q_1=-\infty}^{+\infty} \sum_{\alpha=1}^{M} \frac{Z_\alpha}{|\vec{r}_1 - \vec{R}_\alpha^{q_1}|} \tag{4}$$

Here, for instance $\chi_s^q \equiv \chi(\vec{r}_1 - \vec{r}_s - q\vec{a}) = \chi_s^q(\vec{r}_1)$ stands for the sth AO in the qth cell, m is the number of orbitals and M the number of nuclei in the cell, Z_α is the charge (in atomic units) of the α th nucleus, $\vec{r}_1$ is the position vector of the electron, $\vec{R}_\alpha^{q_1}$ that of the αth nucleus in the qth cell, and finally a is the elementary translation. Further the four-center integral

$$\langle \chi_r^{o}(1)\,\chi_u^{q_1}(2)\,|\,\chi_s^{q}(1)|\chi_v^{q_2}(2)\rangle \equiv \int \chi_r^{o}(\vec{r}_1)^{*}\,\chi_u^{q_1}(\vec{r}_2)^{*}$$

$$\frac{1}{r_{12}}\,\chi_s^{q}(\vec{r}_1)\,\chi_v^{q_2}(\vec{r}_2)\,d\vec{r}_1\,d\vec{r}_2 \tag{5}$$

and the matrix elements of the generalized charge-bond order matrix are defined as (2)

$$P(q_1-q_2)_{u,v} = \frac{a}{2\pi}\int_{-\pi/a}^{\pi/a} 2\sum_{h=1}^{n_f} d(k)_{h,u}^{*}\,d(k)_{h,v} \times$$

$$e^{ik(q_1-q_2)a}dk \tag{6}$$

where n_f denotes the number of filled bands. Finally, the elements of the overlap matrices $\underline{\underline{S}}(q)$ are defined by the equation

$$[\,\underline{\underline{S}}(q)\,]_{r,s} = \langle \chi_r^{o}\,|\,\chi_s^{q}\rangle \equiv \int \chi_r^{o}(\vec{r}_1)^{*}\,\chi_s^{q}(\vec{r}_1)\,d\vec{r}_1 \tag{7}$$

To derive the PPP CO method we take into account explicitly, as in the case of molecules, only the π electrons of the system. Putting $\underline{\underline{S}}(k) = \underline{\underline{1}}$ we have instead of (1) the matrix eigenvalue equation

$$\underline{\underline{H}}(k)\underline{\tilde{d}}(k)_h = \tilde{\varepsilon}(k)_h \underline{\tilde{d}}(k)_h \tag{8}$$

Introducing, as the PPP method (3) does in the case of molecules, the neglect of differential overlap and taking only first neighbours interactions, we have for the elements of the matrices $\underline{\underline{H}}(q)$ instead of (3) the expressions

$$[\underline{\underline{H}}(o)]_{r,r} = <\chi_r^o(1)|\hat{H}^{eff.}|\chi_r^o(1)> + \sum_{u=1}^{m_\pi} \sum_{q_1=-1}^{+1} P(o)_{u,u} \times <\chi_r^o(1)\chi_u^{q_1}(2)|\chi_r^o(1)\chi_u^{q_1}(2)> - \frac{1}{2} P(o)_{r,r} <\chi_r^o(1)\chi_r^o(2)| \chi_r^o(1)\chi_r^o(2)> \tag{9a}$$

$$[\underline{\underline{H}}(q)]_{r,s} = <\chi_r^o(1)|\hat{H}^{eff.}|\chi_s^q(1)> - \frac{1}{2} P(-q)_{r,s} <\chi_r^o(1)\chi_s^q(2)| \chi_r^o(1)\chi_s^q(2)>$$

$$(q = -1, o, 1 \ ; \text{ if } q = o : r \neq s) \tag{9b}$$

where χ_r^o denotes now the rth π orbital in the reference cell and

$$\hat{H}^{eff.} = \hat{H}^N + \sum_{q_1=-1}^{+1} \sum_{\alpha=1}^{m_\pi} V_{el.\alpha}^{eff.q_1} \tag{10}$$

Here $V_{el.\alpha}^{eff.q_1}$ stands for the potential of the inner-shell and σ electrons of the αth atom with a π orbital in the cell characterized by q_1, $\hat{H}^N$ is given by (4), but now q_1 runs only from -1 to +1 and m_π is the number of π orbitals in the unit cell. Introducing the usual approximations and notations of the PPP MO method generalized for our case (3, 4, 5) :

$$<\chi_r^o \ \chi_s^{q_1} | \chi_r^o \ \chi_S^{q_1}> = \gamma(q_1)_{r,s} \tag{11a}$$

$$< \chi_r^o \chi_r^o | \chi_r^o \chi_r^o > = I_r - E_r \tag{11b}$$

$$< \chi_r^o | \hat{H}^{eff.} | \chi_r^o > = -I_r - \sum_{q_1=-1}^{+1} \sum_{s=1}^{m_\pi} Z_s \gamma(q_1)_{r,s} \tag{11c}$$

$$(s \neq r \text{ if } q_1 = 0)$$

$$< \chi_r^o | \hat{H}^{eff.} | \chi_s^q > = \beta(q)_{r,s} \tag{11d}$$

we can write

$$[\,\underline{\underline{H}}(o)\,]_{r,r} = -I_r + \frac{1}{2} P(o)_{r,r} (I_r - E_r) + \sum_{\substack{s=1\\(s\neq r)}}^{m_\pi} [\, P(o)_{s,s} - Z_s] \, \gamma(o)_{r,s} + \sum_{s=1}^{m_\pi} [\, P(o)_{s,s} - Z_s\,] \, [\, \gamma(1)_{r,s} + \gamma(-1)_{r,s}\,] \tag{12a}$$

$$[\,\underline{\underline{H}}(q)\,]_{r,s} = \beta(q)_{r,s} - \frac{1}{2} P(-q)_{r,s} \; \gamma(q)_{r,s} \tag{12b}$$

$$(\text{if } q = 0 : r \neq s)$$

Here I_r and E_r are the ionization potential and electron affinity, respectively, of the rth atom in its appropriate valence state (their values can be taken from Hinze and Jaffé (6) and the generalized charge-bond orders have been defined previously by (6). The Coulomb integrals $\gamma(q)_{r,s}$ have been approximated in the actual calculations with the aid of the expression

$$\gamma(q)_{r,s} = \frac{1}{a_{r,s} + R(q)_{r,s}} \quad , \quad \frac{1}{a_{r,s}} = \frac{1}{2} [\, I_r + I_s - E_r - E_s] \tag{13}$$

given by Mataga and Nishimoto (7), where $R(q)_{r,s}$ is the distance between atom r in the reference cell and atom s in the qth cell.

For the core integrals $\beta(o)_{r,s}$ between atoms within the same molecule the usual PPP values (8) have been used, whereas the intercell $\beta(1)_{r,s}$ core integrals have been taken proportional to the corresponding overlap integrals (5). In calculating and storing the occuring quantities the relations $\underline{\underline{R}}(-1) = \underline{\underline{R}}(1)^{tr}$, $\underline{\underline{\gamma}}(-1) = \underline{\underline{\gamma}}(1)^{tr}$, $\underline{\underline{P}}(-1) = \underline{\underline{P}}(1)^{tr}$, $\underline{\underline{\beta}}(-1) = \underline{\underline{\beta}}(1)^{tr}$ have been used which means that it was not necessary to calculate these quantities separately for $q = -1$

Since the k-dependence of the eigenvector-components $\tilde{d}(k)_{h,r}$ is not known analytically, the integration (6) were performed numerically in each iteration step to obtain the elements of the matrices $\underline{\underline{P}}(q)$. To be able to do this the eigenvalue problem of the Hermitian complex matrix

$$\underline{\underline{H}}(k) = \underline{\underline{H}}(o) + \underline{\underline{H}}(1)\, e^{ika} + \underline{\underline{H}}(-1)\, e^{-ika} \tag{14}$$

had to be solved in each iteration step by different values of k.[1.)] The SCF criteria used in the calculations of periodic DNA models were

$$|P(o)^{(n+1)}_{r,s} - P(o)^{(n)}_{r,s}| \leqslant 10^{-3}$$

$$|P(1)^{(n+1)}_{r,s} - P(1)^{(n)}_{r,s}| \leqslant 10^{-3} \tag{15}$$

simultaneously for all r and s.[2.)]

In the case of the CNDO/2 CO scheme for the valence electrons of one-dimensional periodic systems we can write (8) with (14) for $\underline{\underline{H}}(k)$ if we apply again the first neighbours interactions approximation. In this case the expressions for the elements of the matrices $\underline{\underline{H}}(q)$ (which we can derive again from the corresponding _ab initio_ equations introducing the neglections and approximations of the CNDO/2 method (9)) will be (10)

$$[\underline{\underline{H}}(o)]_{r,r} = U_r + [P(o)_{A,A} - \frac{1}{2} P(o)_{r,r}]\ \gamma(o)_{A,A} +$$

1) In the course of the actual calculations it has been found that applying seven different k values in the interval $(0,\frac{\pi}{a})$ one could obtain consistent results.

2) The number of necessary iteration steps was by the smaller periodic DNA models 5 - 7, whereas in the case of two interacting base pairs as elementary cell 10 - 11.

$$\sum_B [P(o)_{B,B} - Z_B]\ \gamma(o)_{A,B} \quad +$$

$$\sum_B [P(o)_{B,B} - Z_B][\gamma(1)_{A,B} + \gamma(-1)_{A,B}]^{(B\neq A)}_{(r\in A)} \qquad (16a)$$

$$[\underline{\underline{H}}(q)]_{r,s} = \beta^{o}_{A,B}\, S(q)_{r,s} - \frac{1}{2} P(q)_{r,s}\ \gamma(q)_{A,B}$$

$$(\text{if } q = 0,\ r \neq s\ ;\ r \in A\ ,\ s \in B) \qquad (16b)$$

Here

$$P(o)_{A,A} = \sum_{r\in A} P(o)_{r,r} \qquad (17)$$

for the expressions of $P(q)_{r,s}$ see again (6) and with a generalization of the integral expression occuring in the CNDO method

$$\gamma(q)_{A,B} = \int |\chi^{o}_{A}(1)|^{2}\ \frac{1}{r_{12}}\ |\chi^{q}_{B}(2)|^{2}\ dV_1\, dV_2 \quad (q = -1,o,1) \qquad (18)$$

where χ^{o}_{A} is an appropriate valence s orbital centered on atom A in the reference cell. Further the parameter $\beta^{o}_{A,B} = \frac{1}{2}(\beta^{o}_{A} + \beta^{o}_{B})$, where the constants β^{o}_{A} and β^{o}_{B} are given numerically for the elements of the first two rowes of the periodic table (11), Z_B is the core charge of atom B and the core integrals $U_r = \langle \chi^{o}_{r} | \hat{H}^{eff.}_{A,o} | \chi^{o}_{r} \rangle$ can be calculated with the aid of the valence state ionisation potentials and electron affinities of the atom A ($r\in A$) and of its positive and negative ions. Finally $\hat{H}^{eff.}_{A,o} = -\frac{1}{2}\Delta + V^{eff.}_{A,o}$ where $V^{eff.}_{A,o}$ is the effective potential of the atomic core (nucleus and inner shell electrons) A in the reference cell.

In the two-dimensional case we have instead of (8)

$$\underline{\underline{H}}(k_1,k_2)\underline{d}(k_1,k_2)_h = \varepsilon(k_1,k_2)_h \underline{d}(k_1,k_2)_h \qquad (19)$$

Introducing the second neighbours interactions approximation we can write

$$\underline{\underline{H}}(k_1,k_2) = \underline{\underline{H}}(o,o) + \underline{\underline{H}}(1,o)\, e^{ik_1 a} + \underline{\underline{H}}(-1,o)\, e^{-ik_1 a} \quad +$$

$$\underline{\underline{H}}(o,1)\ e^{ik_2a_2} + \underline{\underline{H}}(o,-1)\ e^{-ik_2a_2} +$$

$$\underline{\underline{H}}(1,1)\ e^{i(k_1a_1+k_2a_2)} + \underline{\underline{H}}(1,-1)\ e^{i(k_1a_1-k_2a_2)} +$$

$$\underline{\underline{H}}(-1,1)\ e^{-i(k_1a_1-k_2a_2)} + \underline{\underline{H}}(-1,-1)\ e^{-i(k_1a_1+k_2a_2)} =$$

$$\underline{\underline{H}}(0,0) + \underline{\underline{\tilde{H}}}^{1.nb.} + \underline{\underline{\tilde{H}}}^{2.\,nb.} \tag{20}$$

where $\underline{\underline{\tilde{H}}}^{1.nb.}$ stands for the sum of the 2nd - 5th terms in (20) and $\underline{\underline{\tilde{H}}}^{2.\,nb.}$ for the sum of the 6th - 9th terms.

Introducing again the neglections and parametrizations characteristic for the CNDO/2 method (9) we can write for the elements of the matrices $\underline{\underline{H}}(q_1,q_2)$ in (20)

$$[\underline{\underline{H}}(0,0)]_{r,r} = U_r + [P(0,0)_{A,A} - \tfrac{1}{2}P(0,0)_{r,r}]\ \gamma(0,0)_{A,A}$$

$$+ \sum_{B\neq A} [P(0,0)_{B,B} - Z_B]\ \gamma(0,0)_{A,B}$$

$$+ \sum_{B} \{[P(0,0)_{B,B} - Z_B]\ [\sum_{q_1=-1}^{+1}\ \sum_{q_2=-1}^{+1}\ \gamma(q_1,q_2)_{A,B}$$

$$- \gamma(0,0)_{A,B}]\}_{(r\in A)} \tag{21a}$$

$$[\underline{\underline{H}}(q_1,q_2)]_{r,s} = \beta^{o}_{A,B}\ S(q_1,q_2)_{r,s} - \tfrac{1}{2}P(q_1,q_2)_{r,s}$$

$$\times\ \gamma(q_1,q_2)_{A,B} \tag{21b}$$

Now

$$P(0,0)_{A,A} = \sum_{r\in A} P(0,0)_{r,r} \tag{22a}$$

$$P(q_1,q_2)_{r,s} = \frac{a_1a_2}{(2\pi)^2} \int_{-\pi/a_1}^{\pi/a_1} \int_{-\pi/a_2}^{\pi/a_2} \sum_{h=1}^{n_f} d(k_1,k_2)^{*}_{h,r}$$

$$\times \; d(k_1,k_2)_{h,s} \; e^{i(k_1 q_1 a_1 + k_2 q_2 a_2)} \; dk_1 \, dk_2 \tag{22b}$$

(a_1 and a_2 are the elementary translations in the two different directions) and

$$\gamma(q_1,q_2)_{A,B} = \int | \chi_A^{0,0} (1) |^2 \frac{1}{r_{12}} | \chi_B^{q_1,q_2} (2) |^2 \, dV_1 \, dV_2 \tag{23}$$

where $\chi_B^{q_1,q_2}$ is the appropriate valence s Slater orbital centered on atom B in the cell characterized by

$$\vec{q} = q_1\vec{a}_1 + q_2\vec{a}_2$$

Instead of the CNDO/2 parametrization we can write down the elements of the matrices $\underline{\underline{H}}(q_1,q_2)$ also in the MINDO/2 parametrization (12). We obtain in this way (13)

$$[\underline{\underline{H}}(0,0)]_{r,r} = U_r + \frac{1}{2} P(0,0)_{r,r} < \chi_r^o \chi_r^o | \chi_r^o \chi_r^o > + \sum_{\substack{\sigma \in A \\ \sigma \neq r}} P(0,0)_{\sigma,\sigma} [< \chi_r^o \chi_r^o | \chi_\sigma^o \chi_\sigma^o > - \frac{1}{2} < \chi_r^o \chi_\sigma^o | \chi_r^o \chi_\sigma^o >] + \tilde{\sum_B} \; (r \in A) \tag{24a}$$

$$[\underline{\underline{H}}(0,0)]_{r,s} = P(0,0)_{r,s} [\frac{3}{2} < \chi_r^o \chi_s^o | \chi_r^o \chi_s^o > - \frac{1}{2} < \chi_r^o \chi_r^o | \chi_s^o \chi_s^o >]$$

$$(r \neq s \; ; \; r,s \in A). \tag{24b}$$

$$[\underline{\underline{H}}(q_1,q_2)]_{r,s} = \Omega_{A,B} (I_r + I_s) \, S(q_1,q_2)_{r,s} - \frac{1}{2} P(q_1,q_2)_{r,s} \; \gamma(q_1,q_2)_{A,B}$$

$$(r \in A \; ; \text{ if } q_1 = q_2 = 0 : s \in B) \tag{24c}$$

Here the term $\tilde{\sum_B}$ in (24a) stands for all the expressions sumed over B in (21a), for the one-center core integrals U_r the valence state ionization potential I_r and for the parameters $\Omega_{A,B}$ appropriate values have been given by Baird and Dewar (12). All the one-center two electron integrals

$$<\chi_r^o \chi_s^o \mid \chi_r^o \chi_s^o> \quad \text{and} \quad <\chi_r^o \chi_r^o \mid \chi_s^o \chi_s^o>$$

have been calculated with the aid of the appropriate valence-electron Slater orbitals, whereas for the two-center Coulomb integrals $\gamma(q_1,q_2)_{A,B}$ the Ohno-Klopman expressions (14)

$$\gamma(q_1,q_2)_{A,B} = \frac{1}{\sqrt{\left[R(q_1,q_2)_{A,B}\right]^2 + a_{A,B}^2}} \tag{25a}$$

$$\frac{1}{a_{A,B}} = \frac{1}{2}\left[I_A + I_B - E_A - E_B\right] \tag{25b}$$

has been used, where I_A and E_A is the valence s electron ionization potential and electron affinity of atom A, respectively. All other quantities have been defined before.

The corresponding formulae for the MINDO/2 CO method for one-dimensional periodic systems in the first neighbours interactions approximation we can obtain from the above expressions very easily, we have to substitute only into (14) the equations (24) suppressing in them everywhere the variable q_2 (15).

In the actual all-valence electron calculations again the SCF criteria (15) were used and convergence has been reached in 12 - 20 iteration steps (the MINDO/2 CO methods requiring the smaller number of iteration steps).

The geometry used in the case of the periodic DNA models is due to Spencer (16) and Langridge et al (17), respectively, whereas for the two-dimensional polyglycine network a simplified model geometry (with 90° valence angles) and the parallel chain pleated sheet conformation of β polyglycine (18) has been used.

DISCUSSION OF THE RESULTS

A. Periodic DNA Models

The band structures of the periodic DNA models obtained either with the aid of the PPP CO method can be divided into two groups. The five homopolynucleotides (including also poly U) 19 and the two poly (base pairs) (poly (A-T) and poly (G-C)) (20) have rather broad bands (the widths of the valence bands are 0.2 - 0.3 eV, those of the conduction bands $\sim$ 0.1 eV, and the widths of the lowest filled bands are in most cases $\sim$1 eV). On the other hand for the more complicated periodic DNA models, where we have two different base pairs in the unit cell, like poly $\binom{A-T}{G-C}$ etc. (21), very narrow bands (the widths of the valence bands are usually 0.01 - 0.03 eV, those of the conduction bands are $\sim$ 0.01 eV and the widths of the lowest filled bands are $\sim$ 0.1 eV). The CNDO/2 CO calculations of the homopolynucleotides (22) and of the sugar-phosphate chain of DNA (23) give again rather broad bands (valence band widths between 0.15 and 0.50 eV for the homopolynucleotides and 0.05 eV for poly (SP) and conduction band widths 0.1 - 0.25 eV for both kinds of systems), whereas the MINDO/2 CO results for the homopolynucleotides indicate somewhat less broad bands (widths of valence bands 0.03 - 0.30 eV and those of the conduction bands 0.03 - 0.17 eV). Further, both (CNDO/2 and MINDO/2) all-valence-electron band structure calculations have given very broad lowest filled bands (1-2 eV) for the homopolynucleotides, whereas the CNDO/2 results [3.)] for the poly (SP) chain give only 0.1 eV as the width of the lowest filled band, but some other lowlying filled bands have widths 0.4 -1.2 eV (23). (For further details of the PPP and all-valence-electron CO calculations of different periodic DNA models we refer to the original paper (19 - 23)).

On the basis of perturbation theory it is easy to understand that if the same bases on base pairs are repeated (homopolynucleotides or the poly (A-T), poly (G-C) systems) the splitting of the levels of the constituent single systems is much larger in the polymer than if different subsystems are alternating in it (as is the case in all periodic DNA models for which narrow bands have been obtained).

We can conclude on the basis of the obtained band structures that for the complicated narrow band periodic DNA models a delocalized, Bloch-type description is not very correct and therefore the application of the hopping model seems to be more realistic. On the other hand, the Bloch picture is justified in the cases of periodic DNA models with broad bands. These conclusions are supported also by the calculation of the mean free path and mobility values of the different periodic DNA models starting from their band structures (24).

For the width of the forbidden band between the valence and conduction bands a value of $\sim$ 6.1 eV has been found from the PPP CO

3) In the lack of published parametrization of the MINDO/2 method for second row atoms no MINDO/2 CO calculations could be executed for the sugar-phosphate chain of DNA.

calculation of the poly $(\frac{A-T}{G-C})$ system (21). On the other hand, the u.v. absorption maximum of DNA lies at ∿ 4.5 eV. Therefore the PPP CO method has given a seemingly too high gap. The reason for this difficulty is (for a detailed discussion see (20)) that in any closed shall Hartree-Fock scheme (so also in the PPP scheme) the excitation energy is not just the difference of the corresponding one-electron energies, but

$$^{1}\Delta E_{i\to j} = \varepsilon_j(k) - \varepsilon_i(k) - J_{i,j}(k) + 2K_{i,j}(k) \qquad (26)$$

Here the $J_{i,j}(k)$ Coulomb and $K_{i,j}(k)$ exchange integrals disappear if the number of cell, $N\to\infty$(20). (If we take into account all neighbours'interactions they disappear as lnN/N (25)). Thus. eq. (26) reduces to the difference of one-electron energies, giving too high results. The physical explanation of the problem is that using equ. (26) we treat the motion of the promoted electron and the remaining hole as completely uncorrelated which is not correct. To overcome this difficulty we have to take into account the correlation between them. Perhaps the best way to do this is to assume an excitation occurs also in a solid primarily locally and to mix to the local excitations also some charge transfer excited states. In other words, a proper form of the intermediate exciton theory could be used. Starting from the delocalized Blich functions as linear combinations of them localized Wannier functions can be formed and the elements of the matrix (whose eigenvalues give the excitation energies in the framework of the intermediate exciton theory) can be calculated with the aid of these Wannier functions. Preliminary calculations for the excitation energies of a polyene chain using this method have given rather good agreement with experiment (26).

As first step in investigating the effect of impurities on the electronic structure of DNA the band structure of poly (G-C) has been computed in the PPP CO approximation assuming that one or two water molecules are bound by hydrogen bonds to the NH_2 group of each C molecule or/and to the C = O group of each G molecule (27). According to the results obtained for these systems the additional π-orbital of the H_2O molecules produced an extra π-band between the lowest filled bands, while the other bands remained practically unchanged (27).

As next step the band structures of poly (G-C) and poly (A-T) were recalculated in the presence of Mg^{2+} ions using again the PPP CO method modified suitably to account for the presence of charged ions (for the details see (27, 28))[4]. The calculations have been

4) The ability of divalent metal ions, especially Mg^{2+} ions, to react with a variety of electron-donor sites of the polynucleotide chains was experimentally demonstrated (29).

performed for the poly (G-C) and poly (A-T) systems with all possible types of Mg^{2+} attachments to the heterocyclic bases (including both in the plane of the base pairs and out of plane positions), taking one Mg^{2+} per unit cell. According to the results obtained for these model calculations the presence of the Mg^{2+} ions changes drastically the band structures : the bands become generally broader by a factor 2-3 and also their positions change considerably. This causes great changes in the band gap, for instance if the Mg^{2+} ion is attached to the NH_2 nitrogen of C in the G-C base pair it decreases from $\sim$ 6.0 eV. to $\sim$ 2.0 eV. This means that the exciton band (which lies always below the conduction band) will probably overlap with the valence band in this case which may give rise to the possibility of a phase transition to the excitonic insulator state.

For the interpretation of the conduction properties of DNA (which will not be discussed here) besides the width of the forbidden band (which as we could see just above is strongly dependent on the charged impurities in DNA) also the possibility has to be taken into account that in DNA a charge transfer reaction may occur from the poly (base pairs) to the poly (SP) chain (for the detailed discussion see (30)). If this really occurs the base pair region of DNA would show a hole and the poly (SP) chain an electronic conduction.

Concluding this discussion it seems rather probable that the band structure of the base pairs part of the real aperiodic DNA lies nearer to the narrow-band case than to the case of the broad-band periodic models. On the other hand, the experimentally found nearest-neighbour frequencies of the nucleotide bases in DNA (31) indicate a preference to have the same base repeated. This fact together with the above-described band-broadining effect of charged impurities (and the same expected effect of the non-zero resulting inhomogeneous electric field due to the PO_4^- - K^+ charged double layers at the outer part of the DNA double helix) will increase at least in some segments of the macromolecule the delocalization of the π electrons. For the description of the behavior of the electrons in the other segments of the DNA molecule the hopping model seems to be more correct.

B. Simple Periodic Protein Models

In the case of one-dimensional polyformamide chains, if the interactions were taken along the hydrogen bonds the widths of the valence and conduction bands were $\sim$ 0.2 and $\sim$ 0.3 eV, respectively, in the CNDO/2 case, whereas the MINDO/2 CO calculations have provided smaller widths ($\sim$ 0.1 eV and $\sim$ 0.05 eV,respectively) (13). On the other hand, if the interactions were taken along the main chain, the corresponding band widths were much larger (width of the valence band ($\delta\varepsilon_V$) $\sim$ 2.6 eV, width of the conduction band ($\delta\varepsilon_C$)

$\sim$0.6 eV in the CNDO/2 CO case, whereas the MINDO/2 CO results are $\delta\varepsilon_V \approx 0.9$ eV, $\delta\varepsilon_C \approx 2.5$ eV). Finally, taking into account both types of interactions, the CNDO/2 CO results for the two-dimensional polyformamide network were $\delta\varepsilon_V \approx 2.6$ eV, $\delta\varepsilon_C \approx 0.8$ eV and the MINDO/2 CO ones $\delta\varepsilon_V \approx 2.0$ eV, $\delta\varepsilon_C \approx 1.6$ eV indicating that as periodic protein models really two-dimensional networks have to be treated (13).

In a subsequent calculation of a two-dimensional polyglycine model network 5.) it was demonstrated that the band structure are essentially different (for obvious geometrical reasons), if we take into account instead of only first neighbours'interactions also the effect of the second neighbours (13) ($\delta\varepsilon_V$: $\sim$2.1 eV $\rightarrow$ $\sim$ 1.5 eV ; $\delta\varepsilon_C$: $\sim$ 2.6 eV$\rightarrow$ $\sim$3.1 eV in the CNDO/2 case ; $\delta\varepsilon_V$:$\sim$ 2.7 eV$\rightarrow$ $\sim$1.8 eV; $\delta\varepsilon_C$: $\sim$3.0 eV$\rightarrow$ $\sim$3.5 eV if we go from 1st neighbours'interactions to second neighbours'interactions).

Finally, the MINDO/2 CO calculation of the two-dimensional parallel chain pleated sheet β-polyglycine network (two glycine molecules in the unit cell) using second neighbours'interactions have resulted in $\delta\varepsilon_V \approx 1.2$ eV ; $\delta\varepsilon_C \approx 1.7$ eV and a forbidden band width of $\sim$ 4.8 eV (32) showing that the choice of the geometry influences the band structure considerably. Further it should be mentioned that while in the cases of the polyformamide and of the model polyglycine network we had planar structures and so we could define σ-and π -electron bands (in both cases the valence band was a π, and the conduction band a σ band), this classification becomes impossible in the case of the parallel chains pleated sheet conformation of β-polyglycine because it is not planar.

ACKNOWLEDGEMENT

The author should like to express his sincere gratitude to the NATO Scientific Affairs Division and to the University of Namur whose sponsorship made it possible to organize the Advanced Study Institute on Electronic Structure of Polymers and Molecular Crystals.

He is further very much indebted to Professor G.L Hofacker whose interest and support has given him the opportunity to continue his research.

REFERENCES

(1) J.D.H. Watson and F.H.C. Crick, Nature 171, 737, 964 (1953) ; F.H.C. Crick and J.D. Watson, Proc. Roy. Soc. (London) A223, 80 (1954).

5) To be able to place only one glycine molecule in the unit cell it valence angles have been distorted to rectangular ones.

(2) J. Ladik in "Electronic Structure of Polymers and Molecular Crystals", ed. J.-M. André and J. Ladik, Plenum Press, London, (1975).

(3) R. Pariser and R. G. Parr, J; Chem. Phys. 21, 466, 707 (1953); J. A. Pople, Trans. Far. Soc. 49, 1375 (1953).

(4) J. Ladik, Acta Phys. Acad. Sci., Hung, 18, 185 (1965).

(5) J. Ladik, D. K. Rai, and K. Appel, J. Mol. Spectr. 27, 72 (1968).

(6) J. Hinze and J. Jaffé, J. Amer. Chem. Soc. 84, 540 (1962).

(7) N. Mataga and K. Nishimoto, Z. Physik. Chemie 13, 140 (1957).

(8) J. Ladik and K. Appel, Theor. Chim. Acta 4, 132 (1966).

(9) J.A. Pople, D.P. Santry and G. A. Segal, J. Chem. Phys. 43, 129 (1965) ; J. A. Pople and G. A. Segal, 43, 136 (1965).

(10) J. Ladik and G; Biczó, Acta Chim. Acad. Sci. Hung. 67, 397 (1971) ; H. Fujita and A. Imamura, J. Chem. Phys. 53, 4555 (1970) ; K. Morokuma, J. Chem. Phys. 54, 962 (1971).

(11) J. A. Pople and D. C. Beveridge, Approximate Molecular Orbital Theory, McGraw-Hill Publ. Co., New York, (1970).

(12) N. C. Baird and M.J.S. Dewar, J. Chem. Phys. 50, 1262 (1967).

(13) S. Suhai and J. Ladik, Theor. Chimica Acta 28, 27 (1972); S. Suhai and J. Ladik, Acta Phys. Acad. Sci. Hung. (accepted).

(14) K. Ohno, Theor. Chim. Acta 2, 219 (1964); G. Klopman, J. Amer. Chem. Soc. 87, 3300 (1965).

(15) S. Suhai and J. Ladik, Int. J. Quant. Chem. 7, 547 (1973).

(16) M. Spencer, Acta Cryst. 12, 59, 66 (1959).

(17) R. Langridge, J. Marvin, W. Seeds, H. R. Wilson, C. W. Hooper, M.H.F. Wilkins, and L. D. Hamilton, J. Mol. Biol. 2, 38 (1960).

(18) L. Pauling and R. B. Corey, Proc. Natl. Acad. Sci. USA 39, 253 (1953).

(19) See (5).

(20) J. Avery, J. Packer, J. Ladik and G, Biczó, J. Mol. Spectr. 29, 194 (1969).

(21) B. F. Rozsnyai, F. Martino, J. Ladik, J. Chem. Phys. 52, 5708 (1970).

(22) S. Suhai and J. Ladik, Int. J. Quant. Chem. 7, 547 (1973).

(23) S. Suhai, Biopolymers (accepted).

(24) S. Suhai, J. Chem. Phys. 57, 5599 (1972).

(25) A. Bierman and J. Ladik (unpublished result).

(26) M. Kertèsz, to be published.

(27) B. F. Rozsnyai and J. Ladik, J. Chem. Phys. 52, 5711 (1970).

(28) B. F. Rozsnyai and J. Ladik, J. Chem. Phys. 53, 4325 (1970).

(29) G. L. Eichhorn and Y. A. Shin, J. Am. Chem. Soc. 90, 7323 (1970); P. Cheng, D. Honbo and J. Rozsnyai, Biochem. 1, 239 (1969).

(30) J. Ladik, Int. Quant. Chem. S8, (accepted).

(31) J. Josse, A. D. Kaiser, and A. Kornberg, J. Biol. Chem. 236, 864 (1961).

(32) S. Suhai, Theor. Chim. Acta (submitted).

PROTEIN ENERGY CONVERTERS

K. Laki and J. Ladik *
National Institute of Arthritis, Metabolism and Digestive Diseases, National Institutes of Health
Public Health Service
U. S. Department of Health, Education and Welfare
Bethesda, Maryland 20014

SUMMARY

On the basis of experimental evidence, it seems plausible to assume that actin, the main constituent of the thin filaments in muscle, analogously to the coupling factors of mitochondria and chloroplasts, serves as an energy transformer. An approximate stereostructure of actin has been built up on the basis of its known amino acid sequence and the statistical rules of Chou and Fassman. Using analogies to other nucleotide binding proteins, the probable ATP binding site of actin has been pointed out. In the neighborhood of this site, the polypeptide chain is in random coil allowing many different conformations with slightly different energies. Thus this part of the actin molecule could store and release energy *via* conformational changes. To treat theoretically the conformational changes of a polypeptide chain, a new quantum chemical method for the calculation of the interactions between several molecules is outlined.

INTRODUCTION

We are well aware of the achievements of biochemistry. Just to consider our knowledge of hormones and how these help in health and desease, or the progress made in our understanding of protein structure, is enough to realize the progress made. Similarly, molecular biology unravelled the structure of DNA and genetic engineering is around the corner as a possibility to create fatal havoc in the living world.

---

* Visiting Scientist

But if we direct our attention to biological energy converters, we see a much less advanced picture. Living matter connot exist without the operation of the energy-producing organelles. Here I need to mention three of them : the chloroplast, the mitochondria and the motile organelles.

The chloroplasts utilize photons to reduce CO_2 and produce ATP. The mitochondria use oxygen to burn reducing compounds in order to produce ATP. In their normal functioning, the motile organelles use ATP to produce mechanical work, but as A. V. Hill (1) pointed out, the motile organelle in muscle also can convert mechanical energy into chemical energy (synthesis of ATP).

According to the current view, these organelles were once free-living prokaryotes and became organelles after a primitive cell captured them (2). With these prokaryotes, we go back in time when oxygen appeared in the atmosphere (mitochondria) or even further to the time when light reached the earth (choroplasts). The motile organelle in the form of flagellum might even be older going back in time when not yet photons reached the surface of the earth.

MUSCLE AS AN ENERGY TRANSFORMER

My purpose in this presentation is to discuss the motile organelle as it is exemplified in muscle. One would think that this is the simpler of the three converters, but as I see it, some of the features of the other energy converters can guide us in our understanding of the mechano-chemical energy converter.

As you know, muscle consists of muscle fibers, often several centimeters long and 30-100 microns wide. A branch of the nerve innervating the muscle goes to every fiber. For the sake of simplicity, let us imagine that every fiber is a pile of tiny discs about 2 microns thick and 30-100 microns diameter. This disc is actually a collection of smaller discs (two microns thick and about one micron wide) put side-by-side separated by mitochondria and tubules of a reticular systemn like sewer lines in a city. In the fiber, these tiny discs are also piled up on top of each other, thus creating long tiny rods, the myofibrils. Lets us now concentrate on the tiny discs. These, on their top and bottimn are bordered by the so-called Z-membrane (about two microns apart).

Electron microscopy tells us that thin filaments, a thousand of them, emanate from the Z-membrane toward each other. In the middle of the tiny discs, we find the so-called thick filaments. When the tiny discs shorten, the thin filaments ecomes attachesd to the thick filaments and slide along them with the result that in the shortened disc, the thin filaments are located among the thick filaments.

The thick filaments are orderly aggregates of myosin molecules. The current view is that the thick filaments have protrusions which attach to the thin filaments and when energy is liberated from ATP these protrusions swing and pull in the attached thin filaments and with them the Z-membrane. In this view, the thin filaments are inert rods. In this paper, this view will not be followed ; instead, the alternative that the thin filaments are not inert rods but actively propel themselves into the domain of the thick filaments will be examined.

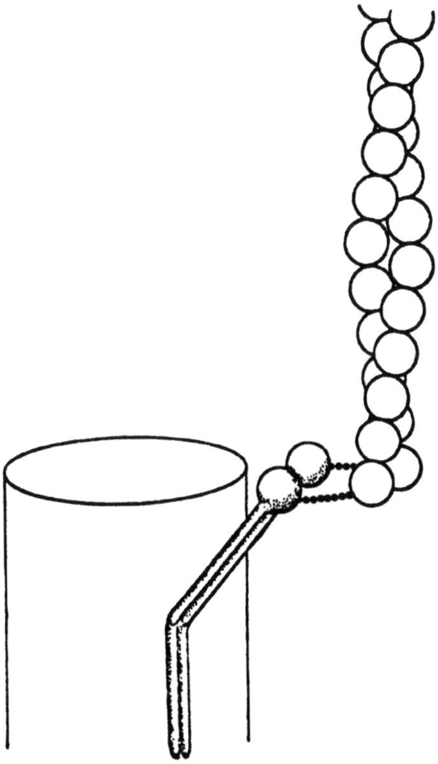

Fig. 1. Schematic representation of the attachment of thin filament to a protrusion of the thick filament in muscle.

I will cite two reasons why we should concentrate our attention on actin rather than on myosin. Actin filaments have been seen and actin has been isolated from many primitive organisms (for review, see ref. 3) where actin is involved in various kinds of motions. Yet in these organisms there are no thick filaments. If myosin is found, it is present only in small quantities. The sliding filament mechanism cannot be invoked in these instances.

Muscle fibers, as Szent-Györgyi has show decades ago, may be kept in concentrated glycerol solution in which the excitability of the membranes is destroyed, yet these fibers contract when ATP is added to them. Oplatka and his co-workers (4) have shown very recently that from the glycerol-treated fibers myosin may be entirely extracted (ghost fibers); nevertheless, such fibers contract with ATP if the globular fragment of myosin (S_1) is added to them. The globular fragment of myosin does not aggregate into filaments, thus these experiments clearly demonstrate that myosin organized into thick filaments is not needed for contraction. The only structural organization in these 'ghost fibers" is the thin filaments anchored in the Z-membranes. Clearly, muscle must be a special case of a more general mechanism where thick filaments are not needed, not even intact myosin.

The thin filament basically is a double row of about 400 actin molecules to which tropomyosin and troponin molecules are attached in such a way that a unit of seven actin molecules are associated with one tropomyosin and one troponin molecule. (Fig. 2.)

PROTEIN ENERGY TRANSFORMERS

We are considering such an actin complex as the unit of energy conversion. This unit has great similarity to the couplint factors (F_1) of mitochondria and chloroplasts.

In chloroplasts, the coupling of P_i to ADP is catalyzed by a coupling factor (CF_1), a high molecular weight protein resembling the coupling factor (F_1) in mitochondria. These coupling factors, in addition to being involved in terminal phosphorylation, also act as ATPase in the presence of Mg^{2+} ions.

When the mitochondrial F_1 (mol. wt. 384,000) is subjected to gel electrophoresis in the presence of SDS and mercaptoethanol it falls into smaller pieces which have molecular weights of 54,000, 50,000, 33,000, 16,000 and 11,000 (5). Similar findings are reported for CF_1 by others (6,7). The last of these subunits is an inhibitor of ATPase. These findings may be compared to the structural unit of the thin filaments composed of 3 different proteins, one of which, the troponin, consists of 3 components (TP-I, TP-N, TP-C). The similarity goes even further. The TP-I (inhibitory) component of troponin which inhibits actomyosin ATPase also inhibits the activity of the mitochondrial ATPase. Thus, the TP-I component of troponin is able to substitute for the inhibitor component of the coupling factor F_1 (8). This is a remarkable finding and further indicates the relationship among the energy converters.

Fig. 2. Schematic representation of the thin filament structure in muscle. The glabules are actin, the long rod tropomyosin, and the small circle stands for the three components of troponin.

When chloroplasts are illuminated in the presence of tritiated water, the coupling factor (CF_1) isolated in the dark contained incorporated tritium. The explanation for the light-induced tritium exchange is that conformational change has taken place during illumination, exposing part of the polypeptide chain to tritium. These experiments strongly suggest that part of the energy of the photon in the chloroplasts is eventually stored in a conformational change of the coupling factor and this stored energy is utilized in the synthesis of ATP from ADP and P_i.

I have previously proposed that when ATP bound to actin shifts to its low energy form, its excess energy becomes stored in a labile helix (3). Or conversely, when work is applied to active muscle the work may be converted into conformational change and from there used for the resynthesis of ATP. I believe that recent developments concerning nucleotide binding proteins permit us to specify and locate this labile helix on the actin structure.

THE STRUCTURE OF ACTIN

It is now becoming increasingly apparent that the folding of the peptide chain in many nucleotide binding proteins is similar in spite of the great variety in their amino acid sequences(9,10,11)

Lactate dehydrogenase (LDH), malate dehydrogenase (MDH), liver alcohol dehydrogenase (LADH), and glyderaldehyde-3-phosphate dehydrogenase (GAPDH) bind nicotinamide adenine dinucleotide. Flavodoxin binds flavine mononucleotide (FMN). (For references , see 9.)

The three-dimensional structure from X-ray studies and the amino acid sequence of these nucleotide binding proteins is available. Quite recently, the structure of adenyl kinase, which binds ATP, ADP or AMP, has also been established (12,13).

In dehydrogenases, the nucleotide binding portion of the molecule consists of a β-pleated sheet. The β-folds in the sheet are termed βA, βB, βC, βD, βE and βF. Helical-folds αB and αC connect the strands βA with βB and βB with βC.

Proceeding along the peptide chain, βA, αB, βB, αC and βC form the adenine nucleotide binding fragment of the molecule. The rest of the pleated sheet and associated helices are involved in forming the nicotinamide binding portion which is related to the other by a twofold axis. Apparently adenyl kinase exhibits similar structural features. The dominant feature of adenyl kinase is a five-stranded pleated sheet (12).

Of the two structural elements, the pleated sheet, not the helix, appears strongly conserved in these proteins. For example, in LDH and MDH, one of the helices is absent and instead, there is a flexible loop which undergoes large conformational change during the functioning of the enzyme (11). It has been proposed that the conformational change takes place in LDH when the phosphate residue of the nucleotide interacts with the guanidinum group of an arginine residue (Arg 101) in the loop.

This pleated sheet structure with the adjoining helices or loops is now considered to be one of the earliest and most common architectural elements of proteins.

Actin is one of the most ancient nucleotide binding proteins (3) ; therefore, it seems reasonable to apply to actin the basic structural features of the nucleotide binding proteins. In constructing the three-dimensional fold of actin, we utilized the amino acid sequence of rabbit actin (Table 1), (14) on which we first

Table 1 Amino acid sequence of actin. Arrows indicate the first three β-folds.

Asp	Glu	Thr	Glu	Asp	Thr	Ala	Leu	Val	Cys	10
Asp	Asp	Gly	Ser	Gly	Leu	Val	Lys	Ala	Gly	20
Phe	Ala	Gly	Asp	Asp	Ala	Pro	Arg	Ala	Val	30
Phe	Pro	Ser	Ile	Val	Gly	Arg	Pro	Arg	His	40
Gln	Gly	Val	Met	Val	Gly	Met	Gly	Gln	Lys	50
Asp	Ser	Tyr	Val	Gly	Asp	Glu	Ala	Gln	Ser	60
Lys	Arg	Gly	Ile	Leu	Thr	Leu	Lys	Tyr	Pro	70
Ile	Glu	His(τMe)	Trp	Gly	Ile	Ile	Thr	Asn	Asp	80
Asp	Met	Glu	Lys	Ile	Trp	His	His	Thr	Phe	90
Tyr	Asn	Glu	Leu	Arg	Val	Ala	Pro	Glu	Glu	100
His	Pro	Thr	Leu	Leu	Thr	Glu	Ala	Pro	Leu	110
Asn	Pro	Lys	Ala	Asn	Arg	Glu	Lys	Met	Thr	120
Gln	Ile	Met	Phe	Glu	Thr	Phe	Asn	Val	Pro	130
Ala	Met	Tyr	Val	Ala	Ile	Gln	Ala	Val	Leu	140
Ser	Leu	Tyr	Ala	Ser	Gly	Arg	Thr	Thr	Gly	150
Ile	Val	Leu	Asp	Ser	Gly	Asp	Gly	Val	Thr	160
His	Asn	Val	Pro	Ile	Tyr	Glu	Gly	Tyr	Ala	170
Leu	Pro	His	Ala	Ile	Met	Arg	Leu	Asp	Leu	180
Ala	Gly	Arg	Asp	Leu	Thr	Asp	Tyr	Leu	Met	190
Lys	Ile	Leu	Thr	Glu	Arg	Glv	Tyr	Ser	Phe	200
Val	Thr	Thr	Ala	Glu	Arg	Glu	Ile	Val	Arg	210
Asp	Ile	Lys	Gln	Lys	Leu	Cys	Tyr	Val	Ala	220
Leu	Asp	Phe	Glu	Asn	Glu	Met	Ala	Thr	Ala	230
Ala	Ser	Ser	Ser	Leu	Glu	Lys	Ser	Tyr	Glu	240
Leu	Pro	Asp	Gly	Glu	Val	Ile	Thr	Ile	Gly	250
Asn	Glu	Arg	Phe	Arg	Cys	Pro	Glu	Thr	Leu	260
Phe	Gln	Pro	Ser	Phe	Ile	Gly	Met	Glu	Ser	270
Ala	Gly	Ile	His	Glu	Thr	Thr	Tyr	Asn	Ser	280
Ile	Met	Lys	Cys	Asp	Ile	Asp	Ile	Arg	Lys	290
Asp	Leu	Tyr	Ala	Asn	Asn	Val	Met	Ser	Gly	300
Gly	Thr	Thr	Met	Tyr	Pro	Gly	Thr	Ala	Asp	310
Arg	Met	Gln	Lys	Glu	Ile	Thr	Ala	Leu	Ala	320
Pro	Ser	Thr	Met	Lys	Ile	Lys	Ile	Ile	Ala	330
Pro	Pro	Glu	Arg	Lys	Tyr	Ser	Val	Trp	Ile	340
Gly	Gly	Ser	Ile	Leu	Ala	Ser	Leu	Ser	Thr	350
Phe	Gln	Gln	Met	Trp	Ile	Thr	Lys	Gln	Glu	360
Tyr	Asp	Glu	Ala	Gly	Pro	Ser	Ile	Val	His	370
Arg	Lys	Cys	Phe							

localized helical folds, β-folds and β-turns utilizing the prescription of Fassman (15). After localizing these structural elements, we reconstructed a portion of the β-pleated sheet from the first three β-folds, patterned according to the adenyl kinase as seen schematically on Fig. 3.

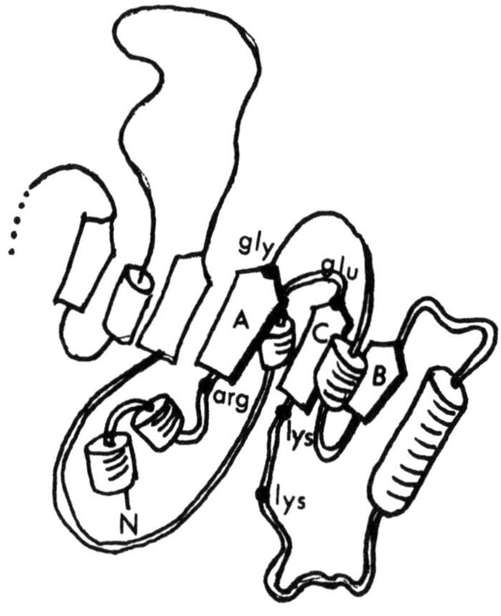

Fig. 3. Schematic representation of the nucleotide binding portion of actin. The peptide chain represented by arrows forms a portion of the β-pleated sheet. The cylinders illustrate helical regions. For details, see text.

Here I would like to dwell on a number of the features of this model which strengthen my belief that this model is a reasonable approximation of the first nucleotide binding portion of actin.

The chain in actin, just like in adenyl kinase, starts with helical regions. These containing basic residues lead into the first β-fold (βA). It is characteristic for all the βA folds in all the nucleotide binding proteins that βA ends with glycine (Table 2)

Table 2

AMINO ACID SEQUENCE OF THE βA FOLD IN VARIOUS PROTEINS*

Dogfish LDH	N K I T V V G
Pig GAPDH	V K V G V D G
Lobster GAPDH	S K I G I D G
Yeast GAPDH	V R V A I D G
Horse LADH	S T C A V F G
Bovine Glu DH	K T F A V Q G
Subtilisin	D V I N M S L
Rabbit actin	G V M V G M G
Aldenyl kinase	K I I F V V G

*For the one letter notation of amino acids see : M. O. Dayhoff, Atlas of Protein Sequence and Structure 1972. Nat. Biomed. Res. Found., Georgetown Univ. Med. Center, Washington, D.C. The data in this table, except for those on actin and adenyl kinase, are taken from Ref. 9. (G=Gly, V=Val, M=Met, I=Ile, and so on).

Glycine is recognized to be conserved in these proteins at this positions because more bulky residue would hinder the binding of those ribose ring of ATP. The β-region next to βA in actin and adenyl kinase is βC and it should contain at its end aspartate or glutamate to be involved in the binding of O2' atom of the ribose ring (9). Adenyl kinase and our model contain glutamate in this position. In LDH and Glu DH, the strand βB is closest to βA and their ends, these contain aspartate or glutamate residues to bind the ribose ring (9).

As could be expected, the nucleotide binding region should contain basic side chains to interact with the phosphate groups of the nucleotide. As Schulz et al.(12) point out, there are several basic groups in the binding area of adinyl kinase. If these residues have importance, as they should in adenyl kinase, and also in actin because the phosphates could be expected to be properly localized, this area should be conserved. Table 3 shows that the sequence of the peptide chain going into βC in actin and adenyl kinase has similar positions for the lysine residues. The situation is also similar in the sequence leading into βA except that in actin, one finds arginine rather than lysine. It is hard to imagine that these correspondences would be accidental. In fact, we have independent evidence that some of these basic residues are indeed involved in phosphate binding.

Table 3 For explanation, see text.

THE SEQUENCES OF ACTIN AND ADENYL KINASE PRECEDING THE βC FOLD

Actin	(113) Lys	Ala	Asn	Arg	Glu	Lys	βC
Adenyl kinase	(83) Lys	Val	Asp	Thr	Ser	Lys	βC

We found that the limited protease, thrombin, specific for arginine and to a lesser extent, lysine side chains splits actin at Arg (28) at Arg (39) and Lys (113) residues (16). It turned out that thrombin split ADP-actin faster than ATP-actin. Our interpretation is that the terminal phosphate of ATP in ATP-actin binds to these two residues and thereby hinders the specificity site of thrombin to attach to these residues in ATP-actin. There are many other basic residues in the actin molecule which could bind the terminal phosphate of ATP. Our identification of the two residues is based on indirect evidence. Nevertheless, the fact that His (40), which is involved in ATP-actin → ADP-actin transformation (17) is also in the nucleotide binding region, suggests that our identification of Arg (39) and Lys (113) may be correct.

It can be expected that when the terminal phosphate of ATP binds to Arg (39) and to Lys (113) otherwise widely seperated in the sequence introduces a constraint in the peptide chain with the result that a labile, energy-rich conformation develops locked in place by weak hydrogen bonds. It is noteworthy that the lysine residue (Lys 83) in adenyl kinase corresponding to Lys (113) in actin is located in a helical segment. In contrast, the residues of the segment in actin containing Lys (113) have low helix potential, but when constraint is introduced, these may form a labile helix.

ACTIN AS ENERGY TRANSDUCER

The central idea of the mechanism I have proposed is that ATP bound to actin can exist in two forms, a high energy form and a low energy form (3). This implies that when ATP in the low energy form becomes hydrolyzed, the heat released and the free energy change should correspond to that observed during the hydrolysis of ordinary phosphate esters. This expectation is borne out from experiments. When ATP-actin changes to ADP-actin, several kcalories less heat develops than expected from ATP hydrolysis (18). This is also true for the globular portion of myosin. When it dephosphorylates bound ATP to ADP, the heat expected released from this reaction remains trapped in a local conformational change of the protein (19). The free energy change of this reaction turned out to be -2.8 kcal/mole rather being -7.5 kcall/mole (20). These findings clearly support the idea that bound ATP can exist in two states.

This conclusion may imply two bindings sites on the actin molecule. The three dimensional structure of actin advanced in this paper actually provides two nucleotide sites, in analogy with the other nucleotide binding proteins. It is quite conceivable that ATP attached to one site is in the high energy form ; in the other site, it is in the low energy form. This would require a movement from one site to the other, a possibility that may eventually have to be considered.

The high energy, labile helix may be created in two ways : by stretching active muscle or permitting the bound ATP to go into its low energy state. So the collapse of the labile helix can cause either a shortening of the molecule thereby carrying out work, or re-energization of ATP from its low energy state to the high energy state. Normally in muscular contraction, the role of ATP is to recreate the high energy conformation. Our proposition is that this is done in the following way : When the terminal phosphate of ATP binds to Arg (39) and Lys (113), the constraint introduced (with the cooperation of a nearby nucleation center) may put a segment of the peptide chain containing residues of low helix forming potential into a helical conformation or some other energetically different conformation of the chain can be reached. The chain then becomes locked in this conformation with weak hydrogen bonds. The extra energy of ATP thus appears as conformational energy. In this state, actin is energized and elongated. How these events may be utilized for muscular contraction has been dealt with in earlier publications.

From all these considerations it is clear that it would be very helpful if the different energies associated with the different conformations of the nucleotide binding parts of the protein could be calculated with the aid of quantum chemical methods.

THEORY OF INTERACTIONS BETWEEN SEVERAL MOLECULES

To obtain a deeper insight into the energetically different states associated with the different conformations of a polypeptide chain, one has to calculate the interactions between the side chains of the different amino acid residues and between these side chains and segments of the main polypeptide chain at different geometrical configurations. Since the energy stored and released during muscle action _via_ conformational changes of the polypeptide chain is only 4-5 kcal/mole, one would need a rather accurate method for this calculation. In principle, the best thing would be to treat _ab initio_ at least six amino acid residues in actin around Arg 28 or around Lys 113 (to which two sites, most probably the terminal phosphate group of ATP is bound [1.]) as a single supermolecule with a large basis set taking into account a fair amount of correlation and performing the calculation for a number of propable geometrical configurations. Such a calculation, however, is today still out of the question even if the largest computers would have been used. On the other hand, the usual perturbation theoretical method to describe the van der Waals interactions have proven to be rather inaccurate, at the biologically interesting medium intermolecular distances of 3-4 A°, if _ab initio_ wave functions of the constituents have been used as unperturbed wave functions (21). Further the perturbation theoretical expression for the dispersion interaction correct to the second order (22) is hopelessly complicated to be executed already for two interacting medium size molecules. Therefore, in practical applications for the different terms in the expression

$$\Delta E = E_{el.st.} + E_{pol.} + E_{disp.} + E_{exch.} + E_{ch.t.} \qquad (1)$$

of the interaction energy, approximate formulae are used in many cases experimentally measurable quantities (like polarizabilities, ionization potentials, etc.) of the interacting molecules are in-

1.) An approximative investigation of the α-helical and β-sheet regions of rabbit actin based on its known amino acid sequence (14) using the rules proposed by Chou and Fassman (15) indicates flexible random coil segments before and after Arg 28 and Lys 113 which can easily undergo into different conformational states with slightly different energies.

corporated into the theory (23). Here $E_{el.st}$ is the electrostatic interaction energy, $E_{pol.}$ stands for the polarization term, E_{disp} is the dispersion energy and $E_{exch.}$ and $E_{ch.t.}$ give the exchange repulsion and the charge transfer contributions, respectively. The latter two terms disappear at large intermolecular separation, but at medium distances, where the overlap of the electron clouds of the interacting molecules is non-negligible, they have to be taken into account. These semiempirical interaction energy schemes are usually parametrized in such a way that they give acceptable interaction energies between two molecules of certain types starting from semiempirical unperturbed molecular wave functions obtained in a certain approximation (for instance, iterative extended Hückel wave functions ; see (23) and reference cited there).

The same can be said about the still more empirical interaction potential function approach of Scheraga (24) and others (25). The basic disadvantage of these theories is that besides being neither general nor accurate enough, they are unable to handle correctly the simultaneous interactions of more than two molecules.

To handle the problem of interactions between several molecules (like in a polypeptide chain), one clearly needs a theory which is

1) accurate enough to account for small conformational energy changes ;
2) can be executed with the present day computers ; and
3) takes into account the simultaneous interactions of several molecules not just adding pair interactions (energies computed between two molecules in the absence of the further ones).

Here we try to outline briefly a method which most probably will fulfill these three conditions. The details, together with some computational results, will be published elswhere (26) after the implementation of the necessary programs has been finished.

Let us assume we have N molecules, n_i should stand for the number of electrons and M_i for the number of nuclei of the i-th molecule. If the molecules are not interacting, the total energy of the system is simply

$$E_1 = \sum_{i=1}^{N} E^{(i)} \tag{2}$$

where $E^{(i)}$ stands for the total energy of the i-th molecule including correlation,

$$E^{(i)} = E^{(i)}_{HF} + E^{(i)}_{corr.} \tag{3}$$

$E^{(i)}$ is dependent on the potential due to its nuclei and electronic charge distribution,

$$E^{(i)} = E^{(i)}(V^{(i)}); \quad V^{(i)}(\vec{r}) = - \sum_{d_i=1}^{M_i} \frac{Z_{d_i}}{|\vec{r} - \vec{r}_{d_i}|} + \int \frac{\rho^{(i)}(\vec{r}\,')}{|\vec{r} - \vec{r}\,'|}\, d\vec{r}\,' \tag{4}$$

where $\rho^{(i)}(\vec{r}')$ is the electronic density of the i-th molecule.

In the case the molecules interact with each other, usually one one writes for the total energy

$$E_2 = \sum_{i=1}^{N} E^{(i)} + \sum_{i<j}^{N} \Delta E^{(i,j)} \tag{5}$$

where $\Delta E^{(i,j)}$ stands for the interaction energy between molecules i and j. If one starts from the wave functions of the free (not interacting) molecules and applies perturbation theory, one obtains the van der Waals terms of eqn. (1). Instead of this, we propose to find the mutually consistent total energies of all the interacting molecules in the potential field of each other ; that is, to take into account that the total energy of the i-th molecule is not only a function of its own potential, but also of those of all the other molecules,

$$\tilde{E}^{(i)}(V^{(i)} ; V^{(1)}, V^{(2)}, \ldots\ldots, V^{(j)}, \ldots, V^{(N)}) \quad (j \neq i) \tag{6}$$

To avoid the numerically difficult treatment of all interacting molecules as one supermolecule, we set up all the potentials in the simple form of (4) neglecting the exchange contributions. In that case, $\tilde{E}(i)$ can be written as

$$\tilde{E}^{(i)} = \langle \psi^{(i)}_{corr.} | \hat{\tilde{H}}^{(i)} | \psi^{(i)}_{corr.} \rangle = \langle \psi^{(i)}_{corr.} | \sum_{k=1}^{n_i} \hat{\tilde{H}}^{(i)}_k + \sum_{k<l}^{n_i} \hat{H}^{(i)}_{k,l} | \psi^{(i)}_{corr.} \rangle \tag{7}$$

with the modified one-electron operator

$$\hat{\tilde{H}}^{(i)}_k = -\frac{1}{2}\Delta_k - \sum_{d_i=1}^{M_i} \frac{Z_{d_i}}{|\vec{r}^{(i)}_k - \vec{r}_{d_i}|} - \sum_{\substack{j=1 \\ j\neq i}}^{N} \Big[\sum_{d_j=1}^{M_j} \frac{Z_{d_j}}{|\vec{r}^{(i)}_k - \vec{r}_{d_j}|} - \int \frac{\rho^{(j)}(\vec{r}')}{|\vec{r}^{(i)}_k - \vec{r}'|} d\vec{r}' \Big] \tag{8}$$

Here $\psi^{(i)}_{corr.}$ is the normalized correlated many electrons wave function of the i-th molecules and $\hat{H}^{(i)}_{k,l}$ stands for $\frac{1}{|\vec{r}^{(i)}_k - \vec{r}^{(i)}_l|}$. With the help of (7) we can write for the total energy of the N interacting molecules instead of (5)

$$\tilde{E}_2 = \sum_{i=1}^{N} \tilde{E}^{(i)} + \sum_{i<j}^{N} \Delta\tilde{E}^{(i,j)} ,$$

where

$$\Delta\tilde{E}^{(i,j)} = \sum_{d_i=1}^{M_i} \sum_{d_j=1}^{M_j} \frac{Z_{d_i} Z_{d_j}}{|\vec{r}_{d_i} - \vec{r}_{d_j}|} + \int \frac{\tilde{\rho}^{(i)}_{corr.}(\vec{r})\, \tilde{\rho}^{(j)}_{corr.}(\vec{r}\,')}{|\vec{r}-\vec{r}\,'|} d\vec{r}d\vec{r}\,' \tag{9}$$

$$- \left(\sum_{d_i=1}^{M_i} Z_{d_i} \int \frac{\tilde{\rho}^{(j)}_{corr.}(\vec{r}\,')}{|\vec{r}_{d_i}-\vec{r}\,'|} d\vec{r}\,' + \sum_{d_j=1}^{M_j} Z_{d_j} \int \frac{\tilde{\rho}^{(i)}_{corr.}(\vec{r}\,')}{|\vec{r}_{d_j} - \vec{r}\,'|} d\vec{r}\,' \right) +$$

$$E^{(i,j)}_{exch.} + E^{(i,j)}_{ch.t.} \tag{10}$$

Here $\tilde{\rho}^{(i)}_{corr.}$ is the correlated charge density of the i-th molecule in the Coulomb field of all the others (mutually consistent correlated charge density), $\vec{r}_{d_i}$ stands for the position vector of the d_i-th nucleus in the i-th molecule and for the exchange and charge transfer terms $E^{(i,j)}_{exch.}$ and $E^{(i,j)}_{ch.t.}$ in first approximation the usual van der Waals expressions could be used. It is clear from this discussion that if electronic correlation of the single molecule can be treated in an acceptable approximation, the first four terms of (10) will give besides the electrostatic term automatically also the polarization and dispersion contributions [2.)] and only the exchange and charge transfer terms have to be calculated separately.

Comparing (9) with (2), we obtain for the interaction energies of the N molecules the expression

$$\tilde{E}_2 - E_1 = \sum_{i=1}^{N} (\tilde{E}^{(i)} - E^{(i)}) + \sum_{i<j} \Delta\tilde{E}^{(i,j)} \tag{11}$$

which especially in the case of more than two interacting molecules may give considerably different numerical results from $\sum_{i<j} \Delta E^{(i,j)}$ if the latter quantities are calculated with the aid of (1)

Preliminary mutually consistent calculations for two interacting HF and CH_2O molecules, respectively, at the Hartree-Fock level using the simple (and rather crude) monopole approximation for the potentials (4) of the partner molecule, have given a somewhat better approximation for the sum of the electrostatic and polarization terms, than the usual van der Waals treatment (26). One would expect that if one would use instead of the monopole approximation a multiple expansion of the potential or still better, would approximate the function $\rho^{(i)}(\vec{r})$ in (4) with the aid of a

2) This statement is, of course, only then true when a mutually consistent solution of the problem has been achieved.

large number of point charges in a three-dimensional mesh and would perform the integration in (4) numerically, the method would provide still beter results

In the above mentioned calculation, the mutually consistent solutions have been obtained by building into the Fock operators of the interacting molecule the potential of the partners. Instead of this, one could try to minimize the total energy (9) of the system directly. In the case of the spinprojected extended Hartree-Fock (EHF) method (27), which gives a considerable part of the correlation energy if the number of electrons in the single molecules is not large, the total energy can be rewritten in a comparatively simple form and its minimum can be found with the help of deepest descent in a rather straightforward way (28). The programs necessary for these calculations have been implemented and the numerical results could be obtained in about twice the time of the conventional ab initio HF procedure (28). The method of deepest descent requires an expression of the variation of the total energy, which was derived by Rosenfeld and Martino (28). One can modify the expression for the case of N interacting molecules (29) and look for

$$\delta \tilde{E}_2 = \sum_{i=1}^{N} \delta \tilde{E}^{(i)} + \sum_{i<j}^{N} \delta (\Delta \tilde{E}^{(i,j)}) = 0 \qquad (12)$$

In this way the minimum of the total energy of N interacting molecules can be obtained in a mutually consistent way at the EHF level providing besides the electrostatic interactions also the polarization and dispersion energies. Using as first step again the monopole approximation to describe the charge distributions of the partner molecule and only the first four terms of $\Delta\tilde{E}^{(i,j)}$ (see equ.10), the expression of $\delta \tilde{E}_2$ for N interacting molecules has been derived and the corresponding modification of the EHF programs is in progress (29). Here again, better approximation than the monopole could be used and for $\Delta\tilde{E}^{(i,j)}$ the full expression (10) including exchange and charge transfer could be applied. This is planned as a further step of the investigations although especially the inclusion of $E^{(i,j)}_{exch.}$ and $E^{(i,j)}_{ch.t.}$ in the expression subject to variation makes both the algebra and the programming rather involved. Since, however, as we have seen the stakes are high, it would be worthwhile to perform with time also these rather complicated calculations.

CONCLUDING REMARKS

To complete the picture and to emphasize the importance of the mechanochemical energy converter, I have to say a few words about the flagellum and the mitotic spindle. In spite of their differing roles, the two organelles show great similarity, so much so that many researchers believe the mitotic spindle is a variant of the flagellum. As an example of the flagellum, we can take the tail of a spermium. The working anatomical element of the flagellum is a group of some 20 microtubules arranged circularly in pairs close to the boundary of the flagellum. The building blocks of the microtubules are protein molecules very similar to actin molecules but containing GTP rather than ATP bound to them. The operating elements of the mitotic spindle are also microtubules. It is generally assumed that contraction of the proteins along one side of the microtubules causes it to bend. Out of such bending motions, the propelling waves of the flagellum can be reconstructed.

The separated flagellum or the mitotic spindle can be kept in glycerol solution and just like the glycerinated muscle fiber, the flagellum starts to beat and the mitotic spindle contracts when ATP is supplied to them.

It is apparent that understanding of the action of the mitotic spindle is vital to understanding of cell division. It is the mitotic spindle whose microtubules when attached to the chromosomes pull them into the two poles of the dividing cell so that after separation, the two daughter cells contain one set of the chromosomes of the original cell. Uncontrolled cell division is at the heart of the cancer problem and perhaps if we understand how mitotic spindle operates, we can get a better idea of how it may be regulated.

REFERENCES

(1) A. V. Hill, Science 131, 897 (1960).
(2) L. Margulis in "Origin of eukaryotic cells", Yale Univ. Press, New Haven and London, (1970).
(3) K. Laki, J. Theor. Biol. 44, 117 (1974).
(4) A. Oplatka, H. Gadasi, R. Tirosh, Y. Lamed, A. Mühlrad and N. Liron, J. Mechanochem. Cell Motility 2, 295 (1974).
(5) I. A. Kozlov and H. N. Mikelsaar, FEBS Letters 43, 212 (1974).
(6) A. F. Knowles and H. S. Penefsky, J. Biol. Chem. 247, 6617 (1972).
(7) W. A. Catterall and P. L. Pedersen, J. Biol. Chem. 246, 4987 (1971).
(8) Y. Tamaura, S. Yamazaki, S. Hirose and Y. Inada, Biochem. Biophys. Res. Commun. 53, 673 (1973).
(9) M. G. Rossmann, D. Moras and K. W. Olsen, Nature 250, 194 (1974).
(10) M. Buehner, G. C. Ford, D. Moras, K. W. Olsen and M. G. Rossman Proc. Nat. Acad. Sci. USA 70, 3052 (1973).
(11) C. I. Bränden, H. Eklund, B. Nordström, T. Boiwe, G. Söderlund, E. Zeppezauer, I. Ohlsson and A. Akeson, Proc. Nat. Acad. Sci. USA 70, 2439 (1973).
(12) G. E. Schulz, M. Elzinga, F. Marx and R. H. Schirmer, Nature 250, 120 (1974).
(13) G. E. Schulz and R. H. Schirmer, Nature 250, 142 (1974).
(14) M. Elzinga, J. H. Collins, W. M. Kuehl and R. S. Adelstein,

Proc. Nat. Acad. Sci. USA 70, 2687 (1973).
(15)P. Y. Chou and G. F. Fassman, Biochemistry 13, 222 (1974).
(16)L. Muszbek, J. A. Gladner and K. Laki, Arch. Biochem. Biophys. (submitted).
(17)G. Hegyi, G. Premecz, B. Sain and A. Mühlrad, Eur. J. Biochem. 44, 7 (1974).
(18)K. Laki, Actin. In "Contractile Proteins and Muscle", Ed. K. Laki. M. Dekker, New York, 1971, p. 97.
(19)T. Yamada, H. Shimizu and H. Suga, Biochim. Biophys. Acta 305, 642 (1973).
(20)R. G. Wolcott and P. Boyer, Biochem. Biophys. Res. Commun. 57, 709 (1974).
(21)H. Lischka, Chem. Phys. 2, 191 (1973).
(22)E. F. Haugh, J.O. Hirschefelder, J. Chem. Phys. 23, 1778 (1955).
(23)For a review of the simplified interaction schemes, see R. Rein in "Electronic Structure of Polymers and Molecular Crystals", ed. J.-M. André and J. Ladik, Plenum Press, London, (1975).
(24)H. A. Scheraga, Adv. Phys. Org. Chem. 6, 103 (1968).
(25)See, for instance, G. N. Ramachandran and V. Sasisekharan, Adv. in Prot. Chem. 23, 283 (1968).
(26)P. Otto and J. Ladik, Chem. Phys. (sumitted).
(27)F. Martino and J. Ladik, J. Chem. Phys. 52, 2269 (1970) ; I. Mayer, J.Ladik and G. Biczo, Int. Quant. Chem. 7, 583 (1973).
(28)M. Rosenfeld and F. Martino (to be published).
(29)J. Ladik, F. Martino and M. Rosenfeld (to be published).

SUBJECT INDEX

MIX
Papier aus verantwortungsvollen Quellen
Paper from responsible sources
FSC® C105338

If you have any concerns about our products,
you can contact us on
ProductSafety@springernature.com

In case Publisher is established outside the EU,
the EU authorized representative is:
Springer Nature Customer Service Center GmbH
Europaplatz 3, 69115 Heidelberg, Germany

Printed by Libri Plureos GmbH
in Hamburg, Germany